Integer Programming and Network Models

Springer
Berlin
Heidelberg
New York
Barcelona
Hong Kong
London
Milan
Paris
Singapore
Tokyo

H. A. Eiselt · C.-L. Sandblom

Integer Programming and Network Models

With Contributions by
K. Spielberg, E. Richards, B. T. Smith,
G. Laporte, B. T. Boffey

With 165 Figures
and 43 Tables

Springer

Prof. H. A. Eiselt
University of New Brunswick
Faculty of Administration
P.O. Box 4400
Fredericton, NB E3B 5A3
Canada

Prof. C.-L. Sandblom
Dalhousie University
Department of Industrial Engineering
P.O. Box 1000
Halifax, Nova Scotia
Canada B3J 2X4

ISBN 3-540-67191-9 Springer-Verlag Berlin Heidelberg New York

Library of Congress Cataloging-in-Publication Data
Die Deutsche Bibliothek – CIP-Einheitsaufnahme
Eiselt, Horst A.: Integer programming and network models/H. A. Eiselt; C.-L. Sandblom. – Berlin; Heidelberg; New York; Barcelona; Hong Kong; London; Milan; Paris; Singapore; Tokyo: Springer, 2000
ISBN 3-540-67191-9

Springer-Verlag Berlin Heidelberg New York
a member of BertelsmannSpringer Science+Business Media GmbH

Printed in Germany

Hardcover-Design: Erich Kirchner, Heidelberg

SPIN 10724054 42/2202-5 4 3 2 1 0 – Printed on acid-free paper

PREFACE

The purpose of this book is to provide readers with an introduction to the very active field of integer programming and network models. The idea is to cover the main parts of the field without being too detailed or too technical. As a matter of fact, we found it somewhat surprising that most—especially newer—books are strongly algorithmically oriented. In contrast, the main emphasis of this book is on models rather than methods. This focus expresses our view that methods are tools to solve actual problems and not ends in themselves. As such, graduate (and with some omissions, undergraduate) students may find this book helpful in their studies as will practitioners who would like to get acquainted with a field or use this text as a refresher.

This premise has resulted in a coverage that omits material that is standard fare in other books, whereas it covers topics that are only infrequently found elsewhere. There are some, yet relatively few, prerequisites for the reader. Most material that is required for the understanding of more than one chapter is presented in one of the four chapters of the introductory part, which reviews the main results in linear programming, the analysis of algorithms, graphs and networks, and dynamic programming, respectively. Readers who are familiar with the issues involved can safely skip that part. The three main parts of the book rely on intuitive reasoning and examples, whenever practical, instead of theorems and proofs. This does, however, not mean that rigor is sacrificed: main results are formally stated and mostly proved.

This volume was originally intended to be a lightly edited book with all contributions written by specialists in the field. While individual chapters were written by different colleagues, few of the contributions are close to their original form. In order to ensure that all chapters were similar in terms of coverage, style, and notation, it was necessary to modify, delete, and add explanations, examples, background material, and cross-references. We do apologize to the contributors if some of them are no longer able to recognize their contribution. We hope, though, that they will agree that the final product is readable, coherent, and—the ultimate goal—useful. It is our pleasant duty to thank our contributors: Dr. K. Spielberg who supplied us with material for Chapters 4 and 6 of Part I, Professor E. Richards, who drafted Chapter 7 of Part I, Professor B. Smith who contributed material for Chapters 1 and 2 of Part II, Professor G. Laporte who drafted and corrected the revision of Chapters 3 and 4 of Part II, and Dr. B.T. Boffey, who wrote the first draft of Chapters 1, 3, 4, and 5 of Part III. We express our sincere thanks to them. We are also grateful to Drs. J. Janssen and T. Larsson who kindly provided comments on parts of the manuscript. Many generations of assistants have worked on the project, but the bulk of the work was done by #7, a.k.a. G. Murphy, and #8, alias A. Lewczuk. We wish to express our gratitude to them for their meticulous typing and copyediting of the material, and the long hours they have put into this work. All mistakes remain, of course, our responsibility.

H.A. Eiselt C.-L. Sandblom

CONTENTS

INTRODUCTION: BASIC DEFINITIONS AND RESULTS

The chapters in this preliminary part will provide background material for the rest of the book. This organization of the coverage will allow a reader with the prerequisite knowledge of fundamentals to proceed directly to the main parts of the book. At the same time, it makes possible a unified treatment of the basic definitions and results which will be needed later on. The material in this part is not repeated in the later parts, but referred to when needed.

Chapter a presents fundamental concepts of linear programming, including the primal simplex method, duality and the dual simplex method, as well as a classification of changes to a linear programming problem and a description of some special structured problems. *Chapter b* deals with the analysis of time complexity of algorithms. The concepts of worst case and average case behaviour of algorithms are explained and a number of examples of algorithms of different time complexity are given. The classes *P*, *NP* and *NPC* and their relationships to each other are also discussed. *Chapter c* provides an introduction to graph theory with definitions of basic concepts and properties of graphs and networks. Several examples are given, the representation and storage of graphs are discussed, reachability and connectivity are covered, and a variety of special graphs and networks are presented. Finally, dynamic programming is treated in *Chapter d*, describing the basic ideas, giving a general algorithm and illustrating the approach by various models and examples.

CHAPTER a LINEAR PROGRAMMING

This chapter will sketch the basic notions and essential results of linear programming that are needed in subsequent chapters of this book. For a full treatment, the reader is referred to the pertinent literature.

a.1 Fundamental Concepts and the Simplex Method

Definition 1: Let $\mathbf{A}$ be an $[m \times n]$-dimensional matrix where $n \geq m$. Any collection of m linearly independent columns of $\mathbf{A}$ is said to be a *basis* of $\mathbf{A}$ and the square submatrix $\mathbf{B}$ formed by these columns is a *basis matrix*.

We may partition $\mathbf{A}$ into $\mathbf{A} = [\mathbf{B}, \mathbf{N}]$, possibly after reordering the columns, where the $[m \text{ x } (n - m)]$-dimensional matrix $\mathbf{N}$ consists of all columns of $\mathbf{A}$ that are not in the basis; the vector of variables $\mathbf{x}$ is partitioned accordingly into $\mathbf{x} = \begin{bmatrix} \mathbf{x_B} \\ \mathbf{x_N} \end{bmatrix}$, where the components of the m-vector $\mathbf{x_B}$ are called the *basic variables* and the components of the $(n - m)$-vector $\mathbf{x_N}$ the *nonbasic variables*. The system $\mathbf{Ax} = \mathbf{b}$ can then be written as $\mathbf{Bx_B} + \mathbf{Nx_N} = \mathbf{b}$, and setting $\mathbf{x_N} = \mathbf{0}$, we obtain the solution $\mathbf{x_B} = \mathbf{B}^{-1}\mathbf{b}$, $\mathbf{x_N} = \mathbf{0}$.

Definition 2: Let $\mathbf{a}_{i\cdot}$ denote row i of a matrix $\mathbf{A}$, and let $\mathbf{Ax} = \mathbf{b}$ be a given system of linear equations. Then the set of points

- $\{\mathbf{x} : \mathbf{a}_{i\cdot}\mathbf{x} = b_i\}$ is called a *hyperplane*;
- $\{\mathbf{x} : \mathbf{a}_{i\cdot}\mathbf{x} \leq b_i\}$ is called a *closed halfspace*;
- $\{\mathbf{x} : \mathbf{a}_{i\cdot}\mathbf{x} \geq b_i\}$ is called a *closed halfspace*;

- $\{\mathbf{x} : \mathbf{a}_{i\bullet}\mathbf{x} < b_i\}$ is called an *open halfspace*;
- $\{\mathbf{x} : \mathbf{a}_{i\bullet}\mathbf{x} > b_i\}$ is called an *open halfspace.*

Definition 3: A point $\mathbf{y} \in \mathbb{R}^n$ is said to be a *linear convex combination* (*lcc*) of the points $\mathbf{x}^i \in \mathbb{R}^n$, $i = 1, ..., k$, if there exist $\lambda_i \geq 0$, $i = 1, ..., k$ with $\sum_{i=1}^{k} \lambda_i = 1$ such that $\mathbf{y} = \sum_{i=1}^{k} \lambda_i \mathbf{x}^i$. The *convex hull* of the points $\mathbf{x}^i$ is the set of all linear convex combinations of them (see Definition I.4).

Definition 4: A *polytope* is the intersection of any finite number of hyperplanes and/or closed halfspaces. A bounded polytope is a *polyhedron.*

Equivalently, a polyhedron can be defined as the convex hull of a finite number of points. Note that some authors reverse the definitions of polytope and polyhedron.

Definition 5: Let H_i be the set of points satisfying $\mathbf{a}_{i\bullet}\mathbf{x}\ R_i\ b_i$, where R_i is a relation with $R_i \in \{\leq, =, \geq\}$ and $i = 1, ..., m$. If $\bigcap_{i=1}^{m} H_i = \bigcap_{\substack{i=1 \\ i \neq k}}^{m} H_i$, then the k-th relation is called *redundant*; otherwise it is called *essential* (*nonredundant*). If $\mathbf{a}_{k\bullet}\tilde{\mathbf{x}} = b_k$ for some $\tilde{\mathbf{x}} \in \mathbb{R}^n$, the k-th relation is said to be *binding* (*tight*) at $\tilde{\mathbf{x}}$.

Definition 6: A *basic point* in $\mathbb{R}^n$ is the intersection of at least n hyperplanes at one point; a *basic feasible point* is a basic point satisfying all given relations. A basic feasible point which is not a linear convex combination of at least two other basic feasible points is called an *extreme point.*

Definition 7: Given an n-dimensional vector of coefficients $\mathbf{c}$, an m-vector $\mathbf{b}$ and an $[m \times n]$-dimensional matrix $\mathbf{A}$ as well as the variable scalar z and an n-vector $\mathbf{x}$ of variables, then the problem

$$\text{P: } \max_{\mathbf{x}}\ z = \mathbf{cx} \qquad \text{objective (function)}$$

s.t.	$\mathbf{Ax} \leq \mathbf{b}$	(structural) constraints
	$\mathbf{x} \geq \mathbf{0}$	nonnegativity constraints,

is called a *linear programming (LP) problem in canonical form.* If the structural constraints are $\mathbf{Ax} = \mathbf{b}$ rather than $\mathbf{Ax} \leq \mathbf{b}$, we say that the problem is in *standard form.* Constraints in the canonical form $\mathbf{Ax} \leq \mathbf{b}$ can by the addition of a nonnegative *slack variable* column m-vector $\mathbf{S}$ be brought to the standard form: $\mathbf{Ax} + \mathbf{S} = \mathbf{b}$. In this context we refer to the x_j, $j = 1, \ldots, n$ as *decision variables* to distinguish them from the slack variables S_i, $i = 1, \ldots, m$. For other transformations between the canonical, standard and other forms of the linear programming problem, see, e.g., Eiselt *et al.* (1987).

The Simplex Method is an algorithm for solving linear programming problems in standard form. It does this by repeatedly transforming the coefficient matrices of the system of equations

$$\mathbf{Ax} = \mathbf{b} \text{ and}$$

$$z - \mathbf{cx} = z_0$$

where z_0 initially is equal to zero, and which for convenience is expressed in a tableau $\mathbf{T}'$ as follows.

$\mathbf{T}'$:

z	$\mathbf{x}$	1	← column headings row
0	$\mathbf{A}$	$\mathbf{b}$	← m rows
1	$-\mathbf{c}$	z_0	← objective function row
↑ z-column	↑ n columns	↑ right-hand side column	

In $\mathbf{T}'$ we drop the z-column, rename the $-\mathbf{c}$ coefficients by $\mathbf{c}$, and call the right-hand side column "1". This transformation results in the simplified tableau $\mathbf{T}$:

$\mathbf{T}$:

$\mathbf{x}$	1
$\mathbf{A}$	$\mathbf{b}$
$\mathbf{c}$	z_0

We now do a sequence of transformations of tableau **T**, but will always refer to the *current* set of coefficients as **A**, **b**, **c** and z_0 which will therefore no longer necessarily coincide with the original problem data. A tableau is said to correspond to a *basic solution* if the columns of its **A**-matrix contains the m unit vectors $\mathbf{e}_i$, $i = 1, ..., m$; these columns are referred to as *basic columns* (*variables*).

Definition 8: A basic solution with $(\mathbf{x}, \mathbf{S}) \geq \mathbf{0}$ is called a *basic feasible solution* (*BFS*).

The regular simplex method performs transformations between tableaus which correspond to basic feasible solutions. A basic feasible solution corresponds to an extreme point of the feasible polytope and a tableau transformation in the simplex method corresponds geometrically to a move between two adjacent extreme points of the feasible polytope. In the regular primal simplex method we will never move to a point with an inferior objective function value compared to the current point. There are only finitely many extreme points, and if the subjective function is bounded from above and the algorithm does not cycle, i.e., return to a previously visited basic solution after a sequence of steps with non-changing z-values, we will then arrive at an optimal point after a finite number of steps.

The tableau transformations are also done in such a way that $c_j = 0$ for all basic columns. Phase 2 of the regular primal simplex algorithm starts from any basic feasible solution; arriving at such a solution is the purpose of Phase 1 briefly outlined below. We will now summarize the Phase 2 method in algorithmic form.

The Primal Simplex Algorithm: Phase 2

Step 0: Start with a tableau that represents a basic feasible solution.

Step 1: Is $c_j \geq 0 \ \forall j = 1, ..., n$?
If yes: Stop; the current solution is optimal.
If no: Go to Step 2.

Step 2: Select any nonbasic variable x_s as entering variable, so that $c_s < 0$. The s-th column is called the *pivot column*.

Step 3: Is there any positive element in the pivot column $a_{is} > 0$, $i = 1, ..., m$?
If yes: Go to Step 4.
If no: Stop, unbounded optimal solutions exist.

Step 4: Determine the r-th row as *pivot row*, i.e., find r, such that

$$\frac{b_r}{a_{rs}} = \min_{i=1,\ldots,m} \left\{ \frac{b_i}{a_{is}} : a_{is} > 0 \right\}.$$

The variable which is in the basis in the r-th row leaves the basis. The element $a_{rs} > 0$ is called the *pivot element* or simply *pivot.*

Step 5: With a_{rs} as the pivot element, do a tableau transformation as follows:

$$a_{ij} := \begin{cases} \left.\begin{array}{l} 1 \text{ if } i=r \text{ and } j=s \\ 0 \text{ if } i \neq r \text{ and } j=s \end{array}\right\} \text{(pivot column)} \\ \dfrac{a_{ij}}{a_{rs}} \text{ if } i=r \text{ and } j \neq s \text{ (pivot row)} \\ a_{ij} - \dfrac{a_{rj}\, a_{is}}{a_{rs}} \text{ if } i \neq r \text{ and } j \neq s \text{ (all other elements),} \end{cases}$$

$$b_i := \begin{cases} \dfrac{b_r}{a_{rs}} \text{ if } i=r \quad \text{(pivot row)} \\ b_i - \dfrac{b_r\, a_{is}}{a_{rs}} \text{ if } i \neq r \text{ (all other rows),} \end{cases}$$

$$c_j := \begin{cases} 0 \text{ if } j=s \text{ (pivot column)} \\ c_j - \dfrac{a_{rj}\, c_s}{a_{rs}} \text{ (all other columns), and} \end{cases}$$

$$z_0 := z_0 - \frac{b_r c_s}{a_{rs}}$$

Go to Step 1.

If the algorithm stops with an optimal solution, z_0 will be the optimal value of the objective function.

When no initial basic feasible solution is readily available to start Phase 2, a Phase 1 procedure must be used. The idea is to add *artificial variables* A_i to form the constraints $\mathbf{a}_{i\bullet}\mathbf{x} + A_i = b_i$ and start the algorithm with the artificial variables in the initial basis. Phase 1 drives the artificial variables out of the basis by minimizing the artificial objective function which is the sum of the nonnegative artificial variables. If this sum is still positive at the termination of Phase 1, the initial problem has no feasible solution; otherwise an initial basic feasible solution has been identified, the artificial variables can be dropped and Phase 2 begin. For details, see, e.g., Eiselt *et al.* (1987).

a.2 Duality and Postoptimality Analysis

An important concept in linear programming is that of duality. A problem dual to a given linear programming problem is defined as follows.

Definition 9: Given a linear programming problem P in canonical form:

$$\text{P:}\quad \underset{\mathbf{x}}{\text{Max}}\ z = \mathbf{cx}$$

$$\text{s.t}\qquad \mathbf{Ax} \leq \mathbf{b}$$

$$\mathbf{x} \geq \mathbf{0}$$

In this context, the problem P is referred to as the *primal problem*. With each primal problem P we associate a *dual problem* P_D. This dual problem can then be written as

$$\text{P}_D\text{:}\quad \underset{\mathbf{u}}{\text{Min}}\ z_D = \mathbf{ub}$$

$$\text{s.t.}\qquad \mathbf{uA} \geq \mathbf{c}$$

$$\mathbf{u} \geq \mathbf{0}$$

Note that while the primal problem has n variables and m structural constraints, the dual problem has m variables and n structural constraints. There is a one-to-one correspondence between each primal variable and dual constraint, and between each primal constraint and dual variable. We now state (without proof) the following four main results.

Theorem 10 (Weak duality): If **x** is feasible for P (i.e., **x** is primal feasible) and **u** feasible for P_D (i.e., **u** is dual feasible), then $\mathbf{cx} \leq \mathbf{ub}$.

Theorem 11 (Strong duality): Exactly one of the following three statements is true:

- P and P_D have finite optimal solutions $\bar{\mathbf{x}}$ and $\bar{\mathbf{u}}$, respectively, with $\mathbf{c}\bar{\mathbf{x}} = \bar{\mathbf{u}}\mathbf{b}$.
- Neither P nor P_D has a feasible solution.
- One of the problems has feasible, but no finite optimal solutions, and the other has no feasible solution.

Theorem 12 (Weak complementary slackness): Assume that $(\bar{\mathbf{x}}, \bar{\mathbf{u}})$ is a pair of feasible solutions for P and P_D. Then $(\bar{\mathbf{x}}, \bar{\mathbf{u}})$ is a pair of optimal solutions if and only if

$$\bar{\mathbf{u}}(\mathbf{A}\bar{\mathbf{x}} - \mathbf{b}) = \mathbf{0}, \text{ and } (\bar{\mathbf{u}}\mathbf{A} - \mathbf{c})\bar{\mathbf{x}} = \mathbf{0}$$

Theorem 13 (Strong complementary slackness): Assume that feasible solutions exist for both P and P_D. Then there exists a pair $(\bar{\mathbf{x}}, \bar{\mathbf{u}})$ of optimal solutions such that

$$\bar{\mathbf{u}} + (\mathbf{b} - \mathbf{A}\bar{\mathbf{x}}) > \mathbf{0}, \text{ and } \bar{\mathbf{x}} + (\bar{\mathbf{u}}\mathbf{A} - \mathbf{c}) > \mathbf{0}$$

From the above four theorems a number of different primal-dual relationships and optimality conditions can be derived. It is worth mention that a tableau in each iteration of the primal simplex method (Phase 2) corresponds to a feasible solution for the primal problem, a solution to the dual problem which is not yet feasible, but such that weak complementary slackness holds. If dual feasibility holds as well, optimality is reached. For the dual simplex method, each tableau corresponds to a dual feasible solution and a primal not yet feasible solution, but weak complementary slackness holds. When primal feasibility is reached, the tableau is optimal.

As opposed to the case for the primal simplex algorithm, the pivot row in the *dual simplex algorithm* is selected first and then the pivot column is determined. To describe the algebraic procedure of the dual simplex method we assume that the problem is given in canonical form. Here, we allow that one or more $b_i < 0$.

The Dual Simplex Algorithm

Step 0: Set up the initial tableau as for the primal simplex algorithm. Is there any $c_j < 0$?

If yes: Add the constraint $\sum_{j=1}^{n} x_j + S^* = M$ as the $(m + 1)$-st row in the tableau (referred to as $\mathbf{T}_0$), select $a_{m+1,s} = 1$ as pivot, where $c_s = \min_{j:c_j<0} \{c_j\}$ and perform a tableau transformation. Go to Step 1.

If no: Go to Step 1.

Step 1: Is there any $b_i < 0$?

If yes: Go to Step 3.

If no: Go to Step 2.

Step 2: Is the value of the objective function z a function of M?

If yes: Stop; unbounded "optimal" solutions exist.

If no: Stop; the current solution is optimal.

Step 3: Select the r-th row as the pivot row, such that $b_r < 0$.

Step 4: Is there at least one element $a_{rj} < 0, j = 1, ..., m + n$?

If yes: Go to Step 5.

If no: Stop; the problem has no feasible solution.

Step 5: Select the s-th column as pivot column, where

$\frac{c_s}{|a_{rs}|} = \min\left\{\frac{c_j}{|a_{rj}|} : a_{rj} < 0\right\}$. Perform one tableau transformation with the pivot $a_{rs} < 0$ (see Step 5 in the primal simplex algorithm) and go to Step 1.

The dual simplex method is quite important for cutting plane and branch-and-bound methods for integer programming; see Chapters 5 and 6 of Part I.

The study of how the optimal solution to a linear programming problem is affected by changes in the original problem structure and data is usually called *postoptimality analysis*. It turns out that many important postoptimality questions can readily be answered by utilizing the final tableau of the simplex procedure. The various cases that are usually considered are shown in Figure 1.

We generally distinguish between *sensitivity analysis* and *parametric programming*. In sensitivity analysis we consider the range over which a given parameter can change without changing the optimal basis or solution; in parametric programming we study what happens if we move outside this range.

The simplex tableau is the key element in linear programming postoptimality analysis and its usefulness in this regard is an important *raison d'être* for the simplex method.

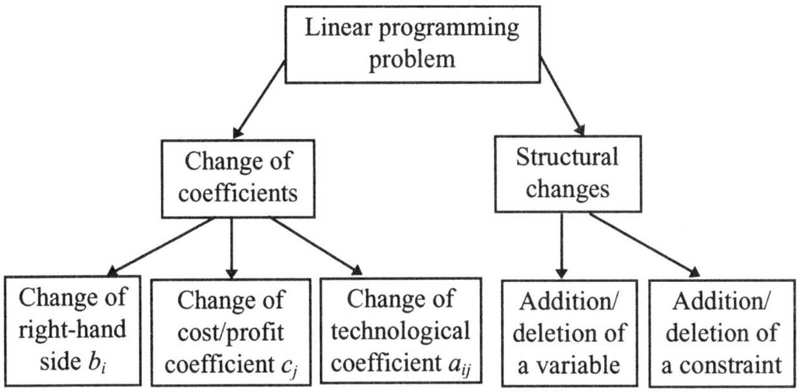

Figure 1

a.3 Problems with Special Structures

A number of linear programming problems have special structures which can be exploited for finding specialized solution algorithms. Such specially structured problems often arise as subproblems of larger and more complex linear programming problems.

The first special structure to be addressed is referred to as the *Transportation Problem*. Consider a situation in which there is a supply s_i, $i = 1, ..., m$ of a certain commodity at each of m locations, called *origins*, and a demand d_j, $j = 1, ..., n$ at each of n locations, called *destinations*. Assume that total supply equals total demand, $\sum_{i=1}^{m} s_i = \sum_{j=1}^{n} d_j$, and that the cost of transporting one unit of the commodity from origin i to destination j is c_{ij}, $i = 1, ..., m$; $j = 1, ..., n$. The transportation problem consists in finding a minimum total cost plan of sending all the supply from the origins to the destination points in such a way that all demand is met. This is achieved by letting x_{ij} be the amount transported from origin i to destination j. The transportation problem can then be formulated as follows.

$$\text{P: Min } z = \sum_{i=1}^{m}\sum_{j=1}^{n} c_{ij}x_{ij}$$

$$\text{s.t.} \quad \sum_{j=1}^{n} x_{ij} = s_i, i = 1,\ldots,m \qquad \text{(supply availability)}$$

$$\sum_{i=1}^{m} x_{ij} = d_j, j = 1,\ldots,n \qquad \text{(demand requirement)}$$

$$x_{ij} \geq 0,\ i = 1, \ldots, m; j = 1, \ldots, n$$

The transportation problem arises naturally in a transportation context, but linear programming problems with this particular structure arise in other situations as well. There are special-purpose methods of the simplex method that are more efficient than the regular simplex method applied to this type of problem. A variety of generalizations of the basic transportation problem exist. They include bounds on the flow between nodes and other additional constraints. For a detailed discussion, see Chapter I.4 and Part III.

Another special structure is the so-called *Assignment Problem*. It can be described as follows. Consider a situation in which n workers are to be assigned to n jobs in such a way that exactly one job has to be assigned to each worker and vice versa. The cost of assigning the i-th worker to the j-th job is given as c_{ij}. The problem of finding an assignment of workers to jobs which minimizes the overall cost can be formulated as a linear programming problem by defining variables x_{ij} that assume a value of one, if the i-th worker is assigned to the j-th job, and equal to zero otherwise.

It is apparent that the assignment (or *bipartite weighted matching*) *problem* is a special case of the transportation problem with $m = n$ and $s_i = d_j = 1\ \forall\ i, j$ and zero-one variables x_{ij}. Although any method that solves the transportation problem can be used to solve the assignment problem, more efficient specialized techniques exist. The first such method to be described in the literature is the so-called *Hungarian method*. More efficient techniques are based on network flow techniques, see, e.g., Part III or Ahuja *et al.* (1993). The assignment problem arises as a subproblem in a variety of important models such as the *traveling salesman problem* that is discussed in Part II.

CHAPTER b ANALYSIS OF ALGORITHMS

b.1 Algorithms and Time Complexity Functions

In this chapter we are not concerned with any specific model but will have a look at some common features shared by many solution techniques or algorithms. In order to avoid confusion, we will use the term *problem* to denote a general mathematical formulation such as a linear, nonlinear, integer etc., programming model, whereas an *instance* of a problem will be any specific numerical case or realization of this problem.

Some problems have so-called *closed form solutions*, i.e., formulas that solve the problem. An example of a problem that does have a closed form solution are quadratic equations. Typically, any problem that is not tiny or elementary will not have a closed form solution. Such problems, and these are virtually all problems of practical interest, require a mathematical method or algorithm for their solution.

An *iterative algorithm* is a computational procedure for finding a solution to every instance of a specific problem. It does so by starting with an initial solution that is either arbitrary or has some prespecified features. This solution is then checked for the desired set of properties Π, such as optimality or feasibility. If the current solution satisfies the properties in Π, then the procedure stops; otherwise a new solution is generated and the procedure is repeated. In order to formalize matters, let k be an iteration counter and denote by $\mathbf{x}^1$ the initial solution. We will write $\mathbf{x}^k \in \Pi$ if the solution $\mathbf{x}^k$ satisfies the properties included in Π. Finally, let $\tau(\mathbf{x}^k)$ be a transformation that generates a new solution $\mathbf{x}^{k+1}$ from the given solution $\mathbf{x}^k$. The basic structure of an iterative algorithm is outlined in Figure 2.

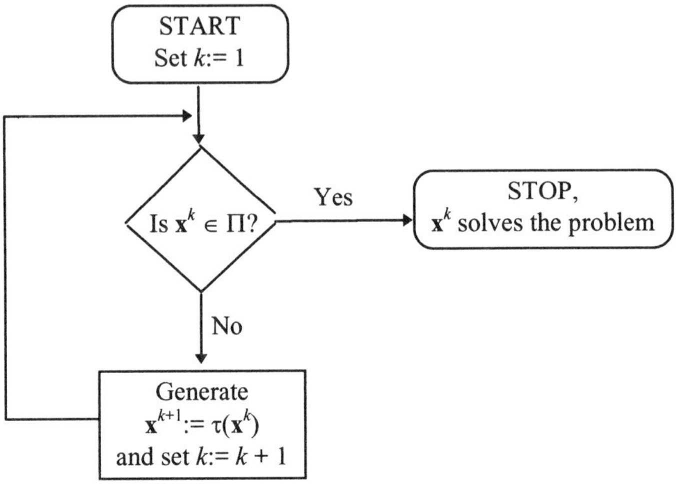

Figure 2

As a well-known example, consider the MODI method for solving the classical transportation problem; for details see, e.g., Eiselt *et al.* (1987). The solution $\mathbf{x}^1$ corresponds to some initial transportation plan, obtained by any of the well-known methods. Then the set Π includes the optimality criterion, namely the nonnegativity of the opportunity costs, associated with a transportation plan. If Π is satisfied, the procedure terminates and the current solution is optimal; otherwise a new solution is generated, using the transformation function τ to improve the solution.

Many algorithms do require a more sophisticated structure than the one displayed in the above flowchart. As an example, the primal two-phase simplex algorithm requires two rather than only one sets of properties Π_1 and Π_2 where Π_1 includes feasibility conditions and Π_2 comprises optimality conditions. In general, stop criteria are usually associated with the two major problems in mathematical programming: feasibility and optimality. In principle, after an algorithm is designed, two properties will have to be proved, *viz.*,

- the *validity* of the procedure, and
- its *convergence*.

As far as the validity is concerned, we must prove that the solution produced by the procedure satisfies the properties in Π, and that, if an optimal solution does exist, it is found by the algorithm. The proof of convergence for exact methods is usually a proof of finiteness of the procedure, something that may not be required for certain approximation algorithms. As an example, consider again the two-phase primal simplex algorithm. Its validity is proved by demonstrating that the solution found by the method is indeed optimal (e.g., by using duality and complementary slackness arguments) and that if an optimal solution exists, it will be found with the method (usually proved by using the equivalent statement: if no optimal solution is found, then no optimal solution exists). Finite convergence was proved for all simplex methods showing that

- at least one optimal solution (if any) must be located at an extreme point of the given polytope,
- each extreme point of the polytope corresponds to at least one basic feasible solution,
- the number of basic feasible solutions is finite, and
- each basic solution is generated at most once.

Once validity and finiteness are proved, the method can be implemented and applied. Over the years, however, scientists and practitioners alike have found that requiring finiteness of a procedure is not sufficient. For example, what good does it do if one can prove that an algorithm is finite if, even when applied to relatively small models, it would take hours, days or even years of time to solve a given instance of the problem on a computer? This argument may seem far-fetched at first sight, but we will see in this chapter that problems of this type do indeed exist. Consequently, we need a better criterion than finite convergence and, in general, parameters that evaluate or classify an algorithm with respect to its performance.

Since all real world models will be solved on computers, we will focus on the requirements of algorithms as they are implemented on a computer. Typically, two criteria are used as performance criteria for algorithms: the amount of time required for the solution of a given model and the storage space requirements in the solution process. Clearly, the time requirement does not exclusively depend on the algorithm and the model itself but also on the specific computational equipment used, just as the required storage space for a model also depends on the way in which the model is coded. Since it is rather obvious that there is no general and definite answer to questions like: "How long will it take to solve any linear programming problem with the primal simplex algorithm?" even if the type of computational equipment were specified, the analyst may rephrase the question and either ask: "What is the expected time (or storage space) requirement?" or: "How much time (or storage space) will be required in the worst case?" if a specific problem is to be solved. The various types of analyses of an algorithm can be visualized in Figure 3.

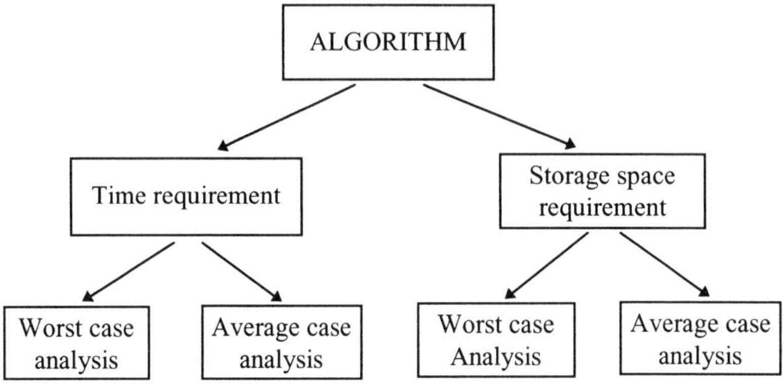

Figure 3

In order to avoid discussing many details concerning the implementation of a model on a computer, we will restrict ourselves mostly to the worst case time analysis of algorithms. In general, while it is generally possible to develop a theoretical upper bound for the time requirement in the worst case, an analysis of the average case either requires a probabilistic analysis using some statistical distribution for the numerical values of the problem coefficients, or an empirical study that involves solving a test series of either collected practical or random-generated instances of a problem. Examples for worst-case as well as probabilistic analyses are provided below.

In order to calculate the time requirements, it is usually beneficial to decompose the term "time" as

$$[\text{time}] = \left[\frac{\text{time}}{\text{elementary operations}}\right]\left[\frac{\text{elementary operations}}{\text{iterations}}\right][\text{iterations}]$$

The above term "elementary operation" is used for addition, subtraction, multiplication, division and comparison (i.e., finding the maximum or minimum of any two given numbers), and it is assumed that all elementary operations require the same constant time on a given computer, regardless of the magnitude of the numbers involved. This is obviously a simplification since a multiplication takes considerably longer than an addition or a comparison but this simplification does not change the principle of the argument. The time requirement for one elementary operation clearly depends on the specific computer model; currently it lies in the vicinity of a nanosecond, i.e., a modern computer is able to perform about a billion elementary operations per second. As the time requirement per iteration has nothing to do with the algorithm but rather with the state-of-the-art technology, we will just count the operations required for an algorithm and the number of iterations. If need be, the speed of the computer will be specified on an ad hoc basis geared towards the specific argument being pursued. The last two factors in the above time function do not represent any fixed numbers that can be associated with a specific algorithm. They are functions depending on the size of the model or, in the above terms, the *size of the instance of the problem*. Since it is assumed that a digital computer is used for the solution of the model, the length of a *binary encoding* of the given instance is a reasonable measure of the size of the model. This yardstick does, however, sometimes prove to be rather awkward, so that we will usually resort to other "natural" measurements of size. As an example, consider a linear programming problem. There the number of variables together with the number of constraints provide a reasonable description of the size of the model. If a string of n numbers is given that are to be sorted or otherwise manipulated, n is again used as a description of the size of the instance of the problem. On the other hand, the magnitude of the numbers in the model is *not* a reasonable measurement of the size, at least not as far as digital computers are concerned. Such machines use binary encodings in which any given number b, e.g., the largest number in the model, is represented by $\log_2 b$ rather than b digits; for an introduction, see Lenstra and Rinnooy Kan (1979). Often, the number of elementary operations per iteration is easy to determine whereas the required number of iterations is either more difficult to compute or its only known bound may be exceedingly large.

Suppose now that the size of any instance of a problem can be described by n, and let $f(n)$ denote the maximal number of elementary operations required to solve any instance of the problem using the algorithm concerned. Then the function $f(n)$ is called the *time complexity function* of the algorithm. As an example, let the time complexity function of some algorithm be $f(n) = ½n^5 + 20n^4 + 2{,}000\ n$. Since we are specifically interested in the behavior of an algorithm applied to large-scale problems, i.e., problems with $n >> 0$, one can easily observe that the term $½n^5$ increases at the fastest rate whereas the other terms become relatively insignificant. Asymptotically, the number of elementary operations increases with n^5 and therefore an algorithm with the above time complexity function will be called of order n^5 or $O(n^5)$ for short. Loosely speaking, $O(f(n))$ includes only the fastest growing term of $f(n)$. As will be seen below, $O(f(n))$ is frequently much easier to determine than $f(n)$ itself. We will call any algorithm for which $f(n)$ is bounded from above by a polynomial function in n a *polynomially bounded algorithm*, whereas all other algorithms are termed *exponentially bounded algorithms* whether $f(n)$ is an exponential function in the strong sense or not. Edmonds (1965) was the first to use the term "good" for polynomially bounded algorithms. The reason to consider polynomially bounded algorithms good and exponentially bounded algorithms poor performers (in the worst case) becomes clear considering Table 1. There it is assumed that the given computer requires one nanosecond (10^{-9} seconds) to perform an elementary operation. Columns 2 – 5 in Table 1 show the computation times for a variety of polynomial and exponential algorithms, and column 6 provides values of n_{hour}, defined as the largest problem size n that can be solved within one hour by the algorithm under consideration. Finally, N denotes the size of problems that can be solved in the same time as a problem of size n, given a computer that is 1,000 times as fast.

Table 1

	Time to solve a problem of size					
O(•)	$n = 10$	$n = 20$	$n = 50$	$n = 100$	n_{hour}	N
$O(n^2)$	$(1)10^{-7}$ sec	$(4)10^{-7}$ sec	$(2.5)10^{-6}$ sec	$(1)10^{-5}$ sec	$n \approx 1.9$ million	$31.6n$
$O(n^3)$	$(1)10^{-6}$ sec	$(8)10^{-6}$ sec	$(1.3)10^{-4}$ sec	$(1)10^{-3}$ sec	$n \approx$ 15.326	$10n$
$O(n^5)$	$(1)10^{-4}$ sec	$(3.2)10^{-3}$ sec	0.31 sec	10 sec	$n \approx 325$	$3.98n$
$O(2^n)$	$(1)10^{-6}$ sec	$(1)10^{-3}$ sec	13 days	$(4)10^{13}$ sec	$n < 41$	$n + 9.97$
$O(3^n)$	$(5.9)10^{-5}$ sec	3.5 sec	22.8 million years	$(1.6)10^{31}$ years	$n = 26$	$n + 6.29$
$O(n!)$	$(3.61)10^{-3}$ sec	77 years	$(9.6)10^{47}$ years	$(3)10^{141}$ years	$n = 15$	$n + 3$ for $n = 10,20$ $n + 2$ for $n = 50,100$
$O(n^n)$	10 sec	$(3.3)10^{9}$ years	$(2.8)10^{68}$ years	$(3.2)10^{183}$ years	$n = 11$	$n + 2.0$ for $n = 10$ $n + 1.7$ for $n = 20$ $n + 1.4$ for $n = 50$ $n + 1.2$ for $n = 100$

So far, we have discussed complexity functions which are assigned to specific *algorithms*. It is also possible to associate a worst case complexity function with every given *problem*. The complexity of a problem is the same as the complexity function of the best known algorithm that solves the problem. As an example, consider the assignment problem with n elements; for details see, e.g., Eiselt *et al.* (1987). It is well known that there exists an $O(n^3)$ algorithm that solves the assignment problem. Clearly, one could also simply enumerate all $n!$ solutions (an obviously vastly inferior "method") so that if only these two solution techniques were known for the assignment problem, one would conclude that the assignment problem has a worst case complexity of $O(n^3)$. This indicates that since a good, i.e., polynomially bounded, algorithm exists, the assignment problem can be considered well solved. This does, however, not necessarily mean that the complexity function cannot be reduced further or that future research in this area is not needed. Clearly, an algorithm with a lower degree polynomial would be preferable to one with a higher degree complexity function, but the improvement which results from a new, say, $O(n^2)$ algorithm where an $O(n^3)$ existed before is minor compared to the improvement of finding a polynomially bounded algorithm where only an exponentially bounded method was previously known. While over the last ten years more efficient algorithms have been found, it is also clear that there is a point where no further improvements are possible. In some of the examples below we will specify a lower bound on the worst case complexity function.

The concept of polynomial algorithms introduced above can be refined somewhat. Again let n be an indication of the size of the problem, let b symbolize the largest number included in the problem, and assume that $f(\bullet)$ is a polynomial function of its argument. We can take an algorithm with a time complexity function of $f(n)$ as *strongly polynomial*, an algorithm with a complexity function of $f(n, \log b)$ as *polynomial*, and an algorithm with $f(n, b)$ as *pseudopolynomial*. Dynamic programming algorithms described in Chapter d are typically pseudopolynomial.

b.2 Examples of Time Complexity Functions

In this section we will provide some simple algorithms whose worst case complexity can easily be determined. The principle in selecting these algorithms has been that they should be simple and not anticipate concepts to be described later in this volume. This may make the exposition fairly elementary, but it should make the principle sufficiently clear. The methods are mostly chosen from the areas of linear algebra and file organization in data processing.

Example 1: Solution of systems of simultaneous linear equations.

Problem: Given a matrix $\mathbf{A} \in \mathbb{R}^n \times \mathbb{R}^n$ and a vector $\mathbf{b} \in \mathbb{R}^n$, find a solution $\mathbf{x}$ to the system $\mathbf{Ax} = \mathbf{b}$ with pivoting.

Complexity: Let $\mathbf{A}^k = (a_{ij}^k)$ be the matrix before the k-th iteration is performed, i.e., $\mathbf{A}^1 = \mathbf{A}$. Suppose that in the k-th iteration the element a_{rs} has been selected as the pivot element. Then the rule requires the following elementary operations:

- Pivot column. Here, we set $a_{rs}^{k+1} := 1$, and $a_{is}^{k+1} := 0 \; \forall i \neq r$. As these elements are set equal to zero and one, respectively, and no elementary operations are required.

- Pivot row. $a_{rj}^{k+1} := a_{rj}^{k} / a_{rs}^{k} \; \forall \, j$. One division is needed for each element in the pivot row.
- All other elements. $a_{ij}^{k+1} := a_{ij}^{k} - (a_{is}^{k} a_{rj}^{k}) / a_{rs}^{k}$, i.e., three elementary operations each. This number can be reduced using the fact that a_{rj}^{k} / a_{rs}^{k} has already been calculated. Hence one can write $a_{ij}^{k+1} := a_{ij}^{k} - a_{is}^{k} a_{rj}^{k+1}$, saving one division per element.

For simplicity, suppose that in the k-th iteration the element a_{kk}^{k} is chosen as pivot element. This means that for the first k columns no operations are necessary, which leaves the remaining $(n - k + 1)$ columns including the right-hand side. Also, the elements in the first k rows do not have to be calculated. Assuming that the matrix **A** has full rank, exactly n iterations will be necessary. Then the time complexity function is

$$f(n) = \sum_{k=1}^{n} (n - k + 1) + 2\sum_{k=1}^{n} (n - k + 1(n - k) = \tfrac{2}{3}n^3 + \tfrac{1}{2}n^2 - \tfrac{1}{6}n,$$

i.e., the complexity of the algorithm is of order $O(n^3)$. This result could have been obtained much easier considering the fact that in each iteration the transformation of a single element takes constant time; given n^2 elements in the matrix, each iteration has a complexity of $O(n^2)$ and the fact that n iterations are needed leads to the above result. Note that since each of the n^2 elements given in the problem has to be considered and used at least once in an elementary operation during any solution method, an obvious lower bound for the complexity function of the problem is $O(n^2)$.

Example 2: Evaluation of a determinant.

Problem: Given a matrix $\mathbf{A} \in \mathbb{R}^n \times \mathbb{R}^n$, evaluate the determinant of **A** denoted by det **A**.

Complexity: The first method pivots in each row and in each column no more than once, and then the product of the pivots is det **A**. As in example 1, $O(n^2)$ elementary operations are required in one iteration, and no more than n iterations (less if **A** is singular) are needed, resulting in $O(n^3)$. In addition, the product of the pivots is calculated in $O(n)$ multiplications, yielding an overall complexity of $O(n^3)$.

The second method is the Laplace expansion. It evaluates the determinant by the i-th row, by computing

$$\det \mathbf{A} = \sum_{j=1}^{n} (-1)^{i+j} \det \mathbf{A}_{ij} ,$$

where $\mathbf{A}_{ij}$ denotes the submatrix of $\mathbf{A}$ obtained from $\mathbf{A}$ by deleting its i-th row and its j-th column. In the worst case $a_{ij} \neq 0 \ \forall \ j = 1, ..., n$ so that det $\mathbf{A}$ is now expressed as the sum of n matrices of dimension $[(n-1) \times (n-1)]$ each of which, in turn, can be expressed as the sum of $(n-1)$ matrices of dimension $[(n-2) \times (n-2)]$ and so forth. The procedure continues until the determinant is expressed in terms of determinants of $[2 \times 2]$ matrices which can be evaluated in three elementary operations each. The total number of elementary operations is roughly $3\prod_{1=2}^{n} i$ which is $O(n!)$. Hence this technique has an exponential complexity and is computationally inferior to the pivoting method above, at least as far as the worst case is concerned. This method is still quite instructional and useful for some proofs.

Example 3: Multiplication of two matrices.

Problem: Given two matrices $\mathbf{A}, \mathbf{B} \in \mathbb{R}^n \times \mathbb{R}^n$, determine $\mathbf{C} = \mathbf{AB}$, using

$$c_{ij} = \sum_{k=1}^{n} a_{ik} b_{kj} \ \ \forall \ i, j = 1, ..., n.$$

Complexity: Each element c_{ij} is the inner (or scalar) product of the n-vector $\mathbf{a}_{i\bullet}$ and the n-vector $\mathbf{b}_{\bullet j}$. The calculation of each element c_{ij} thus requires n multiplications and $(n-1)$ additions, i.e., $O(n)$ elementary operations. Since $\mathbf{C}$ includes n^2 elements, n^2 inner products have to be calculated resulting in a complexity of $O(n^3)$. Note that since each of the $2n^2$ elements of $\mathbf{A}$ and $\mathbf{B}$ is used in at least one operation, a lower bound of $O(n^2)$ exists for the complexity. The reader is referred to the decomposition principle of Strassen (1969) with improved complexity for this problem.

Example 4: Initial solution to the classical transportation problem with m origins and n destinations. For simplicity, we assume that $m = n$.

Problem: Find a feasible solution to the system

$$\sum_{j=1}^{n} x_{ij} = s_i \ \ \forall \ i = 1, ..., n$$

$$\sum_{i=1}^{n} x_{ij} = d_i \ \ \forall \ j = 1, ..., n$$

$$x_{ij} \geq 0 \quad \forall \; i, j = 1, \ldots, n.$$

Complexity: It is known that no more than $(2n - 1)$ variables will have a value different from zero. All exisitng methods differ only in the selection of the variables they assign positive transportation quantities to. For a detailed description of the rules see, e.g., Eiselt *et al.* (1987). Other than that, each iteration requires a single comparison $x_{ij} := \min \{s_i, d_j\}$, and the subsequent updating of supply $s_i := s_i - x_{ij}$ and demand $d_j := d_j - x_{ij}$. In other words, two operations are needed for each of the $2n - 1$ nonzero variables.

In the *Northwest Corner Rule*, no other than the stated elementary operations are necessary, so that the complexity is $O(n)$.

The *Least Cost Method* requires to first scan the given $[n \times n]$-dimensional cost matrix in order to find its minimal element, a task that requires $O(n^2)$ comparisons. Then we make the assignment described above and the procedure is repeated. Even though in the next step the search is only in an $[n \times (n - 1)]$-dimensional matrix (and after that in an $[n \times (n - 2)]$ or in an $[(n - 1) \times (n - 1)]$-dimensional matrix and so forth), the required number of comparisons is still of order $O(n^2)$. This procedure is repeated $(2n - 1) = O(n)$ times, and thus its computational complexity is $O(n^3)$.

In *Vogel's Approximation Method*, the calculation of the 1-minimum and the 2-minimum requires n and $(n - 1)$ comparisons respectively. Since the penalty for each row and column is the difference between the 1-min and the 2-min, $n + (n - 1) + 1 = 2n = O(n)$ elementary operations are needed to determine the penalty for each row and column. This has to be repeated for all $2n = O(n)$ rows and columns, so that the calculation of all penalties requires $O(n^2)$ operations. After these preliminary calculations, the steps start in which quantities are assigned to routes. For that purpose the maximal penalty is sought in $O(n)$ comparisons. In the row or column it belongs to, the 1-min is either determined again (requiring $O(n)$ comparisons) or found by a previously created pointer (in constant time). Then the assignment is made on that element, and the updating commences. Suppose that in the row in which the assignment is made the supply is depleted and no further assignments are possible. Then the penalties of all other rows remain unchanged, and only the penalties of the columns have to be recalculated. If this is done from scratch, $O(n^2)$ operations are needed as shown above. This procedure is repeated $(2n - 1) = O(n)$ times, and therefore its computational complexity is $O(n^2) + n\, O(n^2) = O(n^3)$.

Example 5: Search methods.

Problem: Given a file with n numbers a_j, $j = 1, ..., n$ such that $a_1 \le a_2 \le ... \le a_n$ and each number is allocated the same amount of space on the file. Given some number a^* (the search argument), find j, such that $a_j = a^*$.

Complexity: The *Sequential Scanning Method* compares a^* with a_1, if $a^* = a_1$, stop; otherwise compare a^* with a_2 and so forth. Clearly, in the worst case $a^* = a_n$ and n comparisons are necessary; hence the complexity is $O(n)$. Let $P(a_j)$ now denote the probability that $a^* = a_j$. If $P(a_i) = P(a_j) = 1/n \ \forall \ i \ne j$, then the expected number of comparisons (the average case analysis) is

$$\sum_{j=1}^{n} P(a_j) j = (n+1)/2$$

which is still $O(n)$. However, if numbers that are frequently requested are placed at the beginning of the file, the average time savings are significant.

For the *Bisection Search*, suppose that $n = 2^k$ for some $k \in \mathbb{N}_0$. Compare a^* with $a_{n/2}$; if $a^* \le a_{n/2}$ then discard all a_j with $j > n/2$ and if $a^* > a_{n/2}$, discard all a_j with $j \le n/2$. Then the process is repeated in the remaining interval. It can be seen that in each iteration the relevant file length is cut in half; since only one comparison is required per iteration, it follows that the complexity is of order $O(\log_2 n)$. As an illustration of the above, let $n = 16$ and suppose that $a^* = a_{11}$. Then the arrows in Figure 4 show the steps in the bisection search procedure.

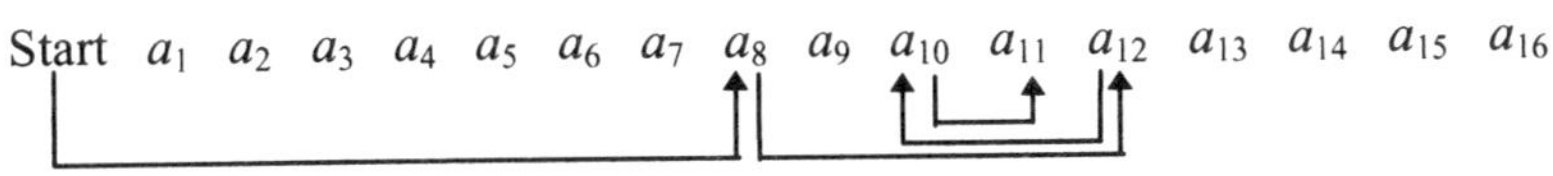

Figure 4

The bisection search procedure is extremely efficient; even in files with one billion numbers require no more than thirty comparisons. It should also be mentioned that with $P(a_i) = P(a_j) = 1/n \ \ \forall \ i \ne j$, the average case complexity is almost identical to the worst case complexity.

In case of *Matrix Search*, suppose that the numbers a_j, $j = 1, ..., n$ are ordered in a $\left[\sqrt{n} \times \sqrt{n}\right]$-dimensional matrix so that $a_{11} \le a_{21} \le ... \le a_{m1} \le a_{12} \le a_{22} \le ... \le a_{m2} \le a_{13} \le ...$, i.e., the ordered file is broken down into $\sqrt{n}$ -dimensional subfiles which are the columns of the matrix. The procedure is to scan the elements a_{11}, a_{12}, ..., $a_{1\sqrt{n}}$ (i.e., the first row of the matrix) until for some k, $a^* > a_{1k}$ for the first time or

$a_{1\sqrt{n}}$ has been reached, in which case we set $k:=1+\sqrt{n}$. Then sequential scanning continues in the $(k-1)$-st column. In the worst case all $\sqrt{n}$ elements in the first row and the $\sqrt{n}$ elements in column $(k-1)$ have to be scanned yielding a worst case complexity of $2\sqrt{n} = O(\sqrt{n})$. Note that the average complexity is $\sqrt{n}$. A possible extension is to conduct a binary, rather than sequential, search in the row and later the column of the matrix.

Search in an unordered set. Although sequential scanning is by far inferior to the two other search methods described above, it is the only one which can also be used for a search in an unordered file. Its complexity does not change whether the file is ordered or not. In many applications, however, the probabilities $P(a_j)$ are not uniformly distributed. Consider, for example, a file of customer numbers. Each time a customer places an order, its number has to be sought, and from there a pointer leads to a customer's file, e.g., his address, most recent orders, etc. Clearly, there will be customers who order frequently, as well as customers whose orders come in infrequently. Since the numbers of "good" customers have to be sought much more frequently, it seems sensible enough to order the file, such that $P(a_1) \leq P(a_2) \ldots \leq P(a_n)$. This may dramatically reduce the average number of necessary comparisons, although it does not change the worst case complexity as the following example may illustrate. Let the customers of a company be divided into three groups, so that group one comprises ten percent of the customers, group two consists of twenty percent of the customers and group three includes the remaining seventy percent. On average, each customer in group one orders five times as often as any customer in group two who, in turn, places ten times as many orders as a customer in the third group. Assume that $a_1, \ldots, a_{n/10}$ are the customer numbers for group one; $a_{n/10+1}, \ldots, a_{3n/10}$ are the numbers for members of group two and the last $(3n/10)$-1 numbers are associated with customers from group three. Then the probability distribution is

$$P(a_j) = \begin{cases} 500/77n & \text{for } j \leq n/10 \\ 100/77n & \text{for } j \in]n/10; 3n/10] \\ 10/77n & \text{for } j > 3n/10 \end{cases}$$

The expected number of comparisons is then

$$\sum_{i=1}^{n/10}(500/77n)\,i + \sum_{i=n/10+1}^{3n/10}(100/77n)\,i + \sum_{i=3n/10+1}^{n}(10/77n)\,i = (131n+680)/1{,}540$$

which approaches $0.085n$ for large values of n. For $n = 1000$ customers, rather than an average of 500.5 comparisons for a completely unordered file, only 85.5 comparisons are needed.

Example 6: Sorting methods.

Problem: Order a file of n numbers $a_1, a_2, ..., a_n$, so that $a_1 \leq a_2 \leq ... \leq a_n$.

Complexity: The simplest technique is *Sorting by Selection*. Find the minimum among all elements on the file (requiring O(n) comparisons), replace it by some very large number $M >> 0$, and write it on a separate file. After repeating this procedure n times, the new file includes all numbers in nondecreasing sequence as desired, whereas the original file consists of only M's. Hence the overall complexity is O(n^2).

The *Pairwise Exchange Method* first compares the first and second, the third and fourth, the fifth and sixth, etc. numbers with each other and exchange them if they are not in nondecreasing order. In the next step, compare the second with the third, the fourth with the fifth, the sixth with the seventh element, etc. and rearrange again if necessary. All odd-numbered iterations are identical to the first, and all even-numbered iterations are the same as the second iteration. The procedure terminates as soon as no two numbers are exchanged in two successive iterations. In each iteration, ½n comparisons are required, and no more than O(n) iterations are necessary. This becomes intuitively clear by considering the worst case in which the smallest number which should appear in the first position on the file is currently at the very end of the file. In each iteration, this number moves one step towards the beginning of the file. This yields an overall complexity of O(n^2) of the procedure. As an illustration, consider the following example in Figure 5 where round brackets indicate the pairs of numbers that are considered for a possible change, and an asterisk next to the bracket shows that a pairwise exchange was made.

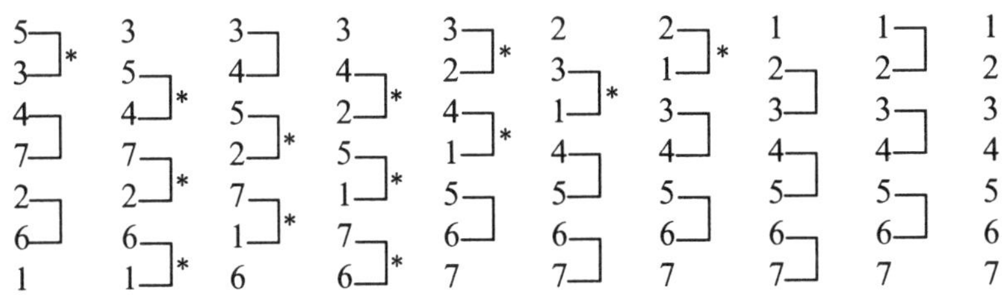

Figure 5

Example 7: Linear Programming.

Problem: Find a vector $\mathbf{x}$ which maximizes $z = \mathbf{cx}$ subject to the constraints $\mathbf{Ax} \leq \mathbf{b}, \mathbf{x} \geq \mathbf{0}$ with $\mathbf{c} \in \mathbb{R}^n$, $\mathbf{b} \in \mathbb{R}^m$, and $\mathbf{A} \in \mathbb{R}^m \times \mathbb{R}^n$.

Complexity: Suppose that any of the simplex methods is used for the solution of a given linear programming problem with n variables and m constraints. Since the tableau including slack and artificial variables is of size $[m \times (n + m)]$, the number of elementary operations for one iteration is of order $O(m(m + n))$. The only known upper bound for the number of iterations is, however, the number of distinct bases of the problem. Recall that a basis in the above problem is an $[m \times m]$-dimensional submatrix of full rank, the maximal number of bases equals the number of combinations of $(m + n)$ columns, taken m at a time. This number is $\binom{m+n}{m}$; clearly not part of or bounded by any polynomial function. In most cases, only a tiny fraction of these bases will be examined by the simplex method employed. For some pivot column selection rules, however, Klee and Minty (1972) have constructed pathological examples that require the examination of an exponential number of bases. According to the knowledge at this point in time, none of the simplex methods has been proved to find a solution after examining a polynomial number of bases. In contrast, Borgwardt (1982) and Smale (1982) have demonstrated that on average, the simplex methods perform very well with the number of iterations being linear in n, so that their respective versions require only a polynomial number of operations on average. On the other hand, there are the ellipsoid and Karmarkar's (1984) methods which all have been proved to converge in polynomial time in the worst case. Among these, only the last appears to be slowly gaining ground on the simplex methods as far as average performance is concerned.

For all problems outlined above, polynomially bounded algorithms exist or have been described. In this book we are concerned with combinatorial optimization problems, many of which deal with the selection of a subset of a given set of elements, satisfying certain criteria as well as with finding certain sequences of the given elements. For many of these problems the only sure way to find an optimal solution is to go through a partial enumeration procedure. The number of solutions to be enumerated, i.e., the number of iterations, is then bounded above only by the number of solutions to the problem which is typically "astronomically finite", such as 2^n, $n!$ or n^n.

b.3 Classes of Problems and Their Relations

In this section we will define classes of problems and explore some of the relations between them. In order to do so, it is necessary to rephrase the known *optimization problems* as *recognition* or *decision problems* for which the result is a "yes" or "no" answer. As an example, consider a linear programming problem. The optimization problem is: What are $\bar{z}$ and $\mathbf{x}$, such that

$$\bar{z} = \max \mathbf{cx}$$
$$\text{s.t.} \quad \mathbf{Ax} \leq \mathbf{b}$$
$$\mathbf{x} \geq \mathbf{0} \qquad ?$$

The corresponding decision problem can be expressed as follows.
Problem: Given $\mathbf{c} \in \mathbb{R}^n$, $\mathbf{b} \in \mathbb{R}^m$, $\mathbf{A} \in \mathbb{R}^m \times \mathbb{R}^n$ and a constant $\bar{z}$.
Question: Does there exist a vector $\mathbf{x} \in \mathbb{R}^n$, such that $\mathbf{Ax} \leq \mathbf{b}$, $\mathbf{x} \geq \mathbf{0}$ and $\mathbf{cx} \geq \bar{z}$?

Given the complexity function $O(f(n))$ defined in the previous section, we write

Definition 14: The class *P* is the class of problems for which there exists an algorithm with a complexity function $O(f(n))$ that is polynomial in the problem's input parameter(s).

Similarly, we can write

Definiton 15: The class *NP* is the class of decision problems for which a guessed solution can be verified in polynomial time.

As an example, consider the problem of finding a feasible solution to a system of simultaneous linear inequalities. Any guessed solution could be verified in polynomial time, thus the problem is in *NP*.

The above definition immediately leads to

Lemma 16: $P \subseteqq NP$.

One of the most important unsolved problems is whether the above inclusion is a proper one or not. This issue will be further elaborated upon below. At this point, it is convenient to state the following result, assuming that *M* is an instance of a given problem (or model).

Defining $\overline{M}$ as the complement of *M*, the following results are known.

Lemma 17: $\overline{M} \in P$ if and only if $M \in P$.

If, however, $M \in NP$ implies that $\overline{M} \in NP$ is not known; it is believed to be not true.

In order to demonstrate this asymmetry, consider the following sequencing problem. Let there be n distinct items $I_1, I_2, \ldots, I_n$ to be manufactured and let c_{ij} denote the setup cost for item I_j if it is processed directly after item I_i. In an optimization problem one would like to determine the sequence that minimizes the sum of the setup costs. In the context of a decision problem, the question is: "Is there a sequence with costs not exceeding a given constant B?" If the answer is "yes", one could easily verify it by adding the corresponding $(n-1)$ parameters in $O(n)$, i.e., in linear time and hence the above scheduling problem is clearly a member of the class *NP*. On the other hand, if the answer to the above question is "no", then there is no easy way to verify its truth.

The next issue addressed here is the reduction of one problem to another. Formally we can state

Definition 18: There is a *polynomial reduction* from a problem M_1 to problem M_2 (in symbols: $M_1 \propto M_2$), if every instance of M_1 can be transformed into an instance of M_2 in polynomial time. M_1 and M_2 are said to be *polynomially equivalent*, if $M_1 \propto M_2$ *and* $M_2 \propto M_1$. In other words, if $M_1 \propto M_2$, then one can view M_1 as a special case of M_2; moreover every algorithm solving M_2 is then also able to solve M_1.

This implies

Lemma 19: If $M_2 \in P$ and some $M_1 \propto M_2$, then $M_1 \in P$.

The polynomial reduction of Lemma 19 is *transitive*, i.e., if $M_1 \propto M_2$ and $M_2 \propto M_3$, then $M_1 \propto M_3$. As a simple example of Lemma 19, let M_2 be a linear programming problem and let M_1 be a classical transportation problem. Since M_2 can be solved in polynomial time (e.g., by Karmarkar's method), $M_2 \in P$. As the classical transportation problem is a special case of linear programming, $M_1 \propto M_2$. Hence, the transportation problem can also be solved in polynomial time. Note, however, that conversely no polynomial reduction from a general linear program to the standard transportation problem exists.

Definition 20: A problem M_1 is called *NP-complete*, if $M \propto M_1$ for every $M \in NP$, indicating that M_1 is at least as hard as any problem in *NP*. The class of all *NP-complete* problems is called *NPC*.

If any decision problem is in *NPC*, then the corresponding optimization problem is *NP-hard* because the existence of a polynomial algorithm for the optimization problem would imply that the decision problem can also be solved in polynomial time. The relations between *P*, *NP*, and *NPC* can be visualized as follows (assuming that $P \neq NP$)

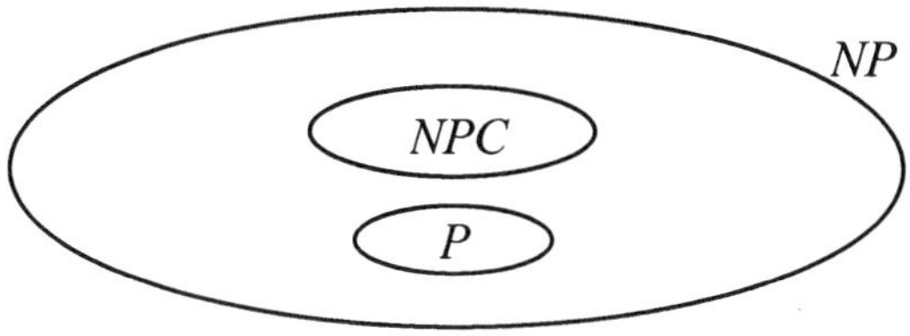

Figure 6

Problems in the class *NPC* are the hardest in *NP*. The first problem known to be in *NPC* was a logical expression called SATISFIABILITY, or SAT for short. The proof was established by Cook (1971) who also outlined a "master reduction", indicating how SAT can be reduced to other problems in *NP*. For a proof of "SAT $\in$ *NPC*" the reader is referred to Cook's paper. The SATISFIABILITY problem can be described as follows. Let n variables x_j be given, where each variable can assume only one or two values, namely true or false, called its *truth value*. If x_j is assigned a truth value of "true", then its complement $\bar{x}_j$ is automatically "false" and if x_j is "false", then $\bar{x}_j$ is "true". The variables x_j and $\bar{x}_j$ are called *literals*. These literals are combined in m expressions C_1, C_2, ..., C_m, called *clauses*. If within each clause the literals are combined by disjunctions whereas the clauses are combined by conjunctions, the resulting expression is called a *boolean expression* in *conjunctive normal form*. An example of such an expression is

$$\underbrace{(\bar{x}_1 \vee x_2 \vee x_3)}_{C_1} \wedge \underbrace{(\bar{x}_2 \vee x_4 \vee x_5)}_{C_2} \wedge \underbrace{(x_2 \vee \bar{x}_3 \vee x_4 \vee \bar{x}_5)}_{C_3}.$$

Such an expression is called *satisfiable*, if there exists at least one assignment of truth values to the literals, such that in each clause at least one literal is "true". Then we can formally state the problem as follows.

SATISFIABILITY: Given a boolean expression in conjunctive normal form, is there at least one satisfiable truth assignment?

In the above example, if x_1, x_2 and x_5 are "true" and x_3 and x_4 are "false" (or, equivalently, $\bar{x}_1, \bar{x}_2, \bar{x}_5$ are "false" and $\bar{x}_3$ and $\bar{x}_4$ are "true") the above expression is "true". Hence, in this example the answer to the above question is yes. It should be noted that various other truth assignments also satisfy the above expression.

It is no problem to prove that SAT $\in$ *NP*. Since each variable can assume only the values "true" or "false", one has "only" to enumerate all 2^n possible truth assignments, and every "yes" answer, which is then accompanied by a truth assignment, can easily be checked in polynomial time which establishes the problems membership in *NP* as per Lemma 16.

After Cook (1971) established the membership of SAT in *NPC*, he provided a means to show that various other problems are also in *NPC*. This is summarized in

Theorem 21: For a given problem M, if
(1) $M \in NP$, *and*
(2) $M_1 \propto M$ for some $M_1 \in NPC$,
then $M \in NPC$.

Usually the first part of the above theorem is easy to establish for a new problem M whereas the second part is frequently more difficult. Starting with SAT being the only element of *NPC* at that time, Karp (1972), Garey and Johnson (1979) and many others have largely increased the number of problems known to be in using the above theorem. Today, there is an ever growing-number of problems known to be in *NPC*. An extensive collection from various areas of mathematics can be found in Garey and Johnson (1979); this collection is constantly updated by the authors in the *Journal of Algorithms*. It is clear from the above that all problems in *NPC* can directly or indirectly be polynomially reduced to SAT. In the following we will describe one problem and show that it is in *NPC*. One problem whose applications and solution techniques are discussed in this book is

0-1 INTEGER: Given a system of simultaneous linear inequalities $\mathbf{Ay} \geq \mathbf{b}$, where $\mathbf{A} \in \mathbb{R}^m \times \mathbb{R}^n$ and $\mathbf{b} \in \mathbb{R}^m$. Does the system have a solution $\mathbf{y} \in \{0, 1\}^n$?

First we will establish the membership of 0-1 INTEGER in *NP*. Suppose that there is a solution $\mathbf{y}^* \in \{0, 1\}^n$ which satisfies $\mathbf{Ay}^* \geq \mathbf{b}$. In order to verify the validity of this answer, the variables y_i are replaced by their values y_j^* and the above "yes" answer can be checked in O(mn) time, i.e., in polynomial time, thus establishing the fact that 0-1 INTEGER $\in$ *NP*. Now a polynomial reduction of SAT to 0-1 INTEGER will be performed. For any given instance of SAT with n variables and n clauses, the literal x_j is replaced by a 0-1 variable y_j and the literal $\bar{x}_j$ is replaced by $(1 - y_j)$. Moreover, each clause C_i is replaced by the inequality

$$\sum_{j:x_j \in C_i} y_i - \sum_{j:\bar{x}_j \in C_i} y_i \geq 1 - \left|\{j : \bar{x}_j \in C_i\}\right| \tag{1}$$

Now each satisfiable truth assignment to SAT also solves the corresponding zero-one integer programming problem; hence 0-1 INTEGER $\in$ *NPC*. The above transformation may be explained by the following example.

SAT: $(x_1 \vee \bar{x}_2 \vee x_3) \wedge (\bar{x}_1 \vee x_2 \vee \bar{x}_3 \vee \bar{x}_4) \wedge (x_2 \vee x_3 \vee x_4)$, where the three clauses contain one, three and no $\bar{x}_j$-literals, respectively. Using the transformation (1), we then obtain the following zero-one integer programming problem:

$$\begin{aligned} y_1 - y_2 + y_3 \quad\quad &\geq 0 \\ -y_1 + y_2 - y_3 - y_4 &\geq -2 \\ y_2 + y_3 + y_4 &\geq 1 \\ y_1,\ y_2,\ y_3,\ y_4 &= 0 \vee 1 \end{aligned}$$

Then the truth assignment $x_1 = x_2 =$ "false" and $x_3 = x_4 =$ "true" corresponds to $\mathbf{y} = [0, 0, 1, 1]^T$ and both solutions satisfy their respective problems. For a variety of transformations such as the above, see Garey and Johnson (1979) and Lenstra and Rinnooy-Kan (1979). Using the transitivity of polynomial reductions, a problem $M \in NPC$ could now be shown to be in *NPC* if either SAT or 0-1 INTEGER can be polynomially reduced to it. In many proofs which establish the membership of some problem *M* in *NPC*, it is more convenient to reduce some problem in *NPC* other than SAT to *M*. One of those problems is 3SAT, a satisfiability problem with no more than three literals per clause. Since 3SAT $\in NPC$ and SAT $\propto$ 3SAT, 3SAT $\in NPC$ follows and again the transitivity of polynomial reductions can be used to establish membership of a problem *M* in *NPC* by showing that $M \in NP$ and 3SAT $\propto M$. It should be mentioned that, as opposed to the "hard" problem 3SAT, the problem 2SAT which contains at most two literals per clause is easy, i.e., 2SAT $\in P$.

As already mentioned above, the fundamental question today is the relation between *P* and *NP*, i.e., $P = NP$ or $P \subsetneq NP$? It is believed that *P* is a proper subset of *NP*, but it seems that the existing mathematical tools are not sufficient to prove or disprove this claim. Since it has been shown that all problems in *NPC* are polynomially equivalent, the existence of a polynomial algorithm for any problem in *NPC* would imply that $P = NP$. On the other hand, Ladner (1977) has shown that $P \neq NP$ implies that there exists at least one problem *M*, such that $M \notin P$ *and* $M \notin NPC$, i.e.: $NP \supsetneq P \cup NPC$.

In general, one can distinguish between four classes of problems.

- *Easy problems*, i.e., problems for which polynomially bounded algorithms exist. Problems in this category are in class *P* and examples of problems in this class are provided in the previous section.

- *Probably difficult problems*, i.e., problems for which at this point only exponentially bounded algorithms exist, but it is not known whether or not polynomial algorithms exist for these problems. This is the class *NPC*; examples are SAT, 0-1 INTEGER and many other combinatorial problems.

- *Intractable problems* are provably difficult problems, i.e., problems for which proof exists that no polynomial algorithms can possibly exist for any problem in this class.

- *Undecidable problems*, i.e., problems for which we have proof that no algorithm can possibly exist to solve them. Examples for problems in this class are the question whether or not a computer comes to a halt after a finite number of steps for any given program and any given input string, the problem of finding integer solutions to a set of polynomial equations (Hilbert's tenth problem), or nonlinear integer programming, see Jeroslow (1974).

Assume now that a real-world situation has been modeled as a mathematical problem and it is known or can be proved that this problem may be in *NP* but it is not in *P*. As discussed above, it is very unlikely that a polynomially bounded algorithm exists for the problem. Only if the model is very small one may still go ahead and solve it with one of the known exponential algorithms. Unfortunately, however, almost all realistic problems are large so that this approach is unlikely to be applicable. This leaves the user with basically two approaches: either approximate the problem, i.e., simplify it until the problem is obtained for which a polynomial algorithm exists, or take the problem as is and use an approximation algorithm. Clearly, neither technique will necessarily result in an optimal solution to the original problem. There is nothing much that can be said in general about the approximation of a problem; it largely depends on the specific situation and on how much information the decision maker is willing to sacrifice. On the other hand, there is the large class of *approximation methods* or *heuristics* which are, at least in most cases, also tailored for the problem but have certain common features. All heuristics are or at least should be designed to find good solutions quickly. The quality of a solution is clearly measured by its deviation from the true optimum; as far as heuristic methods are concerned, we have either empirical data about the average quality of the solutions (from past test series) or sometimes also a proven error bound. An error bound ε for a minimization problem is defined as $\varepsilon = (\tilde{z} - \bar{z})/\bar{z}$, where $\bar{z}$ is the objective function value of the optimal solution and $\tilde{z}$ denotes the objective function value found by a given heuristic method in the worst case. As an example, if the objective value of a minimization problem at optimum is $\bar{z} = 250$ and the solution found by the heuristic method finds a solution with objective function value of no more than 500, then $\varepsilon = 1$, which is called a 100 percent error bound (or sometimes also $1/(1 + \varepsilon) = 0.5$-optimality). Some heuristics with proven error bounds are discussed in this book, for example the traveling salesman problem, bin packing problem and others. One of the advantages of using a heuristic method is that the decision maker quickly obtains a feasible, i.e., implementable, solution which hopefully is not too far from optimum. It should also be noted that it is not always necessary to design a heuristic for a given problem; sometimes one can use an exact technique and abort the calculations prematurely. Clearly, this approach is only possible if the solution obtained in this way is feasible. (In the context of linear programming, dual methods are not useful for this purpose since the first feasible solution is the optimal one, so that one would have to "go all the way"). This approach is justified by the fact that in many instances the exact algorithm finds reasonably good solutions quickly but may fail to find an optimal solution within a reasonable amount of time. Unfortunately, it usually takes very long to establish optimality for the solution or to find a slightly better one. For a survey of non-exact methods and a proposal for their classification, the reader is referred to Müller-Merbach (1981). A variety of heuristics is surveyed in Chapter 7 of Part I.

Finally a word with respect to the evaluation criteria discussed in this chapter and the criterion of worst case time analysis in particular. It was shown at the beginning of this chapter that a good worst case time complexity is a crucial feature of a successful algorithm. Consider, however, any of the simplex algorithms on the one hand and the ellipsoid method on the other. With respect to worst case time analysis, the simplex methods are poor performers whereas the ellipsoid method performs well. Still, most linear programming problems in practice are solved by one of the simplex methods because their average performance is much better than that of the other approaches for linear programming problems. Moreover, the simplex methods have a distinct advantage over the other technique as far as storage space requirement is concerned. But even a combination of all four criteria, worst case, and average case analysis in terms of time and storage space requirement is not sufficient to evaluate an algorithm. It must also be possible to easily take advantage of special structures (such as block-angular or staircase structures or transportation and assignment-type structures in the simplex methods) and, very important, the method should allow the user to perform sensitivity analyses efficiently, a factor which certainly is one of the deciding factors that has contributed to the success of the simplex methods.

CHAPTER c GRAPH THEORY

c.1 Basic Definitions and Examples

In this chapter we introduce some basic definitions and properties of graphs. As opposed to many other areas of combinatorial optimization or mathematical programming in general, the terminology in the field of graphs and networks is not a universally accepted one. In order to make the wide range of books and articles more accessible to the reader, we will—whenever possible—at least mention some of the more frequently used alternative terms even if they are not used in this book.

The first treatment of a problem related to graphs was that by the Swiss mathematician Leonard Euler in 1736, who then described and solved the famous Königsberg bridge problem which is discussed in some detail in Part II of this volume. Subsequently, various scientists from a large variety of disciplines reduced a variety of problems to graph structures. Examples are Cayley (1875) who worked on structures of chemical components, particularly isomers; another one is Sir Hamilton (1856) whose game "around the world" is a mathematical puzzle; and Kirchhoff (1847) with the theory of flows in electrical networks, just to mention a few. Finally, two hundred years after Euler's original work, the Hungarian mathematician Denès König published the first unified approach to graph theory in his book *Theorie der endlichen und unendlichen Graphen* (Theory of finite and infinite graphs), which was published in 1936. A detailed account of the history of graph theory is given by Biggs *et al.* (1976) and a short but interesting list of graph-related problems can be found in Gondran and Minoux (1984). The revived interest in graph and network problems in the 1950s culminated in books by Ford and Fulkerson (1962) and Berge (1958 and 1962). Since then, hundreds of papers in the field have appeared in the literature, and some journals such as *Networks* or *Theory of Graphs* deal exclusively with graph and network problems. To introduce the subject matter, we first define a graph.

Definition 22: A *graph* is a tuple $G = (N, A)$, where $N = \{n_1, n_2, ..., n_n\}$ is a set of elements called *nodes* and $A = \{a_{ij} : n_i, n_j \in N\}$ is a set of binary relations between these nodes. An element of A is called an *arc*.

In the literature one frequently finds the terms *vertices*, *points*, and *elements* instead of nodes and the terms *edges*, *lines*, *branches*, and *links* for arcs. Apart from the above definition with nodes and arcs, graphs denoted by $G = (V, E)$ with vertex set V and edge set E appear most frequently. Sometimes a graph with n nodes is said to be "of order n". If the relations between the nodes are no longer binary, a generalized structure called a *hypergraph* emerges. Hypergraphs were first introduced by Berge (1970 and 1973).

Each arc a_{ij} in a graph is either directed or undirected. An undirected arc is often termed an edge. A directed arc is shown by an arrow which points from its initial node (sometimes called *tail*) to its terminal node (also called *head*). A directed arc $a_{ij} = (n_i, n_j)$ can only be traversed from n_i to n_j whereas an undirected edge $e_{ij} = (n_i, n_j)$ can be traversed from n_i to n_j as well as from n_j to n_i. At times it is convenient to number the given arcs consecutively by $a_1, a_2, ..., a_m$. Then the initial node of a_k is denoted by $\alpha\ (a_k)$ whereas the terminal node of a is written as $\omega\ (a_k)$. A graph that includes only undirected arcs is called an *undirected graph*, a graph with only directed arcs is a *directed graph* or *digraph*, and a graph with undirected as well as directed arcs is referred to as a *mixed graph*.

In the following paragraphs we will describe three types of situations that can be modeled in terms of graphs. We should emphasize that the problems shown here represent only a small proportion of the large variety of models which use graph structures.

Example 1: Graphs in which arcs represent transportation lines.
Consider a graph whose nodes correspond to certain cities and whose arcs represent the existing flights between these cities for a certain airline. Similarly, the train connections in a given area may be modeled by representing all stations, switches, level crossings and ends of tracks by nodes and the existing directed connections between them by directed or undirected arcs. Road and street systems can be modeled in the same way, so that each intersection is represented by a node. Note that non-level crossings such as bridges and underpasses are not represented by nodes as they do not allow a right or left turn. Communication networks can also be modeled as graphs where the nodes represent relay stations, switchboards and end users. In this context we should also mention the pioneering work of Kirchhoff (1847) on electrical networks; another classical reference is Dennis (1959).

Another example demonstrates how a highway crossing in the shape of a clover leaf could be modeled as a graph. Figure 7 shows the physical appearance of the crossing.

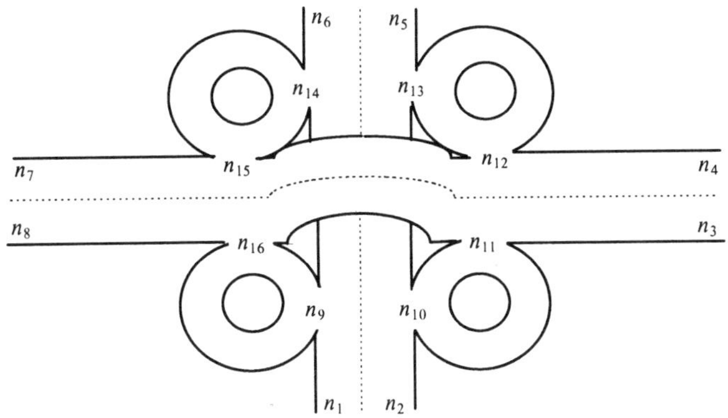

Figure 7

Assume that driving is on the right side of the road and all connections are one way. Then nodes n_1 to n_{16} can be assigned to all junctions and intersections with n_1 to n_8 indicating points on the highway that are located outside the direct vicinity of the intersection and the nodes n_9 to n_{16} represent points at which entering or exiting a highway is possible. Then the clover leaf of Figure 7 can be represented by the directed graph shown in Figure 8.

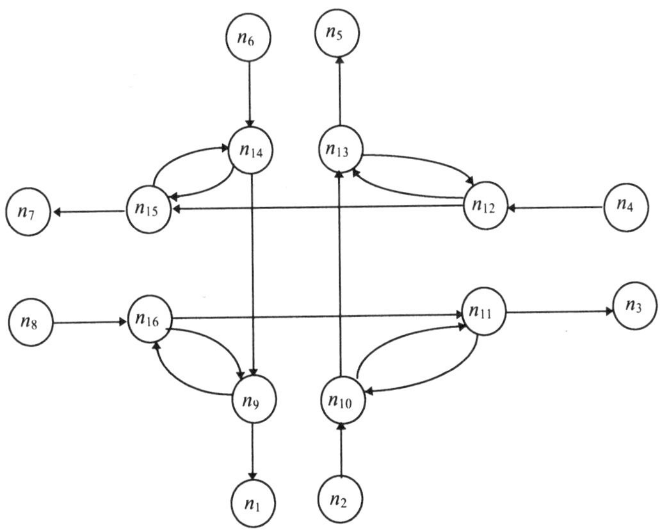

Figure 8

As an example, driving northbound intending to go east, our trip would commence at n_2 and then follow the arcs connecting n_2, n_{10}, n_{11}, n_3. On the other hand, having missed the exit at n_{10}, we would have to follow n_2, n_{10}, n_{13}, n_{12}, n_{15}, n_{14}, n_9, n_{16}, n_{11}, and n_3.

Example 2: Graphs in which nodes represent states of a system.

As one possible example, consider the game of chess. Each node corresponds to a particular positioning of the chess pieces of the two players. A directed arc leading from n_i to n_j indicates that it is possible to reach situation n_j from situation n_i in one move. The modeling of the game of chess in this form is theoretically possible but not practical as it requires a finite but astronomically large number of nodes. In a much simplified setting, consider a machine that cuts given rods into certain lengths. The state of the machine at any point in time may be characterized according to the lengths of the resulting rods as extremely accurate, very accurate, ..., inoperable. Each of these states is represented by a node, and a directed arc from n_i to n_j indicates that there is a possibility of the machine transferring from the i-th state of accuracy to the j-th state within a specified period of time. In addition, we could also assign the probability for such a transition to that arc. Another example in this category can be found in the theory of automata. An automaton is an apparatus which—starting in an initial state—moves into certain

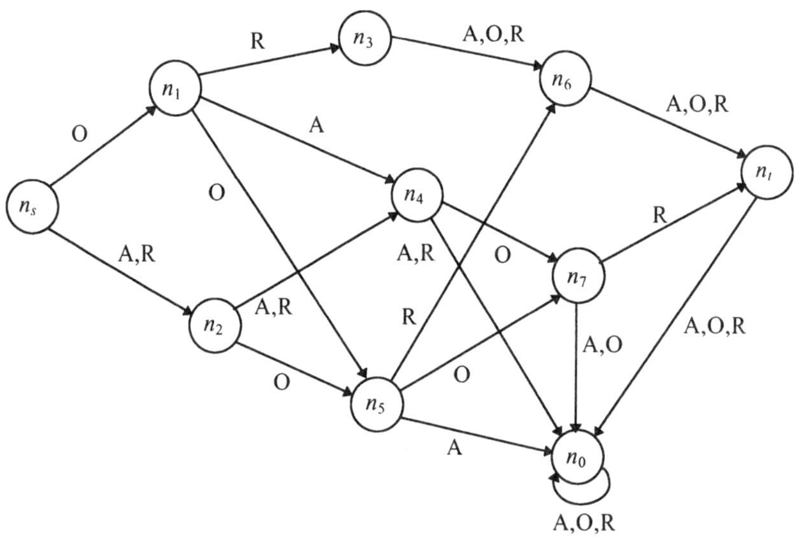

Figure 9

states depending on the input symbol it reads. Having read the input consisting of a string of symbols, the automaton halts in a state which either symbolizes acceptance or rejection of the string. In the following example, n_s denotes the initial state, n_0 is the overflow, the nodes n_1 to n_7 are intermediate states and n_t is the state which, if reached after all symbols have been read, denotes acceptance of the string. If the automaton terminates at any node other than n_t, the work or string will be considered rejected. Suppose that the input alphabet consists of the letters A, O, and R. Let now the letter(s) next to an arc a_{ij} cause the automaton to move from state n_i to state n_j.

It is easy to verify that the automaton described by the graph in Figure 9 accepts all four-letter words with the two letter "OR" appearing anywhere in the word in direct sequence.

Example 3: Graphs with nodes representing physical elements.

An example of this class of models can be found in the hierarchy of an organization. In such a model the nodes represent positions such as President, Chief Executive Officer, Vice-President Operations, Vice President Finance, Operations Research Division, Data Processing Department, etc. Depending on the specific application, the arcs may represent flows of information or they may indicate who reports to whom. Other examples in the same category include the study of molecules, or parts lists of materials. Another interesting application is found in the field of sociology. The basic idea of the model is as follows. Let each node in a graph represent a student in a class. Then a "sympathy graph" is constructed by introducing an arc a_{ij}, if student n_i likes student n_j. Such a situation is depicted in Figure 10.

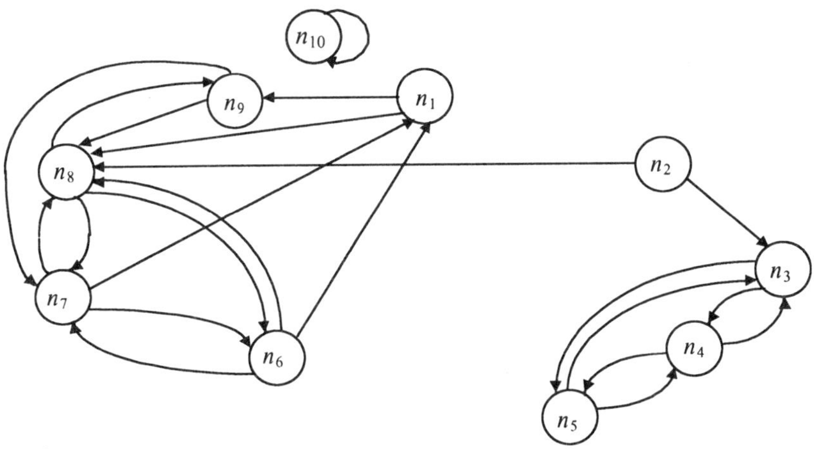

Figure 10

From the (admittedly extreme) situation in Figure 10, one will be able to draw important conclusions, specifically as far as students n_2 and n_{10} and the cliques (n_1,

n_6, n_7, n_8, n_9) and (n_3, n_4, n_5) are concerned. Here, the term "clique" is used here with its usual meaning as a word of the English language, not as a technical term, well-defined in graph theory (see Definition 45).

Many other systems can be modeled as graphs. A variety of such problems can be found in Parts II and III of this book. To facilitate the ensuing discussion, some further definitions are needed.

Definition 23: A *chain* is an alternating sequence of nodes and arcs of the type $(n_{i_1}, a_{i_1 i_2}, n_{i_2}, a_{i_2 i_3}, n_{i_3}, \ldots, n_{i_{t-1}}, a_{i_{t-1} i_t}, n_{i_t})$. A *path* is a chain where $\alpha(a_{i_{k-1}, i_k}) = n_{i_{k-1}}$ and $\omega(a_{i_{k-1}, i_k}) = n_{i_k}$ for all arcs, i.e., all arcs on the path point in the same direction. A *cycle* is a chain with $n_{i_1} = n_{i_t}$ and a *circuit* is a path with $n_{i_1} = n_{i_t}$.

Sometimes it is more convenient to denote chains (and similarly paths, cycles, and circuits) by sequences of arcs $(a_{i_1 i_2}, a_{i_2 i_3}, \ldots, a_{i_{t-1} i_t})$ or sequences of nodes $n_{i_1}, n_{i_2}, \ldots, n_{i_t}$ if this can be done without ambiguity. The above concepts may be explained by the graph in Figure 11.

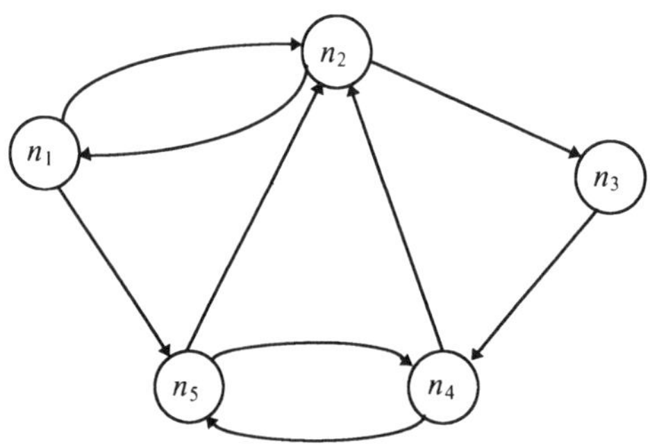

Figure 11

Here, the sequence (n_1, a_{12}, n_2, a_{52}, n_5, a_{54}, n_4, a_{34}, n_3) is a chain but not a path. The sequence (n_1, a_{12}, n_2, a_{23}, n_3, a_{34}, n_4) is a chain and also a path. Similarly, (n_1, a_{12}, n_2, a_{52}, n_5, a_{15}, n_1) is a cycle but not a circuit and (n_2, a_{23}, n_3, a_{34}, n_4, a_{45}, n_5, a_{52}, n_2) is a cycle and also a circuit. Chains, paths, cycles and circuits are *elementary*, if their sequence includes each node no more than once. The chain, path, cycle, and circuit in the above example are all elementary whereas the sequence (n_1, a_{12}, n_2, a_{52}, n_5, a_{54}, n_4, a_{24}, n_2 a_{23} n_3) represents a nonelementary chain since node n_2 is used twice in this chain. If not otherwise stated, the term "chain" and, similarly, path, cycle and circuit will denote an elementary chain in the remainder of this book. It should be mentioned that in undirected graphs, every chain is also a path and every cycle is also a circuit. If not defined otherwise, the *length of a chain* (path cycle, circuit) denotes the number of arcs included in it. Readers should be cautioned as some authors define chains, paths, cycles, and circuits differently. Finally, a graph without circuits is called *acyclic*, whereas a graph with at least one cycle is called *cyclic*.

Definition 24: An arc a_{ij} is called a *loop* or *self loop*, if $i = j$. Two arcs a_{ij} and $a_{k\ell}$ are said to be *parallel*, if $\alpha(a_{ij}) = \alpha(a_{k\ell})$ and $\omega(a_{ij}) = \omega(a_{k\ell})$. A graph with parallel arcs is sometimes called a *multigraph*. A graph without loops and parallel arcs is called a *simple graph*.

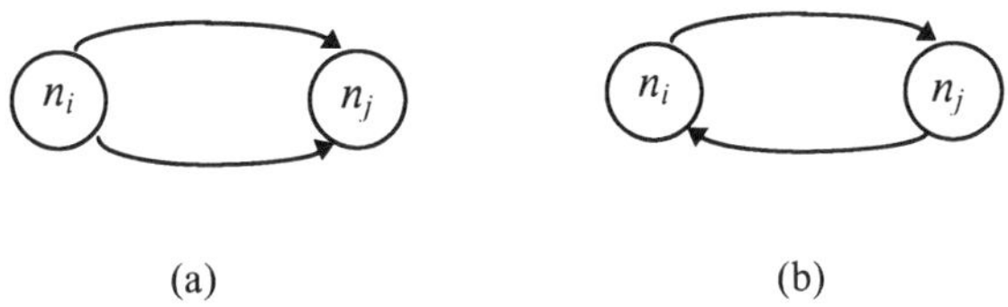

Figure 12

The graph in Figure 10 includes a loop that involves the node n_{10}. The graph in Figure 12a includes parallel arcs whereas that in Figure 12b does not, as the directions of the two arcs are opposite to each other. If a graph is not a multigraph like the one in Figure 12a, then every path and circuit can be referred to by its sequence of nodes or arcs as mentioned above. This short notation is ambiguous for chains and cycles as long as pairs of mutually inverse arcs (such as the ones in Figure 12b) or parallel arcs (such as the ones in Figure 12a) exist. If we refer to just any chain or path starting at some node n_i and terminating at some node n_j, we will call it an (n_i, n_j) *chain* (*path*).

In the following we take a closer look at the relations between any specified node and all the other nodes in the graph. In any given graph, define $\mathcal{N}(n_i)$ as the set of neighbors of a node n_i, formally $\mathcal{N}(n_i) = \{n_j : a_{ij} \in A\}$. In directed graphs the set of neighbors may be decomposed into two subsets, $\mathcal{S}(n_i)$ which is the set of *successors* of node n_i and $\mathcal{P}(n_i)$, which is the set of *predecessors* of node n_i. Formally, $\mathcal{S}(n_i) = \{n_j : a_{ij} \in A\}$ and $\mathcal{P}(n_i) = \{n_j : a_{ji} \in A\}$. Then we can state

Definition 25: In a given graph $G = (N, A)$ the degree of a node n_i, denoted by $\delta(n_i)$, equals the number of neighbors of n_i, i.e., $\delta(n_i) = |\mathcal{N}(n_i)|$. If all nodes of an undirected graph have the same degree, say $\delta(n_i) = r \; \forall \; n_i \in N$, then G is said to be *regular of degree r*. In a directed graph $G = (N, A)$ the indegree of a node n_i, $\delta^+(n_i)$ is the number of its predecessors, i.e., $\delta^+(n_i) = |\mathcal{P}(n_i)|$. Similarly, the *outdegree* of a node n_i, $\delta^-(n_i) = |\mathcal{S}(n_i)|$. If a node n_i has no neighbors then it is called an *isolated node*.

As examples for the above concepts, consider the graphs G_1, G_2, and G_3 in Figures 13a, b, and c, respectively.

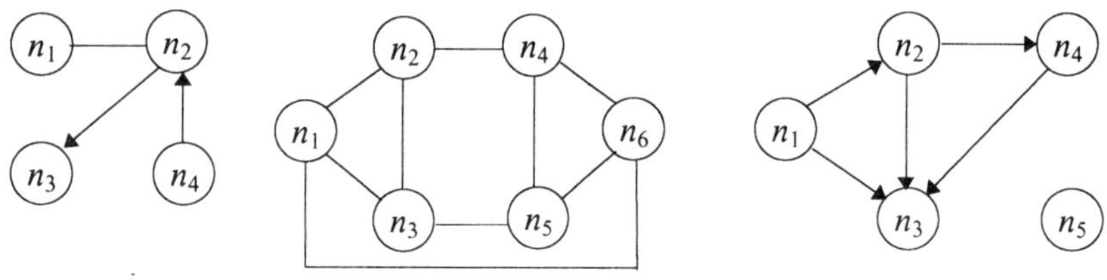

(a) (b) (c)

Figure 13

In the mixed graph G_1, $\mathcal{N}(n_1) = \{n_2\}$, $\mathcal{N}(n_2) = \{n_3, n_4\}$, $\mathcal{N}(n_3) = \{n_2\}$, and $\mathcal{N}(n_4) = \{n_2\}$ so that $\delta(n_1) = 1$, $\delta(n_2) = 2$, $\delta(n_3) = 1$, and $\delta(n_4) = 1$. In G_2, $\delta(n_i) = 3 \; \forall \; i = 1, ..., 6$, so that G_2 is regular of degree 3. Finally, in G_3, $\mathcal{P}(n_1) = \emptyset$, $\mathcal{S}(n_1) = \{n_2, n_3\}$, $\mathcal{P}(n_2) = \{n_1\}$, $\mathcal{S}(n_2) = \{n_3, n_4\}$, $\mathcal{P}(n_3) = \{n_1, n_2, n_4\}$, $\mathcal{S}(n_3) = \emptyset$, $\mathcal{P}(n_4) = \{n_2\}$, $\mathcal{S}(n_4) = \{n_3\}$, and $\mathcal{P}(n_5) = \mathcal{S}(n_5) = \emptyset$. Hence $\delta^+(n_1) = 0$, $\delta^-(n_1) = 2$, $\delta^+(n_2) = 1$, $\delta^-(n_2) = 2$, $\delta^+(n_3) = 3$, $\delta^-(n_3) = 0$, $\delta^+(n_4) = \delta^-(n_4) = 1$, $\delta^+(n_5) = \delta^-(n_5) = 0$. In general, we can prove the following

Lemma 26: For any given graph G, $\sum_{n_i \in N} \delta(n_i) \equiv 0 \pmod 2$. If G is directed, then

$$\sum_{n_i \in N} \delta^+(n_i) = \sum_{n_i \in N} \delta^-(n_i).$$

The lemma can be proved by starting with the graph $G = (N, 0)$ and then introducing the arcs $a_{ij} \in A$ one at a time. Each arc a_{ij} introduced into G increases $\delta^-(n_i)$ as well as $\delta^+(n_j)$ by one, and hence the sum of degrees of all nodes increases by two. In the directed case, the introduction of an arc a_{ij} increases $\delta^-(n_i)$ as well as $\delta^+(n_i)$ by one. Initially, all degrees were zero in both cases, thus the lemma. A direct consequence of the first part of the above lemma is

Corollary 27: In any given graph G the number of nodes with odd degree is even.

It should also be mentioned that if G is regular of degree r, then the number of arcs in G is $m = ½rn$, and hence at least one of r and n must be even for G to exist.

Definition 28: A *subgraph* of a given graph $G = (N, A)$ induced by a set $N_s \subset N$ is a graph $G = (N_s, A_s)$ where $A_s = \{a_{ij} : n_i, n_j \in N_s, a_{ij} \in A\}$. A *partial graph* of a given graph $G = (N, A)$ induced a set $A_p \subset A$ is a graph $G_p = (N, A_p)$ with a given set A_p. A *partial subgraph* induced by sets $N_{ps} \subset N$ and $A_{ps} \subset A$ is a graph $G_{ps} = (N_{ps}, A_{ps})$ with $A_{ps} = \{a_{ij} : n_i, n_i \in N_{ps} : a_{ij} \in A_{ps}\}$.

An example for the above definitions, consider a map of the world where the set of nodes indicates the destinations of all existing airline carriers and the set or arcs show the routes of all carriers. A subgraph induced by all North American destinations would show all flights between all North American destinations. A partial graph induced by one carrier, say Cathay Pacific, would show all flights (worldwide) of that carrier, and the partial subgraph induced by North American destinations and Cathay Pacific would include all flights of Cathay Pacific between North American destinations.

c.2 Representation and Storage of Graphs

In this section we will explore various ways of capturing the information given in a graph without relying on its graphical representation. We then discuss some possibilities to store this information efficiently on a digital computer.

Definition 29: The *adjacency matrix* of a graph $G = (N, A)$ is an $[n \times n]$-dimensional matrix $\mathbf{A}^*(G) = (a^*_{ij})$, where a^*_{ij} denotes the number of arcs connecting n_i and n_j. If G is not a multigraph, then $a^*_{ij} = 1$, if $a_{ij} \in A$, and 0 otherwise.

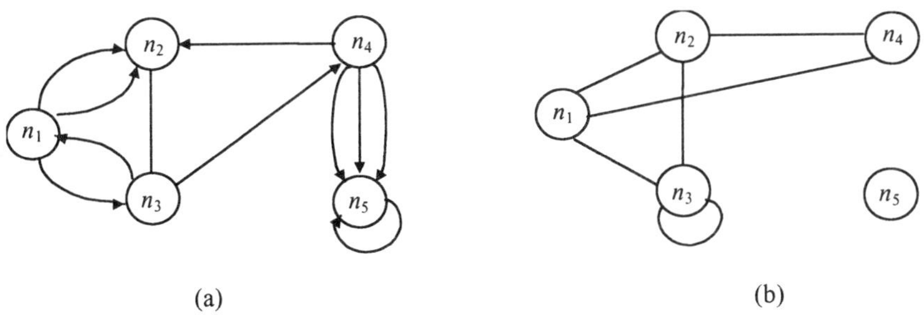

Figure 14

As an example for the above definition, consider the two graphs G_1 and G_2 in Figure 14.

The corresponding adjacency matrices are

$$\mathbf{A}^*(G_1) = \begin{bmatrix} 0 & 2 & 1 & 0 & 0 \\ 0 & 0 & 1 & 0 & 0 \\ 1 & 1 & 0 & 1 & 0 \\ 0 & 1 & 0 & 0 & 3 \\ 0 & 0 & 0 & 0 & 1 \end{bmatrix} \qquad \mathbf{A}^*(G_2) = \begin{bmatrix} 0 & 1 & 1 & 1 & 0 \\ 1 & 0 & 1 & 1 & 0 \\ 1 & 1 & 1 & 0 & 0 \\ 1 & 1 & 0 & 0 & 0 \\ 0 & 0 & 0 & 0 & 0 \end{bmatrix}$$

Some properties of adjacency matrices are:

- If n_i is an isolated node, then $\mathbf{a}^*_{i\bullet} = \mathbf{a}^*_{\bullet i} = 0$.
- If G is undirected, then $a^*_{ij} = a^*_{ji}$, i.e., $\mathbf{A}^*(G)$ is symmetric.
- If G has no loops, then $a^*_{ii} = 0 \ \forall\ i = 1, ..., n$.
- If G is directed, then the sum of elements in the i-th row equals the outdegree of node n_i, i.e., $\mathbf{e}\mathbf{a}^*_{i\bullet} = \sum_{j=1}^{n} a^*_{ij} = \delta(n_j)\ \ \forall\ j = i, ..., n$, where $\mathbf{e} = [1, 1, ..., 1]$ is the summation vector. Similarly, the sum of elements in the j-th column equals the indegree of node n, i.e., $\mathbf{e}\mathbf{a}^*_{\bullet j} = \sum_{i=1}^{n} a^*_{ij} = \delta^+(n_j)\ \ \forall\ j = 1, ..., n$. Also, $\sum_{i=1}^{n}\sum_{j=1}^{n} a^*_{ij} = m$.
- If G is an undirected graph without loops, then the sum of elements in the i-th row and—due to the symmetry of $\mathbf{A}^*(G)$—the sum of elements in the i-th column equals the degree of note n_i, i.e., $\mathbf{a}^*_{i\bullet}\mathbf{e} = \mathbf{e}\,\mathbf{a}^*_{\bullet j} = \sum_{i=1}^{n} a^*_{ij} = \sum_{j=1}^{n} a^*_{ji} = \delta(n_i), i = 1, ..., n$; also $\sum_{i=1}^{n}\sum_{j=1}^{n} a^*_{ij} = 2m$.

As far as storage space is concerned, n^2 nonnegative integers must be stored if a graph G is to be stored by its adjacency matrix $\mathbf{A}^*(G)$. If G is not a multigraph, this can be done more efficiently as only zeroes and ones have to be stored. A generalization of the concept of an adjacency matrix may be used if values are assigned to the nodes and/or arcs of the given graph.

Definition 30: A *valued graph* is a graph $G = (N, A; c_1(n_i), c_2(n_i), \ldots, c_q(n_i); c_1(a_j), c_2(a_j), \ldots, c_p(a_j))$, where the values $c_1(n_i), c_2(n_i), \ldots, c_q(n_i)$ are associated with the nodes n_i and the values $c_1(a_j), c_2(a_j), \ldots, c_p(a_j)$ are assigned to the arcs a_j.

As an example of a valued graph, consider the graph in Figure 15. The meaning of the node and arc values depends on the specific problem under consideration. The graph in Figure 15 may represent a road network where the nodes stand for the intersections of some roads, possibly with traffic lights, and the arcs represent certain directions of the roads. Then the values associated with the nodes, $c_1(n_i)$, may indicate the probability of the traffic light being red on arrival, and $c_2(n_i)$ could symbolize the expected waiting time at node n_i after arrival. The arc values $c_1(a_j)$ could denote the posted speed limit, $c_2(a_j)$ may symbolize the expected travel time, and $c_3(a_j)$ may specify the probability of getting caught for speeding on the road segment a_j.

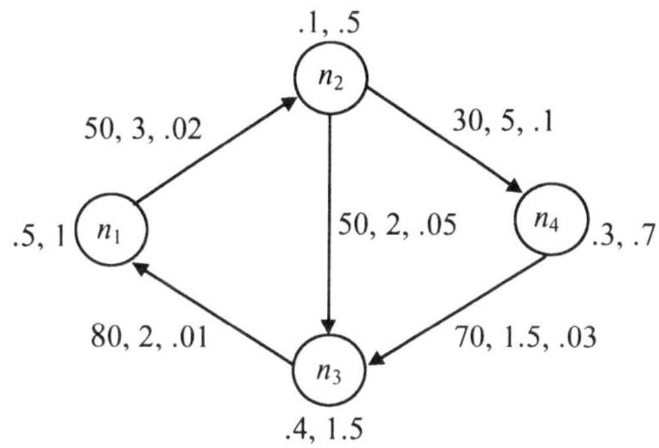

Figure 15

If G does not contain parallel arcs, then this information can be stored in various vectors and matrices. In general, q node vectors of length n and p *valued adjacency matrices* $\mathbf{VA}_1^*(G), \mathbf{VA}_2^*(G), \ldots, \mathbf{VA}_3^*(G)$, each of dimension $[n \times n]$, are required. Hence a total of $qn + pn^2$ numbers will have to be stored. For the above example we obtain two valued node vectors are [0.5, 0.1, 0.4, 0.3] and [1, 0.5, 1.5, 0.7], respectively, and the three respective valued adjacency matrices are

$$\mathbf{VA}_1(G) = \begin{bmatrix} - & 50 & - & - \\ - & - & 50 & 30 \\ 80 & - & - & - \\ - & - & 70 & - \end{bmatrix} \qquad \mathbf{VA}_2(G) = \begin{bmatrix} - & 3 & - & - \\ - & - & 2 & 5 \\ 2 & - & - & - \\ - & - & 1.5 & - \end{bmatrix}$$

$$\mathbf{VA}_3(G) = \begin{bmatrix} - & .02 & - & - \\ - & - & .05 & .1 \\ .01 & - & - & - \\ - & - & .03 & - \end{bmatrix}$$

Note that in each specific application it has to be determined whether the "–" entries in the valued adjacency matrices are set to 0, ∞, or any other specific value. Here they just denote the nonexistence of the corresponding arc. We will address this problem whenever a valued adjacency matrix is used. It should also be pointed out that if G has a special structure, it may be sufficient to store only part of the valued adjacency matrix without losing any information. This topic is further elaborated upon later when special graph structures are discussed.

Adjacency matrices are, however, by no means the only way to store the structure of a graph. While the matrix introduced here is—at least in its crude form—not immediately applicable for graph storage, it does possess some appealing properties that will be exploited when this matrix is used in mathematical formulations. For convenience assume that the arcs of the given graph have been arbitrarily numbered $a_1, a_2, \ldots, a_m$.

Definition 31: The *node-arc incidence matrix* $\mathbf{I}^*(G) = (I^*_{k\ell})$ associated with a directed graph G is defined by

$$(I^*_{k\ell}) = \begin{cases} 1, \text{if } n_k = \alpha(a_\ell) \\ -1, \text{if } n_k = \omega(a_\ell) \\ 0 \text{ otherwise.} \end{cases}$$

If the given graph G is undirected, the node-arc incidence matrix is defined as

$$(I^*_{k\ell}) = \begin{cases} 1, \text{if } n_k \text{ is one of the end nodes of } a_\ell \\ 0 \text{ otherwise.} \end{cases}$$

Observe that each column of the $[n \times m]$-dimensional matrix $\mathbf{I}^*(G)$ includes exactly two nonzero entries, a "+1" and a "–1" if the corresponding arc is directed and two "+1"'s if the arc is undirected. As an example consider the mixed graph in Figure 16.

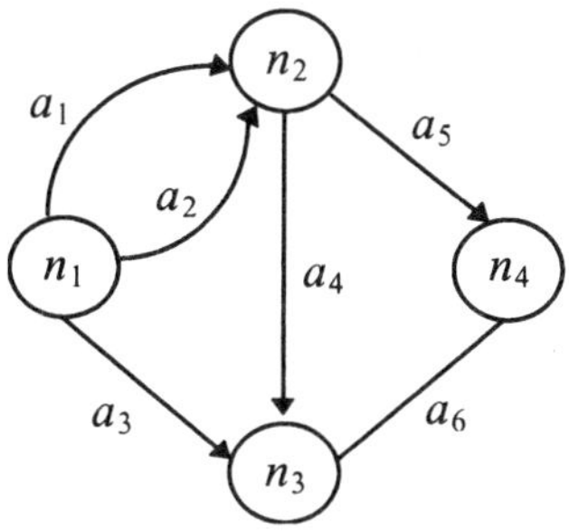

Figure 16

The node-arc incidence matrix of the mixed graph G of Figure 16 is then

$\mathbf{I}^*(G)$:

	a_1	a_2	a_3	a_4	a_5	a_6
n_1	1	1	1	0	0	0
n_2	-1	-1	0	1	1	0
n_3	0	0	-1	-1	0	1
n_4	0	0	0	0	-1	1

Note that parallel arcs (given that they are both directed or both undirected) have identical columns in $\mathbf{I}^*(G)$. For directed graphs, the number of + 1 (–1) entries in a row of $\mathbf{I}^*(G)$ equals the outdegree (indegree) of the node associated with that row. Another property of the node-arc incidence matrix of a directed graph should also be mentioned. Let P_{ij} denote a path from n_i to n_j in a directed graph and let $\mathbf{e}_\ell$ denote a vector [0, ..., 0, 1, 0, ..., 0] with the "1" in the ℓ-th position. Then the sum of all columns of arcs on that path is a column vector with a "1" in the i-th and a "–1" in the j-th position, i.e., $\sum_{k:a_k \in P_{ij}} I^*_{\bullet k} = \mathbf{e}_i - \mathbf{e}_j$. Similarly, the sum of columns of all arcs on a circuit equals the zero vector. Finally, there is yet another possibility to describe part of the structure of a graph G. For that purpose, two distinct nodes of G receive special labels; one is called the *source* n_s of G and the other is called the *sink* n_t of G. While the particular meaning of source and sink depends on the specific situation the graph is used to model, we can think of units being routed from source to sink, e.g. messages in communication networks or traffic in street networks. Once a source and a sink have been located, all elementary paths from source to sink are enumerated. Let the arcs of the given graph again be numbered from a_1 to a_m and suppose that the paths from n_s to n_t have also been (arbitrarily) numbered as $P_1, P_2, ..., P_r$. Then we can define

Definition 32: An *arc-path incidence matrix* $\tilde{\mathbf{I}}(G) = (\tilde{I}_{k\ell})$ associated with a graph G is defined as

$$\tilde{I}_{k\ell} = \begin{cases} 1, \text{if } a_k \in P_\ell \\ 0 \text{ otherwise} \end{cases}$$

As an example, consider again the graph in Figure 16. If $n_s = n_1$ and $n_t = n_4$, then there are five distinct paths from n_s to n_t, *viz.*, $P_1 = (a_1, a_5)$, $P_2 = (a_2, a_5)$, $P_3 = (a_1, a_4, a_6)$, $P_4 = (a_2, a_4, a_6)$, and $P_5 = (a_3, a_6)$.

The arc-path incidence matrix is then

$$\tilde{\mathbf{I}}(G) = \begin{bmatrix} 1 & 0 & 1 & 0 & 0 \\ 0 & 1 & 0 & 1 & 0 \\ 0 & 0 & 0 & 0 & 1 \\ 0 & 0 & 1 & 1 & 0 \\ 1 & 1 & 0 & 0 & 0 \\ 0 & 0 & 1 & 1 & 1 \end{bmatrix}$$

For most graphs, $r >> m$ meaning that many more paths (columns) than arcs (rows) exist. The excessive number of paths prohibits the explicit use of arc-path incidence matrices for all but the smallest graphs. Also, if in the above graph n_4 is chosen as source and n_3 as sink, then only one single path $\mathrm{P}_1 = (a_6)$ exists, so that the corresponding arc-path incidence matrix captures only a very small part of the structure of the entire graph. This problem could at least partially be overcome by extending the concept to an *arc-chain incidence matrix* which is similar to the above arc-path incidence matrix except that a "+1" ("–1") entry indicates an arc which is used in forward (backward) direction on a chain from a specified source to a sink. Another problem related to the arc-path incidence matrix is that exactly one such matrix does belong to a given graph but not vice versa. Consider the graph in Figure 17, which corresponds to the same arc-path incidence matrix as the graph in Figure 16.

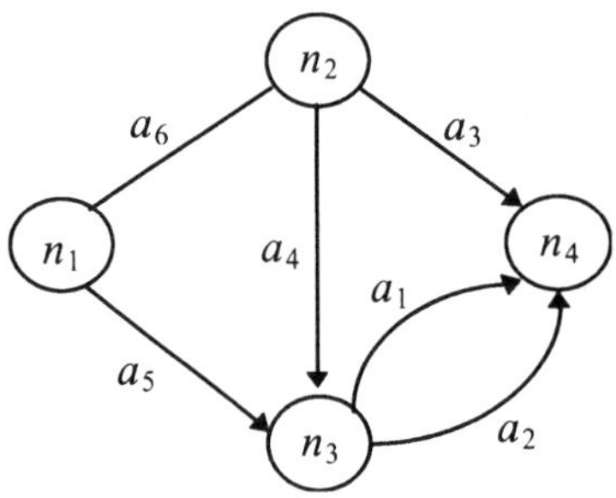

Figure 17

The reason for this non-uniqueness lies in the fact that the sequence of the arcs on the paths is not captured in the matrix. But even without the problem of non-uniqueness of the graph representation, the construction of the graph from a given arc-path incidence matrix is not a trivial one.

All of the above methods of capturing the structure of a graph in a matrix have in common that they are—at least in the form outlined here—rather inefficient. In other words, we will be needing $O(n^2)$, $O(mn)$, and $O(nr)$ units of storage space. Much more efficient types of storage exist, specifically for *sparse graphs*, i.e., graphs with $m << n^2$. Two of the more popular of these types are described below.

Arc list. In principle, the storage of a graph by means of an arc list is nothing more than a listing of all arcs of a graph by their initial and terminal nodes. If sequentially stored, an arc list can be written as

$$//n/m//\alpha\,(a_1)/\omega\,(a_1)//\;\alpha\,(a_2)/\;\omega\,(a_2)//\;...\;//\;\alpha\,(a_m)/\;\omega\,(a_m)//.$$

This list can also conveniently be stored in two blocks, one consisting of all initial nodes and the other consisting of all terminal nodes of the arcs in the same sequence, i.e., the first block is then $\alpha(a_1)$, $\alpha(a_2)$, ..., $\alpha(a_m)$ and the second block is $\omega(a_1)$ $\omega(a_2)$, ..., $\omega(a_m)$.

For simplicity, we write the nodes as 1, 2, 3, ..., n, rather than n_1, n_2, ..., n_n. Then the arc list for the graph in Figure 17 consists of a block with the initial nodes 3, 3, 2, 2, 1, 1, and a block with the terminal nodes 4, 4, 4, 3, 3, 2. As the last elements of the two lists correspond to an undirected arc, the "1" and "2" entries in the initial and terminal lists, respectively, could have been reversed. The space requirement for this type of storage is about $2m$.

It is now easy to add any number of arc values $c_1(a_j)$, $c_2(a_j)$, ..., $c_p(a_j)$ by attaching additional lists to the bottom of the above two strings, i.e., $c_1(a_1)$, $c_1(a_2)$, ..., $c_1(a_m)$; $c_2(a_1)$, $c_2(a_2)$, ..., $c_2(a_m)$; ...; $c_p(a_1)$, $c_p(a_2)$, ..., $c_p(a_m)$. The space requirement for a valued graph with p arc values is $(p + 2)m$. Values associated with the nodes can also be added in lists similar to the above arc lists.

Neighbor list. In some applications it is more efficient to use a different type of storage that allows the user to quickly scan all neighbors (or successors or predecessors) of a node. In principle, such a list can be written as

$$//n//\;\delta\,(n_1)/\;\mathcal{N}(n_1)//\;\delta\,(n_2)/\;\mathcal{N}(n_2)//\;\;...\;\;//\;\delta(n_n)/\;\mathcal{N}(n_n)//.$$

For easy reference, the neighbor list could be stored as follows. In a first sequential list, all nodes n_1, n_2, ..., n_n are stored. Suppose now that the arcs adjacent to some node n_k are denoted by $a_{k_1}, a_{k_2}, \ldots, a_{k_{\delta(n_k)}}$. Then the second sequential list enumerates all arcs adjacent to n_1, followed by those adjacent to n_2, n_3, ..., and n_n. A pointer leads from a node in the first (node) list to the first arc adjacent to it. In this way the arcs adjacent to a node can easily be scanned by starting at a given node, following the pointer to the first arc adjacent to it and continuing the sequential scan until the terminal node of another pointer is encountered; this is where the neighbors of the next node start being enumerated. Finally, a third list is added so that there exists a one-to-one correspondence between its entries and the elements of the second list. In particular, if the second list includes an arc a_{k_j} (which by definition is a neighbor of n_k), then the third list indicates the node other than n_k to which a_{k_j} is adjacent, i.e., $\omega(a_{k_j})$.

As an example consider the graph in Figure 18.

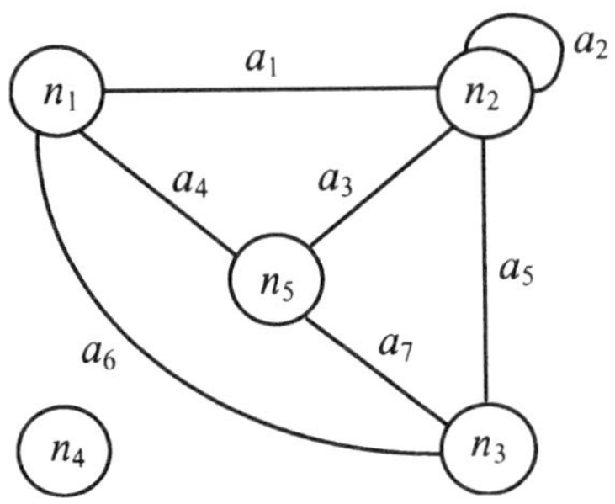

Figure 18

The neighbor list is then as follows:

Nodes:	1	2	3	4	5									
Adjacent arcs:	1	4	6	1	2	2	3	5	5	6	7	3	4	7
Terminal nodes:	2	5	3	1	2	2	5	3	2	1	5	2	1	3

As each arc is now listed twice in such a storage form, the space requirement for neighbor lists is about $4m + n$. Again, lists of all node and arc values can be added.

Similarly, predecessor and successor lists for directed graphs are set up. Their storage requirements are smaller as each arc is only stored once.

c.3 Reachability and Connectivity

In this section we will discuss some problems related to the connectivity of nodes in a given graph. Again, it will be necessary to distinguish between directed and undirected graphs. First, consider undirected graphs.

Definition 33: Two nodes n_i and n_j, $i \neq j$, of an undirected graph are said to be *connected*, if there exists at least one path connecting n_i and n_j. An *undirected graph G* is connected if each pair of its nodes is connected.

Suppose now that G is not connected. Then G consists of a set of subgraphs G_1, G_2, ..., G_s, where $G_k = (N_k, E_k)$, $k = 1, ..., s$; such that $\bigcup_{k=1}^{s} N_k = N$ and there exists no edge e_{ij} with $n_i \in N_k$ and $n_j \in N_\ell$ for any $k \neq \ell$. If each of the subgraphs G_k is connected, they are called *connected components*. The number of connected components is called the *connectivity number* of G.

The connectivity of a given graph G can be tested by means of the following algorithm which is initialized with the set of nodes $N = \{n_1, n_2, ..., n_n\}$, the set of labeled nodes $J = \{n_i\}$ with arbitrary $n_i \in N$, the set of currently scanned nodes $S = \emptyset$ and the adjacency matrix $\mathbf{A}^*(G) = (e^*_{ij})$.

Connectivity of Undirected Graphs

Step 1: Set $S := S \cup \{n_i\}$ as well as $J := J \cup \bigcup_{j=1}^{n} \{n_j : e^*_{ij} > 0\}$.

Step 2: Is $J = N$? If yes: Stop; the graph G is connected.

If no: Go to Step 3.

Step 3: Is $J = S$? If yes: Stop; the graph G is not connected.

If no: Select some $n_i \in J \setminus S$ and go to Step 1.

Example: The above algorithm may be illustrated by the graph G^1 in Figure 19a.

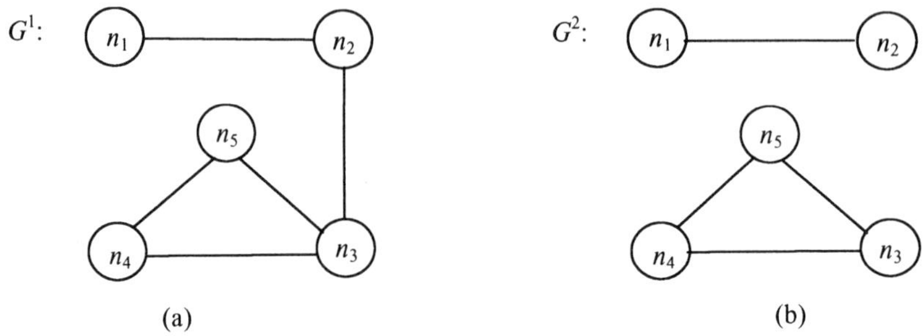

Figure 19

The adjacency matrix of the graph is

$$\mathbf{A}^*(G^1) = \begin{bmatrix} 0 & 1 & 0 & 0 & 0 \\ 1 & 0 & 1 & 0 & 0 \\ 0 & 1 & 0 & 1 & 1 \\ 0 & 0 & 1 & 0 & 1 \\ 0 & 0 & 1 & 1 & 0 \end{bmatrix}$$

The procedure is initialized with n_1, $S = \emptyset$ and $J = \{n_1\}$. As $e^*_{12} = 1$, the two sets are redefined as $S = \{n_1\}$ and $J = \{n_1, n_2\}$ and as $S \underset{\neq}{\subset} J \underset{\neq}{\subset} N$ another iteration is required. Proceeding with n_2, with edges e_{21} and e_{23} in E, the sets are redefined as $S = \{n_1, n_2\}$ and $J = \{n_1, n_2, n_3\}$. In the next iteration we choose $n_3 \in J \setminus S$, and the sets are redefined as $S = \{n_1, n_2, n_3\}$ and as e_{32}, e_{34} and e_{35} are in E, $J = \{n_1, n_2, n_3, n_4, n_5\} = N$; thus G is connected.

The same procedure can be applied to graph G^2 in Figure 19b. Again, the first step yields $S = \{n_1\}$ and $J = \{n_1, n_2\}$, so that $n_2 = J \setminus S$ is selected to proceed. As only the edge e_{21} connects the node n_2 to any other node, the redefined sets are $S = \{n_1, n_2\} = J \underset{\neq}{\subset} N$ and thus G^2 is unconnected. Apart from the message that G^2 is not connected, the algorithm has also produced one connected component, *viz.*, $S = J$. In case we are interested in *all* connected components, the rows and columns of the nodes in S are deleted from the adjacency matrix and the procedure is repeated, starting again with an arbitrary node. The algorithm terminates if each node is included in exactly one connected component (with the possibility of some connected components including only a single node in which case these nodes are isolated). It should be mentioned that the efficiency of the above procedure can be enhanced by working with a neighbor list rather than the adjacency matrix of the given graph. As in the above algorithm, each row is scanned no more than once and each row consists of n elements, the algorithm is of complexity $O(n^2)$. Using Tarjan's (1972) implementation, the complexity can be reduced to $O(m)$ which is optimal as each arc has to be considered at least once.

The concept of connectivity can also be defined for directed and mixed graphs. For that purpose we define

Definition 34: Two nodes n_i and n_j in a directed or mixed graph G are called *strongly connected*, if there exists at least one path from n_i to n_j *and* a path from n_j to n_i. A graph is called strongly connected if each pair of its nodes is strongly connected. Two nodes n_i and n_j in a directed or mixed graph G are *weakly connected* if there exists at least one chain between n_i and n_j. A graph G is called weakly connected if each pair of its nodes is weakly connected.

As an example of the above definition, consider the three graphs in Figure 20.

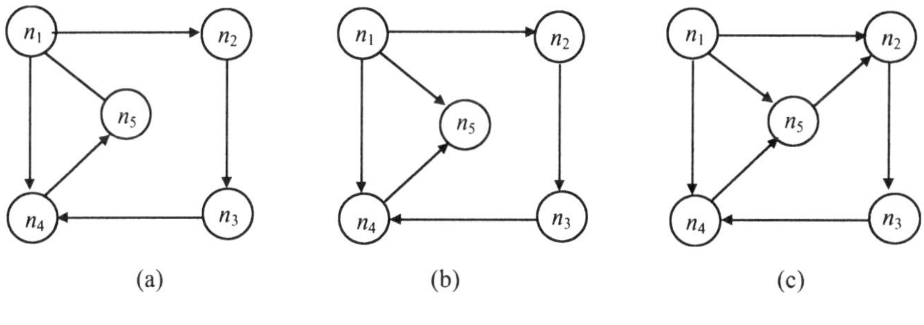

Figure 20

The mixed graph in Figure 20a is strongly connected as each of its nodes can be reached from each other node which is easily verified by inspection. On the other hand, the graph in Figure 20b is not strongly connected. For instance, the node n_1 cannot be reached from any other node. A similar argument reveals that the graph in Figure 20c is also not strongly connected.

Definition 35: A subgraph of G_k of a given graph G that is strongly (weakly) connected so that no supergraph of G_k which is also a subgraph of G has that property is referred to as a *strongly* (*weakly*) *connected component.*

In other words, all pairs of nodes in a strongly (weakly) connected component are strongly (weakly) connected but no node in some component G_k is strongly (weakly) connected to any node not in G_k. Inspection of the above Figure 21b reveals that no pair of nodes is strongly connected, thus there exist five strongly connected components. The graph in Figure 20c consists of two strongly connected components $G_1 = (N_1, A_1)$ and $G_2 = (N_2, A_2)$ with $N_1 = \{n_1\}$ and $N_2 = \{n_2, n_3, n_4, n_5\}$. On the other hand, all of the graphs in Figure 20 are weakly connected.

In order to determine all strongly connected components of a graph, the algorithm determining connected components of undirected graphs could be suitably modified. The basic idea is as follows. The above method is initialized with some starting node, say n_i. After being applied, the method terminates with some set J_i that includes all nodes that can be reached from n_i. Select now some node $n_j \in J_i$ and determine J_j. If $n_i \in J_j$, then n_i and n_j are strongly connected. This procedure is then repeated for all nodes in J_i resulting in the strongly connected component that includes n_i. Now all nodes from this strongly connected component are deleted from the given graph and the algorithm is repeated until each node is included in exactly one strongly connected component. If $J_i = N$ and $n_i \in J_j \ \forall \ n_j \in N$, then the given graph is strongly connected.

A simpler and somewhat more systematic method for the determination of strongly connected components is the *Simple Triple Algorithm* described below. It is initialized with the adjacency matrix $\mathbf{A}^*(G)$ of the given graph G where parallel arcs and loops have been deleted as they do not contribute towards the connectedness. Furthermore, set $k := 0$ and compute $\mathbf{A}^k = \mathbf{A}^*(G) + \mathbf{I}$ where $\mathbf{I}$ denotes the $[n \times n]$-dimensional identity matrix.

The Simple Triple Algorithm

Step 1: Is $k = n$? If yes: Go to Step 3.

If no: Set $k := k + 1$ and go to Step 2.

Step 2: Determine the matrix $\mathbf{A}^{k+1} = (a_{ij}^{k+1})$, so that $a_{ij}^{k+1} = \max\left\{a_{ij}^{k}; a_{kj}^{k} a_{ik}^{k}\right\} \forall$ i, j. Go to Step 1.

Step 3: All nodes whose rows (or columns) are identical in $\mathbf{A}^{n+1}$ belong to the same strongly connected component.

The computational complexity of the simple triple algorithm is easily determined. As the adjacency matrix is $[n \times n]$-dimensional, the loop involving steps 3 and 4 is executed no more than n times. Within each such loop the values of no more than n^2 elements a_{ij}^{k+1} are computed. Each such computation requires two elementary comparisons, hence the determination of the strongly connected components from $\mathbf{A}^{n+1}$ with the simple triple algorithm is of complexity $O(n^3)$. The determination of the strongly connected components from $\mathbf{A}^{n+1}$ requires additional computational effort of no more than $O(n^3)$ making the entire method an $O(n^3)$ algorithm.

It should also be noted that in the k-th iteration the elements in the k-th row and in the k-th column do not change as for $i = k$, $a_{kj}^{k+1} = \max\{a_{kj}^{k}; a_{kj}^{k} a_{kk}^{k}\} = a_{kj}^{k}$ since $a_{kk}^{k} = 1$ by construction. A similar argument applies to $j = k$. For simplicity of the manual computations, the k-th row and columns are circled in the succeeding numerical example. Consider the graph in Figure 21.

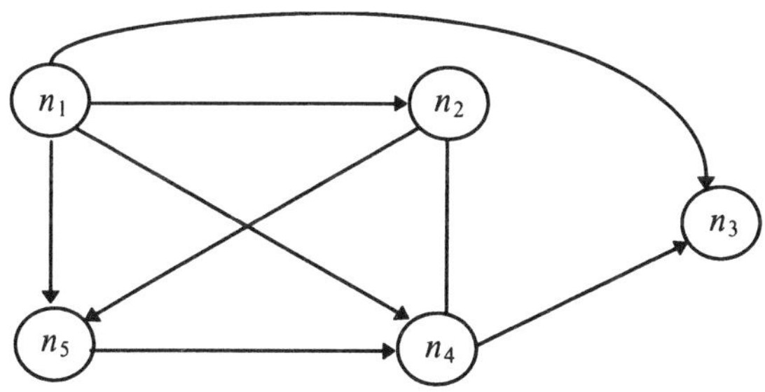

Figure 21

The adjacency matrix of G is

$$\mathbf{A}^*(G) = \begin{bmatrix} 0 & 1 & 1 & 1 & 1 \\ 0 & 0 & 0 & 1 & 1 \\ 0 & 0 & 0 & 0 & 0 \\ 0 & 1 & 1 & 0 & 0 \\ 0 & 0 & 0 & 1 & 0 \end{bmatrix}. \quad \mathbf{A}^0 = \begin{bmatrix} 1 & 1 & 1 & 1 & 1 \\ 0 & 1 & 0 & 1 & 1 \\ 0 & 0 & 1 & 0 & 0 \\ 0 & 1 & 1 & 1 & 0 \\ 0 & 0 & 0 & 1 & 1 \end{bmatrix}$$

is obtained by adding the identity matrix $\mathbf{I}$ to $\mathbf{A}^*(G)$. As $\mathbf{a}_{\bullet 1} = \mathbf{e}_1$ (where $\mathbf{e}_i$ denotes the i-th unit vector, i.e., the i-th column of **I**), there are no changes in the first iteration, so that we obtain

$$\mathbf{A}^1 = \begin{bmatrix} 1 & 1 & 1 & 1 & 1 \\ 0 & 1 & 0 & 1 & 1 \\ 0 & 0 & 1 & 0 & 0 \\ 0 & 1 & 1 & 1 & 0 \\ 0 & 0 & 0 & 1 & 1 \end{bmatrix}$$

. In the second iteration the only change in $\mathbf{A}^2$ occurs in the element a_{45}. The matrix $\mathbf{A}^2$ is then

$$\mathbf{A}^2 = \begin{bmatrix} 1 & 1 & 1 & 1 & 1 \\ 0 & 1 & 0 & 1 & 1 \\ 0 & 0 & 1 & 0 & 0 \\ 0 & 1 & 1 & 1 & 1 \\ 0 & 0 & 0 & 1 & 1 \end{bmatrix}. \quad \mathbf{A}^3 = \begin{bmatrix} 1 & 1 & 1 & 1 & 1 \\ 0 & 1 & 0 & 1 & 1 \\ 0 & 0 & 1 & 0 & 0 \\ 0 & 1 & 1 & 1 & 1 \\ 0 & 0 & 0 & 1 & 1 \end{bmatrix} = \mathbf{A}^2$$

as $\mathbf{a}^2_{3\bullet} = \mathbf{e}_3$. The changes in the fourth iteration occur in the elements a^3_{23}, a^3_{52}, and a^3_{53}. Now

$$\mathbf{A}^4 = \begin{bmatrix} 1 & 1 & 1 & 1 & 1 \\ 0 & 1 & 1 & 1 & 1 \\ 0 & 0 & 1 & 0 & 0 \\ 0 & 1 & 1 & 1 & 1 \\ 0 & 1 & 1 & 1 & 1 \end{bmatrix} = \mathbf{A}^5$$

as there are no further changes in the last iteration and the procedure terminates.

As the first and third rows (and columns) each are different from any other row (and column) in the matrix $\mathbf{A}^5$, hence they are each in a strongly connected component by themselves. On the other hand, the second, fourth, and fifth rows and columns are identical, so that n_2, n_4, and n_5 are members of the same strongly connected component. In other words, the strongly connected components are the subgraphs G^1, G^2, and G^3 with $N_1 = \{n_1\}$, $N_2 = \{n_3\}$, and $N_3 = \{n_2, n_4, n_5\}$. It is now also easily possible to show the adjacency relations between the strongly connected components. For that purpose one representative node is chosen from each component and we delete all rows and columns of the original adjacency matrix $\mathbf{A}^*(G)$ which do not belong to one of these representative nodes. What results are the reduced graph G^{red} and its adjacency matrix $\mathbf{A}^*(G^{\text{red}})$. Choosing $n_1 \in G_1$, $n_3 \in G_2$, and $n_4 \in G_3$ in the above example, the reduced graph G^{red} and its adjacency matrix $\mathbf{A}^*(G^{\text{red}})$ are shown in Figure 22.

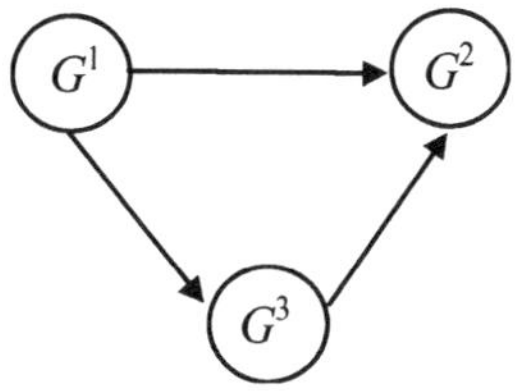

$A^*(G^{red})$:

	n_1	n_3	n_4
n_1	0	1	1
n_3	0	0	0
n_4	0	1	0

Figure 22

It is obvious that no two nodes in the reduced graph can be strongly connected since in such a case all nodes in those distinct connected components would be included in one component.

The problem of determining the weakly connected components of a directed or mixed graph G^d can be solved by replacing each given arc a_{ij} in the original graph by two inversely directed arcs a_{ij} and a_{ji}. It is then easy to prove that each strongly connected component in the new graph corresponds to a weakly connected component in the original graph. We can also prove

Lemma 36: Each weakly connected component of a directed or mixed graph G is a union of some strongly connected components of G.

c.4 Graphs with Special Structures

In this section we will introduce some specific types of graphs frequently encountered in a variety of applications.

Definition 37: A *network* is a directed weakly connected graph without loops.

The graph in Figure 20b is an example of a network whereas the graph in Figure 20a is not as it is mixed.

Definition 0.38: A *forest* is an undirected graph G without cycles. If a forest is connected, then it is said to be a *tree.*

An example of a forest is shown in Figure 23a and a tree is depicted in Figure 23b. The forest in Figure 23a consists of two trees with nodes $\{n_1\}$ and $\{n_2, n_3, n_4, n_5, n_6\}$, respectively.

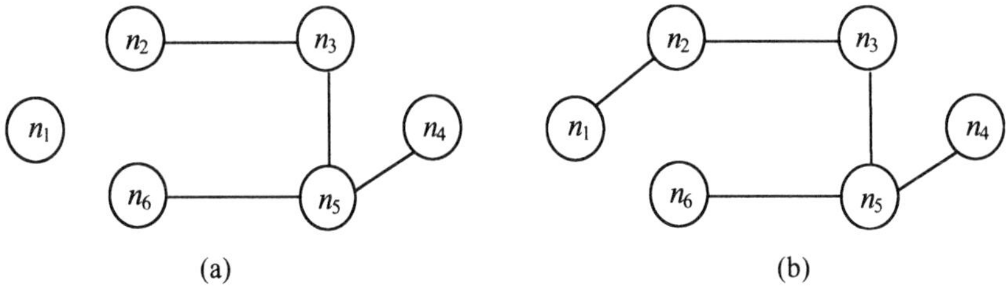

Figure 23

Due to their importance, we will state a theorem that includes a number of equivalent characterizations of trees.

Theorem 39: An undirected graph $G = (N, E)$ is a tree if and only if any one of the statements (1)-(5) below is true:

(1) G is connected and $|E| = n - 1$.

(2) G is acyclic and $|E| = n - 1$.

(3) There exists exactly one elementary path between each pair of nodes n_i, n_j with $i \neq j$.

(4) G is minimal with respect to connectivity, i.e., there exists no connected partial graph G_p with $E_p \subset E$. In particular, $G \setminus e = (N, E \setminus \{e\})$ is disconnected for each edge $e \in E$.

(5) G is acyclic and if any two nonadjacent nodes of G are joined by a new edge e then $G + e = (N, E \cup \{e\})$ has exactly one cycle.

Definition 40: An *arborescence* is a directed graph with the following property. Given some node n_r (the *root*), each node n_i of G can be reached from n_r on exactly one path $(n_r, ..., n_i)$. A *branching* is a collection of disconnected graphs, each of which is an arborescence.

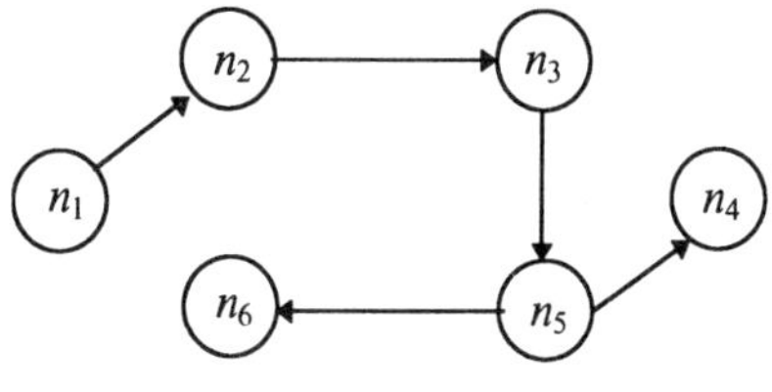

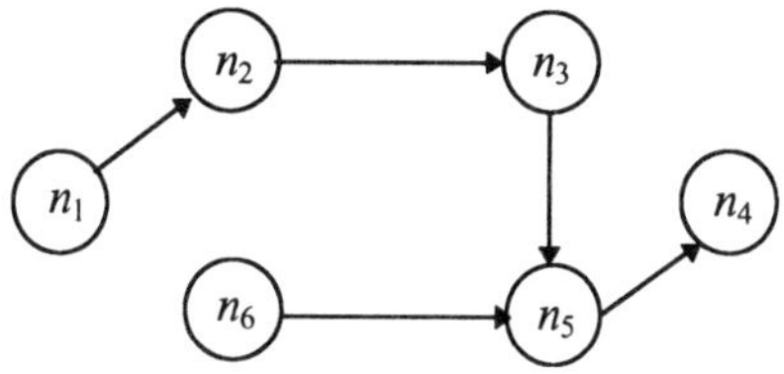

Figure 24

Clearly, arborescences and branchings are the directed equivalents of trees and forests, respectively. The graph in Figure 24a is an arborescence with root n_1 whereas the graph in Figure 24b is not an arborescence since $\delta^-(n_1) = \delta^-(n_6) = 0$, so that both nodes cannot be reached from any other node and thus both would have to be roots, thus violating the definition of an arborescence.

Definition 41: A graph $G = (N, A)$ is called *bipartite*, if N can be decomposed into two sets N_1 and N_2 with $N_1 \cup N_2 = N$ and $N_1 \cap N_2 = \varnothing$, so that $a_{ij} \in A$ implies that either $n_i \in N_1$, $n_j \in N_2$ or $n_i \in N_2$, and $n_j \in N_1$. A graph is called *complete bipartite* if there exists at least one arc for each $n_i \in N_1$, and $n_j \in N_2$.

It is now possible to prove

Lemma 42: A graph is bipartite if and only if each of its cycles is of even length.

The idea of the proof is this. Start a cycle at a node in N_1. Then by definition any arc with one end in N_1 will have its other end in N_2. As we are shuttling between N_1 and N_2 and, in order to close the cycle, we have to move back to N_1, each cycle is of length 2, 4, 6, ...; thus the lemma.

The special structures of bipartite graphs also shows in their adjacency and incidence matrices. In particular, the following structures are encountered.

A*(G):

	N_1	N_2
N_1	**0**	**(0, 1)**
N_2	**(0, 1)**	**0**

In the adjacency matrix $\mathbf{A}^*(G)$, $(\mathbf{0}, \mathbf{1})$ denotes some submatrix of zeroes and ones. In case a bipartite graph G is undirected, then—due to the symmetry of its adjacency matrix—the structure of G can be stored by the matrix formed by the rows of N_1 and the columns of N_2 (or, equivalently, the part that includes the rows of N_2 and the columns of N_1). If G is directed, the symmetry property is lost and both those parts have to be stored. The incidence matrix $\mathbf{I}^*(G)$ of a bipartite graph also has a special structure. A column representing an arc a_{ij} is $(\mathbf{e}_i - \mathbf{e}_j)$ with exactly one nonzero element in a row assigned to N_1 and the other in a row assigned to N_2. This specific structure has a very useful property that occurs in a variety of important problems described in Parts II and III. Also note that an undirected complete bipartite graph without parallel arcs has $|N_1|\,|N_2|$ arcs. Setting $p = |N_1|$ and $q = |N_2|$, it is frequently referred to as $K_{p,q}$ graph, some details of which are discussed below.

Definition 43: A directed graph $G = (N, A)$ is called *symmetric*, if $a_{ij} \in A$ implies that $a_{ji} \in A$. A graph is called *antisymmetric* if $a_{ij} \in A$ implies that $a_{ji} \notin A$.

Examples of symmetric and antisymmetric graphs are shown in Figures 25a and b, respectively. Note that antisymmetric graphs cannot have any loops.

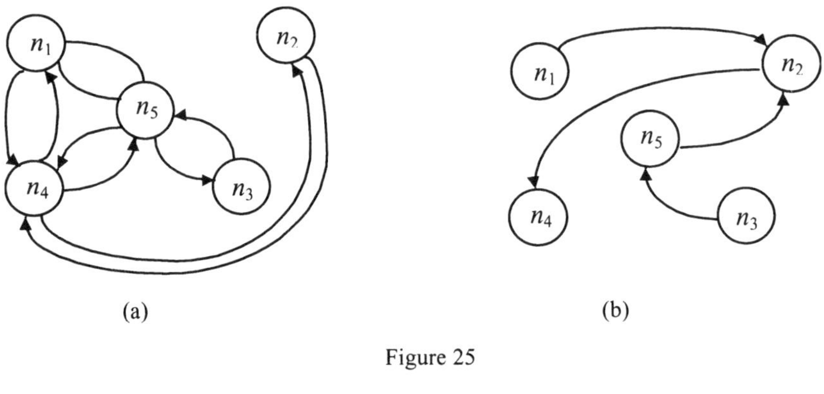

Figure 25

Definition 0.44: A graph $G = (N, A)$ is called *complete*, if for each pair of nodes $n_i, n_j \in N$, either $a_{ij} \in A$ or $a_{ji} \in A$. A complete undirected graph with n nodes and without parallel arcs is usually referred to as a K_n.

According to the above definition, an undirected complete graph without parallel arcs and loops will contain exactly $\frac{1}{2}n(n-1)$ arcs whereas a mixed or directed complete graph without parallel arcs and loops contains at least $\frac{1}{2}n(n-1)$ arcs.

The graphs in Figure 26a and b are examples for an undirected complete graph and a mixed complete graph, respectively.

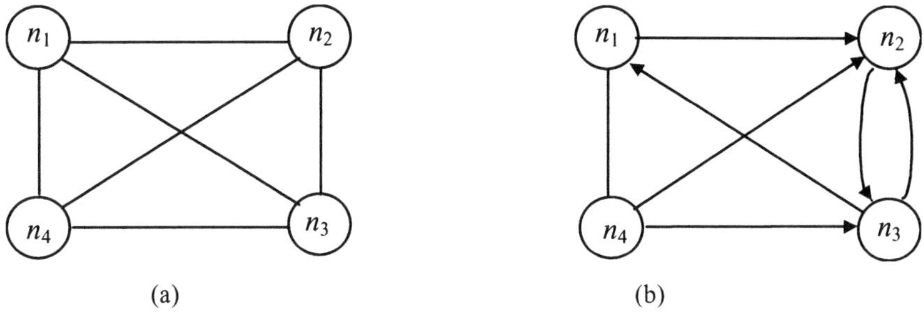

Figure 26

It is also possible to combine the notion of symmetry and completeness. Symmetric and antisymmetric complete graphs are shown in Figures 27a and b.

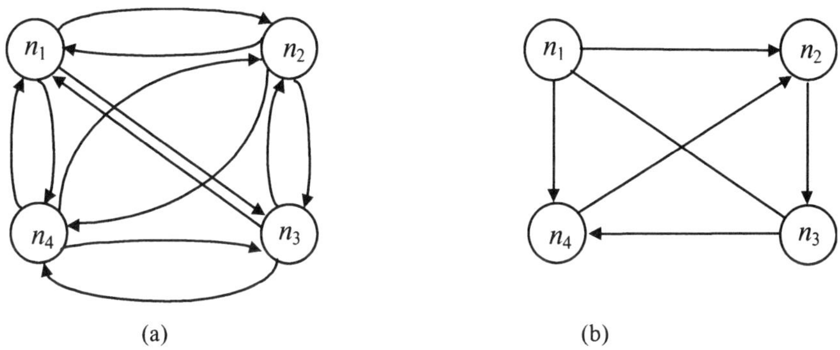

Figure 27

The concept of complete graphs is also used in

Definition 45: A *clique* of a graph G is the largest complete subgraph of G.

Another important concept in graph theory is that of planarity. This concept is defined as follows.

Definition 46: The drawing of a graph G in the two-dimensional plane is called a *planar representation of* G (or *plane graph*) if no two arcs of G intersect. A graph G is called *planar* if it has at least one *planar representation*.

In the following we will state a necessary and sufficient condition for any given graph G to be planar. For that purpose all nodes with degree 2 are eliminated from G by replacing a node with $\delta(n_i) = 2$ as well as its adjacent arcs a_{ij} *and* a_{ik} by the arc a_{jk}. The graph derived from G in such a way is called G'. Furthermore, we will be using the complete graph with five nodes K_5 and the complete bipartite graph $K_{3,\,3}$. The latter is popularly known from the "houses and utilities" problem: given three houses and three utilities (e.g., gas, electricity, and water works), is it possible to connect the houses and the utilities, so that each house receives all three supplies and none of the connecting links cross in the plane? The answer to this question is provided in the following theorem due to Kuratowski (1930).

Theorem 47: A graph G is planar if and only if G' does not contain K_5 or $K_{3,3}$ as a subgraph.

For the proof of this theorem we refer to the original work by Kuratowski or to Busacker and Saaty (1965). This theorem answers the above "houses and utilities" problem in the negative. Hopcroft and Tarjan (1974) have developed an algorithm that indicates whether or not a given graph G is planar and if so, it finds a planar representation in O(n) time. The concept of planarity has a variety of important applications. Among those is the production of printed circuits where intersections

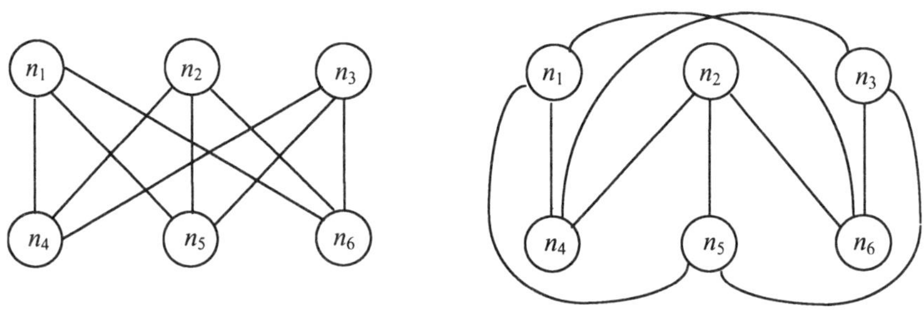

(a) (b)

Figure 28

of wires are costly and to be avoided whenever possible. This leads immediately to a new problem: given that a graph G is not planar, find a representation of G in the plane which minimizes the number of intersections of its arcs.

The $K_{3,3}$ graph displayed in Figure 28a has nine intersections whereas the representation in Figure 28b of the very same graph has only one (which, by definition, must be minimal). Also connected with the concept of planarity is that of a region.

Definition 48: A *region* of a plane graph G is a set of points in the plane bounded by a chain of G.

This definition includes the "exterior region" which extends to infinity in all directions. We can now prove what is known as *Euler's theorem.*

Theorem 49: The planar representation of a connected graph with n nodes, m arcs and r regions satisfies

$$n + r - m = 2.$$

Proof: The theorem will be proved by constructing the given graph G from an initial graph which consists of only one node. Thus, initially $n = 1$, $r = 1$ (the exterior region) and $m = 0$ satisfying the above relation. Now arcs are introduced one at a time. Due to the assumption of connectivity of G, at any stage of the construction it is possible to find an arc which either connects two already existing nodes or one existing and one new node. First consider the case of a new arc a_{ij} which connects two nodes that already exist. Based on the connectivity of the graph G (which is retained throughout the additions of arcs) there is a chain from n_i to n_j. The addition of a_{ij} closes the cycle and thus produces a new region. In summary, n remains unchanged, and m and r increase by one each. Hence the left-hand side of Euler's relation does not change. On the other hand if a new node is introduced together with an arc, then no new regions develop, n and m increase each by one while r remains unchanged; hence again the relation remains valid if it was so before. As the condition was satisfied initially, the theorem is proved.□

The above theorem can be generalized by allowing the graph to consist of a number of components. If c denotes the number of connected components of G, then one can state

Corollary 50: A plane graph with n nodes, m arc, r regions and c connected components satisfies

$$n + r - m - c = 1.$$

The proof is similar to the one above and is omitted here.

CHAPTER d DYNAMIC PROGRAMMING

d.1 Basic Ideas

Richard Bellman is universally recognized as the father of dynamic programming. His research in the 1950's led to the publication of his book *Dynamic Programming*; see Bellman (1957). Unfortunately, the name that he gave to this approach to optimization is somewhat misleading as a variety of static problems can be solved with it, but as it is generally accepted we will use it in this book. On the other hand, the term dynamic programming conveys some of the essence of the approach used in this very general technique. The term is used in the sense of *recursive* or *multistage* optimization, since the decomposition of a decision problem into interrelated stages lies at the heart of the dynamic programming principle. Unlike other areas of optimization, it is difficult to establish a "canonical" or "standard" form of dynamic programming into which all problems to be solved by this method can be cast. Rather, we may think of dynamic programming as a general computational approach. First a single-stage subproblem is solved, and then successively larger subproblems are solved, recursively, until finally a solution to the original problem has been found.

In order to explain the basic principles of dynamic programming, we will solve a small illustrative example in the next section. In doing so, we also introduce some general notation. The third section of this chapter outlines the basic principles of dynamic programming together with an algorithm and a numerical example. In the fourth section we consider a few different problems that are particularly amenable to dynamic programming.

In addition to Bellman's (1957) original work, significant contributions to the field are those by Bellman and Dreyfus (1962), Hadley (1964), Howard (1960), Jacobson and Mayne (1970), and Nemhauser (1966). Later treatments can be found in Dreyfus and Law (1977) and in Denardo (1982). Standard textbooks in operations research usually include dynamic programming; see for example Hillier and Lieberman (1995), Ravindran, Phillips and Solberg (1987) and Wagner (1975). The important field of probabilistic dynamic programming is beyond the scope of this short introduction.

The classical way to introduce dynamic programming is by way of the so-called *stagecoach problem*, first mentioned by Wagner (1969). A traveler is planning his journey by stagecoach from city n_1 to city n_{11} via some of the cities n_2 to n_{10}. A graph that shows all available links is shown in Figure 29 where the numbers next to the arcs represent the travel time between cities. Some direct connections are not available, such as those between n_2 and n_7, n_3 and n_5, and n_6 and n_{10}. The problem is now to find a route from city n_1 to city n_{11} through the network in such a way that the total travel time is minimized.

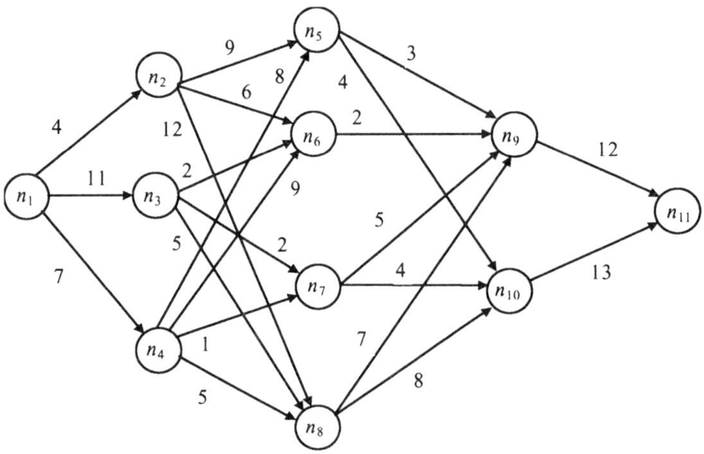

Figure 29

One way to solve this problem would be to enumerate all possible routes from n_1 to n_{11}, evaluate the length of each of them and then select the best. In our example this would not be too difficult as there are no more than 17 possible routes that can easily be evaluated and compared. However, as the number of cities grows this *total enumeration* approach quickly becomes impractical. A problem with, say, 100 cities would easily keep even the fastest of today's computers busy for years in evaluating the astronomically many routes that would be possible. Dynamic programming is an *implicit enumeration* technique that avoids total enumeration and enables us to solve the problem by decomposing it into subproblems or stages which are then solved sequentially.

At this point, we have the choice between *forward recursion* and *backward recursion*. As the names imply, the forward recursion commences with node n_1 and successively computes the shortest distances to successor nodes, until the shortest route to node n_{11} is found. The backward recursion proceeds in a reverse direction, starting at n_{11} and successively finds shortest paths from predecessors to n_{11}, until the node n_1 is reached. The important point is that it is not required to store entire paths; knowledge about the latest optimal decision and the length of the shortest path found so far are sufficient. The forward recursion can be illustrated by using a simple labeling procedure that is very similar to labeling techniques to be introduced in parts II and III of this book. The label of a node n_j consists of two parts: the first part indicates the direct predecessor of n_j on the shortest path, and the second part is the length of the shortest path from n_1 to n_j. In our example, the label of n_1 is (–, 0). Consider now all nodes in the next stage, i.e., n_2, n_3, and n_4. As n_2 can only be reached from n_1 directly, its label is (n_1, 4) with the second part of the label indicating the total distance from n_1 to n_2. Similarly, we label n_3 and n_4 with (n_1, 11), and (n_1, 7), respectively. Consider now labeling n_5. The node can be reached from n_2 and n_4. By way of n_2, the length of the shortest path equals 4 (by virtue of the second part of the label of n_2) plus 9 (the direct distance between n_2 and n_5. Similarly, if we were to label n_5 from n_4, the length of the shortest path from n_1 to n_5 via n_4 is 7 (by virtue of the label of n_4) plus 8 (the length of the direct distance from n_4 to n_5). Hence, the label of n_5 is either (n_2, 13) or (n_4, 15). Since we are interested in the shortest path, we choose (n_2, 13) as the label of n_5. Similarly, we label the nodes n_6, n_7, and n_8 as (n_2, 10), (n_3, 13), and (n_4, 12), respectively. In the next stage, we label n_9 and n_{10}. For n_9, we either label the node from n_5 with (n_5, 13 + 3) = (n_5, 16), from n_6 with (n_6, 10 + 2) = (n_6, 12), from n_7 with (n_7, 13 + 5) = (n_7, 18), or from n_8 with (n_8, 12 + 7) = (n_8, 19). The shortest path is by way of n_6 (notice that all we know at this point is the best direct predecessor of the node n_9, but *not* the entire path from n_1 to n_9), so that the label n_9 is (n_6, 12). Similarly, the node n_{10} is labeled as (n_5, 17) or (n_7, 17). In the last stage, the terminal node n_{11} is labeled either from n_9 with the label (n_9, 12 + 12) = (n_9, 24) or from n_{10} with the label (n_{10}, 17 + 13) = (n_{10}, 30). The shortest distance is 24, so that node n_{11} is labeled (n_9, 24).

At this point, we have already determined the length of the shortest path (i.e., the objective value of the problem), but not the course of actions that leads to that solution. In order to determine which route to follow in order to reach n_{11} from n_1 in 24 hours, a *backward pass* is required. Starting at n_{11}, it moves backwards by scanning the labels. Here, the best predecessor of n_{11} is n_9. The label of n_9 indicates that its optimal predecessor is n_6 whose optimal predecessor, in turn, is n_2. Finally, the optimal predecessor of n_2 is n_1 which, in reverse order, then gives the unique optimal set of decisions (i.e., the shortest path) as $n_1 - n_2 - n_6 - n_9 - n_{11}$. The shortest path could also have been found using backward recursion in an analogous fashion by starting with node n_{11} and proceeding backwards until node n_1 is reached.

This simple example, while quite primitive, shows the main features of dynamic programming: reliance of an optimal policy on the immediate processor in a sequential structure. It is this sequential structure that gives the method its name. Clearly, a number of simplifications were made: the example does not include any links that lead from a node in a certain stage to another node that is not in the next, but in some further stage (such cases can be dealt with by introducing intermediate "dummy cities"), and backward arcs that would introduce circuits in the graph are not included. It is also worth mention that some structural changes in the problem can be dealt with without requiring solving the entire problem all over again. This is possible, as changes in, say, stage k do not impact the process in the preceding k stages. This feature allows for simple postoptimality analyses.

d.2 A General Algorithm

Using the above stagecoach problem as an example, some general concepts and notation common to all dynamic programming problems will now be described.

(1) The problem can be divided into a number of different *stages* $k = 1, ..., K$. At each stage k there are a number of alternative *states* y_k that indicate the current status of the system. The set of feasible states y_k in stage k is called the *state space* Y_k. The state y_k is determined at the end of stage k.

(2) At each stage k, a decision $x_k \in X_k$ is taken, where X_k denotes the set of feasible decisions of stage k.

(3) Given a state y_{k-1} and a decision x_k, the system is transformed to a state y_k via the expression $y_k = t_k(y_{k-1}, x_k)$, where t_k is the *stage transition* or *stage transformation* function. It is also useful to define the inverse function $y_{k-1} = t_k^{-1}(y_k, x_k)$.

(4) Starting from any current state, an optimal policy for the remaining stages is independent of the policy adopted in the previous stages. (This is one version of Bellman's famous *principle of optimality*).

(5) In each state, an optimal policy for the remaining stages can be found from a recursive relationship in which $f_k(y_{k-1}, x_k)$ denotes the contribution of the decision x_k to the objective function in stage k, given that the system is in state y_{k-1}. The term $\overline{F}_k(y_k)$ expresses the cumulative optimal objective function value in the first k stages given that the system is in state y_k in stage k.

(6) Using the recursive relationship, an optimal solution to the original problem is calculated by moving from stage to stage in a forward fashion (backwards in

case a backward recursion is used), starting from the initial stage and ending in the terminal stage. Formally, $F_k(y_k, x_k) = \overline{F}_{k-1}(y_{k-1}) + f_k(y_{k-1}, x_k)$, where $\overline{F}_{k-1}(y_{k-1})$ denotes the optimal contribution up to (and including) stage $k - 1$, and $f_k(y_{k-1}, x_k)$ is again the contribution to the objective function in stage k. In order to optimize x_k in stage k we determine $\overline{F}_k(y_k)$ by solving the problem $\underset{x_k \in X_k}{\text{Max}}\, F_k(y_k, x_k)$.

We are now ready to state a dynamic programming algorithm in general form. Recall that the problem to be solved has been decomposed as a K-stage maximization problem, x_k denote the *decision variables* and y_k the *state variables.* We assume that $K > 2$. The algorithm is initialized by setting $y_0 := 0$, $\overline{F}_0(0) := 0$ and $k := 1$.

A Dynamic Programming Algorithm

Step 1: Determine Y_k and X_k.

Step 2: For each pair (y_k, x_k) with $y_k \in Y_k$, $x_k \in X_k$, and $y_{k-1} \in Y_{k-1}$, determine $F_k(y_k, x_k) = \overline{F}_{k-1}(y_{k-1}) + f_k(y_{k-1}, x_k)$.

Step 3: (Optimization): Determine $\overline{F}_k(y_k) = F_k(y_k, \overline{x}_k) = \underset{x_k \in X_k}{\text{Max}}\, \{F_k(y_k, x_k)\}$. Set $k := k + 1$.

Step 4: Is $k > K$? If yes: Find a solution to $\underset{y_k \in Y_k}{\text{Max}}\, \overline{F}_K(y_K)$ and go to Step 5.
If no: Go to Step 1.

Step 5: (Backward sweep). The optimal value of the objective function is $\overline{z} = \overline{F}_K(\overline{y}_K) = F_K(\overline{y}_K, \overline{x}_K) = \underset{y_K \in Y_K}{\max}\, \{F_K(y_K, \overline{x}_K)\}$. For $k = K - 1, K - 2, \ldots, 1$, determine $\overline{y}_k$, the corresponding $\overline{x}_k$ and $\overline{y}_{k-1}$ values using the inverse function $y_{k-1} = t_k^{-1}(y_k, x_k)$. The values $\overline{x}_1, \overline{x}_2, \ldots, \overline{x}_K$ then constitute an optimal solution.

In the above algorithm we refer to steps 1 to 4 as the forward recursion phase and to step 5 as the *backward substitution* phase or *backward sweep*. For a minimization problem we simply replace "max" in the algorithm above by "min". Alternatively, one might make use of the fact that Min $f(y)$ = Max $(-f(y))$ for any function $f(y)$. We also point out that the procedure for optimizing the functions is not specified here; this can be accomplished by any appropriate method depending on the particular problem under consideration. The above dynamic programming algorithm can also be stated in a backwards recursion version. We will restrict our treatment to the intuitively more appealing forward recursion formulation.

Example: Consider the following nonlinear integer optimization problem

$$\begin{array}{llcl} \text{P: Max } z = & 6x_1 + 2^{x_2} + x_3^3 & & \\ \text{s.t.} & 3x_1 + 4x_2 + 2x_3 & \le & 12 \\ & x_1 & \in & [0; 3] \\ & \quad\quad x_2 & \in & [1; 4] \\ & \quad\quad\quad\quad x_3 & \in & [0; 2] \\ & x_1, \quad x_2, \quad x_3 & \in & \mathbb{N}_0 \end{array}$$

In order to apply dynamic programming, the problem is artificially subdivided into stages, so that each stage deals with exactly one variable. This is possible as the problem is additively separable, i.e., functions of individual variables are connected solely by addition and subtraction. The constraint can be written as $\sum_j a_j x_j \le b$ with $a_1 = 3$, $a_2 = 4$, $a_3 = 2$, and $b = 12$, where the right-hand side can be interpreted as a resource that is consumed by the activities x_j. Defining the state variable y_k as the amount of the resource allocated to all steps up to and including stage k, the amount of the resource that is consumed in stage k is then $a_k x_k$. The transition function therefore expresses the resource consumption so far, i.e., $y_k = y_{k-1} + a_k x_k$. Define now the state space Y_k, $k = 1, ..., n$ as the set of possible values that the state variable y_k can assume. It is crucial to define the state spaces as tightly as possible as their cardinalities tend to grow exponentially even if only a few stages are included in the model. Here, we notice that as $x_2 \ge 1$, that variable will consume at least $a_2 x_2 = 4x_2 \ge 4$ resource units, so that only $12 - 4 = 8$ resource units are really available for the other two variables if they assume values above their lower bounds.

Initially, $y_0 = 0$ and for $k = 1$, $y_1 = y_0 + a_1 x_1 = 0 + 3x_1$, or $y_1 = 3x_1$. The decision variable x_1 ranges from 0 to 3; however, as each unit of x_1 consumes 3 resource units and only 8 are available, this range can be tightened to [0; 2]. Given now that $x_1 \in X_1 = \{0, 1, 2\}$, it follows from $y_1 = 3x_1$ that $y_1 \in Y_1 = \{0, 3, 6\}$. All pairs ($y_1$, x_1) for which $y_1 \ne 3x_1$ are shown with "–". These considerations result in Table $\mathbf{T}_1$.

$\mathbf{T}_1$:

y_1 \ x_1	$F_1(y_1, x_1) = \bar{F}_1(y_0) + f_1(y_0, x_1)$: 0	1	2	$\bar{F}_1(y_1) = F_1(y_1, \bar{x}_1)$	$\bar{x}_1$
0	0	-	-	0	0
3	-	6	-	6	1
6	-	-	12	12	2

Set now $k = 2$. The transition function is then $y_2 = y_1 + a_2 x_2 = y_1 + 4x_2$ or $y_1 = y_2 - 4x_2$.

The range of x_2 can be reduced by considering the fact that $x_2 = 4$ would require 16 resource units while only 12 are available. Hence, $x_2 \in X_2 = \{1, 2, 3\}$. As the transition function is $y_2 = y_1 + 4x_2$ and $y_1 \in Y_1 = \{0, 3, 6\}$, we obtain $Y_2 = \{4, 7, 8, 10, 11, 12\}$. This allows us to solve the second stage of the problem. Again, we indicate by "–" pairs (y_2, x_2) for which $y_2 - 4x_2 \notin Y_1$. The cells of all other pairs in the table will include a number in the top left corner, indicating the value of y_1 that belongs to the (y_2, x_2) pair in this cell. As an example, the pair $(y_2, x_2) = (8, 2)$ leads to $y_1 = y_2 - 4x_2 = 8 - 8 = 0$ which is shown in the appropriate cell in the table below. The partial objective values in the cells (y_2, x_2) are computed as the objective value achieved so far for the value of y_1 shown in the corner plus the objective contribution by the appropriate value of x_2. As an example, consider the pair $(y_2, x_2) = (10, 1)$ with $y_1 = 6$. The partial objective value $\overline{F}_1(y_1) = \overline{F}_1(6) = 12$ as shown in Table $\mathbf{T}_1$, while the contribution $f_2(y_1, x_2)$ in the objective function of problem P equals $2^{x_2} = 2^1 = 2$. The other values are computed similarly. For each value of y_2, we then find the value of x_2 that maximizes $F_2(y_2, x_2)$. In Table $\mathbf{T}_2$, this means finding the largest value in each row and the corresponding x_2-value. This information is stored in the two rightmost columns of the Table $\mathbf{T}_2$.

$\mathbf{T}_2$:

	$F_2(y_2, x_2) = \overline{F}_1(y_1) + f_2(y_1, x_2)$				
y_2 \ x_2	1	2	3	$\overline{F}_2(y_2) = F_2(y_2, \overline{x}_2)$	$\overline{x}_2$
4	[0] 0 + 2 = 2	–	–	2	1
7	[3] 6 + 2 = 8	–	–	8	1
8	–	[0] 0 + 4 = 4	–	4	2
10	[6] 12 + 2 = 14	–	–	14	1
11	–	[3] 6 + 4 = 10	–	10	2
12	–	–	[0] 0 + 8 = 8	8	3

Consider now the final stage $k = 3$. The transition function is $y_3 = y_2 + a_3x_3 = y_2 + 2x_3$, or $y_2 = y_3 - 2x_3$. Given the range of $x_3 \in X_3 = \{0, 1, 2\}$, the transition function results in $Y_3 = \{4, 6, 7, 8, 9, 10, 11, 12\}$. Note that even in this small example with tightly bounded x_k-values, the state spaces Y_k (and with them the decision tables) have grown to sizable proportions. It is easy to imagine what a right-hand side value $b = 10{,}000$ with no individual explicit bounds on the variables x_k would do to the state spaces Y_k. The results of our computations are summarized in Table $\mathbf{T}_3$.

$\mathbf{T}_3$:

y_3 \ x_3	0	1	2	$\overline{F}_3(y_3) = F_3(y_3, \overline{x}_3)$	$\overline{x}_3$
	$F_3(y_3, x_3) = \overline{F}_2(y_2) + f_3(y_2, x_3)$				
4	4 | 2 + 0 = 2	–	–	2	0
6	–	4 | 2 + 1 = 3	–	3	1
7	7 | 8 + 0 = 8	–	–	8	0
8	8 | 4 + 0 = 4	–	4 | 2 + 8 = 10	10	2
9	–	7 | 8 + 1 = 9	–	9	1
10	10 | 14 + 0 = 14	8 | 4 + 1 = 5	–	14	0
11	11 | 10 + 0 = 10	–	7 | 8 + 8 = 16	16	2
12	12 | 8 + 0 = 8	10 | 14 + 1 = 15	8 | 4 + 8 = 12	15	1

This terminates the forward recursion. We are now able to determine $\overline{y}_3$, i.e., the value that maximizes $\overline{F}_3(y_3)$ by finding the maximal value in the $\overline{F}_3(y_3)$ column. Here, the maximum is $\overline{F}_3(y_3) = 16$, which is found in the row with $\overline{y}_3 = 11$. The corresponding $\overline{x}_3$-value is found in the last column; here $\overline{x}_3 = 2$. At this point, the transition function enables us to move to the previous step and table. In the cell $(\overline{y}_3, \overline{x}_3) = (11, 2)$, we find $\overline{y}_2 = 7$. Back in the Table $\mathbf{T}_2$, the row with $\overline{y}_2 = 7$ shows that $\overline{x}_2 = 1$ and the element $(\overline{y}_2, \overline{x}_2) = (7, 1)$ indicates that $\overline{y}_1 = 3$. Back in Table $\mathbf{T}_1$, the row with $\overline{y}_1 = 3$ shows that $\overline{x}_1 = 1$ which completes the backward sweep. In summary, the unique optimal solution of the problem is $\overline{x}_1 = 1$, $\overline{x}_2 = 1$, and $\overline{x}_3 = 2$ with an optimal value of the objective function $\overline{z} = 16$.

It should be pointed out that sensitivity analyses on the right-hand side values can easily be performed. For instance, if $b = 10$ rather than 12, all we have to do is eliminate the last two rows in table $\mathbf{T}_3$. Now the maximum objective value is $\overline{z} = \overline{F}_3(10) = 14$ and $\overline{x}_3 = 0$. The backward sweep then reveals that $\overline{y}_2 = 10$, $\overline{x}_2 = 1$, $\overline{y}_1 = 6$, and $\overline{x}_1 = 2$, or simply $\overline{x}_1 = 2$, $\overline{x}_2 = 1$, and $\overline{x}_3 = 0$. Notice that the reduction of the right-hand side value by 2 units has not resulted in a minor modification of the previous solution, but a completely different optimal solution.

d.3 Various Examples

To show the versatility of the dynamic programming approach we present a number of different problems that can be cast into the mold of and solved as dynamic programming problems. Possible modifications and extensions to the basic model are indicated wherever appropriate.

Example 1: *A Production-Inventory Planning Model*
Consider a facility that produces, stores, and sells a particular item. Production constraints and costs are assumed to be known and given, as are inventory constraints and costs as well as contracted demands for a 4-month period. The objective is to find an optimal, i.e., least cost, production-inventory plan. In particular, assume that the demand for electric transformers is 30, 40, 20, and 30 for the months of May, June, July, and August, respectively. The costs of producing the transformers are shown in Table 2.

Table 2

Production level	0	10	20	30	40
Production costs (in \$1,000)	0	7	9	10	11

The production of the transformers takes place during a month, and all deliveries are made at the end of the month. Transformers that are not delivered in the same month that they are produced may be kept in inventory at a cost of \$100 per month, per transformer. Inventory costs are based on the number of transformers in stock at the beginning of each month, or, equivalently at the end of the previous month. The maximum storage capacity is 30 transformers. The opening inventory on May 1 is 20 units, and the closing inventory on August 31 must be zero. Only batches of ten units are considered for storage, production and delivery purposes. The task is to determine an optimal production-inventory plan that simultaneously meets the demand and minimizes total costs.

In order to obtain a dynamic programming formulation of this problem we first define the stages to be the months. Using the forward recursion, let $k \in K = \{1, 2, 3, 4\}$ correspond to the months of May, June, July, and August, respectively. Furthermore, let the decision variable x_k denote the production level in month k, and let the state variable y_k indicate the inventory level at the end of month $k = 1, 2, 3, 4$. As the opening inventory in May is 20 units, $y_0 = 20$, and as the inventory level at the end of August must be zero, $y_4 = 0$. Defining d_k as the demand for transformers in month k, the inventory balancing equations can then be written as $y_k = y_{k-1} + x_k - d_k$ for $k = 1, 2, 3, 4$. In other words, the inventory at the end of month $k-1$ plus the production in month k minus the demand in month k must equal the inventory at the end of month k; this is the stage transformation equation $y_k = t_k(y_{k-1}, x_k)$. Additionally, feasible inventory levels are $y_k \in Y_k = \{0, 10, 20, 30\}$ and possible production capacities are $x_k \in X_k = \{0, 10, 20, 30, 40\}$ for $k = 1, 2, 3, 4$. Finally, let $c(x_k)$ be the production cost, so that the total cost in each month is the sum of production and inventory cost, i.e., $f_k(y_{k-1}, x_k) = c(x_k) + 0.1y_{k-1}$ in month k.

In the first stage, $y_0 = 0$ and $\overline{F}_0(y_0) = 0$ and in stage 1 we solve the problem

$\operatorname*{Min}_{x_1 \in X_1} F_1(y_1, x_1) = \overline{F}_0(y_0) + f_1(y_0, x_1) = c(x_1) + 0.1y_0$ where $y_1 = y_0 + x_1 - d_1$ or $y_1 = 20 + x_1 - 30$ so that $y_1 = x_1 - 10$. We then obtain the tableau $\mathbf{T}_1$:

T$_1$: $F_1(y_1, x_1) = c(x_1) + 0.1y_0$

y_1 \ x_1	0	10	20	30	40	$\overline{F}_1(y_1) = F_1(y_1, \overline{x}_1)$	$\overline{x}_1$
0	-	7+2=9	-	-	-	9	10
10	-	-	9+2=11	-	-	11	20
20	-	-	-	10+2=12	-	12	30
30	-	-	-	-	11+2=13	13	40

In stage 2 we solve

$\operatorname*{Min}_{x_2 \in X_2} F_2(y_2, x_2) = \overline{F}_1(y_1) + f_2(y_1, x_2) = \overline{F}_1(y_1) + c(x_2) + 0.1y_1$ where $y_2 = y_1 + x_2 - d_2 = y_1 + x_2 - 40$. The tableau $\mathbf{T}_2$ is then

T$_2$: $F_2(y_2, x_2) = \overline{F}_1(y_1) + c(x_2) + 0.1y_1$

y_2 \ x_2	0	10	20	30	40	$\overline{F}_2(y_2)$	$\overline{x}_2$
0	-	13+7+3 =23	12+9+2 =23	11+10+1 =22	9+11+0 =20	20	40
10	-	-	13+9+3 =25	12+10+2 =24	11+11+1 =23	23	40
20	-	-	-	13+10+3 =26	12+11+2 =25	25	40
30	-	-	-	-	13+11+3 =27	27	40

and in stage 3 we solve

$\underset{x_3 \in X_3}{\text{Min}}\ F_3(y_3, x_3) = \overline{F}_2(y_2) + f_3(y_2, x_3) = \overline{F}_2(y_2) + c(x_3) + 0.1y_2$, where $y_3 = y_2 + x_3 - d_3 = y_2 + x_3 - 20$, so that the tableau $\mathbf{T}_3$ is

$\mathbf{T}_3$: $F_3(y_3, x_3) = \overline{F}_2(y_2) + c(x_3) + 0.1y_2$

y_3 \ x_3	0	10	20	30	40	$\overline{F}_3(y_3)$	$\overline{x}_3$
0	25+0+2 =27	23+7+1 =31	20+9+0 =29	-	-	27	0
10	27+0+3 =30	25+7+2 =34	23+9+1 =33	20+10+0 =30	-	30	0,30
20	-	27+7+3 =37	25+9+2 =36	23+10+1 =34	20+11+0 =31	31	40
30	-	-	27+9+3 =39	25+10+2 =37	23+11+1 =35	35	40

Finally, in stage 4 we solve

$\underset{x_4 \in X_4}{\text{Min}}\ F_4(y_4, x_4) = \overline{F}_3(y_3) + f_4(y_3, x_3) = \overline{F}_3(y_3) + c(x_4) + 0.1y_4$ where $y_4 = y_3 + x_4 - d_4 = y_3 + x_4 - 30$. Since $y_4 = 0$, the tableau $\mathbf{T}_4$ is

$\mathbf{T}_4$: $F_4(y_4, x_4) = \overline{F}_3(y_3) + c(x_4) + 0.1y_3$

y_4 \ x_4	0	10	20	30	40	$\overline{F}_4(y_4)$	$\overline{x}_4$
0	35+0+3 =38	31+7+2 =40	30+9+1 =40	27+10+0 =37	-	37	30

As now $k = 5 > 4 = K$, we set $\overline{z} := 37$ and proceed with the backward sweep by successively finding $\overline{y}_4 = 0$, $\overline{x}_4 = 30$, and then, since $\overline{y}_4 = \overline{y}_3 + \overline{x}_4 - 30$, that $\overline{y}_3 = 0$. From $\mathbf{T}_3$ we obtain $\overline{x}_3 = 0$, so that, since $\overline{y}_3 = \overline{y}_2 + \overline{x}_3 - 20$, $\overline{y}_2 = 20$ results. The tableau $\mathbf{T}_2$ yields $\overline{x}_2 = 40$ and, with $\overline{y}_2 = \overline{y}_1 + \overline{x}_2 - 40$, $\overline{y}_1 = 20$. Finally, $\mathbf{T}_1$ then implies that $\overline{x}_1 = 30$. Recall that $\overline{y}_0 = 20$ was predetermined. Therefore, the unique optimal solution is $(\overline{y}_1, \overline{x}_1) = (20, 30)$, $(\overline{y}_2, \overline{x}_2) = (20, 40)$, $(\overline{y}_3, \overline{x}_3) = (0, 0)$, $(\overline{y}_4, \overline{x}_4) = (0, 30)$ and $\overline{z}$ = \$37,000. By coincidence, already in tableau $\mathbf{T}_2$ we know that the optimal production of transformers in June will be $\overline{x}_2 = 40$, since this value is optimal for all possible values of y_2. The unique optimal plan is summarized in Table 3.

Table 3

Month	Opening inventory level	Production level	Demand level	Closing inventory level
May	20	30	30	20
June	20	40	40	20
July	20	0	20	0
August	0	30	30	0

Example 2: A Staff Assignment Model
A company is operating four outlets for which the staffing is to be optimized. Profits at each outlet depend on the number of employees assigned to it. The net profits per day are shown in Table 4.

Table 4

Number of employees	Outlet 1	Outlet 2	Outlet 3	Outlet 4
2	195	-	165	-
3	245	215	220	-
4	270	230	240	250
5	275	245	255	260
6	265	245	-	270

As can be seen in Table 4, outlet 1 needs at least two employees, outlet 2 at least three, outlet 3 at least two but no more than five, and outlet 4 at least four employees. Assuming that 14 employees are available, the staff assignment is to be optimized.

Assign one stage to each of the four outlets and let the decision variable x_k denote the number of employees assigned to outlet k. Furthermore, let y_k be the state variable denoting the cumulative number of staff assigned to outlets $j \leq k$, $k = 1, 2, 3, 4$. Clearly, $y_0 = 0$.In stage $k = 1$, we obtain $y_1 = y_0 + x_1 = 0 + x_1 = x_1 \in X_1 = \{2, 3, 4, 5, 6\}$ so that $Y_1 = \{2, 3, 4, 5, 6\}$ and with $\overline{F}_0(y_0) = 0$ we find that $F_1(y_1, x_1) = \overline{F}_0(y_0) + f_1(y_0, x_1) = f_1(y_0, x_1)$. Table $\mathbf{T}_1$ is then

$\mathbf{T}_1$:

	$F_1(y_1, x_1) = f_1(y_0, x_1)$						
y_1 \ x_1	2	3	4	5	6	$\overline{F}_1(y_1)$	$\overline{x}_1$
2	195	-	-	-	-	195	2
3	-	245	-	-	-	245	3
4	-	-	270	-	-	270	4
5	-	-	-	275	-	275	5
6	-	-	-	-	265	265	6

In stage $k = 2$, the transition function is $y_2 = y_1 + x_2$ with $x_2 \in X_2 = \{3, 4, 5, 6\}$, so that $Y_2 = \{5, 6,..., 12\}$.The tableau $\mathbf{T}_2$ is

$\mathbf{T}_2$:

y_2 \ x_2	$F_2(y_2, x_2) = \overline{F}_1(y_1) + f_2(y_1, x_2)$: 3	4	5	6	$\overline{F}_2(y_2)$	$\overline{x}_2$
5	410	-	-	-	410	3
6	460	425	-	-	460	3
7	485	475	440	-	485	3
8	490	500	490	440	500	4
9	480	505	515	490	515	5
10	-	495	520	515	520	5
11	-	-	510	520	520	6
12	-	-	-	510	510	6

In stage 3 the transition function is $y_3 = y_2 + x_3$ and with $x_3 \in X_3 = \{2, 3, 4, 5\}$, we obtain $Y_3 = \{7, 8,..., 14\}$, since no more than 14 employees are available. The tableau $\mathbf{T}_3$ is then

$\mathbf{T}_3$:

y_3 \ x_3	$F_3(y_3, x_3) = \overline{F}_2(y_2) + f_3(y_2, x_3)$: 2	3	4	5	$\overline{F}_3(y_3)$	$\overline{x}_3$
7	575	-	-	-	575	2
8	625	630	-	-	630	3
9	650	680	650	-	680	3
10	665	705	700	665	705	3
11	680	720	725	715	725	4
12	685	735	740	740	740	4,5
13	685	740	755	755	755	4,5
14	675	740	760	770	770	5

Finally, in stage 4 the transition function is $y_4 = y_3 + x_4$ and as exactly 14 employees are to be assigned, $Y_4 = \{14\}$ and the tableau $\mathbf{T}_4$ is

$\mathbf{T}_4$:

y_4 \ x_4	$F_4(y_4, x_4) = \overline{F}_3(y_3) + f_4(y_3, x_4)$: 4	5	6	$\overline{F}_4(y_4)$	$\overline{x}_4$
14	955	940	900	955	4

As now $k = 5 > 4 = K$, $\overline{z} = 955$ and with $\overline{y}_4 = 14$ it then successively follows that

$\bar{x}_4 = 4$, $\bar{y}_3 = \bar{y}_4 - \bar{x}_4 = 10$, $\bar{x}_3 = 3$, $\bar{y}_2 = \bar{y}_3 - \bar{x}_3 = 7$, $\bar{x}_2 = 3$, $\bar{y}_1 = \bar{y}_2 - \bar{x}_2 = 4$, $\bar{x}_1 = 4$. In other words, four employees should be assigned to outlet 1, three to outlet 2, three to outlet 3 and four to outlet 4. The total net profit with this staff assignment is then \$955 per day.

Postoptimality analysis can easily be carried out by appropriately using the tableaus constructed during the solution procedure of a problem. As an example, assume that only 13 rather than 14 employees were available. Based on the tableaus of the first three stages we would then obtain the new stage 4 tableau

$\mathbf{T}'_4$:

y_4 \ x_4	$F_4(y_4, x_4) = \bar{F}_3(y_3) + f_4(y_3, x_4)$ 4	5	6	$\bar{F}_4(y_4)$	$\bar{x}_4$
13	930	890	845	930	4

With $\bar{z} = 930$, $\bar{y}_4 = 13$ and $\bar{x}_4 = 4$ we then successively find $\bar{y}_3 = 9$, $\bar{x}_3 = 3$, $\bar{y}_2 = 6$, $\bar{x}_2 = 3$, $\bar{y}_1 = 3$, and $\bar{x}_1 = 3$. Consequently, in case 13 employees were available, three of them should be assigned to outlet 1, three to outlet 2, three to outlet 3 and four to outlet 4, for a total net profit of \$930 per day.

Example 3: *A Budget Allocation Model*

A firm has \$100,000 available for the purchase of new machinery. Three types of machines are considered, *viz.*, M_1, M_2, and M_3. Each machine of type M_1 costs \$20,000 to purchase and will generate a net profit of \$3,000 in its first year of operation. For each machine M_2, the corresponding figures are \$20,000 and \$2,000, respectively. Each machine M_3 costs \$30,000 to purchase, but unlike the other machines, the profits generated are nonlinearly related to the number of machines bought. Due to economies of scale, if j machines of type M_3 are bought, the total first-year net profits generated by them are $j\,(2j - 1)$. The question is how many machines of each type should be purchased in order to maximize the total first-year net profits.

To set this problem up for dynamic programming, we first define the stages to correspond to the different machines, and denote by x_k the number of machines of type M_k to be purchased. Let y_k denote the amount of funds (in \$1,000) invested in the stages $j \leq k$, $k = 1, 2, 3$. Scaling the budget constraint to $2x_1 + 2x_2 + 3x_3 \leq 10$, the problem can be written as

$$\begin{aligned} \text{P: Max } z = \; & 3x_1 + 2x_2 + 2x_3^2 - x_3 \\ \text{s.t.} \quad & 2x_1 + 2x_2 + 3x_3 \leq 10 \\ & x_1, x_2, x_3 \in \mathbb{N}_0. \end{aligned}$$

It is apparent that this formulation is a single-constraint problem with a nonlinear objective function.

Using the dynamic programming algorithm of Section d.3 we find that $F_1(y_1, x_1) = \overline{F}_0(y_0) + f_1(y_0, x_1) = f_1(y_0, x_1) = 3x_1$ where $y_1 = y_0 + 2x_1$. As $y_0 = 0$, $X_1 = \{0, 1, 2, 3, 4, 5\}$, and we obtain $y_1 = 2x_1$. The first table is then

T$_1$:

y_1 \ x_1	0	1	2	3	4	5	$\overline{F}_1(y_1)$	$\overline{x}_1$
	$F_1(y_1, x_1) = f_1(y_0, x_1) = 3x_1$							
0	0	-	-	-	-	-	0	0
2	0	3	-	-	-	-	3	1
4	0	3	6	-	-	-	6	2
6	0	3	6	9	-	-	9	3
8	0	3	6	9	12	-	12	4
10	0	3	6	9	12	15	15	5

In stage 2, the transition function is $y_2 = y_1 + 2x_2$ and we solve $\underset{x_2 \in X_2}{\text{Max}} \ F_2(y_2, x_2) = \overline{F}_1(y_1) + f_2(y_1, x_2) = \overline{F}_1(y_1) + 2x_2$. As $2x_2 \le 10$, $X_2 = \{0, 1, ..., 5\}$ and Table **T**$_2$ is then

T$_2$:

y_2 \ x_2	0	1	2	3	4	5	$\overline{F}_2(y_2)$	$\overline{x}_2$
	$F_2(y_2, x_2) = \overline{F}_1(y_1) + 2x_2$							
0	0	-	-	-	-	-	0	0
2	3	2	-	-	-	-	3	0
4	6	5	4	-	-	-	6	0
6	9	8	7	6	-	-	9	0
8	12	11	10	9	8	-	12	0
10	15	14	13	12	11	10	15	0

Note that $\overline{x}_2 = 0$ regardless of the value of y_2.

In stage 3 the transition function is $y_3 = y_2 + 3x_3$. As $3x_3 \le 10 = b$, $X_3 = \{0, 1, 2, 3\}$ and the objective is $\underset{x_3 \in X_3}{\text{Max}} \ F_3(y_3, x_3) = \overline{F}_2(y_2) + f_3(y_2, x_3) = \overline{F}_2(y_2) + 2x_3^2 - x_3$. The tableau is

T$_3$:

y_3 \ x_3	0	1	2	3	$\overline{F}_3(y_3)$	$\overline{x}_3$
	$F_3(y_3, x_3) = \overline{F}_2(y_2) + 2x_3^2 - x_3$					
0	0	-	-	-	0	0
1	-	-	-	-	-	-
2	3	-	-	-	3	0
3	-	1	-	-	1	1
4	6	-	-	-	6	0
5	-	4	-	-	4	1
6	9	-	6	-	9	0
7	-	7	-	-	7	1
8	12	-	9	-	12	0
9	-	10	-	15	15	3
10	15	-	12	-	15	0

At this point, the process terminates with $\overline{z} = 15$. Backtracking reveals that $\overline{y}_3 = \arg\max_{y_3 \in Y_3} \overline{F}_3(y_3) = 10$ with $\overline{x}_3 = 0$ or 3. Since $\overline{y}_3 = \overline{y}_2 + 3\overline{x}_3$, $\overline{y}_2 = 10$ or 0. In both cases $\overline{x}_2 = 0$. Finally, as $\overline{y}_2 = \overline{y}_1 + 2\overline{x}_2 = 10$ or 0 so that $\overline{y}_1 = 10$ or 0 and $\overline{x}_1 = \arg\max_{x_1 \in X_1} \overline{F}_1(\overline{y}_1) = \arg\max_{x_1 \in X_1} \overline{F}_1(10) = 5$ or $\overline{x}_1 = \arg\max_{x_1 \in X_1} \overline{F}_1(1) = 0$.

In summary, there are two optimal solutions. One is $\overline{x}_1 = 5$, $\overline{x}_2 = 0$, and $\overline{x}_3 = 0$, and the other is $\overline{x}_1 = 0$, $\overline{x}_2 = 0$, and $\overline{x}_3 = 3$. Both solutions have the objective value $\overline{z} = 15$. In other words, the decision maker should either buy five machines of type M_1 and none of the other two types, or buy three M_3 machines and none of the others, for a maximal total first-year net profit of \$15,000. Actually, while the first of these two optimal policies uses the entire \$100,000 budget, the second policy only needs \$90,000, as is seen in our stage 3 tableau.

Example 4: *An Equipment Replacement Model*
A machine needs to be replaced periodically; possible replacement times are at the beginning of calendar years. We assume that the cost of a new machine is c and that the salvage value of a t-year old machine is $b - t$ for $t \leq b$ and zero otherwise; c and $b \geq 1$ are some given constants. The annual profit from a machine which is t years old at the beginning of the year is $b^2 - t^2$ for $t \leq b$ and zero otherwise, and the planning horizon is K years. We first formulate the problem of finding an optimal replacement policy for the machine as a dynamic programming model and then solve a numerical problem.

First, define the stages k to be the number of the years and then let y_k be the age of the machine at the end of stage k. The decision x_k = "keep" or "replace" is made at the beginning of stage $k = 1,2, ..., K$ after the age of the machine is determined. A year after a decision concerning replacement was made, the machine is either 1 year old in case it was replaced, or $y_{k-1} + 1$ years old in case it was not replaced. Formally, we can write $y_k = y_{k-1} + 1$ if x_k is "keep", and 1 if x_k is "replace" for $k = 1, ..., K$. This relationship will be the stage transformation function. Finally, let $f_k(y_{k-1}, x_k)$ be the net profit contribution from stage k, i.e., $f_k(y_{k-1}, \text{keep}) = \max\{b^2 - y_{k-1}^2, 0\}$, and $f_k(y_{k-1}, \text{replace}) = b^2 + \max\{b - y_{k-1}, 0\} - c$ for $k = 1, ..., K-1$. In the last year, we have $f_K(y_{K-1}, \text{keep}) = \max\{b^2 - y_{K-1}^2, 0\} + \max\{b - y_{K-1} - 1, 0\}$, and $f_K(y_{K-1}, \text{replace}) = b^2 + \max\{b - y_{K-1}, 0\} - c + b - 1$.

Consider now a specific case with $K = 5$ years, $b = 4$ years, $c = \$9$ and $y_0 = 3$. For year $k = 1$ we solve the problem $\max_{x_1 \in X_1} = F_1(y_1, x_1) = \bar{F}_0(y_0) + f_1(y_0, x_1) = f_1(y_0, x_1)$. Here, $f_1(y_0, \text{keep}) = \max\{16 - y_0^2, 0\} = 16 - 3^2 = 7$, and $f_1(y_0, \text{replace}) = 16 + \max\{4 - y_0, 0\} - 9 = 16 + 1 - 9 = 8$, so that $y_1 = y_0 + 1 = 3 + 1 = 4$ if x_1 is "keep" and $y_1 = 1$ if x_1 is "replace"; therefore $Y_1 = \{1, 4\}$ (since $y_0 = 3$ is given, $Y_0 = \{3\}$). This leads to the tableau

T₁:

y_1 \ x_1	$F_1(y_1, x_1) = \bar{F}_0(y_0) + f_1(y_0, x_1) = f_1(y_0, x_1)$: keep	replace	$\bar{F}_1(y_1)$	$\bar{x}_1$
1	–	8	8	replace
4	7	–	7	keep

In the next stage $k = 2$ and the transition function is $y_2 = y_1 + 1$ if x_2 is "keep" and 1 if x_2 is "replace", so that $Y_2 = \{1, 2, 5\}$ and we solve $\max_{x_2 \in X_2} = F_2(y_2, x_2) = \bar{F}_1(y_1) + f_2(y_1, x_2)$. Evaluating $f_2(y_1, x_2)$ results in $f_2(y_1, \text{keep}) = \max\{16 - y_1^2, 0\} = 16 - y_1^2$ and $f_2(y_1, \text{replace}) = 16 + \max\{4 - y_1, 0\} - 9 = 11 - y_1$. However, if the decision is to replace a machine, we must also record the age y_1 of the machine at the time of the replacement so that the tableau is

T₂:

y_2 \ x_2	$F_2(y_2, x_2) = \bar{F}_1(y_1) + f_2(y_1, x_2)$: keep	replace: $y_1 = 1$	replace: $y_1 = 4$	$\bar{F}_2(y_2)$	$\bar{x}_2$
1	—	8 + 10 = 18	7 + 7 = 14	18	replace
2	8 + 15 = 23	—	—	23	keep
5	7 + 0 = 7	—	—	7	keep

In stage $k = 3$, the transition function is $y_3 = y_2 + 1$ if x_3 is "keep", and 1 if x_3 is "replace" so that $Y_3 = \{1, 2, 3, 6\}$ and we solve $\underset{x_3 \in X_3}{\text{Max}} = F_3(y_3, x_3) = \overline{F}_2(y_2) + f_3(y_2, x_3)$, where $f_3(y_2, \text{keep}) = \max\{16 - y_2^2, 0\}$ and $f_3(y_2, \text{replace}) = 7 + \max\{4 - y_2, 0\}$. Tableau $\mathbf{T}_3$ is then

T$_3$: $F_3(y_3, x_3) = \overline{F}_2(y_2) + f_3(y_2, x_3)$

y_3 \ x_3	keep	replace $y_2 = 1$	replace $y_2 = 2$	replace $y_2 = 5$	$\overline{F}_3(y_3)$	$\overline{x}_3$
1	—	18+10=28	23+9=32	7+7=14	32	replace
2	18+15=33	—	—	—	33	keep
3	23+12=35	—	—	—	35	keep
6	7+0=7	—	—	—	7	keep

In stage $k = 4$, $Y_4 = \{1, 2, 3, 4, 7\}$ and the formulas are analogous to those for $k = 3$. Tableau $\mathbf{T}_4$ is then

T$_4$: $F_4(y_4, x_4) = \overline{F}_3(y_3) + f_4(y_3, x_4)$

y_4 \ x_4	keep	replace $y_3 = 1$	replace $y_3 = 2$	replace $y_3 = 3$	replace $y_3 = 6$	$\overline{F}_4(y_4)$	$\overline{x}_4$
1	—	32+10 =42	33+9 =42	35+8 =43	7+7 =14	43	replace
2	32+15 =47	—	—	—	—	47	keep
3	33+12 =45	—	—	—	—	45	keep
4	35+7 =42	—	—	—	—	42	keep
7	7+0 =7	—	—	—	—	7	keep

Finally, in stage $k = 5$ the transition function is $y_5 = y_4 + 1$ if x_5 is "keep", and 1 if x_5 is replace, so that $Y_5 = \{1, 2, 3, 4, 5, 8\}$ and $f_5(y_4, \text{keep}) = \max\{16 - y_4^2, 0\} + \max\{4 - y_4 - 1, 0\}$ and $f_5(y_4, \text{replace}) = 16 + \max\{4 - y_4, 0\} - 9 + 4 - 1 = 10 + \max\{4 - y_4, 0\}$. The final tableau, where "k" and "r" in the last column refer to "keep" and "replace" respectively, is then

T$_5$:

$y_5 \backslash x_5$	keep	replace $y_4 = 1$	replace $y_4 = 2$	replace $y_4 = 3$	replace $y_4 = 4$	replace $y_4 = 7$	$\overline{F}_5(y_5)$	$\overline{x}_5$
	$F_5(y_5, x_5) = \overline{F}_4(y_4) + f_5(y_5, x_5)$							
1	—	43+13 =56	47+12 =59	45+11 =56	42+10 =42	7+10 =17	59	r
2	43+17 =60	—	—	—	—	—	60	k
3	47+13 =60	—	—	—	—	—	60	k
4	45+7 =52	—	—	—	—	—	52	k
5	42+0 =42	—	—	—	—	—	42	k
8	7+0 =7	—	—	—	—	—	7	k

With $k: = k + 1 = 6 > 5 = K$, we set $\overline{z}$:= 60 and $\overline{y}_5 := \arg\max_{y_5 \in Y_5} \overline{F}_5(y_5) = 2$ or 3 and $\overline{x}_5$ is "keep". The associated value of $\overline{y}_4$ = 1 or 2 and $\overline{x}_4 = \arg\max_{x_4 \in X_4} \overline{F}_4(y_4)$ which is "replace" with $\overline{y}_3$ = 3 or "keep"; therefore $\overline{y}_3$ = 3 or 1 and $\overline{x}_3 = \arg\max_{x_3 \in X_3} \overline{F}_3(y_3)$ which is either "keep" or "replace" with $\overline{y}_2$ = 2. Now in both cases $\overline{y}_2$ = 2, so that $\overline{x}_2 = \arg\max_{x_2 \in X_2} \overline{F}_2(y_2)$ which is "keep". Finally, $\overline{y}_1$ = 1 so that $\overline{x}_1 = \arg\max_{x_1 \in X_1} \overline{F}_1(y_1)$ is "replace".

There are two optimal solutions, *viz.*, $\overline{x}_1$ = replace, $\overline{x}_2$ = keep, $\overline{x}_3$ = keep, $\overline{x}_4$ = replace, and $\overline{x}_5$ = keep; and $\overline{x}_1$ = replace, $\overline{x}_2$ = keep, $\overline{x}_3$ = replace, $\overline{x}_4$ = keep), and $\overline{x}_5$ = keep. Both optimal solutions result in a total profit of \$60 over the 5 years.

Example 5: *A Nonlinear Programming Problem*
Consider the nonlinear programming problem

$$\begin{aligned} \text{P : Max } z &= x_1 x_2 x_3 \\ \text{s.t. } \quad & 2x_1 + 3x_2 + x_3 \leq 5 \\ & x_1, x_2, x_3 \geq 0 \end{aligned}$$

We define the stages to correspond to the three decision variables x_k and define $y_k \geq 0$ as the amount of resources allocated to the stages $j \leq k$, so that $y_0 = 0$, $y_1 = y_0 + 2x_1 = 2x_1$, $y_2 = y_1 + 3x_2$ and $y_3 = y_2 + x_3$ (≤ 5). We obtain the function $f_k(y_{k-1}, x_k) = x_k$, so that $F_k(y_k, x_k) = \overline{F}_k(y_{k-1}) f_k(y_{k-1}, x_k)$ and the contributions towards the objective function at each stage are connected in a multiplicative fashion rather than an additive as in the previous examples. Suitably modifying the dynamic programming algorithm and setting $\overline{F}_0(y_0) = 1$, the problem in stage $k = 1$ is $\underset{x_1 \in X_1}{\text{Max}} = F_1(y_1, x_1) = f_1(y_0, x_1) = x_1$. Given that $2x_1 = y_1$, we obtain $\overline{x}_1 = \frac{1}{2}y_1$ so that $\overline{F}_1(y_1) = \frac{1}{2}y_1$. In stage $k = 2$ we solve $\underset{x_2 \in X_2}{\text{Max}} = F_2(y_2, x_2) = \overline{F}_1(y_1) f_2(y_1, x_2) = (\frac{1}{2}y_1)x_2$, where $y_2 = y_1 + 3x_2 \geq 0$. Simple differential calculus reveals that $\partial F_2(y_2, x_2)/\partial x_2 = 0$ results in $\overline{x}_2 = \frac{1}{6}y_2$, so that $\overline{F}_2(y_2) = \frac{1}{24}y_2^2$. In stage $k = 3$ we solve the problem $\underset{x_3 \in X_3}{\text{Max}} = F_3(y_3, x_3) = \overline{F}_2(y_2) f_3(y_2, x_3) = \frac{1}{24}y_2^2 x_3$ where $y_3 = y_2 + x_3$, so that $F_3(y_3, x_3) = \frac{1}{24}(y_3^2 x_3 - 2y_3 x_3^2 + x_3^3)$. Now $\partial F_3(y_3, x_3)/\partial x_3 = 0$ results in $\overline{x}_3 = \frac{1}{3}y_3$, so that $\overline{F}_3(y_3) = \frac{1}{162}y_3^3$.

The backward sweep reveals that $\overline{y}_3 = \underset{y_3 \in Y_3}{\text{argmax}}\ \overline{F}_3(y_3) = 5$, so that $\overline{x}_3 = \frac{1}{3}\overline{y}_3 = 5/3$ and $\overline{z} = 125/162$. Then $\overline{y}_2 = \overline{y}_3 - \overline{x}_3 = 5 - 5/3 = 10/3$ and $\overline{x}_2 = \frac{1}{6}\overline{y}_2 = 5/9$. Furthermore, $\overline{y}_1 = \overline{y}_2 - 3\overline{x}_2 = 10/3 - 5/3 = 5/3$ and $\overline{x}_1 = \frac{1}{2}\overline{y}_1 = 5/6$. The unique optimal solution is therefore $\overline{x}_1 = \frac{5}{6}$, $\overline{x}_2 = \frac{5}{9}$, $\overline{x}_3 = \frac{5}{3}$ with a value of the objective function of $\overline{z} = 125/162$.

This problem could have been solved using Kuhn-Tucker optimality conditions or geometric programming; see, e.g., Eiselt *et al.* (1987). In this problem the functions $f_k(y_{k-1}, x_k)$ were connected in a multiplicative rather than additive fashion. Dynamic programming can easily handle such objectives by suitably modifying the algorithm. It should also be pointed out that while in this example, the maximizations in stages 2 and 3 above were carried out by way of differential calculus, this is not always necessary. In general, the subproblems may be solved by any suitable optimization technique that is appropriate for the particular problem.

Example 6: *A Linear Programming Problem*

This example demonstrates that dynamic programming is able to solve linear programming problems. However, due to the complexity of the approach and the availability of very fast alternative methods, this is definitely not recommended in practice. Consider the problem

$$\begin{array}{llllll} \text{P : Max } z = & 4x_1 & + 3x_2 & + 5x_3 & & \\ \text{s.t.} & 2x_1 & + \ x_2 & + 4x_3 & \leq 15 & \\ & 3x_1 & + 2x_2 & + 3x_3 & \leq 12 & \\ & x_1, & x_2, & x_3 & \geq 0. & \end{array}$$

Let the stages again correspond to the decision variables x_k, $k = 1, 2, 3$. Considering the two constraints as resource constraints with availability of 15 and 12 units, respectively, the two state variables v_k and w_k denote the allocation of the two resources in the first k stages. To solve this problem, we set $v_0 := w_0 := 0$ and $\bar{F}_0(v_0, w_0) = 0$.

In stage $k = 1$, we can then solve the problem $\underset{x_1 \in X_1}{\text{Max}}\ F_1(v_1, w_1, x_1) = \bar{F}_0(v_0, w_0) + f_1(v_0, w_0, x_1) = 4x_1$ where x_1 is constrained by $2x_1 \leq v_1$, $3x_1 \leq w_1$. Clearly $\bar{x}_1 = \min\{\frac{1}{2}v_1, \frac{1}{3}w_1\}$ since both constraints have to be satisfied, and $\bar{F}_1(v_1, w_1) = 4\bar{x}_1 = 4\min\{\frac{1}{2}v_1, \frac{1}{3}w_1\}$. The problem in stage $k = 2$ is $\underset{x_2 \in X_2}{\text{Max}}\ F_2(v_2, w_2, x_2) = \bar{F}_1(v_1, w_1) + f_2(v_1, w_1, x_2) = \bar{F}_1(v_1, w_1) + 3x_2$, where $v_2 = v_1 + x_2 \geq 0$ and $w_2 = w_1 + 2x_2 \geq 0$. The objective can be rewritten as $F_2(v_2, w_2, x_2) = 4\min\{\frac{1}{2}(v_2 - x_2), \frac{1}{3}(w_2 - 2x_2)\} + 3x_2 = \min\{x_2 + 2v_2, \frac{1}{3}x_2 + \frac{4}{3}w_2\}$, subject to $v_2 - x_2 \geq 0$ and $w_2 - 2x_2 \geq 0$, so that $\bar{x}_2 = \min\{v_2, \frac{1}{2}w_2\}$ results. If now $v_2 \leq \frac{1}{2}w_2$ then $\bar{x}_2 = v_2$ so that $\bar{F}_2(v_2, w_2) = \min\{v_2 + 2v_2, \frac{1}{3}v_2 + \frac{4}{3}w_2\} = \min\{3v_2, \frac{1}{3}(v_2 + 4w_2)\} = \frac{1}{3}v_2 + \min\{\frac{8}{3}v_2, \frac{4}{3}w_2\} = \frac{1}{3}v_2 + \frac{4}{3}\min\{2v_2, w_2\} = 3v_2$. If, on the other hand, $v_2 \geq \frac{1}{2}w_2$, then $\bar{x}_2 = \frac{1}{2}w_2$, so that $\bar{F}_2(v_2, w_2) = \min\{\frac{1}{2}w_2 + 2v_2, \frac{1}{6}w_2 + \frac{4}{3}w_2\} = \min\{\frac{1}{2}w_2 + 2v_2, \frac{3}{2}w_2\} = \frac{1}{2}w_2 + \min\{2v_2, w_2\} = \frac{3}{2}w_2$. In summary, $\bar{F}_2(v_2, w_2) = 3v_2$ if $2v_2 \leq w_2$, and $3w_2/2$ if $2v_2 \geq w_2 = 3\min\{v_2, \frac{1}{2}w_2\}$.

Finally, the problem in stage $k = 3$ is $\underset{x_3 \in X_3}{\text{Max}} = F_3(v_3, w_3, x_3) = \bar{F}_2(v_2, w_2) + f_3(v_2, w_2, x_3) = \bar{F}_2(v_2, w_2) + 5x_3$, subject to $v_3 = v_2 + 4x_3 \geq 0$, and $w_3 = w_2 + 3x_3 \geq 0$. The objective can be written as $3\min\{v_3 - 4x_3, \frac{1}{2}(w_3 - 3x_3)\} + 5x_3 = \min\{3v_3 - 7x_3, \frac{3}{2}w_3 + \frac{1}{2}x_3\}$. It is maximized for $\bar{x}_3 = \frac{1}{5}(2v_3 - w_3)$, with $\bar{F}_3(v_3, w_3) = \frac{3}{2}w_3 + \frac{1}{10}(2v_3 - w_3) = \frac{1}{5}(v_3 + 7w_3)$ which equals 99/5 for $\bar{v}_3 = 15$, $\bar{w}_3 = 12$, so that $\bar{x}_3 = 18/5$ and $\bar{z} = 99/5$. The backward sweep determines that $\bar{v}_2 = \bar{v}_3 - 4\bar{x}_3 = 15 - 72/5 = 3/5$ and $\bar{w}_2 = \bar{w}_3 - 3\bar{x}_3 = 12 - 54/5 = 6/5$, so that $\bar{x}_2 = \min\{\bar{v}_2, \frac{1}{2}\bar{w}_2\} = \min\{3/5, 3/5\} = 3/5$. Finally, $\bar{v}_1 = \bar{v}_2 - \bar{x}_2 = 3/5 - 3/5 = 0$, and $\bar{w}_1 = \bar{w}_2 - 2\bar{x}_2 = 6/5 - 6/5 = 0$, so that $\bar{x}_1 = \min\{\frac{1}{2}\bar{v}_1, \frac{1}{2}\bar{w}_1\} = \min\{0,0\} = 0$. The unique optimal solution is then $\bar{\mathbf{x}} = [\bar{x}_1, \bar{x}_2, \bar{x}_3] = [0, \frac{3}{5}, \frac{18}{5}]$ with $\bar{z} = 99/5$.

PART I INTEGER PROGRAMMING

Optimization problems in which some or all decision variables are required to take only integer values occur in a variety of contexts. Finding feasible solutions is already a difficult task. Except for a few classes of specially structured problems, proving optimality for a given feasible solution is even more difficult. Unlike the case of continuous optimization where no integrality restrictions are imposed on the variables, integer programming problems, even if linear, can in general not be solved in polynomial time, i.e., do not belong to the class P discussed in Chapter b in the Introduction. Furthermore, for most of them there do not exist any polynomial approximation schemes. These matters are quite well understood and pursued in the subject of complexity theory. We will not deal with them here, because it does not appear to contribute much to the subject of solving real problems. Readers are referred to Chapter b for a discussion of such issues.

Chapter 1 presents fundamental concepts of integer programming in a broad overview, with simple examples when possible. Chapter 2 introduces the use of logical variables in formulations, and Chapter 3 describes applications and considers special structures that are important for the overall field. Chapter 4 investigates possibilities to reformulate integer programming problems in order to make them more easily solvable. Finally, Chapters 5, 6 and 7 deal with cutting plane, branch-and-bound, and heuristic methods, respectively, for solving integer programming problems.

CHAPTER 1 THE INTEGER PROGRAMMING PROBLEM AND ITS PROPERTIES

The subject of this part, the solution of mathematical problems in integer numbers, is not as new as the reader might believe. Certain problems were already known to the Greeks, e.g., Euclid (3rd century B.C.) and Diophantos (3rd century A.D.). Their achievement was the determination of the greatest common divisor (g.c.d.) of a set of numbers (accomplished by the Euclidean Algorithm) as well as some answers to the question: when does a given set of equalities have at least one integer solution? We provide some thoughts to the latter problem in the following section. In 1641 Fermat (1601-1665) in his famous "Last Theorem" considered the equation $x^n + y^n = z^n$ and conjectured that this equality has no integer solution x, y, z in the case of $n > 2$. Fermat claimed that he had found a proof of this theorem, but it has never been found. The theorem has defied efforts by mathematicians trying to prove it for 300 years, until it was finally proved in 1993 by Andrew Wiles. The proof is long and complicated and involves the so-called Taniyama-Shimura conjecture. A flaw in the proof was corrected a year later. For a full account of the problem, its long history and final solution, see Singh (1997). In the nineteenth century George Boole (1815-1864) expressed logical connections in terms of zero-one variables, i.e., variables that equal either zero or one. Finally, in the middle of the twentieth century after the development of linear programming, the stage was set for finding integral optimal solutions for practical problems. Some of the most important contributions came from individuals such as Balas, Gomory, Glover, Hammer, and Hu, and later from Edmonds and Lawler. The developments since the mid-sixties demonstrated that there is no strict borderline between integer programming on the one and network models on the other side, as many important network problems can be formulated as integer programming problems.

1.1 Definitions and Basic Concepts

In this section we will explore the relations between certain problems, mostly integer linear programming problems and their linear programming relaxations. To give a glimpse of the difference in complexity between the integer and the continuous problem, consider a single linear equation in two or more variables where the goal is to find continuous and integer solutions to this equation. In the continuous case, all we have to do is set all but one variable to arbitrary values and then solve the equation for the remaining variable. In case an integer solution is required, the equations are commonly referred to as *Diophantine* after the mathematician Diophantos, and a solution may be found by the so-called *Rule of Virgins* due to Euler and Sylvester; see Saaty (1970). The procedure can be summarized as follows. First the given equation is solved for the variable with the smallest coefficient. The solution is expressed in mixed numbers, and all fractional terms, if they exist, are set equal to a new variable. Multiplying this equation by the least common multiple, we obtain a new equation and the procedure starts again with this equation. If there is no more fractional part, the procedure terminates. In this case all given variables are expressed as functions of the variable that was introduced last. Then the enumeration of integer values for this variable generates integer solutions for all given variables. Additional constraints, such as nonnegativity constraints and upper and lower bounds for the given variables, can be included at this stage. The following two examples may explain this procedure.

Example 1: Consider the equation $3x_1 + 5x_2 = 34$ with the integrality requirements, $x_1, x_2 \in \mathbb{N}_0$. As x_1 has the smallest coefficient, we solve for x_1, and can write $x_1 = {}^1/_3(34 - 5x_2) = 11{}^1/_3 - 1{}^2/_3\, x_2 = 11 - 1x_2 + x_3$, where $x_3 = {}^1/_3 - {}^2/_3\, x_2$. This new equation can be written as $2x_2 + 3x_3 = 1$ or, since x_2 has the smallest coefficient, as $x_2 = {}^1/_2 - {}^3/_2 x_3 = -x_3 + x_4$, where $x_4 = {}^1/_2 - {}^1/_2 x_3$. This equation can then be written as $x_3 + 2x_4 = 1$ or, as x_3 has the smallest coefficient, $x_3 = 1 - 2x_4$. At this point, there are no more fractions present in the relation and the process terminates. A simple backward recursion allows us to express the variables x_1 and x_2 as functions of x_4. In particular, we obtain

$$x_1 = 13 - 5x_4,$$
$$x_2 = 3x_4 - 1.$$

Applying the nonnegativity constraints $x_1, x_2 \geq 0$ to these two expressions, we obtain the bounds $x_4 \leq 2{}^3/_5$ and $x_4 \geq {}^1/_3$ which, due to the integrality requirement, can be tightened to $x_4 \leq 2$ and $x_4 \geq 1$. This leaves $x_4 = 1$ and $x_4 = 2$, which, in turn, leads to the two solutions $(x_1, x_2) = (8, 2)$ and $(3, 5)$.

Example 2: Find solutions to $4x_1 + 7x_2 = 17$ with $x_1, x_2 \in \mathbb{N}_0$. Applying the above procedures, we obtain

$$x_1 = 6 - 7x_4,$$
$$x_2 = -1 + 4x_4.$$

If x_1 and x_2 were only required to be integer but not nonnegative, then all solutions could be determined by setting x_4 to arbitrary integer values. However, as $x_1, x_2 \geq 0$ is required, we incorporate these nonnegativity conditions and obtain the bounds

$x_1 \geq 0$ or $x_4 \leq {}^6/_7$ which can be strengthened to $x_4 \leq 0$, and
$x_2 \geq 0$ or $x_4 \geq {}^1/_4$ which can be strengthened to $x_4 \geq 1$,
an obvious contradiction. Hence, the given equation in this example has many solutions in integers but none in nonnegative integers. These two examples may highlight the difference in difficulties of solving the integer and the nonnegative integer problems.

In general, one can prove

Theorem I.1: The linear equation $\sum_{j=1}^{n} a_j x_j = b$; $b, a_j \in \mathbf{Z}\ \forall j$ has a solution $x_j \in \mathbf{Z}$, $j = 1, ..., n$, if and only if the greatest common divisor (gcd) of a_j, $j = 1, ..., n$, divides b in integers.

In both examples above gcd $(a_1, a_2) = 1$, so that it would be guaranteed that at least one integer, but not necessarily nonnegative, solution exists. The above procedure can also be applied to problems involving more than one equality; however, the resulting method is very tedious. The small examples above may illustrate the degree of difficulty added by the integrality requirements.

Back to optimization problems, a significant difference between integer and continuous problems exists regarding the number of feasible solutions. Whereas linear programming problems may have zero, one or an infinite number of solutions, all-integer problems may have any number of solutions. More specifically, to this day most linear programming problems are solved with the simplex method whose popular primal version in its second phase moves from one extreme point to an adjacent extreme point in a feasible direction. In doing so, the method usually finds the optimal solution in a number of steps that is proportional to a relatively small multiple of the number of constraints of a given problem.

Although it might be thought that for m constraints (including the nonnegativity requirements) and n variables, the number of potential extreme points might be the binomial coefficient $\binom{m}{n}$, it turns out to be a gross overestimate; the upper bound is actually $\binom{m - \left\lfloor \frac{n+1}{2} \right\rfloor}{m-n} + \binom{m - \left\lfloor \frac{n+2}{2} \right\rfloor}{m-n}$, see McMullen (1970). Still, with 100 structural constraints and 100 variables (a very small problem by today's standards), so that $m = 200$, $n = 100$, there could be more than 10^{40} extreme points. Even if one takes into account that many of these vertices are actually infeasible, that hardly reduces the numbers in a meaningful manner. Yet we know that commercial linear programming codes often have no difficulty at all coping with much larger problems for which the number of feasible vertices would be astronomically large.

One might think that integer programming is simpler, since one can exclude all points other than those with integral coordinates. Nothing can be further from the truth. Integer programming is, in fact, orders of magnitude more difficult than linear programming, because there are no simple methods for excluding the fractional points or focusing on the set of integer points. As we will see, one may actually have to solve integer programming problems via the execution of (often inordinately) many linear programming problems.

Just to give an idea of the magnitude of the solution space for even small integer problems, consider the following example of a *generalized assignment problem* (*GAP*). In an office building there are 11 floors on which 30 tenants are to be located. Each tenant has given space requirements that may vary with the floor to which he is assigned. If tenant j is assigned to floor i, he needs a_{ij} square feet of floor space. Floor i is assumed to have an area of b_i square feet. The rent that the managing company extracts from its tenants is a fixed proportion of the tenants' revenue. Hence the tenants as well as the managing company are interested in maximizing the tenants' revenue. Suppose that each square foot of space on floor i that is rented will yield c_i dollars per month for the managing company. Defining zero-one decision variables y_{ij} which equal 1 if and only if floor i is assigned to tenant j, and 0 otherwise, then the problem can then be formulated as

$$\text{P:} \qquad \text{Max } z = \sum_{i=1}^{11}\sum_{j=1}^{30} c_i\, a_{ij}\, y_{ij}$$

$$\text{s.t.} \qquad \sum_{j=1}^{30} a_{ij}\, y_{ij} \le b_i \;\forall\, i = 1, \ldots, 11$$

$$\sum_{i=1}^{11} y_{ij} = 1, \forall\, j = 1, \ldots, 30$$

$$y_{ij} = 0 \vee 1 \;\forall\, i, j$$

Here, the objective maximizes the revenue, the first set of constraints models the limitation of floor space, and the second set of constraints ensures that each customer is assigned to exactly one floor.

The number of possible zero-one solutions y_{ij} is $2^{11\times 30} = 2^{330} \cong (2.187)10^{99}$. This is an enormous number, in practice impossible to enumerate even with the most modern computational equipment. But it is possible to simplify the task of enumeration by taking into account the uniqueness of the floor assignment for a given tenant by setting $y_{kj} = 0 \;\forall\, k \neq i$, if a tenant-floor assignment $y_{ij} = 1$ is made. For each tenant, one need only record his floor, which gives $11^{30} \cong (1.745)10^{31}$

possible solutions, about 10^{68} times less than the first time! Yet, suppose that one could evaluate a billion, i.e., 10^9 solutions in a second, and keep the best feasible one, one would still need $(1.745)10^{22}$ seconds, or roughly $(5.53)10^{14}$ years: this still represents about 36,000 times the age of the universe, since the Big Bang! The good news is that with current specialized software, it is usually possible to solve such a problem to optimality within a few minutes on a workstation. While, in general, integer and mixed integer programming problems are *NP-hard*, there are problem types that frequently permit the user to solve instances of even fairly large size within a reasonable amount of time by using modern commercial solvers.

Below, we present a transformation that is useful in many applications. Frequently, there are upper bounds u_j and/or lower bounds ℓ_j on a continuous variable x_j or an integer variable y_j; to impose such bounds as tightly as possible may be a matter of considerable algorithmic and practical importance as we will show in various places below. Furthermore, the case where the integer variables y_j are restricted to be binary, i.e., $y_j = 0$ or 1, occurs most often in practical problems. If on the other hand the integer variables are not binary, one can usually distinguish between two cases:

- The upper bounds for the variables y_j are large. In such a case, it often suffices to replace these integer variables by continuous variables, and to obtain an approximate solution by, say, the standard simplex method of linear programming followed by integer rounding of any fractional values of y_j in the solution.

- The upper bounds for the variables y_j are modest in size. It is then possible to use a binary expansion of the y_j and reduce the problem to an equivalent form P' with a new set of binary variables, although of larger cardinality. Specifically, if the upper bound for y_j is u_j, we need to replace y_j by its binary expansion which requires $n_j = \lfloor \log_2 u_j \rfloor + 1$ binary digits and therefore as many new binary variables. As an example, consider the following linear programming problem

$$\begin{aligned} \text{P: Max } z &= 3y_1 + 7y_2 \\ \text{s.t.} \quad -y_1 + 4y_2 &\le 7 \\ 5y_1 + 2y_2 &\le 20 \\ y_1 \qquad &\le 12 \\ y_2 &\le 70 \\ y_1, \quad y_2 &\in \mathbb{N}_0 \end{aligned}$$

Representing the integer variables y_j by their binary expansions, we obtain $y_j = \sum_{k=0}^{p_j} 2^k y_{j_k}$ with zero-one variables y_{j_k} and a parameter p_j, such that $2^{p_j} < u_j < 2^{p_j+1}$. Here, $y_1 \le u_1 = 12 \in [8; 16[= [2^3; 2^4[$, so that $p_1 = 3$ and we replace y_1 by

$y_{1_0} + 2y_{1_1} + 4y_{1_2} + 8y_{1_3}$. Similarly, as $y_2 \le u_2 = 70 \in [64; 128[= [2^6; 2^7[$, so that $p_2 = 6$, and $y_2 = y_{2_0} + 2y_{2_1} + 4y_{2_2} + 8y_{2_3} + 16y_{2_4} + 32y_{2_5} + 64y_{2_6}$. The problem can then be reformulated as

$$
\begin{aligned}
\text{P}':\ \text{Max}\ z = {} & 3y_{1_0} + 6y_{1_1} + 12y_{1_2} + 24y_{1_3} + 7y_{2_0} + 14y_{2_1} + 28y_{2_2} + 56y_{2_3} + 112y_{2_4} + 224y_{2_5} + 448y_{2_6} \\
\text{s.t.}\ & -y_{1_0} - 2y_{1_1} - 4y_{1_2} - 8y_{1_3} + 4y_{2_0} + 8y_{2_1} + 16y_{2_2} + 32y_{2_3} + 64y_{2_4} + 128y_{2_5} + 256y_{2_6} \le 7 \\
& 5y_{1_0} + 10y_{1_1} + 20y_{1_2} + 40y_{1_3} + 2y_{2_0} + 4y_{2_1} + 8y_{2_2} + 16y_{2_3} + 32y_{2_4} + 64y_{2_5} + 128y_{2_6} \le 20 \\
& y_{1_0} + 2y_{1_1} + 4y_{1_2} + 8y_{1_3} \le 12 \\
& y_{2_0} + 2y_{2_1} + 4y_{2_2} + 8y_{2_3} + 16y_{2_4} + 32y_{2_5} + 64y_{2_6} \le 70 \\
& y_{1_0}, y_{1_1}, y_{1_2}, y_{1_3}, y_{2_0}, y_{2_1}, y_{2_2}, y_{2_3}, y_{2_4}, y_{2_5}, y_{2_6} = 0 \vee 1
\end{aligned}
$$

Clearly, if the upper bounds u_j are very large, the problem P′ may include many more variables than is practical.

To formalize matters, define vectors of variables **x** and **y** and let J and K be two disjoint index-sets, so that the number of variables is $n = |J \cup K|$. Furthermore, let R_i describe one of the relations $\le$, $=$, $\ge$, and define $f(\mathbf{x}, \mathbf{y})$ and $g_i(\mathbf{x}, \mathbf{y})$ as linear functions. Moreover, if $b_i \in \mathbb{R}\ \forall\ i$, then the general integer programming problem can be written as

$$
\begin{aligned}
\text{P:}\quad & \text{Max}\ z = f(\mathbf{x}, \mathbf{y}) \\
& \text{s.t.}\quad g_i(\mathbf{x}, \mathbf{y})\ R_i\ b_i\ \ \forall\ i = 1, \ldots, m \\
& \qquad\quad x_j \in \mathbb{R}_+\ \ \forall\ j \in J \\
& \qquad\quad y_j \in \mathbb{N}_0\ \ \forall\ j \in K.
\end{aligned}
$$

In many applications, integer variables y_j are restricted to assume values of only zero or one. Such zero-one variables are accomodated in this general formulation by including $y_j \le 1$ in the constraints. The nonnegativity constraints and integrality requirements leave then only 0 and 1 as feasible solutions for y_j. Disaggregations of this type replace $y_j = 0 \vee 1$ by $y_j \ge 0$, $y_j \le 1$, and $y_j \in \mathbf{Z}$, are often useful and are further discussed later in this part.

In this volume we will concentrate on integer programming problems with $f(\mathbf{x}, \mathbf{y})$ and $g_i(\mathbf{x}, \mathbf{y})$ being all linear functions, so that the term, "integer programming" is actually shorthand for "integer linear programming". Furthermore, we assume that all coefficients and parameters are known with certainty. This allows us to formally define two classes of integer programming problems.

Definition I.2: A *mixed-integer linear programming problem* (*MILP*) is a linear programming problem with the additional constraint that at least one (but not all) variables are restricted to be integer. An *all-integer linear programming problem*

(*AILP or ILP*) is a linear programming problem with the additional restriction that all variables must be integer.

In terms of the above formulation P, an all-integer linear programming problem has $J = \emptyset$. If not defined otherwise, we assume in the following that the relations R_i are always given as $\leq$ inequalities. If this is not the case, the well-known transformations from linear programming may be applied. Similarly, we often assume that a minimization function is given which can always be obtained. Defining **x** as a vector of continuous variables and **y** as a vector of integer variables, the all-integer and mixed-integer programming problem P_{ILP} and P_{MILP} can then be written as

$$P_{ILP}\text{: Max } z = \mathbf{cy} \quad \text{s.t. } \mathbf{Ay} \leq \mathbf{b}, \quad \mathbf{y} \in \mathbb{N}_0^n$$

$$P_{MILP}\text{: Max } z = \mathbf{c}(\mathbf{x}, \mathbf{y}) \quad \text{s.t. } \mathbf{A}(\mathbf{x}, \mathbf{y}) \leq \mathbf{b}, \quad x_j \in \mathbb{R}_+ \ \forall j \in J, \quad y_j \in \mathbb{N}_0 \ \forall j \in K$$

In order to demonstrate the basic differences between all integer, mixed integer and the corresponding linear programming problems, we consider the constraints

$$3x_1 + 5x_2 \leq 15$$
$$4x_1 + 2x_2 \leq 8.$$

The figures below show the four cases: a standard LP with $x_1, x_2 \geq 0$ in Figure I.1a, an AILP with $x_1, x_2 \in \mathbb{N}_0$ in Figure I.1b, an MILP with $x_1 \in \mathbb{N}_0$, $x_2 \geq 0$ in Figure I.1c, and another MILP with $x_1 \geq 0$, $x_2 \in \mathbb{N}_0$ shown in Figure I.1d.

In case (a), the shaded area is feasible, in case (b) only the indicated eight points are feasible, in cases (c) and (d) all points on the bold lines are feasible. Notice that as soon as at least one integer variable is introduced, the given polytope is no longer convex (cases (b), (c) and (d)).

Consider now the location of optimal solutions. Given that feasible solutions exist, it is well known that at least one optimal solution of a linear programming problem can be found at an extreme point of the polytope described by the constraints of the problem. In contrast, the optimal solution of an all-integer linear programming problem may be located anywhere in the polytope, including its interior. As an

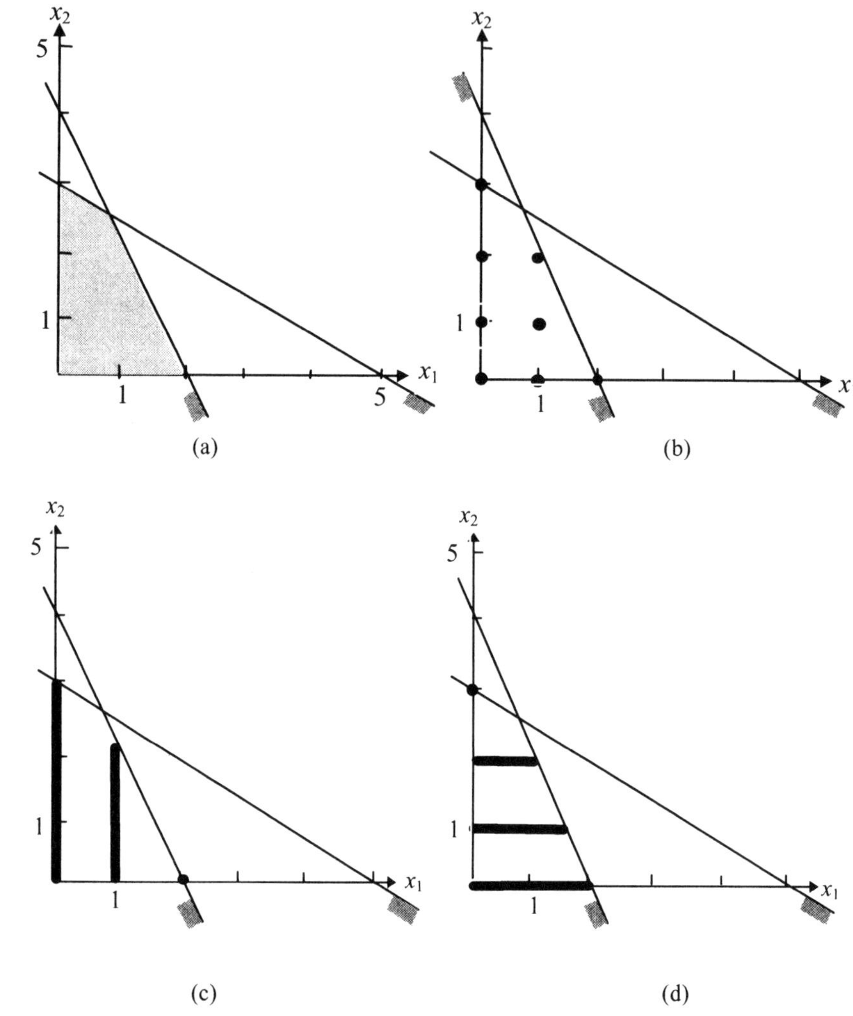

Figure I.1

illustration, consider the following all-integer linear programming problem with the integrality requirements deleted.

$$\begin{array}{lrl} \text{P:} & \text{Max } z = & 2x_1 + 3x_2 \\ & \text{s.t.} & x_1 + 2x_2 \le 5 \\ & & -2x_1 + 3x_2 \ge 0 \\ & & x_1 \;,\; x_2 \ge 0 \end{array}$$

The problem is displayed in Figure I.2 where the shaded area represents the feasible set and the points shown in bold are the all-integer feasible points.

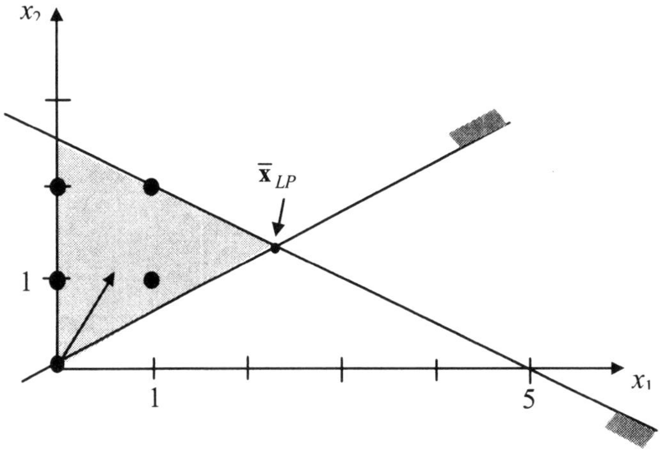

Figure I.2

The optimal solution of the linear programming problem is $\bar{\mathbf{x}} = [2^1/_7, 1^3/_7]$ with $\bar{z} = 8^4/_7$. It is easy to see that rounding will not result in an optimal integer solution, as all four neighboring all-integer points $[x_1, x_2] = [2, 1], [2, 2], [3, 1]$, and $[3, 2]$ are infeasible. Actually, the all-integer solution is $\bar{\mathbf{x}}_{\text{AILP}} = [1, 2]$ with $\bar{z}_{\text{AILP}} = 8$. The mixed-integer solution with the requirements $x_1 \geq 0$ and $x_2 \in \mathbb{N}_0$ is the same. Finally, the solution to the mixed-integer programming problem with $x_1 \in \mathbb{N}_0$ and $x_2 \geq 0$ is $\bar{\mathbf{x}}_{\text{MILP}} = [2, 1\frac{1}{2}]$ with $\bar{z}_{\text{MILP}} = 8\frac{1}{2}$.

Similarly, it is easy to show that while a linear programming problem may have an infinite number of solutions, the addition of integrality constraints may result in an empty feasible set. As an example, consider the constraints

$$\begin{aligned} 2x_1 + 10x_2 &\leq 9 \\ -x_1 + 5x_2 &\geq 1 \\ x_1, \quad x_2 &\geq 0 \end{aligned}$$

The graph in Figure I.3 clearly demonstrates that the feasible set including integrality requirements for both variables is empty.

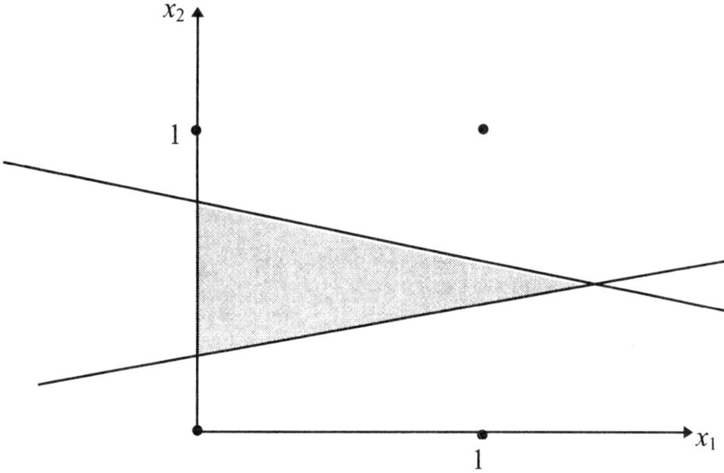

Figure I.3

Lemma I.3: The existence of *continuous feasible solutions* for a linear programming problem does not guarantee the existence of *integral feasible solutions*.

Less dramatically, but very importantly, the optimal solution of a linear programming problem relaxation may be, relatively speaking, very far from the optimal integer solution, and rounding (up or down or some components up and others down) an optimal solution of the linear programming solution can result in values that are very different from those of an optimal solution of the integer programming problem P.

As an example, consider the following problem:

$$\begin{array}{lrl} \text{P:} & \text{Max } z = & (6\alpha + 1)y_1 - \alpha y_2 \\ & \text{s.t.} & -y_1 + 6y_2 \leq 24 \\ & & 6y_1 - \ \ y_2 \leq 30 \\ & & y_1, \ \ y_2 \in \mathbb{N}_0 \end{array}$$

where α is some given positive parameter. The optimal integer solution is $\bar{\mathbf{y}}_{\text{IP}}$ = [5, 0] with $\bar{z}_{\text{IP}} = 30\alpha + 5$, the optimal solution of the linear programming relaxation is $\bar{\mathbf{y}}_{\text{LP}} = [5\frac{29}{35}, 4\frac{34}{35}]$ with $\bar{z}_{\text{LP}} = 30\alpha + 5\frac{29}{35}$. Rounding the linear programming solution results in the solutions $\tilde{\mathbf{y}}$ = [5, 4], [6, 4], [5, 5], and [6, 5], only the first of which is feasible. Its associated objective value is $\tilde{z}$ = 26α + 5. The relative error in the objective function value due to rounding can then be written as

$\dfrac{\bar{z}_{LP} - \tilde{z}}{\bar{z}_{LP} - \bar{z}_{IP}} - 1 = \dfrac{140}{29}\alpha$, which goes to infinity as α approaches infinity. In other words, the relative error of rounding can be arbitrarily large.

Define now $\mathcal{P}_{IP} = \{(\mathbf{x}, \mathbf{y}): \mathbf{A}(\mathbf{x}, \mathbf{y}) \leq \mathbf{b}, y_j \in \mathbb{N}_0, x_j \geq 0\}$ as the set of feasible solutions of the given all-integer or mixed-integer programming problem, let $\mathcal{P}_{LP} = \{(\mathbf{x}, \mathbf{y}): \mathbf{A}(\mathbf{x}, \mathbf{y}) \leq \mathbf{b}; \mathbf{x}, \mathbf{y} \geq \mathbf{0}\}$ and assume that $\mathcal{P}_{IP} \neq \emptyset$. We can then formulate

Definition I.4: The *convex hull* $\mathcal{P}_{IP}^{CH}$ of the set $\mathcal{P}_{IP}$ is a convex polytope that includes all feasible points in $\mathcal{P}_{IP}$, and is such that no proper subset of $\mathcal{P}_{IP}^{CH}$ has this property.

In other words, $\mathcal{P}_{IP}^{CH}$ is the intersection of all convex subsets of $\mathbb{R}^n$, that include $\mathcal{P}_{IP}$. The relations between the sets $\mathcal{P}_{IP}$, $\mathcal{P}_{IP}^{CH}$, and $\mathcal{P}_{LP}$ can be summarized in

Lemma I.5: $\mathcal{P}_{IP} \subseteq \mathcal{P}_{IP}^{CH} \subseteq \mathcal{P}_{LP}$.

The first inclusion holds by definition, and the second inclusion holds by virtue of the convexity of $\mathcal{P}_{LP}$. To illustrate the concept, consider the constraints $x_1 + x_2 \leq 2½$, $x_1 \leq 1½$, and $x_1, x_2 \in \mathbb{N}_0$. The shaded area in Figure I.4 is $\mathcal{P}_{LP}$, the area surrounded by the bold line is $\mathcal{P}_{IP}^{CH}$, and $\mathcal{P}_{IP}$ consists of the five points (0, 0), (1, 0), (0, 1), (1, 1), and (0, 2).

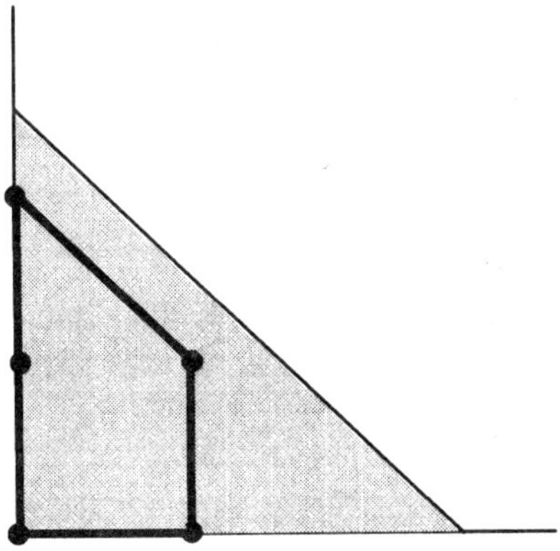

Figure I.4

Given any linear maximization objective function, Lemma I.5 implies

Corollary I.6: $\bar{z}_{IP} \leq \bar{z}_{IP}^{CH} \leq \bar{z}_{LP}$.

The fact that all extreme points in $\mathcal{P}_{\text{IP}}^{\text{CH}}$ are integer for all variables which are required to be integer implies that $\bar{z}_{\text{IP}} = \bar{z}_{\text{IP}}^{\text{CH}}$.

The next two sections deal with the relations between the problems P_{IP} and P_{LP} and their objective values $\bar{z}_{\text{IP}}$ and $\bar{z}_{\text{LP}}$.

1.2 Relaxations of Integer Programming Problems

In this section we explore possibilities that enable the user to solve problems more easily and efficiently, albeit with a loss of accuracy. More specifically, all relaxation methods define sets of feasible solutions that are supersets of the original feasible sets. The way this superset is defined distinguishes the different techniques. Clearly, any relaxation that has the same objective function as the original problem and a superset of the original feasible set as its own set of feasible solutions will have an objective value that, in models with maximization objective, can be no smaller than that of the original problem. In other words, the solution of the relaxed problem provides an *upper bound* for the original problem. This allows relaxations to be used in computational procedures in which the optimal solution is bounded from above and from below: from above, by the solution of a relaxed problem (the optimal solution of which must have an objective value that is no smaller than that of any known solution) and from below by any known feasible solution. The goal of such procedures will be to find upper bounds that are as small as possible and lower bounds that are as high as possible. One such upper-bound generating relaxation is the linear programming relaxation. It can be defined as follows.

Definition I.7: Given an integer programming problem P_{IP}: Max $\mathbf{cx}$, s.t. $\mathbf{Ax} \le \mathbf{b}$, $x_j \in \mathbb{N}_0\ \forall\, j \in K$, $x_j \in \mathbb{R}_+\ \forall\, j \in J$, then the *linear programming relaxation* is P_{LP}: Max $\mathbf{cx}$, s.t. $\mathbf{Ax} \le \mathbf{b}$, $x_j \in \mathbb{R}_+\ \forall\, j \in J \cup K$.

In simple words, the linear programming relaxation simply deletes the integrality requirements of all variables. Denoting again the optimal objective values of the original problem P_{IP} and its linear programming relaxation P_{LP} by $\bar{z}_{\text{IP}}$ and $\bar{z}_{\text{LP}}$ respectively, we can write

Definition I.8: The *absolute integrality gap* is defined as the difference $\bar{z}_{\text{LP}} - \bar{z}_{\text{IP}}$ and for $\bar{z}_{\text{IP}} \neq 0$, the *relative integrality gap* is defined as $(\bar{z}_{\text{LP}} - \bar{z}_{\text{IP}}) / |\bar{z}_{\text{IP}}|$.

The relative integrality gap is a reasonably good measure of the difficulty of a problem. Anything above 10% is usually a fairly difficult problem, and measures of

50% and above indicate really hard problems. Clearly, the gaps can only be computed *after* a problem has been solved to optimality. However, even though $\bar{z}_{\text{IP}}$ may be unknown, often reasonably good bounds for that value may be known that can be substituted for the exact value, giving reasonable approximations.

As a numerical example, consider the following problem.

$$
\begin{array}{llrl}
\text{P:} & \text{Max } z = & 3x_1 + 5x_2 & \\
& \text{s.t.} & 2x_1 + \ \ x_2 & \leq 5 \\
& & 3x_1 + 7x_2 & \leq 14 \\
& & x_1 \qquad & \leq 2 \\
& & x_2 & \leq 1\tfrac{1}{2} \\
& & x_1\,, \quad x_2 & \in \mathbb{N}_0
\end{array}
$$

The optimal solution of P_{IP} is $\bar{\mathbf{x}}_{\text{IP}} = [2,\ 1]$ with objective value $\bar{z}_{\text{IP}} = 11$. Replacing the integrality requirements by nonnegativity constraints, the linear programming relaxation has the solution $\bar{\mathbf{x}}_{\text{LP}} = [1\frac{10}{11}, 1\frac{2}{11}]$ with objective value $\bar{z}_{\text{LP}} = 11\frac{7}{11}$. Hence, the absolute integrality gap is $\frac{7}{11}$ and the relative integrality gap is $(11\frac{7}{11} - 11)/|11| = 7/121 \cong 5.8\%$.

The second type of relaxation is the *Lagrangean relaxation*. It differs from the linear programming relaxation in a number of aspects. Here, we choose a number of the given constraints, multiply them by nonnegative Lagrangean multipliers or dual variables $u_i \geq 0$, $i = 1, \ldots, m$, and subtract them from the objective function. Formally, the Lagrangean relaxation can be defined as follows.

Definition I.9: Given an integer programming problem P_{IP}: Max $\mathbf{cx}$, s.t. $\mathbf{Ax} \leq \mathbf{b}$, $\mathbf{Dx} \leq \mathbf{d}$, $x_j \in \mathbb{N}_0\ \forall j \in K$, $x_j \in \mathbb{R}_+\ \forall j \in J$, the Lagrangean relaxation with any fixed vector $\mathbf{u} \geq \mathbf{0}$ is P_{LR}: Max $\mathbf{cx} - \mathbf{u}\,(\mathbf{Ax} - \mathbf{b})$, s.t. $\mathbf{Dx} \leq \mathbf{d}$, $x_j \in \mathbb{N}_0\ \forall j \in K$, $x_j \in \mathbb{R}_+\ \forall j \in J$. In this relaxation, the constraints $\mathbf{Ax} \leq \mathbf{b}$ are said to have been *dualized*.

We now prove that the objective function value of the Lagrangean relaxation is indeed an upper bound of the optimal objective value of the original problem P_{IP}.

Lemma I.10: $\bar{z}_{\text{LR}} \geq \bar{z}_{\text{IP}}$.

Proof: Arbitrarily choose some $\mathbf{u}^* \geq \mathbf{0}$. Then denote the optimal solution to P_{LR} by $\hat{\mathbf{x}}$ and let $\bar{\mathbf{x}}$ be the optimal solution of P. We now have

$\bar{z}_{\text{LR}} = \mathbf{c}\,\hat{\mathbf{x}} - \mathbf{u}^*(\mathbf{A}\,\hat{\mathbf{x}} - \mathbf{b}) \geq \mathbf{c}\,\bar{\mathbf{x}} - \mathbf{u}^*(\mathbf{A}\,\bar{\mathbf{x}} - \mathbf{b})$ (as $\hat{\mathbf{x}}$ maximizes this objective)

$\geq \mathbf{c}\,\bar{\mathbf{x}}$ (as $\mathbf{A}\,\bar{\mathbf{x}} - \mathbf{b} \leq \mathbf{0}$ due to feasibility and

$= \bar{z}_{\text{IP}}$ $\mathbf{u}^* \geq \mathbf{0}$ by assumption)

For further details, readers are referred to Eiselt *et al.* (1987).□

It is apparent that Langrangean relaxations have many degrees of freedom that require finetuning. First, it must be decided which constraints $\mathbf{Ax} \leq \mathbf{b}$ to dualize. In some applications, relaxing certain classes of constraints leaves a system $\mathbf{Dx} \leq \mathbf{d}$ that has a special structure that makes it easy to solve, such as a network structure. Alternatively, it may be decomposable, again aiding the solution process. However, selecting the constraints to be dualized according to such principles may result in poor bounds. The other choice that has to be made concerns the multipliers. While any set of nonnegative weights is theoretically possible, some will result in poor bounds. As an illustration, consider again the above numerical example and assume that the first two constraints are to be dualized, leaving as constraints only upper bounds and integrality requirements. Choose, for instance, weights $u_1 = 1$ and $u_2 = 3$. We then obtain the Lagrangean relaxation

$$\text{P}_{\text{LR}}\text{: Max } z_{\text{LR}} = 3x_1 + 5x_2 - 1(2x_1 + x_2 - 5) - 3(3x_1 + 7x_2 - 14) = -8x_1 - 17x_2 + 47$$

or simply

$$\begin{aligned} \text{Min } -z_{\text{LR}} &= 8x_1 + 17x_2 - 47 \\ \text{s.t.} \quad & x_1 \leq 2 \\ & x_2 \leq 1\tfrac{1}{2} \\ & x_1, \; x_2 \in \mathbb{N}_0 \end{aligned}$$

The optimal solution is $\bar{\mathbf{x}}_{\text{LR}} = [0, 0]$ with $\bar{z}_{\text{LR}} = 47$, which is a very poor upper bound. However, dualizing the first, third and fourth constraints with weights $u_1 = 1$, $u_3 = 1$, and $u_4 = 2$ leads to

$$\begin{aligned} \text{P}_{\text{LR}}\text{: Max } z_{\text{LR}} &= 3x_1 + 5x_2 - 1(2x_1 + x_2 - 5) - 1(x_1 - 2) - 2(x_2 - 1\tfrac{1}{2}) = 2x_2 + 10 \\ \text{s.t.} \quad & 3x_1 + 7x_2 \leq 14 \\ & x_1, \; x_2 \in \mathbb{N}_0. \end{aligned}$$

The solution is then $\bar{\mathbf{x}}_{\text{LR}} = [1, 2]$ with $\bar{z}_{\text{LR}} = 14$, a considerably better bound.

Finally, we consider surrogate relaxation. Again, a set of weights $\mathbf{u} \geq \mathbf{0}$ is supplied by the user. These weights are then employed to aggregate constraints. Formally, we can write

Definition I.11: Given an integer programming problem P_{IP}: Max $\mathbf{cx}$, s.t. $\mathbf{Ax} \leq \mathbf{b}$, $\mathbf{Dx} \leq \mathbf{d}$, $x_j \in \mathbb{N}_0 \; \forall j \in K$, $x_j \in \mathbb{R}_+ \; \forall j \in J$ and a nonnegative vector $\mathbf{u} \geq \mathbf{0}$, the *surrogate relaxation* is P_{SR}: Max $\mathbf{cx}$, s.t. $\mathbf{Ax} \leq \mathbf{b}$, $\mathbf{uDx} \leq \mathbf{ud}$, $x_j \in \mathbb{N}_0 \; \forall j \in K$, $x_j \in$

$\mathbb{R}_+ \ \forall j \in J$.

As with Lagrangean relaxation, surrogate relaxation offers various degrees of freedom: which constraints are to be aggregated, and what weights are to be used. To explain the concept, consider again the above example and choose weights $u_1 = 1$, $u_2 = 1$, $u_3 = 1$, $u_4 = 2$, so that the resulting problem has only a single constraint. In particular, it is

$$\begin{aligned} \text{P}_{\text{SR}}\text{: Max } z_{\text{SR}} &= 3x_1 + 5x_2 \\ \text{s.t.} \quad 6x_1 &+ 10x_2 \leq 24 \\ x_1, \quad & x_2 \in \mathbb{N}_0. \end{aligned}$$

The unique optimal solution is $\bar{\mathbf{x}}_{\text{SR}} = [4, 0]$ with $\bar{z}_{\text{SR}} = 12$; obviously this solution (though not its objective value) deviates substantially from the optimal solution of the original problem which was $\bar{\mathbf{x}}_{\text{IP}} = [2, 1]$ with objective value $\bar{z}_{\text{IP}} = 11$. On the other hand, aggregating only the second, third and fourth constraints with weights $u_2 = 4$, $u_3 = 8$, and $u_4 = 2$, leads to the surrogate relaxation

$$\begin{aligned} \text{P}_{\text{SR}}\text{: Max } & 3x_1 + 5x_2 \\ \text{s.t.} \quad & 2x_1 + x_2 \leq 5 \\ & 20x_1 + 30x_2 \leq 75 \\ & x_1, \quad x_2 \in \mathbb{N}_0, \end{aligned}$$

whose unique optimal solution is $\bar{\mathbf{x}}_{\text{SR}} = [1^7/_8, 1^1/_4]$ with $\bar{z}_{\text{SR}} = 11^7/_8$, close to the solution of the original problem. Further details concerning aggregation procedures are discussed in Chapter 4.

1.3 Polyhedral Combinatorics

Polyhedral combinatorics deals with the relationships between the sets $\mathcal{P}_{\text{IP}}^{\text{CH}}$ and $\mathcal{P}_{\text{LP}}$. In particular, we are interested under what conditions these two sets are identical. This is an interesting case, as $\mathcal{P}_{\text{IP}}^{\text{CH}}$ is a polyhedron whose extreme points are all integer-valued, so that $\bar{\mathbf{x}}_{\text{IP}} = \bar{\mathbf{x}}_{\text{LP}}$ for any objective function, i.e., the linear programming relaxation solves the all-integer problem. Some cases for which this is true are discussed below.

One of the key aspects of polyhedral combinatorics are min-max results that can also be used to establish relationships between the objective values of closely related problems. Such results are similar in principle to those obtained for pairs of dual linear programs. The most popular example deals with matchings and coverings. A matching in an undirected graph $G = (N, E)$ is defined as a partial graph $G' = (N, E')$ with $E' \subset E$, so that each node in G' is incident to at most one edge. Similarly, a node covering in $G = (N, E)$ is a subgraph $G'' = (N', E)$, with $N' \subset N$ such that each edge is incident to at least one node. A maximal matching maximizes $z_M = |E'|$, while a minimal covering minimizes $z_C = |N'|$. It is apparent that $z_M \le z_C$ and König has proved that $\bar{z}_M = \bar{z}_C$ in case of bipartite graphs, a result later generalized by Edmonds (1965). Results of this type can be used for the bounding of the objective value of a given problem. Another famous (and strong) result is the min cut - max flow result in network flow theory, see Chapter 1 of Part III. For details and a summary of the field, readers are referred to Schrijver (1986, 1995), Nemhauser and Wolsey (1988), Pulleyblank (1989), and Aardal and van Hoesel (1996, 1999).

In this book we restrict ourselves to one specific aspect of polyhedral combinatorics. In particular, we deal with problem formulations whose linear programming relaxations are either always integer or tend to have integer solutions. First we consider problems that always have integer solutions. This integrality property relies on the matrix of coefficients (its numbers and structure) and the right-hand side values. We first restate a well-known theorem from linear algebra that will be used in our discussion.

Assume that $\mathbf{A} = (a_{ij})$ is an $[n \times n]$-dimensional matrix with full rank, i.e., $\mathbf{A}$ has no linear dependencies. Let $\mathbf{A}_j$ be the matrix that is derived from $\mathbf{A}$ by deleting the j-th column of $\mathbf{A}$ and replacing it by a vector $\mathbf{b}$ of the same dimension. Denoting by det $\mathbf{A}$ the determinant of a given matrix $\mathbf{A}$, we can now state Cramer's rule.

Theorem I.12: Let $\mathbf{A}$ be an $[n \times n]$-dimensional nonsingular matrix, let $\mathbf{b}$ be a given $[n \times 1]$-dimensional column vector and denote $\mathbf{x}$ by a given $[1 \times n]$-dimensional column vector of variables. Then the system of simultaneous equations $\mathbf{Ax} = \mathbf{b}$ has a solution $x_j = \dfrac{\det \mathbf{A}_j}{\det \mathbf{A}}$ $\forall j = 1, \ldots, n$.

We can further define

Definition I.13: An $[m \times n]$-dimensional matrix $\mathbf{A}$ with $m \le n$ is said to be *totally unimodular*, if each nonsingular square submatrix $\mathbf{A}^*$ of $\mathbf{A}$ has determinant det $\mathbf{A}^* = \pm 1$.

At this point we note:

- For a matrix $\mathbf{A}$ to be totally unimodular it is necessary that it includes only elements $a_{ij} = 0 \vee 1 \vee -1$. If this is not the case, say $a_{ij} = 2$, then there is one $[1 \times 1]$-dimensional submatrix $\mathbf{A}^* = [2]$ with det $\mathbf{A}^* = 2$.

- The concept of *unimodularity* is a relaxation of the concept of total unimodularity. It only requires that $[m \times m]$ - dimensional solutions of full rank (i.e., *bases*) have determinant ± 1.

The following examples may explain the concepts.

Example 1: The matrix $\mathbf{A} = \begin{bmatrix} 2 & 1 & 1 \\ 1 & 1 & 0 \end{bmatrix}$ is not totally unimodular since $a_{11} = 2$, but it is unimodular as all three bases $\begin{bmatrix} 2 & 1 \\ 1 & 1 \end{bmatrix}$, $\begin{bmatrix} 2 & 1 \\ 1 & 0 \end{bmatrix}$ and $\begin{bmatrix} 1 & 1 \\ 1 & 0 \end{bmatrix}$ have determinants of ± 1.

Example 2: The matrix

$$\mathbf{A} = \begin{bmatrix} 1 & 0 & 1 & 1 \\ 1 & 1 & 0 & 1 \\ 0 & 1 & 1 & 1 \\ 1 & 1 & 1 & 1 \end{bmatrix}$$

is unimodular, because its only basis (the matrix itself) has a determinant of -1. It is, however, not totally unimodular; the submatrix $\mathbf{A}^*$, which includes the first three rows and columns, has a determinant of 2.

Example 3: The matrix

$$\mathbf{A} = \begin{bmatrix} 1 & 1 & 1 & 0 \\ 1 & 1 & 0 & 1 \\ 0 & 1 & 1 & 0 \end{bmatrix}$$

is unimodular as well as totally unimodular. This can be verified by evaluating the determinants of $\binom{4}{3} = 4$ $[3 \times 3]$-dimensional submatrices, $\binom{4}{2}\binom{3}{2} = 18$ $[2 \times 2]$-dimensional submatrices and $\binom{4}{1}\binom{3}{1} = 12$ $[1 \times 1]$-dimensional submatrices. This process also demonstrates how tedious the process is that determines whether or not a given matrix $\mathbf{A}$ is unimodular. Note that if $\mathbf{A}$ is totally unimodular then its transpose $\mathbf{A}^T$ is also totally unimodular. Two theorems concerning unimodularity were proved by Veinott and Dantzig (1968) and Hoffman and Kruskal (1956), respectively. In order to state them, let $\mathbf{A}$ be an $[m \times n]$-dimensional integral matrix ($m \leq n$) with full row rank, i.e., rk $\mathbf{A} = m$, and let $\mathbf{b}$ be some integral $[m \times$

1]-dimensional column vector. The two theorems are

Theorem I.14: All extreme points of the polytope $\{\mathbf{x}: \mathbf{Ax} = \mathbf{b}, \mathbf{x} \geq \mathbf{0}\}$ are integral, if and only if **A** is unimodular.

Theorem I.15: All extreme points of the polytope $\{\mathbf{x}: \mathbf{Ax} \leq \mathbf{b}, \mathbf{x} \geq \mathbf{0}\}$ are integral, if and only if **A** is totally unimodular.

Moreover, Veinott and Dantzig and Hoffman and Kruskal have shown that each basis (i.e., nonsingular submatrix of **A** with full rank) has an integral inverse. Dantzig and Veinott's theorem is proved by employing Cramer's rule (Theorem I.12), according to which $x_j = \dfrac{\det \mathbf{A}_j^*}{\det \mathbf{A}^*}$ given that **A*** is a basis of **A**. Since the vector **b** is integer, $\mathbf{A}_j^*$ is also an integer matrix, implying that $\det \mathbf{A}_j^*$ is integer and as $\mathbf{A}^*$ is, by assumption, totally unimodular, i.e., $\det \mathbf{A}^* = \pm 1$, the theorem follows. Hoffman and Kruskal's theorem is proved similarly by adding a vector of slack variables to **x**, and their identity matrix to **A**.

Clearly, the requirements a matrix must satisfy in order to be totally unimodular are very strong, so that very few practical problems will exhibit this property. However, there are a number of special structures that have totally unimodular coefficient matrices. Among them are the capacitated and uncapacitated transportation problems, the assignment problem, and the maximal flow problem. With some additional considerations, even the cost-minimal feasible flow problems have totally unimodular coefficient matrices and thus *per se* integral solutions, as long as the given capacities (right-hand sides) are integral. The formulations of the above problems can be found in Parts II and III. Finally, we present a sufficient, but not necessary, condition for the total unimodularity of a zero-one matrix **A** that has an interesting application in production planning, provided certain constraints are fulfilled. For that purpose, define

Definition I.16: The ones of a zero-one matrix **A** are said to have the *consecutive ones property* if the rows and columns of **A** can be arranged, so that all 1-entries in a row (column) form an uninterrupted sequence.

Example: Consider the matrix

$$\mathbf{A} = \begin{bmatrix} 1 & 1 & 1 & 0 \\ 0 & 1 & 1 & 0 \\ 1 & 0 & 1 & 1 \end{bmatrix}$$

As it is stated, not all rows or columns of the matrix **A** have uninterrupted sequences of ones; see, in particular, row 3 and column 1. However, rearranging the columns in the order of column 2, 3, 1, and 4, we see that the rearranged rows of the resulting matrix $\tilde{\mathbf{A}}$ now have all ones in an uninterrupted sequence.

$$\tilde{\mathbf{A}} = \begin{bmatrix} 1 & 1 & 1 & 0 \\ 1 & 1 & 0 & 0 \\ 0 & 1 & 1 & 1 \end{bmatrix}$$

Similarly, the rows of the matrix $\tilde{\mathbf{A}}$ can be rearranged in the order 3, 1, and 2, so that now also the columns exhibit an uninterrupted sequence of ones.

The continuous sorting can be used in the following production scheduling problem. Consider a process in which n products are to be processed on m machines M_i, $i = 1,..., m$. The processing time of the j-th product on the i-th machine is given by a_{ij} time units per quantity unit, and it is assumed that the processing times on one machine are equal for all products that need to be processed on that machine. In particular, $a_{ij} = 0 \vee a_i \; \forall \, j$. Moreover assume that the values $a_{ij} \leq 1$ which, if not satisfied originally, can always be achieved by appropriate scaling. In addition, define the unit profit as c_{ij} and let the machine capacities be b_i time units. The goal is now to determine quantity units x_j, defined as the number of processed (and sold) units of the j-th type. This production problem can be formulated as

$$\begin{aligned} \text{P:} \quad & \text{Max } z = \mathbf{cx} \\ & \text{s.t. } \mathbf{Ax} \leq \mathbf{b} \\ & \qquad \mathbf{x} \geq \mathbf{0} \end{aligned}$$

If the machines can be arranged along an assembly line, so that each product is processed in an uninterrupted sequence of machines M_i, and the given machine capacities were integral, then the values for x_j will all be integral at optimum. This

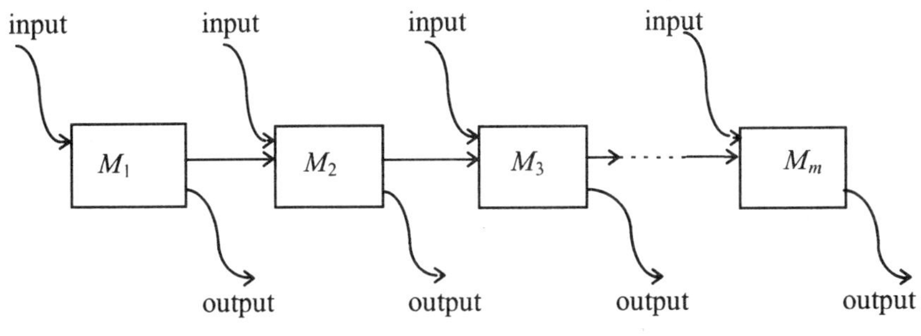

Figure I.5

results from the fact that the matrix of coefficients can be transformed into a column-wise continuously sortable zero-one matrix that, by virtue of our discussion above, is totally unimodular. The general structure of this problem is shown in Figure I.5.

Example: Consider the following matrix of processing times in Table I.1

Table I.1

Machines \ Products	P_1	P_2	P_3	P_4	P_5	b_i
M_1	0	$^1/_2$	$^1/_2$	$^1/_2$	0	10
M_2	$^1/_4$	$^1/_4$	$^1/_4$	0	$^1/_4$	6
M_3	0	$^1/_{12}$	$^1/_{12}$	0	$^1/_{12}$	3
c_j	$^5/_2$	3	$^7/_8$	2	$^3/_2$	

Here, product P_1 is processed only on M_2; P_2 and P_3 are processed on M_1, M_2 and M_3; P_4 is processed only on M_1, and P_5 is processed on M_2 and M_3. Scaling the rows by multiplying their elements by 2, 4, and 12, respectively, and adding slack variables we obtain the tableau

T:

x_1	x_2	x_3	x_4	x_5	S_1	S_2	S_3	1
0	1	1	1	0	1	0	0	20
1	1	1	0	1	0	1	0	24
0	1	1	0	1	0	0	1	36
$-^5/_2$	-3	$-^7/_8$	-2	$-^3/_2$	0	0	0	0

that has consecutive ones property in the columns. The naturally integer optimal solution is $\bar{x}_1 = 24$, $\bar{x}_4 = 20$, and $\bar{x}_2 = \bar{x}_3 = \bar{x}_5 = 0$ with objective value $\bar{z} = 100$. In other words, 24 units of P_1 and 20 units of P_4 are made, resulting in complete use of the capacities of M_1 and M_2, whereas machine M_3 is not used at all.

A much wider class of practically important problems cannot be proved to always have integer solutions, but they tend to be integer or near integer. One example is the class of so-called *set covering* problems

$$\begin{aligned} \text{P: Min } z &= \mathbf{dy} \\ \text{s.t. } \mathbf{Dy} &\geq \mathbf{e} \\ \mathbf{y} &\in \{0, 1\}^n \end{aligned}$$

where **e** is the summation vector of ones, **d** is a nonnegative vector of objective function coefficients, and **D** is a given matrix of zeroes and ones. The idea in this problem is to "cover" the rows of the problem by setting variables y_j equal to one, so that each row of **Dy** equals at least 1. Sometimes, the right-hand side vector consists of integer values larger than 1. A large class of applications consists of location covering problems.

A substantially more difficult problem is the *set partitioning* problem. The problem is similar to set covering, but it includes equality rather than inequality constraints. Set partitioning problems arise in all sorts of scheduling problems of practical importance. One such problem related to set covering and set partitioning is airline crew scheduling, for which the rows represent legs of an airline schedule (defined as a direct connection between two cities), and each column corresponds to one possible crew which could be assigned to that leg. When the problem was first solved on a practical level in the early seventies, the number of rows was in the thousands, and the number of columns in the millions. In the late 1990s, some talk of a number of possible columns in the trillions. All solution methods applied to these problems have a strong heuristic flavor; the given problem is replaced by thousands of subproblems in which thousands or tens of thousands of columns are chosen as to satisfy a variety of requirements on the crews and schedules. Typically, the subproblems are solved as linear programs whose solution values are rounded to integer form. It is apparently possible to almost always find good solutions in this manner for which the difference between the best feasible objective function value and the best lower bound is small. The monetary value of such solutions is enormous. The crew scheduling and fleet scheduling applications routinely save a few percent of the overall costs involved, something that translates into savings in the millions of dollars. Today, almost all airlines have substantial operations research groups which solve these and other problems. An example of the airline crew problem scheduling problem is presented in Section 3.2.2 of this part.

While the airline scheduling application methods are constantly updated, they all continue to rely heavily on the fact that the underlying set covering problems are integer programming problems that are relatively easy to solve. However, it is important to stress that most mixed integer programming problems are in general not easy in the above sense, and it appears unlikely that at some point in the forseeable future general "good" solution methods become available. Yet there has been a steady progress in related systematic approximation procedures, called *heuristic solution methods* or simply *heuristics*. The most important classes of heuristics are discussed in Chapter 7 of this part.

CHAPTER 2 FORMULATIONS IN LOGICAL VARIABLES

In this chapter, we discuss how zero-one variables can be used to model conditions that cannot be modeled with the usual continuous variables. The techniques introduced in this chapter are later employed to model important applications.

2.1 The Modeling of Discrete Variables

A variable is called *discrete* if it can only assume certain prespecified values. As an example, consider a variable representing a quantity of material that can only be purchased in lots or batches, giving it the name *batch size problem*. Specifically, let material j be available only in separate quantities $v_{j_0}, v_{j_1}, \ldots, v_{j_{p_j}}$. Formally, $y_j = v_{j_0} \vee v_{j_1} \vee v_{j_2} \vee \ldots \vee v_{j_{p_j}}$. Defining zero-one variables $y_{j_k} = 1$, if $y_j = v_{j_k}$ and 0 otherwise, we can then express y_j as $y_j = \sum_{k=0}^{p_j} v_{j_k} y_{j_k}$ with the additional constraints $\sum_{k=0}^{p_j} y_{j_k} = 1$, and $y_{j_k} = 0 \vee 1 \; \forall \, k$. The variable y_j is then replaced by this expression wherever it appears in the problem formulation. This transformation is applied to all discrete variables y_j. The obvious drawback of this procedure is the increase in the number of variables, especially if one or more values p_j are large. Since the second constraint forces exactly one variable y_{j_k} to one, this constraint is often called a *multiple choice constraint*. Problems with non-overlapping multiple choice constraints are said to include *special ordered sets* (SOS). Special ordered sets, as introduced by Beale and Tomlin (1970), can be defined as follows.

Definition I.17: A *special ordered set of type 1*, *SOS*1, is a set of variables exactly one of which assumes a positive value.

Special ordered sets can be modeled as *generalized upper bounding* (GUB) constraints in linear programming; for details, see Eiselt *et al.* (1987). Beale and Tomlin (1970) have also introduced another type of special ordered set.

Definition I.18: A *special ordered set of type 2*, *SOS*2, is a set of variables $y_1, \ldots, y_r$ so that $y_\nu = 0 \; \forall \; \nu \neq k, k+1$ for some $k \in [1; r-1]$.

In other words, in a special ordered set of type 2, no more than two adjacent variables may have a value unequal to zero. Such sets are needed for modeling piecewise linear approximations to nonlinear functions similar to the approximation described in Section 2.8. For details and other applications of special ordered sets, the reader is referred to Williams (1993).

Many codes for the solution of integer programming problems incorporate features for special ordered sets, designed to reduce the processing time. As a numerical example for the batch-size problem consider a variable y_j that assumes a value of either 3, 10, 39, or 71. In terms of the parameters defined above, $v_{j_0} = 3$, $v_{j_1} = 10$, $v_{j_2} = 39$, and $v_{j_3} = 71$, and we can replace the original variable by $y_j = 3y_{j_0} + 10y_{j_1} + 39y_{j_2} + 71y_{j_3}$ with the additional constraint $y_{j_0} + y_{j_1} + y_{j_2} + y_{j_3} = 1$ and $y_{j_0}, y_{j_1}, y_{j_2}, y_{j_3} = 0 \vee 1$.

In general, consider the following batch-size problem

$$\text{P: Max } z = \sum_{j=1}^{n} c_j y_j$$

$$\text{s.t.} \quad \sum_{j=1}^{n} a_{ij} y_j \leq b_i \;\; \forall \; i = 1, \ldots, m$$

$$y_j = v_{j_0} \vee v_{j_1} \vee \ldots \vee v_{j_{p_j}} \;\; \forall \; j = 1, \ldots, n.$$

Using the above transformation, the first tableau of the reformulated problem is then

T :

$y_{1_0} \cdots y_{1_{p1}}$	$y_{2_0} \cdots y_{2_{p2}}$	$y_{3_0} \cdots y_{3_{p3}}$	• • •	$y_{n_0} \cdots y_{n_{pn}}$	1
$a_{11}v_{1_0} \ldots a_{11}v_{1_{p1}}$	$a_{12}v_{2_0} \ldots a_{12}v_{2_{p2}}$	$a_{13}v_{3_0} \ldots a_{13}v_{3_{p3}}$	• • •	$a_{1n}v_{n_0} \ldots a_{1n}v_{n_{pn}}$	b_1
$a_{21}v_{1_0} \ldots a_{21}v_{1_{p1}}$	$a_{22}v_{2_0} \ldots a_{22}v_{2_{p2}}$	$a_{23}v_{3_0} \ldots a_{23}v_{3_{p3}}$	• • •	$a_{2n}v_{n_0} \ldots a_{2n}v_{n_{pn}}$	b_2
• •	• •	• •		• •	•
• •	• •	• •		• •	•
• •	• •	• •		• •	•
$a_{m1}v_{1_0} \ldots a_{m1}v_{1_{p1}}$	$a_{m2}v_{2_0} \ldots a_{m2}v_{2_{p2}}$	$a_{m3}v_{3_0} \ldots a_{m3}v_{3_{p3}}$	• • •	$a_{mn}v_{n_0} \ldots a_{mn}v_{n_{pn}}$	b_m
1...1...1					1
	1...1...1				1
		1...1...1	•		1
			•		.
			•	1...1...1	1
$-c_1v_{1_0} \ldots -c_1v_{1_{p1}}$	$-c_2v_{2_0} \ldots -c_2v_{2_{p2}}$	$-c_3v_{3_0} \ldots -c_3v_{3_{p3}}$	• • •	$-c_nv_{n_0} \ldots -c_nv_{n_{pn}}$	

Notice the nonoverlapping sequences of ones in the bottom rows of the body of the tableau. These special ordered sets are further explored in Chapter 3 of this part.

2.2 The Modeling of Fixed Charges

In some problems the cost of an operation is made up of two components, a fixed charge and a cost that is a function of the level of some activity. Examples of fixed charges are the cost of building a new plant, buying a new machine, repairing or refitting a machine that was out of service, and the setup, e.g., cleanup, cost of an operation. The fixed charge may also express a measure other than cost, such as the setup time of getting a machine into service or the fixed elapsed time before a certain operation can commence.

A typical example concerns the shipment of some commodity from a warehouse i to a destination j. The quantity shipped from i to j is denoted by x_{ij}. However, the shipment can only be sent if the i-th warehouse is opened, which can be done at a fixed charge f_i. Defining c_{ij} as the unit transport cost from warehouse i to customer j, the cost for the warehouse and all shipments leading out of it is then $z_i = f_i + \sum_{j=1}^{n} c_{ij} x_{ij}$ if warehouse i operates, and zero if it is decided not to open it. The introduction of a zero-one variable y_i with $y_i = 1$, if warehouse i opens and 0 otherwise will permit us to model this situation by setting $z_i = f_i\, y_i + \sum_j c_{ij} x_{ij}$, and $x_{ij} \le \kappa_{ij}\, y_i\ \forall\, j$ where κ_{ij} denotes an upper bound or capacity on the flow from warehouse i to customer j. Now, if $y_i = 0$, then the warehouse i is closed and the right-hand sides of all constraints are zero, forcing x_{ij} to zero for all j. Clearly, the cost of operation in such a case is $z_i = 0$. On the other hand, if $y_i = 1$, then the constraints are $x_{ij} \le \kappa_{ij}\ \forall\, j$ and the costs equal $z_i = f_i + \sum_j c_{ij} x_{ij}$.

2.3 Disjunctive Variables

Disjunctive variables are variables with the property that if one of them assumes a positive value, the others should equal zero. For instance, consider the case in which only one of a number of available fuel additives may be used in a single blend. Defining the quantities of the additives by $x_1, x_2, \ldots, x_n$, and letting their respective availabilities be denoted by $u_1, u_2, \ldots, u_n$, we define zero-one variables y_j that assume a value of one, if the j-th additive is used in a blend and zero otherwise. We can then write $x_j \le u_j\, y_j$, and $\sum_j y_j\, R\, 1$ where the relation operator R is either "≤", "=" or "≥"; $y_j = 0 \vee 1\ \forall\, j = 1, \ldots, n$, and $x_j \ge 0\ \forall\, j = 1, \ldots, n$. If R is "≤" then *at most* one of the additives will be used. Another situation occurs if R is "=", for *exactly one* of the additives will be used. Similarly, if R is ≥, *at least one* of the additives are to be used. The right-hand sides of all of the first constraints belonging to additives that are not used (i.e., the corresponding $y_j = 0$) force the quantity of the additive to zero. Note that the case of R being "≤" does allow all variables x_j to equal zero, a case in which none of the additives is used. Finally, the constraint $\sum_j y_j R\, k$ models the situation in which at most, exactly, or at least k of the additives have to be used, depending on whether the relation operator R is "≤", "=" or "≥".

2.4 Constraint Selection

Instead of imposing relationships between variables as in Section 2.3 above, we may wish to impose relationships between constraints. One such type of relationship are *dichotomies*, also referred to as *either-or constraints*. In other words, given two constraints, we attempt to formulate them in such a way, that at least one of two given constraints must be satisfied. Formally, we define the two constraints as

$$\mathbf{a}_{i\bullet}\,\mathbf{x}\; R_i\; \mathrm{b}_i$$

$$\mathbf{a}_{k\bullet}\,\mathbf{x}\; R_k\; \mathrm{b}_k$$

where the relations R_i, $R_k \in \{\leq, =, \geq\}$ and where at least one of the two constraints has to be satisfied. As an example, suppose that a company can purchase one of two machines. Clearly, if they purchase the first machine, only the constraint(s) related to that machine must hold, while the constraints relating to the second machine are irrelevant. The basic principle employed here is to ensure that a constraint is made redundant in case it is not required to be satisfied. In case of an inequality, this is achieved by shifting its hyperplane in a parallel fashion away from the feasible solution for a "significant" distance. An equation is dealt with by first replacing it by two inequalities of opposite signs and then shifting these apart in parallel fashion.

Formally we first define a zero-one variable y which equals 1, if the i-th constraint must hold, and 0 otherwise. With a sufficiently large constant $M >> 0$, the original system

$$\mathbf{a}_{i\bullet}\,\mathbf{x} \leq \mathrm{b}_i \text{ or } \mathbf{a}_{k\bullet}\,\mathbf{x} \leq \mathrm{b}_k$$

can be written as

$$\mathbf{a}_{i\bullet}\,\mathbf{x} \leq \mathrm{b}_i + M(1-y)$$

$$\mathbf{a}_{k\bullet}\,\mathbf{x} \leq \mathrm{b}_k + My$$

$$y = 0 \vee 1.$$

The validity of this transformation is readily apparent. If $y = 1$, then the i-th constraint appears in its original form and must hold, whereas the k-th constraint is redundant for sufficiently large values of M. Similarly, if $y = 0$, the i-th constraint is redundant, whereas the k-th constraint appears in its original form and must hold.

If one or both of the original constraints are of the type "$\geq$", the constraint is first multiplied by some negative number in order to transform it to the above form. After adding $M(1-y)$ and My to the respective right-hand sides, we obtain $\mathbf{a}_{i\bullet}\,\mathbf{x} \geq \mathrm{b}_i - M(1-y)$ and $\mathbf{a}_{k\bullet}\,\mathbf{x} \geq \mathrm{b}_k - My$ with $y = 0 \vee 1$. In case at least one of the original constraints is an equation, we first replace the equation $\mathbf{a}_{i\bullet}\,\mathbf{x} = \mathrm{b}_i$ by the inequalities $\mathbf{a}_{i\bullet}\mathbf{x} \leq \mathrm{b}_i$ and $\mathbf{a}_{i\bullet}\,\mathbf{x} \geq \mathrm{b}_i$, before proceeding with the transformations.

2.5 Imposing a Sequence on Variables

In sequencing problems, a priority order among variables may be required. Typically, an activity cannot commence until a preceding activity is fully completed. Similarly, some more expensive resources should only be used if lower-cost resources are fully utilized. As an example, more costly overtime labor should only be used if work during regular hours is insufficient to complete a task. Further examples of applications are:

- Sequencing in time where a manufacturing operation of type j cannot take place until a preceding operation of type $(j-1)$ is completely finished.
- Sequencing related to different objects or materials. Consider two materials A and B are to be used, such that material A must be used up completely before material B may be utilized.
- Piecewise linear cost functions. A typical example are economies of scale, where the first u_1 tons of material are quoted at some unit cost c_1, the next $(u_2 - u_1)$ tons cost c_2 per ton (where usually $c_2 < c_1$), the next $(u_3 - u_2)$ tons cost c_3 $(< c_2)$, etc. A detailed example is provided in Section 2.8.

In all of the above examples, we define new *conditional* zero-one *variables* y_j in addition to the existing continuous activity variables x_j, whose upper bounds are u_j and introduce the following constraints:

$$x_j \leq u_j y_j,\ j = 1, ..., n,$$

$$y_j \geq y_{j+1},\ j = 1, ..., n-1, \text{ and}$$

$$x_j \geq 0; y_j = 0 \vee 1\ \forall j = 1, ..., n.$$

The first set of constraints ensures that an activity is only undertaken if its "indicator variable" y_j equals one, and the second set of constraints guarantees that a variable y_j can only assume a value of one if all variables $y_1, ..., y_{j-1}$ already equal one.

2.6 Imposing a Sequence on Constraints

Similar to the conditional variables discussed in the previous section, we may impose a sequence on some of the constraints. Such *conditional constraints* are of the type "if $\mathbf{a}_{i\bullet}\mathbf{x}\ R_i\ b_i$, then $\mathbf{a}_{k\bullet}\mathbf{x}\ R_k\ b_k$," with any relations R_i and R_k of the type $\leq$, $=$ or $\geq$. Since conditional constraints impose a sequence on constraints, i.e., rows of a problem, they are in some sense dual to conditional variables which impose a sequence on variables, i.e., columns of a problem and which we discussed in Section 2.5 above. Conditional constraints occur in certain applications that involve activities which are arranged in sequence. For example, if a diet in the standard diet problem includes more than a prespecified amount of fat, then its sugar content should not exceed a given bound.

The general idea is best explained graphically. Consider the two constraints $\mathbf{a}_{i\bullet}\mathbf{x} \le b_i$ and $\mathbf{a}_{k\bullet}\mathbf{x} \le b_k$. The condition $\mathbf{a}_{i\bullet}\mathbf{x} \le b_i \rightarrow \mathbf{a}_{k\bullet}\mathbf{x} \le b_k$ then generates the feasible set shown in Figure I.6.

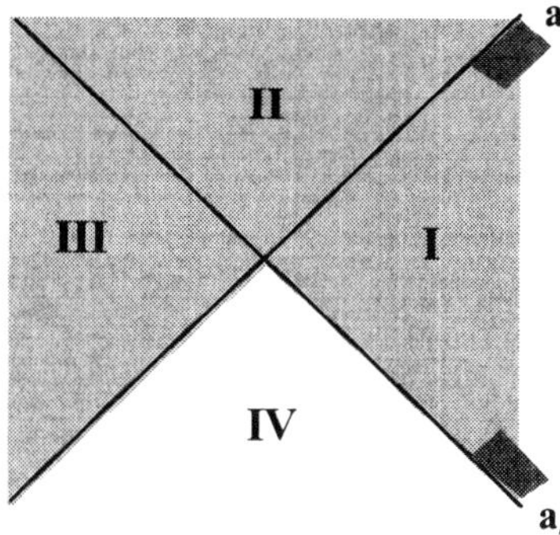

Figure I.6

In case the premise is satisfied and $\mathbf{a}_{i\bullet}\mathbf{x} \le b_i$, then $\mathbf{a}_{k\bullet}\mathbf{x} \le b_k$ must hold as well which is the case in region I in Figure I.6. On the other hand, if the i-th constraint is violated so that $\mathbf{a}_{i\bullet}\mathbf{x} > b_i$, then it is irrelevant whether or not the k-th constraint is satisfied. This situation occurs in the areas II and III. In summary, the entire feasible set is shown as the shaded area. This area is generated by either satisfying the k-th constraint, or by violating the i-th constraint. We can formalize the procedure by defining a zero-one variable y that assumes a value of 1, if the i-th constraint is violated, and 0 otherwise. The condition $\mathbf{a}_{i\bullet}\mathbf{x} \le b_i \rightarrow \mathbf{a}_{k\bullet}\mathbf{x} \le b_k$ can then be written as

$$\mathbf{a}_{i\bullet}\mathbf{x} > b_i + M(1-y)$$

$$\mathbf{a}_{k\bullet}\mathbf{x} \le b_k + My$$

Clearly, if $y = 1$, then $\mathbf{a}_{i\bullet}\mathbf{x} > b_i$, thus violating the i-th constraint, whereas the k-th condition is redundant. On the other hand, $y = 0$ implies redundancy of the first relation, whereas the k-th constraint must hold. This is the desired result.

2.7 Absolute Values of Functions and Nonconcave Objectives

In this section we examine some types of constraints and an objective that can be formulated in zero-one variables. All formulations that do not require integer variables are covered in Eiselt *et al.* (1987).

2.7.1 A Problem with Collective Absolute Values

Consider a constraint that can be written as $|\mathbf{a}_{i\bullet}\mathbf{x}|\ R_i\ b_i$. All cases with $b_i \leq 0$ were covered in Eiselt *et al.* (1987); here we discuss the remaining cases in which $b_i > 0$. Again, if $R_i = \{\leq\}$, a simple linear programming formulation is available. Let now $R_i \in \{=\, ; \geq\}$. Applications of constraints of this type can be found in traffic networks, where the left-hand side expresses the actual traffic flow where $\mathbf{a}_{i\bullet}\mathbf{x} > 0$ denotes a "forward" and $\mathbf{a}_{i\bullet}\mathbf{x} < 0$ a "backward" flow. The right-hand side denotes a lower bound on the traffic flow below which the existence of the traffic link is no longer justified. The two cases considered here are $|\mathbf{a}_{i\bullet}\mathbf{x}| = b_i$ and $|\mathbf{a}_{i\bullet}\mathbf{x}| \geq b_i$ with $b_i > 0$. They are shown in Figures I.7a and b, respectively.

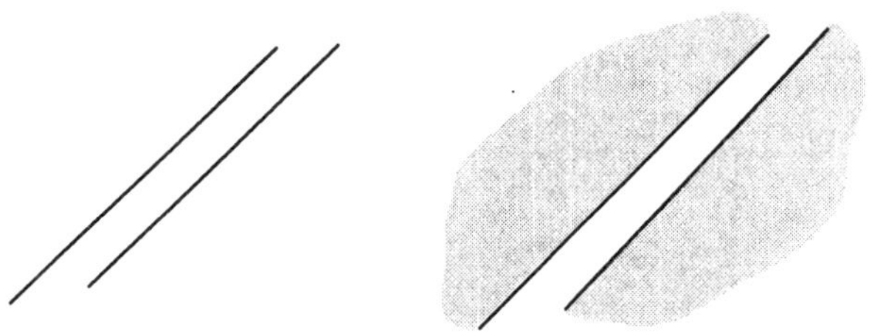

(a) (b)

Figure I.7

The feasible set in the case where $R_i = \{=\}$ is the set of points on the two lines $\mathbf{a}_{i\bullet}\mathbf{x} = b_i$ and $-\mathbf{a}_{i\bullet}\mathbf{x} = b_i$, and for $R_i = \{\geq\}$, the feasible set includes the shaded areas $\mathbf{a}_{i\bullet}\mathbf{x} \geq b_i$ and $-\mathbf{a}_{i\bullet}\mathbf{x} \geq b_i$, i.e., the entire plane without the corridor between the two hyperplanes. Clearly, both sets are nonconvex. Both cases can be formulated requiring that $\mathbf{a}_{i\bullet}\mathbf{x}\ R_i\ b_i$ *or* $-\mathbf{a}_{i\bullet}\mathbf{x}\ R_i\ b_i$, which are either-or constraints. Defining a zero-one variable y which assumes a value of 1, if the former and 0 if the latter constraint is satisfied, and with $M >> 0$ sufficiently large, we can then formulate the system as

$$\mathbf{a}_{i\bullet}\mathbf{x} \leq b_i + M(1-y)$$

$$\mathbf{a}_{i\bullet}\mathbf{x} \geq b_i - M(1-y)$$

$$-\mathbf{a}_{i\bullet}\mathbf{x} \leq b_i + My$$

$$-\mathbf{a}_{i\bullet}\mathbf{x} \geq b_i - My,$$

in case $R_i = \{=\}$. Here the equations were replaced by pairs of equivalent inequalities before applying the formulation for either-or constraints. In case $R_i = \{\geq\}$, we obtain the formulation

$$\mathbf{a}_{i\bullet}\mathbf{x} \geq b_i - M(1-y)$$
$$-\mathbf{a}_{i\bullet}\mathbf{x} \geq b_i - My.$$

2.7.2 A Problem with Individual Absolute Values

Consider a problem in which the left-hand side of a constraint is the sum of absolute values of variables. The general type of constraint under consideration is $\sum_{j=1}^{n} a_{ij}\left|x_j\right| R_i\ b_i$. For convenience, transform the relation, so that $R_i = \{\geq\}$ which is always possible. We then need to discuss the cases in which $b_i > 0$ and $a_{ij} \geq 0$ or $a_{ij} \in \mathbb{R}\ \forall\ j$; all other cases are discussed by Eiselt *et al.* (1987). We begin with the case $a_{ij} \geq 0\ \forall\ j$. There are 2^n possible outcomes for the signs of the x_j - components and accordingly, we use 2^n constraints, specifically $\sum_{j=1}^{n} \{\pm\}^k\ a_{ij}x_j \geq b_i - M(1-y_k)$ for the k-th combination of signs. At least one of the 2^n constraints must be turned on (i.e., $y_k = 1$ for at least one k) so that we also require $\sum_{k=1}^{2^n} y_k \geq 1$, where $y_k = 0 \vee 1\ \forall\ k$.

Example: $2\,|\,x_1\,| + 3\,|\,x_2\,| \geq 6$

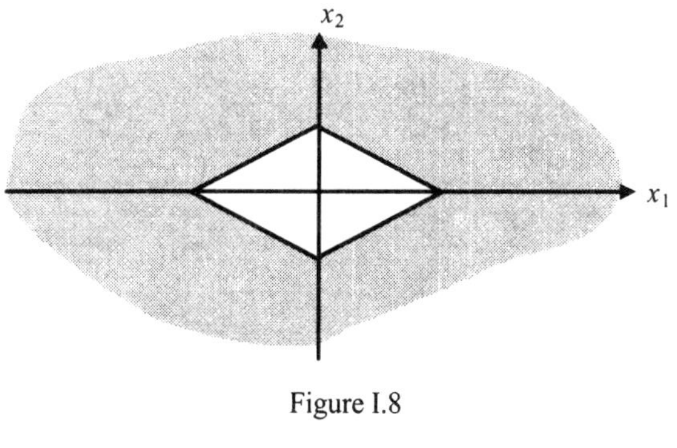

Figure I.8

All constraints form a diamond around the origin. All feasible solutions are located outside of the diamond in Figure I.8. This is formulated as

$$
\begin{aligned}
2x_1 + 3x_2 &\geq 6 - M(1 - y_1) \\
2x_1 - 3x_2 &\geq 6 - M(1 - y_2) \\
-2x_1 + 3x_2 &\geq 6 - M(1 - y_3) \\
-2x_1 - 3x_2 &\geq 6 - M(1 - y_4) \\
y_1 + y_2 + y_3 + y_4 &\geq 1 \\
x_1, x_2 &\in \mathbb{R} \\
y_1, y_2, y_3, y_4 &= 0 \vee 1
\end{aligned}
$$

Consider now the case with $a_{ij} \in \mathbb{R}$. Then we need to add the constraints $\{\pm\}^k x_j \geq -M(1 - y_k)$, partitioning $\mathbb{R}^n$ into its 2^n orthants:

$$
\begin{aligned}
&\sum_{j=1}^{n} \{\pm\}^k a_{ij} x_j \geq b_i - M(1 - y_k) \ \forall\, k = 1, \ldots, 2^n \\
&\{\pm\}^k x_j \geq M(1 - y_k) \ \forall\, j = 1, \ldots, n; \ k = 1, \ldots, 2^n \\
&\sum_{k=1}^{2^n} y_k \geq 1 \\
&y_k = 0 \vee 1, k = 1, \ldots, 2^n
\end{aligned}
$$

Here $\{\pm\}^k x_j$ indicates all 2^n possible sign combinations of the vector $\mathbf{x}$ with components x_j, $j = 1,\ldots, n$.

Example: $2\,|\,x_1\,| - 3\,|\,x_2\,| \geq 6$

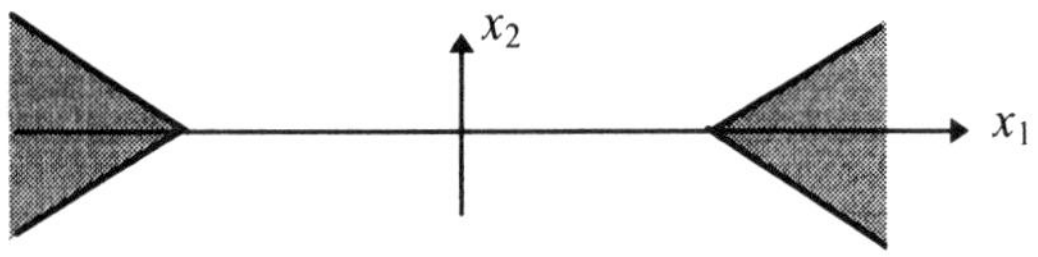

Figure I.9

The feasible region consists of the two disjoint shaded areas in Figure I.9. We obtain

$$
\begin{array}{lll}
2x_1 - 3x_2 \geq 6 - M(1 - y_1), & x_1 \geq -M(1 - y_1), & x_2 \geq -M(1 - y_1), \\
2x_1 + 3x_2 \geq 6 - M(1 - y_2), & x_1 \geq -M(1 - y_2), & -x_2 \geq -M(1 - y_2), \\
-2x_1 - 3x_2 \geq 6 - M(1 - y_3), & -x_1 \geq -M(1 - y_3), & x_2 \geq -M(1 - y_3), \\
-2x_1 + 3x_2 \geq 6 - M(1 - y_4), & -x_1 \geq -M(1 - y_4), & -x_2 \geq -M(1 - y_4)
\end{array}
$$

$$y_1 + y_2 + y_3 + y_4 \geq 1$$

$$x_1, x_2 \in \mathbb{R}$$

$$y_1, y_2, y_3, y_4 = 0 \vee 1$$

For instance, $y_1 = 1$ defines the part of the shaded cone on the right in Figure I.10 that is in the nonnegative quadrant, $y_2 = 1$ defines the part of the shaded cone on the right in the fourth quadrant, etc.

2.7.3 A Problem with a Nonconcave Objective

Finally, consider the following problem with a nonconcave objective function.

$$\text{P: Max } z = \max_j \{c_j x_j\}$$

$$\text{s.t.} \quad \mathbf{Ax} \leq \mathbf{b}$$

$$\mathbf{x} \geq \mathbf{0}$$

The usual transformation applied to maximin objectives would lead to the objective Max z, optimized subject to the given constraints plus n additional constraints $z \geq c_j x_j$ $\forall\, j = 1, \ldots, n$. However, this model would have unbounded "optimal" solutions, rendering such a transformation useless. Consider instead the following procedure. For each $j = 1, \ldots, n$, maximize $z_j = c_j x_j$ (i.e., solve n linear programming problems), and then choose the solution with the highest $\bar{z}_j$ value. This process can be incorporated in a single formulation by using zero-one variables y_j that assume a value of one, if the j-th term in the objective is considered, and zero otherwise.

Clearly, we can deal with one term at a time, so that $\sum_{j=1}^{n} y_j = 1$ is a required constraint. Furthermore, if $y_j = 1$, then the objective value z should be equal to $c_j x_j$, whereas z is not restricted by any of the terms $c_k x_k$, $k \neq j$. This can be formulated as the problem

$$
\begin{aligned}
\text{P}': \ & \text{Max } z \\
& \text{s.t. } \mathbf{Ax} \leq \mathbf{b} \\
& \quad z \geq c_j x_j - M(1 - y_j) \ \ \forall j = 1, \dots, n \\
& \quad z \leq c_j x_j + M(1 - y_j) \ \ \forall j = 1, \dots, n \\
& \sum_{j=1}^{n} y_j = 1 \\
& \quad \mathbf{x} \geq \mathbf{0} \\
& \quad y_j = 0 \vee 1 \ \forall j = 1, \dots, n
\end{aligned}
$$

2.8 Piecewise Linear Functions

Consider a piecewise linear function such as the one shown in Figure I.10.

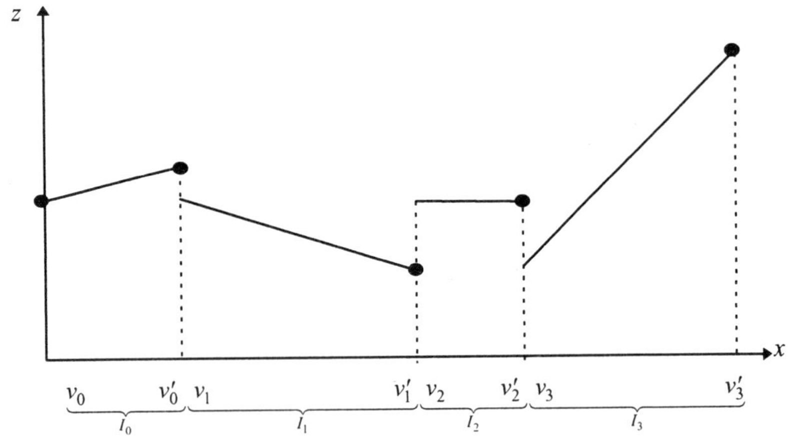

Figure I.10

In general, $z = \sum_j f_j(x_j)$ where we have to assume that each term of the sum depends on no more than a single variable so that we have a separable function. For ease of exposition, we consider only a single variable x and write the piecewise linear function as $z = f(x)$. Introduce now the following notation. Let $f(x)$ be defined on a set of intervals $I_0, I_1, I_2, \ldots, I_p$, with I_k having left and right elemens v_k and v'_k, respectively, and where the interval may be open, half open, or closed. To each interval assign a zero-one variable y_k, such that $y_k = 1$, if $x \in I_k$ and 0 otherwise. If $x \in I_k$, it can be expressed as a linear convex combination of v_k and v'_k, in particular by $v_k \lambda_k + v'_k \lambda'_k$ with nonnegative continous variables $\lambda_k, \lambda'_k \geq 0$, with $\lambda_k + \lambda'_k = 1$. Since x must belong to exactly one interval, we must have $\sum_{k=0}^{p} y_k = 1$, a multiple choice constraint as introduced earlier. In other words, we will express the original variable x as a linear convex combination of the two end points of the interval it is located in, given that x must be in exactly one interval. This can be formulated as

$$x = \sum_{k=0}^{p} v_k \lambda_k + \sum_{k=0}^{p} v'_k \lambda'_k$$

$$\sum_{k=0}^{p} y_k = 1$$

$$\lambda_k + \lambda'_k = y_k \; \forall \; k = 0, \ldots, p$$

$$y_k = 0 \vee 1 \; \forall \; k = 0, \ldots, p$$

$$\lambda_k, \lambda'_k \geq 0 \; \forall \; k = 0, \ldots, p$$

The conditions for λ_k must be strengthened in case of (half) open intervals I_k. In particular, $\lambda_k < 1$ if the interval I_k is left half open, $\lambda'_k < 1$ if the interval I_k is right half open, and $\lambda_k < 1$, $\lambda'_k < 1$ if the interval I_k is open. Since $\lambda'_k = y_k - \lambda_k \; \forall \; k$, the above system, assuming only closed and/or left half open intervals, can be simplified to

$$x = \sum_{k=0}^{p} (v_k - v'_k)\lambda_k + \sum_{k=0}^{p} v'_k y_k$$

$$\sum_{k=0}^{p} y_k = 1$$

$$\lambda_k \in [0; y_k] \ \forall \ k = 0, ..., p$$

$$\lambda_k < 1 \text{ if } I_k \text{ is left half open.}$$

$$y_k = 0 \vee 1 \ \forall \ k = 0, ..., p$$

Note that if the function under consideration is defined on a connected region in $\mathbb{R}$, then $v'_k = v_{k+1} \ \forall \ k = 0, ..., p-1$. This does, however, not simplify the above transformation. Finally, due to the piecewise linearity of $f(x)$, we find that $f(x) =$

$$f(\sum_{k=0}^{p} v_k \lambda_k - \sum_{k=0}^{p} v'_k \lambda_k + \sum_{k=0}^{p} v'_k y_k) = \sum_{k=0}^{p} \left[f(v_k)\lambda_k - f(v'_k)\lambda_k + f(v'_k) y_k \right] =$$

$$\sum_{k=0}^{p} \left[(f(v_k) - f(v'_k))\lambda_k + f(v'_k) y_k \right].$$

Example 1: Consider the piecewise linear function

$$f(x) = \begin{cases} 4 + \frac{1}{4}x & \text{if } x \in [0; 4] \\ 5\frac{1}{3} - \frac{1}{3}x & \text{if } x \in]4; 10] \\ 4 & \text{if } x \in]10; 13] \\ -11 + x & \text{if } x \in]13; 18] \end{cases}$$

Then $v_0 = 0$, $v'_0 = v_1 = 4$, $v'_1 = v_2 = 10$, $v'_2 = v_3 = 13$ and $v'_3 = 18$, so that the variable x can be written as

$$x = -4\lambda_0 + 4y_0 - 6\lambda_1 + 10y_1 - 3\lambda_2 + 13y_2 - 5\lambda_3 + 18y_3,$$

and the constraints are

$$y_0 + y_1 + y_2 + y_3 = 1$$

$$0 \le \lambda_0 \le y_0$$

$$0 \le \lambda_1 \le y_1$$

$$0 \le \lambda_2 \le y_2$$

$$0 \le \lambda_3 \le y_3$$

$$\lambda_1 < 1, \lambda_2 < 1, \lambda_3 < 1$$

$$y_0, y_1, y_2, y_3 = 0 \vee 1$$

Furthermore,

$$f(x) = (4-5)\,\lambda_0 + 5y_0 + (4-2)\,\lambda_1 + 2y_1 + (4-4)\,\lambda_2 + 4y_2 + (4-9)\lambda_3 + 9y_3$$
$$= -\lambda_0 + 5y_0 + 2\lambda_1 + 2y_1 + 4y_2 - 5\lambda_3 + 9y_3.$$

For instance, if $y_3 = 1$ and $\lambda_3 = {}^3/_5$ (implying that $x = 15$), then $f(x) = -5\,({}^3/_5) + 9\,(1) = 6$ which is the desired result.

This transformation can also be used in case a nonlinear function is to be approximated by a piecewise linear function or demonstrated in the following

Example 2: Consider the following mathematical programming problem with nonlinear objective.

$$\text{P}: \text{Max } z = f_1(x_1) + f_2(x_2)$$
$$\text{s.t.} \quad x_1 + x_2 \leq 7$$
$$x_1 \in \,]\,0\,;\,4\,]$$
$$x_2 \in [\,0\,;\,5\,]$$

where $f_1(x_1) = \dfrac{1}{x_1} + 4x_1 - x_1^2$ and $f_2(x_2) = 3x_2$. Assume that it has been decided to approximate $f_1(x_1)$ by the piecewise linear function shown in Figure I.11.

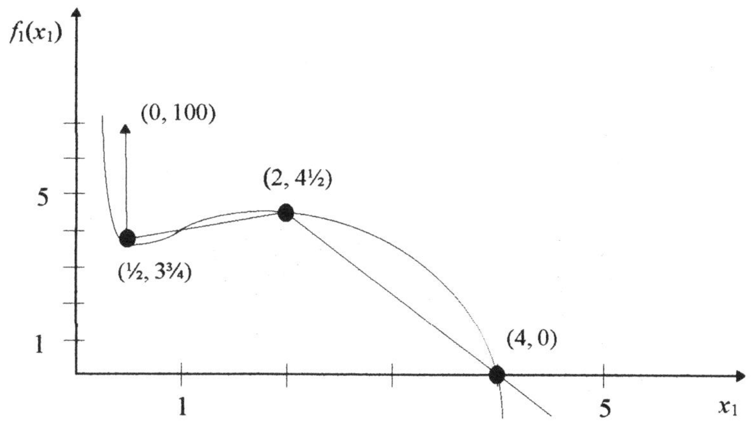

Figure I.11

Here, $v_0 = 0$, and $v_0' = ½$ with $f(v_0) = 100$, and $f(v_0') = 3¾$;

$v_1 = ½$, and $v_1' = 2$ with $f(v_1) = 3¾$, and $f(v_1') = 4½$; and

$v_2 = 2$, and $v_2' = 4$ with $f(v_2) = 4½$, and $f(v_2') = 0$,

Then $x_1 = -½\lambda_0 + ½y_0 - 1½\,\lambda_1 + 2y_1 - 2\lambda_2 + 4y_2$, since I_0 is left half open and I_1 and I_2 are closed,

and $f_1(x_1) = (100 - 3¾)\,\lambda_0 + 3¾\,y_0 + (3¾ - 4½)\,\lambda_1 + 4½\,y_1 + (4½ - 0)\,\lambda_2 + 0y_2$
$= 96¼\,\lambda_0 + 3¾\,y_0 - ¾\,\lambda_1 + 4½\,y_1 + 4½\,\lambda_2.$

The problem can then be written in an equivalent form as

$$
\begin{aligned}
\text{P:}\quad & \text{Max } z = 96¼\,\lambda_0 + 3¾\,y_0 - ¾\lambda_1 + 4½\,y_1 + 4½\,\lambda_2 + 3x_2 \\
\text{s.t.}\quad & -½\lambda_0 + ½\,y_0 - 1½\,\lambda_1 + 2y_1 - 2\lambda_2 + 4y_2 + x_2 \le 7 \\
& -½\lambda_0 + ½\,y_0 - 1½\,\lambda_1 + 2y_1 - 2\lambda x_2 + 4y_2 \in [\,0\,;4] \\
& x_2 \in [\,0\,;5\,] \\
& y_0 + y_1 + y_2 = 1 \\
& 0 < \lambda_0 \le y_0 \\
& 0 \le \lambda_1 \le y_1 \\
& 0 \le \lambda_2 \le y_2 \\
& \lambda_0 < 1 \\
& y_0,\ y_1,\ y_2 = 0 \vee 1
\end{aligned}
$$

There are some special cases of piecewise linear functions in the objective functions where zero-one variables are not needed. Consider the following example.

$$
\begin{aligned}
\text{P:}\quad & \text{Max } z = f_1\,(x_1) + 7x_2 \\
\text{s.t.}\quad & 3x_1 + 6x_2 \le 120 \\
& x_1\,,\quad x_2 \ge 0
\end{aligned}
$$

Suppose now that x_1 denotes the production level of a certain product, and assume that the unit profit for the first five units is \$4, the unit profit for the next seven units is \$3, and the unit profit for the last eight units is \$1. It must now be guaranteed that we do not start making, say, units 6 through 12 before making units 1 through 6. Usually this is done by way of zero-one variables. Here, such variables are unnecessary: it is never wise to increase x_{j_k} from zero by introducing it into the basis before including x_{j_ℓ}, $\ell > k$ as well, as the unit profit of x_{j_ℓ} is not smaller than that of x_{j_k}. In other words, x_j is "filled from zero to its optimal level." In this example we can write

$$f_1(x_1) = \begin{cases} 4x_1 & \text{if } x_1 \in [v_0; v_1] \\ 3x_1 + v_1 & \text{if } x_1 \in [v_1; v_2] \\ x_1 + v_1 + 2v_2 & \text{if } x_1 \in [v_2; v_3] \end{cases}$$

With $v_0 = 0$, $v_1 = 5$, $v_2 = 12$, and $v_3 = 20$, we can write

$$f_1(x_1) = \begin{cases} 4x_1 & \text{if } x_1 \in [0;5] \\ 3x_1 + 5 & \text{if } x_1 \in [5;12] \\ x_1 + 29 & \text{if } x_1 \in [12;20] \end{cases}$$

Plotting $f_1(x_1)$ reveals a piecewise linear concave function with break points at $x_1 = 5$ and $x_1 = 12$. In particular, $f_1(x_1)$ is the lower envelope of the linear functions $4x_1$, $3x_1 + 5$, and $x_1 + 29$. This can be modeled by creating a new variable $x_3 = \min\{4x_1, 3x_1 + 5, x_1 + 29\}$; for details readers are referred to Eiselt *et al.* (1987). This leads to the formulation

$$\begin{array}{llcl} \text{P}': \text{Max } z = & & 7x_2 + x_3 & \\ \text{s.t.} & 4x_1 & - x_3 & \geq 0 \\ & 3x_1 & - x_3 & \geq -v_1 \\ & x_1 & - x_3 & \geq -v_1 - 2v_2 \\ & 3x_1 + 6x_2 & & \leq 120 \\ & x_1, \quad x_2 & & \geq 0, \\ & & x_3 & \in \mathbb{R} \end{array}$$

In general, zero-one variables are not required for maximization problems with concave objective functions or for problems with constraints of type $f_i(x_i) \leq b_i$ with a convex function $f_i(x_i)$. This type of approximation is also useful in the context of special ordered sets of type 2, as defined in Section 2.1.

2.9 Semicontinuous Variables

The modeling of some situations may require the use of so-called *semicontinuous* variables that either take the value of zero or any number above a given positive threshold. Formally, such a variable will satisfy the conditions $x_j = 0$ or $x_j \geq \ell_j$ where $\ell_j > 0$ is some given threshold value. An example for semicontinuous variables occurs in forestry harvesting where x_j indicates the harvesting effort in a logging area. Either the area is not harvested at all, corresponding to $x_j = 0$, or some minimum prescribed level of harvesting effort ℓ_j is carried out, where ℓ_j is determined by economic or other considerations. Above this minimum, any level of x_j can be chosen.

Semicontinuous variables may be modeled in a number of ways. One possibility is to define zero-one variables y_j that assume a value of 1, if $x_j \geq \ell_j$, and 0 if $x_j = 0$. The constraints are then $x_j \leq My_j$, $x_j \geq \ell_j y_j$, $y_j = 0 \vee 1$, and $x_j \geq 0$, where, as usual, $M >> 0$. Clearly, if $y_j = 0$, then constraints require that $x_j \leq 0$ and $x_j \geq 0$, i.e., $x_j = 0$. If $y_j = 1$, then $x_j \leq M$ (a redundant constraint) and $x_j \geq \ell_j$ which is the desired result. Another possibility to model this situation is to define a zero-one variable in the same way as before and then replace the original semicontinuous variable x_j by a new continuous variable x'_j and a zero-one variable y_j, such that $x_j = \ell_j y_j + x'_j$ with $y_j = 0 \vee 1$ and $x'_j \geq 0$. We then add the contraints $0 \leq x'_j \leq (u_j - \ell_j)\, y_j$ where u_j is some appropriately chosen upper bound on the value of x_j. As an example, if a variable x_j must be either zero or take any value between 3 and 21, we can write $x_j = 3y_j + x'_j$ where $y_j = 0 \vee 1$ and $0 \leq x'_j \leq 18y_j$. If $y_j = 0$, then $x'_j = 0$ and hence $x_j = 0$, whereas if $y_j = 1$, then $x'_j \leq 18$. A value of, say, $x'_j = 7$ then leads to $x_j = 3(1) + 7 = 10$. In this way, any feasible value of x_j can be generated. The modeling of semicontinuous variables is reminiscent of the modeling of fixed charges, described in Section 2.2. In that case, the threshold value ℓ_j corresponds to the fixed cost f_i.

CHAPTER 3 APPLICATIONS AND SPECIAL STRUCTURES

The previous chapter demonstrated how zero-one variables can be used to model logical relationships. In this chapter we present a number of examples involving integer variables. The applications are selected to show the wide range and the versatility of integer programming. However, in order to avoid duplication, we will not describe problems related to network models that are treated in detail in Parts II and III of this volume. Although numerous applications of integer programming are reported in the literature, surprisingly few books devote much attention to applications. Notable exceptions are Taha (1997), Williams (1978), and Rardin (1998).

3.1 Applications

3.1.1 A Distribution-Location Problem

Consider a company producing a commodity at each of two factories A and B. The amounts produced can be stored directly at A and/or B, but also in each of the three warehouses D, E, and F. The commodity is delivered free of charge from A, B, D, E and F and sold to eight customers C_1, C_2, ..., C_8. The weekly capacities of the plants and the warehouses A, B, D, E, and F are 3,800; 4,600; 1,100; 2,350;

Table I.2

from \ to	D	E	F
A	4.80	5.20	9.90
B	8.50	5.70	4.90

and 950 tons, respectively, whereas the demands of the customers $C_1, \ldots, C_8$ are 600; 750; 480; 530; 980; 760; 1,030; and 650 tons respectively. The unit transportation costs (in dollars) from factories to warehouses are shown in Table I.2.

Similarly, the unit transportation cost from factories and warehouses to customers are shown in Table I.3 where nonexistent routes are indicated by a dash –:

Table I.3

from \ to	C_1	C_2	C_3	C_4	C_5	C_6	C_7	C_8
A	8.60	9.40	6.30	4.70	–	14.20	16.10	–
B	11.30	–	11.60	5.10	–	–	9.00	4.20
D	6.00	–	11.10	11.90	18.20	21.00	21.50	–
E	16.10	10.50	5.40	4.40	5.80	11.20	16.60	13.10
F	13.30	–	–	13.20	–	11.00	–	4.60

Based on the above data, the objective is to formulate an optimization model that incorporates a cost-minimal distribution plan as well as potential opening, closing, or expansion of warehouses. First consider the components that relate to transportation. For that purpose, define x_{ij} as the amount of the commodity sent from factory/warehouse i to customer j; $i = A, B, D, E, F$, $j = C_1, \ldots, C_8$, and from factory i to warehouse j, $i = A, B$; $j = D, E, F$. Similarly, define c_{ij} as the unit transportation cost from factory/warehouse i to customer j; $i = A, B, D, E, F$, $j = C_1, \ldots, C_8$, and from factory i to warehouse j, $i = A, B$; $j = D, E, F$. Finally, define b_i as the capacity of factory/warehouse $i = A, B, C, D, E, F$; and d_j the demand of customer j for $j = C_1, \ldots, C_8$. The total distribution cost z can then be expressed as the sum of transportation costs from factories to warehouses and the transportation cost to customers, i.e.,

$$z = \sum_{i=A,B} \sum_{j=D,E,F} c_{ij} x_{ij} + \sum_{i=A,B,D,E,F} \sum_{j=C_1,\ldots,C_8} c_{ij} x_{ij}. \qquad (1)$$

In this expression, c_{ij} is set equal to $M >> 0$ for nonexistent routes, e.g., $c_{AC_5} = c_{AC_8} = c_{BC_2} = \ldots = c_{FC_7} = M$. The remaining values of c_{ij} are taken from Tables I.2 and I.3.

Consider now the constraints. There are five different categories of constraints. These are formulated as follows.

$$\sum_{\substack{j=D,E,F\\C_1,\ldots,C_k}} x_{ij} \le b_i\ ,\ i = A, B \tag{2}$$

$$\sum_{i=A,B} x_{ij} \le b_j, j = D,E,F \tag{3}$$

$$\sum_{i=A,B} x_{ij} - \sum_{k=C_1,\ldots,C_8} x_{jk} = 0, j = D, E, F \tag{4}$$

$$\sum_{i=A,B,D,E,F} x_{ij} \ge d_j, j = C_1,\ldots,C_8 \tag{5}$$

$$x_{ij} \ge 0,\ i = A, B, D, E, F;\ j = C_1,\ldots, C_8. \tag{6}$$

The first set of constraints (2) ensures that the material flow out of a factory does not exceed its capacity, constraints (3) guarantee that the warehouse capacities are respected, the flow balancing constraints (4) force the inflow into a warehouse to be equal to its outflow, and constraints (5) force the flow of goods to satisfy customer demand. This model includes 2 + 3 + 3 + 8 = 16 structural constraints.

Consider now the possibility of opening new, and closing or expanding existing warehouses. In particular, suppose that two new warehouses *G* and *H* could be opened, while warehouse *D* could be expanded, and warehouses *E* and/or *F* could be shut down. The capital costs for opening *G* with a capacity of 1,600 would be \$105, for *H* (capacity 1,250) \$93.50, and for expanding *D* (i.e., increasing its capacity from 1,100 to 1,700) \$28.50. The capital cost savings for shutting down *E* would be \$85.00, and \$77.00 for *F*. Unit transportation costs from the factories to the new warehouses and from the new warehouses to the customers are provided in Tables I.4 and I.5.

Table I.4

from \ to	*G*	*H*
A	7.50	7.40
B	3.30	4.10

Table I .5

from \ to	C_1	C_2	C_3	C_4	C_5	C_6	C_7	C_8
G	16.30	17.70	12.60	6.50	4.00	3.50	–	8.50
H	8.60	–	–	11.50	15.40	13.50	10.00	3.50

Finally, it is required that not more than four warehouses should remain in operation, so that if both G and H are opened, at least one of E and F must be shut down.

In order to deal with these additional considerations, the linear programming model formulated above will now have to be extended to a mixed integer linear programming model. We introduce binary zero-one variables y_i as follows. The variable $y_D = 1$, if warehouse D is expanded and 0 otherwise; $y_E = 1$ if warehouse E is kept open and 0 otherwise, and similarly for y_F. Furthermore, $y_G = 1$, if warehouse G is opened and 0 otherwise, and similarly for y_H. We also extend the definitions of the continuous variables x_{ij} and the coefficients c_{ij} and b_{ij} to cover (potential) transportation into and out of G and H.

The new total cost z', consisting of the distribution cost, plus any capital costs, minus any capital savings, is then

$$z' = \sum_{i=A,B} \sum_{j=D,E,F,G,H} c_{ij}x_{ij} + \sum_{i=A,B,D,E,F,G,H} \sum_{j=C_1,\dots,C_8} c_{ij}x_{ij}$$

$$+ 28.5y_D + 85(y_E - 1) + 77(y_F - 1) + 105y_G + 93.5y_H. \qquad (1')$$

The factory capacity restraints are only slightly modified to

$$\sum_{\substack{j=D,E,F,G,H,\\ C_1,\dots,C_8}} x_{ij} \le b_i,\ i = A, B \qquad (2')$$

whereas the other constraint groups are now

$$\begin{cases} \sum_{i=A,B} x_{iD} \le 1{,}100 + 600y_D \\ \sum_{i=A,B} x_{ij} \le b_j y_j,\ j = E,F,G,H \end{cases} \qquad (3')$$

$$\sum_{i=A,B} x_{ij} - \sum_{k=C_1,\dots,C_8} x_{jk} = 0,\ j = D, E, F, G, H \qquad (4')$$

$$\sum_{i=A,B,D,E,F,G,H} x_{ij} \ge d_j,\ j = C_1,\dots,C_8 \qquad (5')$$

$$x_{ij} \geq 0,\, i = A, B, D, E, F, G, H\,; j = C_1, \ldots, C_8, \qquad (6')$$

$$y_j = 0 \vee 1, j = D, E, F, G, H \qquad (6'')$$

where constraints (3′) concern the capacities of the warehouses, constraints (4′) ensure the equality of inflow and outflow at the warehouses, and constraints (5′) guarantee that customer demands are respected. Finally, since no more than four warehouses are allowed to be open, we must have

$$y_E + y_F + y_G + y_H \leq 3. \qquad (7).$$

This mixed integer programming formulation includes 56 continuous and 5 binary variables, and 21 structural constraints. It turns out that at optimum, warehouse D should not be expanded, warehouse E should remain open, while F is shut down, and the new warehouse G should be opened while H should not. The total cost for the original linear programming problem is \$51,417, which decreases to \$34,688 for the mixed integer programming problem; a result of the savings due to the opening of new and closing of existing warehouses.

3.1.2 A Cutting Stock Problem

Assume that a lumber yard has a supply of 10-ft boards which are cut into 3-ft, 4-ft, and 5-ft boards according to customer demand. The 10-ft boards can be cut in six different sensible patterns as shown in Table I.6.

Table I.6

Pattern #	Number of 3-ft boards	Number of 4-ft boards	Number of 5-ft boards	Waste (ft)
1	3	0	0	1
2	2	1	0	0
3	1	0	1	2
4	0	1	1	1
5	0	2	0	2
6	0	0	2	0

There are many other possible but not sensible patterns; for instance, one could cut a 10-ft board into a 3-ft and a 4-ft board, leaving 3-ft as waste. This would not make sense, since the 3-ft waste could be used as a 3-ft board, as in pattern #2. If a customer orders 50 3-ft boards, 65 4-ft boards and 40 5-ft boards, the question is how many 10-ft boards need to be cut and what cut patterns are to be used?

To model this situation, denote by y_j the number of 10-ft boards cut according to pattern j, $j = 1, ..., 6$. Whereas the total customer demand is $50 \times 3 + 65 \times 4 + 40 \times 5 = 610$ ft, the total length of boards actually cut is $10(y_1 + y_2 + y_3 + y_4 + y_5 + y_6)$, and the total waste is therefore $10\sum_{j=1}^{6} y_j - 610$. This implies that when we minimize $\sum_{j=1}^{6} y_j$, the total number of 10-ft boards that need to be cut, we also minimize the total waste, and vice versa. The actual number of 3-ft boards obtained in the cutting procedure is $3y_1 + 2y_2 + y_3$, and therefore $3y_1 + 2y_2 + y_3 \geq 50$ must hold in order to satisfy customer demand for 3-ft boards. Similarly, $y_2 + y_4 + 2y_5 \geq 65$ to satisfy the demand for 4-ft boards and $y_3 + y_4 + 2y_6 \geq 40$ for 5-ft boards. Since the variables y_j must be nonnegative integers, $j = 1, ..., 6$, we can summarize this *cutting stock* problem as:

$$
\begin{array}{llr}
\text{P: Min } z = y_1 + y_2 + y_3 + y_4 + y_5 + y_6 & & \text{(10-ft boards)} \\
\text{s.t.} \quad 3y_1 + 2y_2 + y_3 & \geq 50 & \text{(3-ft boards)} \\
\qquad y_2 + y_4 + 2y_5 & \geq 65 & \text{(4-ft boards)} \\
\qquad y_3 + y_4 + 2y_6 & \geq 40 & \text{(5-ft boards)} \\
\qquad y_j \in \mathbb{N}_0, j = 1, ..., 6 & &
\end{array}
$$

The optimal solution to this problem is to cut a total of 65 10-ft boards, using pattern #2 25 times and patterns #5 and #6 20 times each. The total waste would then be $25 \times 0 + 20 \times 2 + 20 \times 0 = 40$ ft (or, equivalently, $10\sum_{j=1}^{6} y_j - 610 = 10 \times 65 - 610 = 40$ ft).

The above example illustrates a simple instance of the cutting stock or *trim loss problem*, which is widely applicable whenever trim waste is to be minimized. The example above refers to a one-dimensional situation (board length), but two- and three-dimensional applications are also common where two-dimensional patterns or three-dimensional pieces are stamped or cut out of some material which could be wood, metal, or other compounds. The number of cut patterns, six in our example above, can easily run into millions, as reported in the early contributions by Gilmore and Gomory (1961, 1963). Practical solution of such problems requires the use of special techniques such as column generation, leading to subproblems of the knapsack type, see, e.g., Eiselt *et al.* (1987).

A comment concerning the definition of waste is in order. Whereas the six patterns in our example above generate waste only in pieces of 1-ft and 2-ft lengths, it is possible that the difference between board-ft cut and board-ft demanded could include 3-, 4-, or 5-ft pieces as well that are not strictly waste since they could be used for later customer demand. In our example, only patterns 2 and 6 generate no waste in the strict sense, and yet these two patterns produce boards of all three sizes required. If we require that no waste is allowed in the strict sense, the minimum number of 10-ft boards to be cut would then be obtained by solving the reduced problem

$$
\begin{aligned}
\text{P}': \text{Min } z = \; & y_2 + y_6 \\
\text{s.t.} \quad & 2y_2 \qquad \geq 50 \\
& y_2 \qquad \geq 65 \\
& \qquad 2y_6 \geq 40 \\
& y_2, \; y_6 \in \mathbb{N}_0
\end{aligned}
$$

which has an optimal solution $\bar{y}_2 = 65$, $\bar{y}_6 = 20$, $\bar{z} = 85$, so that now 85 instead of 65 10-ft boards have to be cut, yielding 130 3-ft boards, 65 4-ft boards, and 40 5-ft boards, for a surplus of 130 – 50 = 80 unnecessary 3-ft boards. These could be held in stock for future demand.

3.1.3 Examination Timetabling

A university needs to schedule a given set of m examinations E_i, $i = 1,..., m$. The examinations are to be held in specified time slots t, $t = 1,..., T$ during the university examination period. By an *examination timetable* we mean an assignment of the m examinations to the T time slots. Each examination has to be scheduled, but not each time slot must be used; also, it is possible to schedule several examinations in the same time slot. A feasible examination timetable has to satisfy a multitude of restrictions. For example, no conflicts are allowed in the sense that if a student is registered to take two different examinations, they cannot be scheduled in the same time slot. There are also room capacities to consider. To facilitate a discussion of these and other considerations we will first introduce some notation.

Define $y_{it} = 1$ if examination i is scheduled in time slot t, and $y_{it} = 0$ otherwise; $i = 1,..., m$; $t = 1,..., T$. An examination timetable is specified by giving the values of all zero-one variables y_{it}. For a timetable to be feasible, three conditions must be satisfied: each exam must be scheduled exactly once, there can be no conflicts for any registered student, and room capacities are to be respected.

The condition that each exam is scheduled in exactly one time slot is easily formulated as

$$\sum_{t=1}^{T} y_{it} = 1, \; i = 1,\ldots, m.$$

Let row a_{ij} denote the number of students taking examinations E_i and E_j. Clearly, if $a_{ij} > 0$, then E_i and E_j cannot be assigned to the same time slot. This requirement can be modeled by the constraints

$$y_{it} + y_{jt} \le 2 - \left\lceil \frac{a_{ij}}{a_{ij}+1} \right\rceil, \; \forall\, i < j, \forall\, t.$$

This constraint is valid, as the left-hand side is the number of exams E_i and E_j assigned to time slot t, and the right-hand side equals 2, if $a_{ij} = 0$ and 1 for any $a_{ij} > 0$. Note that the right-hand side of this constraint could be replaced by $2 - a_{ij} / (a_{ij} + 1)$, as for $a_{ij} = 0$, the right-hand side value equals zero in both cases, whereas it equals a positive value in the open interval]1; 2[for any $a_{ij} > 0$. As the left-hand side is the sum of two integer variables and hence integer, the effective upper bound is 1 as required. However, the formulation without the ceiling is not as tight which may result in increased computation times. For a discussion of weak and strong formulations, see Chapter 4 of this part.

As far as room capacities are concerned, suppose that b_i students are registered for examination E_i and there is a total of n_t seats available in time slot t, then no more students can be scheduled for examination then there are seats, i.e.,

$$\sum_{i=1}^{m} b_i y_{it} \le n_t, \; t = 1,\ldots, T.$$

An examination timetable that satisfies the above three types of constraints is deemed feasible but might still have undesirable aspects. A desirable feasible timetable might consider time slot preferences and a schedule that spreads out exams as evenly as possible.

First consider time slot preferences. Defining weights $w_{it} \ge 0$ that indicate the disutility or penalty of scheduling exam E_i in time slot t per student, the penalty for scheduling examination E_i in time slot t is therefore $\sum_{t=1}^{T} b_i w_{it} y_{it}$ which takes the value of 0 if E_i is scheduled in a preferred time slot, and $b_i w_{it}$ if E_i is scheduled in an undesirable time slot. The total penalty for the entire examination timetable is then

$$\sum_{i=1}^{m}\sum_{t=1}^{T} b_i w_{it} y_{it}$$

which is used as part of the objective function.

The second aspect concerns the time spread of the scheduled exams. Consider examinations E_i and E_j, both to be taken by a_{ij} students. Even if E_i and E_j are not scheduled at the same time, it is deemed undesirable if E_i and E_j were scheduled closely after each other. With s denoting the number of time periods between the time slots for E_i and E_j in the timetable, we define a penalty weight v_s per student which is large for small s and small for large s. We could now set $v_s = 2^{5-s}$ for $s = 1, \ldots, 5$ and $v_s = 0$, for $s \geq 6$. The total time proximity penalty for scheduling examinations E_i and E_j with $i > j$ exactly s time periods apart would be $a_{ij}v_s$ so that the total timetable time proximity penalty is

$$\sum_{\substack{i,j \\ i>j}}\sum_{s}\sum_{t} (y_{it}y_{j,\,t-s} + y_{it}y_{j,\,t+s})\, a_{ij}v_s$$

which constitutes another part of the objective function. Note that the terms $y_{it}\, y_{j,t-s}$ and $y_{it}\, y_{j,\,t+s}$ are nonlinear, a feature that significantly increases the degree of difficulty of the problem. Among all feasible timetables we now minimize an objective that is constructed as the sum of the total time slot penalties and time proximity penalties.

A model similar to the above was used by Laporte and Desroches (1984) to construct an examination timetable with m = 168 exams to be held and $\sum_{i=1}^{m} b_i = 17{,}682$ exam papers to be written.

3.1.4 Forestry Harvesting

A company is planning the harvesting of a certain forested area. This area is described on a map grid of square cells F_{ij} of equal size. For ease of exposition, we assume that the forested area is rectangular with borders in the north-south and east-west directions, and consisting of $m \times n$ cells F_{ij}, $i = 1,\ldots, m$; $j = 1,\ldots, n$, and with cell $F_{i,\,j+1}$ to the east and cell $F_{i+1,\,j}$ to the south of cell F_{ij}. This pattern is shown in Figure I.18.

	$F_{i-1,j-1}$	$F_{i-1,j}$	$F_{i-1,j+1}$	
	$F_{i,j-1}$	F_{ij}	$F_{i,j+1}$	
	$F_{i+1,j-1}$	$F_{i+1,j}$	$F_{i+1,j+1}$	

Figure I.12

For ecological reasons it is common to prevent large tracts of forest to be clear-cut by requiring that certain harvest adjacency constraints must be satisfied. One such type of requirement is that neighboring cells cannot both be cut. We can formulate this mathematically by introducing the harvest zero-one variable y_{ij} which takes the value of 1 if cell F_{ij} is harvested and 0 if it is not. Furthermore, define N_{ij} as the set of neighbors of F_{ij}, where neighborhood relations may be defined in various ways. The general idea is then to prohibit harvesting of *all* parcels in N_{ij} if F_{ij} is harvested. Formally, $y_{ij} = 1$ would then require that $y_{k\ell} = 0 \ \forall \ F_{k\ell} \in N_{ij}$. A general disaggregated set of constraints would then be assigned to each parcel F_{ij}, such that

$$y_{ij} + y_{k\ell} \leq 1 \ \forall \ F_{k\ell} \in N_{ij}. \tag{8}$$

These constraints would then be aggregated, so that a simple constraint

$$y_{ij} \leq 1 - \frac{1}{|N_{ij}|} \sum_{F_{k\ell} \in N_{ij}} y_{k\ell} \tag{9}$$

would be assigned to each parcel F_{ij}. Clearly, whenever any parcel in the neighborhood of F_{ij} is harvested, then the right-hand side of constraint (9) is strictly less than one, so that the zero-one variable y_{ij} must equal zero.

Consider now the case in which neighbors are defined as parcels that share a common border. In other words, a parcel F_{ij} has four neighbors: $F_{i-1,j}$ to the North, $F_{i,j-1}$ to the West, $F_{i+1,j}$ to the South, and $F_{i,j+1}$ to the East. The disaggregated constraints are

$$\left.\begin{aligned} y_{ij} + y_{i-1,j} &\leq 1 \\ y_{ij} + y_{i,j-1} &\leq 1 \\ y_{ij} + y_{i+1,j} &\leq 1 \\ y_{ij} + y_{i,j+1} &\leq 1 \end{aligned}\right\} \forall \ F_{ij}. \tag{10}$$

Notice the duplication: the constraint for F_{ij} that either it or its neighbor to the north may be harvested (*viz.*, $y_{ij} + y_{i-1,j} \leq 1$) is the same as the constraint for $F_{i-1,j}$

that either it or its neighbor to the south may be harvested (*viz.*, $y_{i-1,j} + y_{ij} \le 1$). Hence, the set of four constraints in (10) can be reduced to the set

$$\left.\begin{aligned} y_{ij} + y_{i,j+1} &\le 1 \\ y_{ij} + y_{i+1,j} &\le 1 \end{aligned}\right\} \forall F_{ij}. \tag{10'}$$

It should be noted that with this neighborhood definition (ignoring boundary constraints), up to half of the entire region may be harvested, similar to all the white squares on a chess board.

Another type of neighborhood constraint requires an unharvested annulus of cells around any harvested cell. This means that for any given cell F_{ij}, the neighborhood N_{ij} includes the neighbors to is North, Northeast, East, Southeast, South, Southwest, West, and Northwest. This means that for each cell F_{ij}, we define eight constraints

$$\left.\begin{aligned} y_{ij} + y_{i-1,j-1} &\le 1 \\ y_{ij} + y_{i-1,j} &\le 1 \\ y_{ij} + y_{i-1,j+1} &\le 1 \\ y_{ij} + y_{i,j-1} &\le 1 \\ y_{ij} + y_{i,j+1} &\le 1 \\ y_{ij} + y_{i+1,j-1} &\le 1 \\ y_{ij} + y_{i+1,j} &\le 1 \\ y_{ij} + y_{i+1,j+1} &\le 1 \end{aligned}\right\} \forall F_{ij} \tag{11}$$

Again, dupliction of constraints means that only four of the constraints in (11) are needed, e.g., the Northeast, East, Southeast, and South of a cell. Another possibility is to simply formulate an aggregated constraint that allows only one of the four cells at F_{ij} or to its East, Southeast, or South, to be harvested. Such a formulation is

$$y_{ij} + y_{i,j+1} + y_{i+1,j} + y_{i+1,j+1} \le 1 \quad \forall F_{ij}. \tag{12}$$

This formulation ensures that an annulus around a harvested parcel remains unharvested. Given such a definition of neighborhood, up to one quarter of the entire region may be harvested, e.g., harvest all parcels F_{ij} with $i, j \equiv 1 \pmod 2$. This neighborhood structure is further discussed by Snyder and ReVelle (1996).

An extension of the model is to also consider the habitat or preservation value h_{ij} of cell F_{ij}. Assume that three mutually exclusive actions may be applied to each cell F_{ij}, *viz.*, harvesting, habitat preservation, or no action. Defining c_{ij} as the monetary value of cutting parcel F_{ij}, the objective is then to maximize both total harvest profit and total preservation value. For that purpose we introduce habitat preservation variables z_{ij} that assume a value of 1 if parcel F_{ij} is selected for habitat preservation and 0 if no action is taken. The multiobjective model can then be written as

$$\text{P: Max } z_1 = \sum_{i=1}^{m} \sum_{j=1}^{n} c_{ij} y_{ij}$$

$$\text{Max } z_2 = \sum_{i=1}^{m} \sum_{j=1}^{n} h_{ij} z_{ij}$$

$$\text{s.t. } y_{ij} + y_{i,j+1} + y_{i+1,j} + y_{i+1,j+1} + z_{ij} \leq 1,\ i = 1, \ldots, m-1; j = 1, \ldots, n-1$$

$$y_{ij}, z_{ij} = 0 \vee 1,\ i = 1, \ldots, m; j = 1, \ldots, n$$

The constraints in the above formulation require that each parcel F_{ij} is either cut, habitat protected or left alone. It also leaves an uncut annulus around every cut cell. Furthermore, if a cell is habitat protected, then no adjacent cell can be harvested, and finally, if a cell is cut then none of the cells in the surrounding annulus may be declared habitat protected.

For further details on the above types of forestry models, see Snyder and ReVelle (1997), and Lockwood and Moore (1992). For a discussion of multiobjective models, the reader is referred to Eiselt *et al.* (1987).

3.1.5 Technology Choice

Consider a situation in which a company has the choice of technologies for the production of a product. For simplicity, assume that the firm manufactures only a single product. Denote the production quantity by x, and let the demand for that product be d. Suppose now that n technologies for the production of the product are available. If the j-th technology is employed, costs of $c_j^f + c_j^v x_j$ are incurrred, where c_j^f (c_j^v) denote the fixed (variable) production cost related to the j-th technology, and x_j denotes the quantity of the product manufactured with this technology. Denote by u_j the capacity of the j-th technology if it is chosen for the production process. The problem is now to choose production processes so as to minimize the production-related costs while respecting production capacities and satisfying customer demand.

For that purpose, define zero-one variables y_j that assume a value of 1, if the j-th production process is chosen, and 0 otherwise. The problem can be written as

$$\text{P: Min } z = \sum_{j=1}^{n} c_j^f y_j + \sum_{j=1}^{n} c_j^v x_j$$

$$\text{s.t.} \quad x_j \leq u_j y_j \ \forall\, j = 1, \ldots, n$$

$$\sum_{j=1}^{n} x_j \geq d$$

$$x_j \geq 0 \ \ \forall\, j = 1, \ldots, n$$

$$y_j = 0 \vee 1 \ \ \forall\, j = 1, \ldots, n$$

The first set of constraints ensures that the production level produced with a certain technology equals zero, if that technology is not chosen. The last structural constraint requires that customer demand is met.

Example: As a numerical illustration, consider three technologies, a low, an intermediate, and a high level technology. Typically, the more advanced a technology, the higher its fixed cost and the lower its variable costs. Here, let the three cost functions be $50 + 6x_1$, $200 + 4x_2$, and $800 + x_3$, respectively. The capacities of the three technologies are 100, 400, and 1,000, respectively, and the total customer demand equals 1,020. Clearly, with the given capacities no single technology will suffice to satisfy the customer demand. The problem can be formulated as

$$\text{P: Min } z = 50y_1 + 200y_2 + 800y_3 + 6x_1 + 4x_2 + x_3$$

$$\text{s.t.} \quad x_1 \leq 100y_1$$

$$x_2 \leq 400y_2$$

$$x_3 \leq 1{,}000y_3$$

$$x_1 + x_2 + x_3 \geq 1{,}020$$

$$x_1, \ x_2, \ x_3 \geq 0$$

$$y_1, \ y_2, \ y_3 = 0 \vee 1$$

The optimal solution to the problem is $\bar{y}_1 = 1$, $\bar{y}_2 = 0$, $\bar{y}_3 = 1$, and $\bar{x}_1 = 20$, $\bar{x}_2 = 0$, $\bar{x}_3 = 1{,}000$ with an objective function value $\bar{z} = 1{,}970$, so that the low and the high, but not the intermediate technology levels will be used. The high level technology production process will operate at full capacity, unlike the low level technology process that operates at only 20% of full capacity.

In addition to the constraints outlined above, we may require that exactly k (or no more than k) technologies may be chosen in the process. This can easily be incorporated by adding the constraint $\sum_{j=1}^{n} y_j = k$ or $\sum_{j=1}^{n} y_j \le k$, respectively.

3.1.6 Political Districting

The fundamental problem in political administration is to define districts, each of which has a certain number of political representatives, so that the districts have certain properties that are either desirable or have to be satisfied by law. The earliest work reported in this field is by Hess *et al.* (1965), some more recent results are presented by Hojati (1996). Basically there are two principles that may be applied: the top-down principle that starts with large population units such as states or provinces and then divides these states into districts; or the bottom-up philosophy which starts with small individual population units such as municipalities and combines these in the procedure.

For illustrative purposes, consider first a very simple model that combines population units to districts in a bottom-up process. Define m as the number of municipalities or population units and let n be the number of electoral districts or ridings. Whereas the population units are known with their location, respective population and other features, nothing is known about the district other than their number. We apply the usual "one man - one vote" principle, and assume that there is exactly one representative per district to be elected. Define now $y_{ij} = 1$, if the i-th population unit is assigned to the j-th district, and 0 otherwise. Let p_i denote the population of the i-th population unit and define q_j as the population of the j-th district. The total population (of the state) is then $p = \sum_{i=1}^{m} p_i = \sum_{j=1}^{n} q_j$. Then state-wide, the average population per representative is p/n whereas in the j-th district it is $q_j/1 = q_j$. The deviation of this number from the state average is $|p/n - q_j|$. One objective is then to minimize the maximal absolute deviation of the population per representative in any district from the state average. This leads to the following generalized minimax assignment problem.

$$\text{P:} \qquad \text{Min } z = \max_j \left\{ \left| \frac{p}{n} - q_j \right| \right\}$$

$$\text{s.t.} \qquad \sum_{i=1}^{m} p_i y_{ij} = q_j \quad \forall j = 1, \ldots, n \qquad (13)$$

$$\sum_{j=1}^{n} y_{ij} = 1, \quad \forall i = 1,...,m \tag{14}$$

$$y_{ij} = 0 \vee 1, \quad \forall\, i = 1,..., m; j = 1,..., n$$

In this formulation, constraints (13) define the population of the j-th district, and constraints (14) ensure that each population unit is assigned to exactly one district.

There are additional constraints that should be respected by any reasonable solution. In order to explain those, consider the following numerical example. There are nine population units that are to be combined so as to form three districts. The location of the units is shown in Figure I.13.

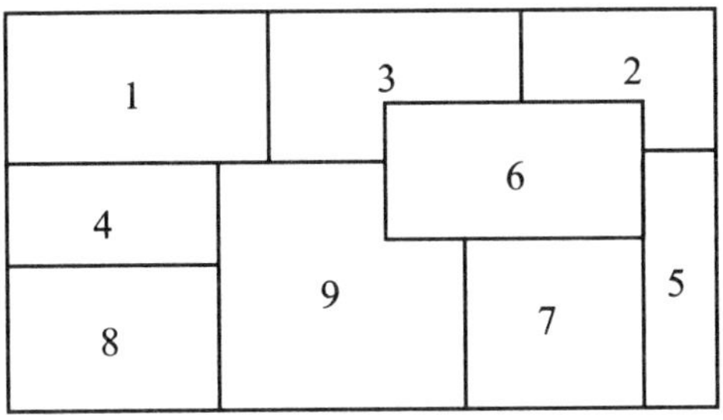

Figure I.13

Suppose now that the population is distributed as shown in Table I.7.

Table I.7

Population unit #	1	2	3	4	5	6	7	8	9
Population (in 1,000s)	3	1	2	3	4	10	2	4	1

The total population is p = 30,000 people. Consequently, the theoretical optimum (that may or may not be achievable in any specific instance) has 10,000 people assigned to each district.

Consider now the following two solutions:

Solution 1: {1, 4, 8}, {2, 3, 5, 7, 9}, {6}

Solution 2: {1, 3, 4, 7}, {2, 5, 8, 9}, {6}

With respect to the minimax objective in the above formulation, both solutions are optimal as they achieve the theoretical limit. However, in Solution 1, population unit six is an enclave surrounded by units 2, 3, 5, 7, and 9, whereas in Solution 2 the district {1, 3, 4, 7} is disconnected. Both features are not desirable and should be excluded. A solution such as {1, 2, 3, 4}, {5, 7, 8, 9} and {6} with 9,000, 11,000 and 10,000 people in the respective districts will be preferred even though the districts are still elongated in shape which may not be desirable. Other requirements, e.g., the incorporation of natural boundaries such as highways, rivers etc., as district boundaries may also be desired. It may be (and usually is) desired to design districts that are homogeneous with respect to population, e.g., annual incomes may be used as surrogate criteria for homogeneity.

3.1.7 Apportionment Problems

Consider a country with a representative system of government. The problem of how to fairly distribute the seats in a legislature or parliament according to the population size of regions, provinces, or number of party votes has haunted politicians and researchers for a long time. Due to the indivisibility of seats, a perfectly fair distribution is almost certainly impossible to achieve and the *apportionment problem* is to find the best compromise solution.

To formalize matters, assume that a country has n states with populations $a_j \in \mathbb{N}$, $j = 1, \ldots, n$ and a legislature with $b \in \mathbb{N}$ seats. Let $y_j \in \mathbb{N}_0$, $j = 1, \ldots, n$ denote the number of seats allocated to the j-th state. An *apportionment* of the b seats among the n states is a vector $\mathbf{y} = [y_1, \ldots, y_n]$ such that $y_j \in \mathbb{N}_0$ and $\sum_{k=1}^{n} y_k = b$. There are usually maximum requirements on y_j as well; for instance, $y_j \geq 1\ \forall\, j$ is required in the United States and $y_j \geq 2\ \forall\, j$ in France; other countries have other lower bound requirements. There may be upper bound constraints on y_j as well. For details, the reader is referred to Balinski and Young (1982).

The question is now how to measure the "fairness" of a given apportionment and, secondly, how to achieve fairness once it has been defined. To answer the first question, we compare the population proportion of the j-th state to the whole country, i.e., $a_j \Big/ \sum_{k=1}^{n} a_k$ with the proportion of seats of the j-th state to the whole legislature, i.e., y_j / b. If these two ratios are equal, i.e., $a_j \Big/ \sum_{k=1}^{n} a_k = y_j / b$, then the j-th state has been fairly treated; otherwise, $\left| \left[a_j \Big/ \sum_{k=1}^{n} a_k \right] - y_j / b \right|$ is a measure of the unfairness of the apportionment that relates to the j-th state. Taking all states into consideration, a measure of the unfairness of the entire apportionment is then $\max_{1 \le j \le n} \left| \left[a_j \Big/ \sum_{k=1}^{n} a_k \right] - y_j / b \right|$ which expresses the worst (largest) unfairness of the apportionment for the whole country. The apportionment problem can then be formulated as

$$\text{P}_1\text{: Min } z_1 = \max_{1 \le j \le n} \left| \left[a_j \Big/ \sum_{k=1}^{n} a_k \right] - y_j / b \right|,$$

$$\text{s.t.} \qquad \sum_{k=1}^{n} y_k = b$$

$$y_k \in \mathbb{N}_0 \; \forall \; k = 1,\ldots, n$$

in addition to any other upper and/or lower bound constraints on the values of y_j. Another formulation of the apportionment problem results if the degree of unfairness is defined as the sum for the unfairnesses for the states. Minimizing this sum results in the problem

$$\text{P}_2\text{: Min } z_2 = \sum_{j=1}^{n} \left| \left[a_j \Big/ \sum_{k=1}^{n} a_k \right] - y_j / b \right|$$

$$\text{s.t.} \qquad \sum_{j=1}^{n} y_j = b$$

$$y_j \in \mathbb{N}_0 \; \forall \; j = 1,\ldots, n$$

For an apportionment to be perfectly fair, we would have to have $a_j \Big/ \sum_{k=1}^{n} a_k = y_j/b$ or $y_j = ba_j \Big/ \sum_{k=1}^{n} a_k \in \mathbb{N}_0$ for all $j = 1,\ldots, n$. Since the size of the legislature b is usually in the hundreds, whereas the parameters a_j are typically in the millions, this requirement is almost certainly never satisfied.

Historically, methods for allocating seats in legislatures have been adopted that attempt to achieve fairness in the above sense. For the United States, the method of Alexander Hamilton was used during the 19th century; it is also known as

Vinton's method. It first calculates the *quota* $ba_j \Big/ \sum_{k=1}^{n} a_k$ for each state and gives to each state the floor of its quota, i.e., $\left\lfloor ba_j \Big/ \sum_{k=1}^{n} a_k \right\rfloor$. The remaining seats are then given to the states having the largest fractional parts of their quotas. A peculiar phenomenon was observed in the 1880s for this method, affecting the state of Alabama, and was accordingly called the "Alabama paradox." We will give an example of this paradox based on a hypothetical country where $b = 100$ seats are to be apportioned among three states, having 45.3%, 44.2%, and 10.5% of the population, respectively. The three quotas for the states are then 45.3, 44.2, and 10.5 seats, respectively. By first allocating the integer parts of these quotas (45, 44, and 10), a total of 99 seats are allocated. Since the remaining fractional parts are 0.3, 0.2, and 0.5, respectively, the smallest state receives the last seat, for an overall allocation of 45, 44, and 11. Assume now that the legislature is to be increased in size, from 100 to 101 seats. The new quotas are $101 \times 0.453 = 45.753$, $101 \times 0.442 = 44.642$, and $101 \times 0.105 = 10.605$, respectively, resulting in a first allocation of 45, 44, and 10, for a total of 99 seats, as before. Two more seats are to be allocated and according to the sizes of the fractional parts of the quotas, the larger states would get one each, for a total allocation of 46, 45, and 10, respectively. The surprising outcome is that increasing the size of the legislature from 100 to 101 seats leads to a smaller allocation for the small state, that decreases from 11 to 10 seats.

The discovery of anomalies for the various apportionment methods has led to a deeper study of the associated problems. For a comprehensive treatment of apportionment problems, readers are referred to Balinski and Young (1982).

3.1.8 Open Pit Mining

A company is mining an area which, if viewed from above, is rectangular. At the surface this area can be subdivided into blocks according to Table I.8 which is displayed on the next page.

In the northwest corner of each block is a triple *ijk*, where *i* is the row number, *j* is the column number and *k* the level, ($k = 1$ at the surface) of the block. Finally, each block, if mined, has a value (in thousands of dollars) indicated in the southeast corner of each block. For instance, block 241 has a value of \$5,300.

Table I.8

111 / 5.1	121 / 5.6	131 / 6.0	141 / 6.1	151 / 5.8	161 / 5.2
211 / 4.3	221 / 4.9	231 / 5.4	241 / 5.3	251 / 5.0	261 / 4.4
311 / 3.8	321 / 4.1	331 / 4.3	341 / 4.4	351 / 4.1	361 / 3.7
411 / 4.7	421 / 4.8	431 / 5.0	441 / 4.9	451 / 3.6	461 / 2.9

Beneath the level 1 blocks are the level 2 blocks arranged according to Table I.9.

Table I.9

112 / 8.9	122 / 9.7	132 / 11.6	142 / 10.3	152 / 9.7
212 / 9.1	222 / 9.9	232 / 12.1	242 / 10.7	252 / 8.9
312 / 7.2	322 / 8.5	332 / 9.2	342 / 8.8	352 / 8.1

Below level 2, there are levels 3 and 4 as shown in Tables I.10a and b.

Table I.10

(a)

113 / 12.7	123 / 14.0	133 / 13.2	143 / 10.1
213 / 11.1	223 / 12.5	233 / 11.7	243 / 9.7

(b)

114 / 11.5	124 / 12.1	134 / 10.8

The way blocks are situated is such that block ijk is located directly beneath the center of the four blocks $ij,\ k-1$; $i, j+1, k-1$; $i+1, j, k-1$; and $i+1, j+1,$

$k - 1$. Clearly, in order to mine block ijk, those four blocks have to be mined first. As an example, block 123 at level 3 cannot be mined until blocks 122, 132, 222 and 232 at level 2 have been mined. Suppose now that mining blocks at the surface costs \$6,000, at level 2 \$9,000, at level 3 \$10,000 and at level 4 \$11,000 per block.

In order to determine which blocks should be mined, we define zero-one variables $y_{ijk} = 1$ if block ijk is mined; and 0 otherwise. Furthermore, let p_{ijk} denote the profit from mining block ijk. For instance, $p_{261} = 4.4 - 6 = -1.6$, i.e., –\$1,600, and $p_{133} = 13.2 - 10 = 3.2$, i.e. \$3,200. The objective function is then

$$\text{Max } z = \sum_{i=1}^{4} \sum_{j=1}^{6} p_{ij1} y_{ij1} + \sum_{i=1}^{3} \sum_{j=1}^{5} p_{ij2} y_{ij2} + \sum_{i=1}^{2} \sum_{j=1}^{4} p_{ij3} y_{ij3} + \sum_{j=1}^{3} p_{ij4} y_{ij4}$$

The only constraints required in this model are constraints that ensure that a block is only mined if the four blocks above it are also mined. This can be expressed by four constraints for each block ijk, $k = 2, 3, 4$:

$$y_{ijk} \le y_{ij,\,k-1}$$

$$y_{ijk} \le y_{i,\,j+1,\,k-1}$$

$$y_{ijk} \le y_{i+1,\,j,\,k-1}$$

$$y_{ijk} \le y_{i+1,\,j+1,\,k-1}$$

Note that there are 104 such constraints; four for each of the 26 blocks on levels 2, 3, 4. Alternatively, the four constraints for each block could be aggregated to $4y_{ijk} \le y_{ij,\,k-1} + y_{i,\,j+1,\,k-1} + y_{i+1,\,j,\,k-1} + y_{i+1,\,j+1,\,k-1}$. Using this approach, we would only need a total of 26 constraints. For computational purposes, however, the model with 104 disaggregated constraints requires less computational effort than that with the 26 aggregated constraints. Other aspects of aggregation and disaggregation are covered in Chapter 4 of this part.

In summary, the model we have constructed to solve the open pit mining problem has 50 binary variables, as many terms in the objective function and either 104 disaggregated or 26 aggregated constraints. It turns out that the maximum total net profit is \$6,500 obtained by mining blocks 121, 131, 141, 151, 221, 231, 241, 251, 321, 331, 341, 351, 122, 132, 142, 222, 232, 242, 123, and 133. Level 4 will not be mined. For further details readers are referred to Williams (1974, 1978).

3.1.9 Bin Packing and Assembly Line Planning

Consider tasks T_1, T_2, ...,T_n that are to be accomplished by a set of workers. It is assumed that all workers have similar abilities and it takes a worker a_j time units to perform task T_j. Given that each worker has a capacity of b time units and that each job must be performed by exactly one worker, what is the smallest number of workers that can perform all tasks? A simple bound can be calculated for this problem by setting the total time required for all jobs which is $\sum_j a_j$ in relation to the capacity of each worker, b. In other words, defining $\underline{m}$ as a lower bound on the number of workers required to perform the jobs, we know that $\underline{m} = \left\lceil \frac{1}{b} \sum_j a_j \right\rceil$. Note that this is only a theoretical bound that may not be achievable. As a simple example, let three tasks with durations of 11, 10, and 14, respectively, be given, and assume that a worker's capacity is 20. Then the theoretical bound on the number of workers is $\underline{m} = \lceil 35/20 \rceil = 2$, whereas it is apparent that three workers are needed to perform the three tasks.

The assignment of tasks to workers has a physical equivalent that is commonly known as the *bin packing* problem. Consider a set of bins that can be used for packing. All bins have a length of b. Suppose now that we have a set of items such as paper rolls that are to be packed in these bins. All rolls have a diameter that fits into the bins, but their widths a_j , $j = 1,..., n$ differ. The rolls are packed into the bins, so that their ends touch. If it is desired to pack the rolls into the smallest number of bins possible, the same problem as that in the task assignment arises.

Formulating the problem, we first define variables y_{ij}, so that $y_{ij} = 1$, if the i-th worker is assigned to the j-th task and 0 otherwise. The problem here is that we do not know beforehand how many workers will be needed to perform all tasks. One possibility is to use the above estimate and add a few workers in case the minimum number is not sufficient. (Should the number of workers still be insufficient, then the solver will indicate that "no feasible solution" exists). In the following we assume that m workers are available to work on the tasks. The constraints can then be formulated as

$$\sum_{j=1}^{n} a_j y_{ij} \leq b \;\forall\; i = 1,..., m \tag{15}$$

$$\sum_{i=1}^{m} y_{ij} = 1 \;\forall\; j = 1,..., n \tag{16}$$

$$y_{ij} = 0 \vee 1 \ \forall\, i = 1,..., m; j = 1,..., n \tag{17}$$

The constraints in (15) guarantee that the capacities of the workers are respected, and the constraints in (16) ensure that each task is assigned to exactly one worker.

Consider now the objective function. If it is desired to minimize the number of workers, we have to define additional variables $y_i = 1$, if the i-th worker is employed with at least one task and 0 otherwise. The objective is then simply

$$\text{Min } z = \sum_{i=1}^{m} y_i \tag{18}$$

and an additional set of constraints is needed to link the new variables to the existing assignment variables. In particular, we write

$$\sum_{j=1}^{n} y_{ij} \leq My_i \ \ \forall\ i = 1,..., m \tag{19}$$

with $M >> 0$. As long as at least one task is assigned to a worker, the left hand side of a relation in constraint set (19) is positive, so that y_i must also be positive, i.e., equal to one, in order to ensure feasibility. Note that the problem has now a total of $mn + m$ variables and $2m + n$ constraints. It is well known that the bin packing problem is *NP-hard*; see Garey and Johnson (1979).

An alternative objective would attempt to balance the existing workload among m workers. Various approaches are available to accomplish that task. One would be to design a measure for the imbalance of the workloads, e.g., the variance or the width of the interval between smallest and highest workload. Another possibility is to minimize the highest workload among the workers. As the workload of the i-th worker is $\sum_j a_j y_{ij}$, this objective would be written as

$$\text{Min } z = \max_i \left\{ \sum_j a_j y_{ij} \right\} \tag{20}$$

The usual transformations, as shown in Eiselt *et al.* (1987), can then be applied to transform the problem into a standard (minisum) mathematical programming problem.

In practice, there are frequently additional constraints. For instance, it may not be possible to perform the tasks in just any order, but certain precedence relations may have to be respected. In particular, a sequence may be imposed upon workers, so that the first worker performs the tasks to be completed first, the next tasks are then performed by the second worker, and so on. It is then required that the k-th worker ($k < \ell$) cannot perform tasks that require another task to be completed that, in turn, is performed by the ℓ-th worker. Such logical constraints are discussed in Chapter 2.

3.2 Problems with Special Structures

3.2.1 Knapsack Problems

Consider a mutual fund company that can choose between a variety of n investment alternatives, e.g., stocks, bonds, real estate, and currencies. The expected return of the j-th investment is denoted by c_j , $j = 1, ..., n$. The risk per dollar of the j-th investment is specified as a_j, where it is assumed that risk is linear in the investment, and is additive. The total risk for all investments combined is not to exceed the prescribed amount b. Defining x_j as the amount of money invested in the j-th alternative, the investment problem can be written as

$$\text{P: Max } z = \sum_{j=1}^{n} c_j x_j \tag{21}$$

$$\text{s.t.} \quad \sum_{j=1}^{n} a_j x_j \leq b \tag{22}$$

$$x_j \geq 0 \ \forall \ j = 1, ..., n \tag{23}$$

This simple *continuous knapsack problem* is nothing but a standard linear programming problem with a single constraint. Dantzig (1957) has shown that the problem has a simple solution: it is optimal to set $\bar{x}_j = b/a_j$ for j, such that $j = \arg\max_{k=1,...,n} \{c_k/a_k\}$ and $\bar{x}_\ell = 0 \ \forall \ \ell \neq j$. In other words, we compute the investment with the highest return per risk unit and allocate the total amount that can be invested to that investment alternative. (The proof is simple: in the nondegenerate case, all constraints in the dual except for one are redundant, and complementary slackness leads to the desired result; see Theorem 12 in Chapter a). The continuous knapsack problem is also called the *cargo loading* problem, describing the situation of a vessel with limited volume or weight capacity which is to be loaded with the most profitable cargo.

Suppose now that rather than limiting the risk in the constraint, we define a_j as the amount of investment required by the j-th alternative and b as the amount of money available for investment. This does not change the basic structure of the problem, as we still attempt to maximize the return subject to a simple budget constraint. Consider, however, a situation in which an investment now requires exactly a_j dollars and it is not possible to purchase partial investments. This restricts our variables to $y_j = 1$ if we do invest in the j-th alternative and 0 if we do not. Formally, this means that constraints (23) are replaced by

$$y_j = 0 \vee 1 \ \forall j=1, \ldots, n \qquad (24)$$

The problem (21), (22), and (24) is commonly referred to as a *zero-one knapsack problem*. While little seems to have changed from the continuous knapsack problem, this is not the case: the zero-one knapsack problem has been proved to be *NP-hard*. It is also easily seen that complete enumeration is not a realistic procedure, as there are 2^n combinations of zeroes and ones, requiring, for example, that more than a trillion solutions be examined in case of only forty variables. It should also be noted that this version of the knapsack problem has given it its name: a hiker may pack any of n given items into his knapsack, where the j-th item weighs a_j and has a utility of c_j. With a carrying capacity of b and the objective of maximizing his utility, the hiker's problem of which items to pack is precisely the zero-one knapsack problem formulated above. Sometimes this problem is also called the *fly-away kit problem*, where an airplane pilot needs to decide which emergency items to take on an airplane. The zero-one knapsack problem is the simplest form of an integer programming problem, yet it includes all basic ingredients of general integer programming problems; see, e.g., Crowder *et al.* (1983) for insightful comments on the likelihood of general integer programming problems being approximately solved by the systematic consideration of their constituent knapsack problems via logical processing procedures.

Some regularity conditions are customarily imposed. For instance, it is often assumed that $a_j, c_j > 0 \ \forall j = 1,\ldots, n$. It is apparent that if $a_j \geq 0$ and $c_j \leq 0$ for some j it would be optimal to set $\bar{x}_j = 0$ as the j-th item apparently consumes a resource but does not contribute to the objective function in a positive way. Similarly, if $a_j \leq 0$ and $c_j \geq 0$, it would be optimal to set $\bar{x}_j = 1$ as the j-th item does not consume the resource but contributes positively to the objective. The remaining case is $a_j < 0$ and $c_j < 0$. Here we can replace the variable x_j by its logical complement $x'_j = 1 - x_j$, so that the new variable x'_j then has positive coefficients as required; see also Section 4.2.1. Furthermore, knapsack problems can safely be assumed to satisfy $\sum_{j=1}^{n} a_j > b$, or otherwise the constraint is not restrictive and the optimal solution is $\bar{x}_j = 1 \ \forall j = 1,\ldots, n$. Similarily, we may assume that $\min_{j=1,\ldots,n} \{a_j\} \leq b$; otherwise the only feasible solution is the trivial one with $x_j = 0 \ \forall \ j = 1, ..., n$.

There exist highly efficient special codes for the zero-one knapsack problem. Most of them order the coefficients in decreasing order of their ratios c_j / a_j, and then solve the relaxed linear programming problem by setting $x_j = 1 \ \forall \ j \leq \mu$ such that $\sum_{j=1}^{\mu} a_j \leq b$ but $\sum_{j=1}^{\mu+1} a_j > b$. In addition, $x_{\mu+1} = \frac{1}{a_{\mu+1}}\left[b - \sum_{j=1}^{\mu} a_j\right]$, and $x_j = 0 \ \forall \ j = \mu + 2, ..., n$. Rounding $x_{\mu+1}$ to zero leads to an integer feasible solution of the knapsack problem. In addition to this property, one can also take advantage of efficient reduction schemes for the zero-one problem to which we will return in Chapter 4 of this part.

The special importance of knapsack problems derives from the fact that there are many problems that can be decomposed into subproblems, some of which are knapsack problems. Three examples of complex applications for which knapsacks feature importantly in suitable decomposition schemes are multicommodity network design problems (Section 4.3 of Part III), the generalized assignment problem, and certain truck scheduling models.

Returning to the investment example, another version of the knapsack problem allows multiple purchases of the same investment. In other words, while partial investments in an alternative are still prohibited, it is now possible to invest in an alternative any integer number of times. This is the so-called *integer knapsack problem* for which constraints (24) are replaced by

$$x_j \in \mathbb{N}_0 \ \ \forall \ j = 1, ..., n \qquad (25)$$

The integer knapsack problem has a number of important applications in addition to the above investment problem, which is also referred to as the *capital budgeting* problem. These include subproblems of the *cutting stock* problem discussed in Section 3.1.2, stocking a warehouse to maximum value with items that come in indivisible units, and designing multiple choice examinations as shown by Feuerman and Weiss (1973). Furthermore, surrogate relaxation of an integer programming problem gives rise to an integer knapsack problem, as described in Section 1.2; the same holds for the aggregation of constraints, which will be covered in Section 4.3 of the next chapter.

In some cases the knapsack constraint is an equation rather than an inequality, corresponding to models where no slack or waste is allowed. This will reduce the number of feasible solutions substantially. If a dynamic programming approach is used to solve the problem, substantial computational savings can be realized. For details, see Salkin (1975), and for computational aspects in general, Martello and Toth (1990).

An important extension of the zero-one knapsack problem is when so-called *multiple choice constraints* are added. In the investment problem this occurs if the investments can be classified according to different types, e.g., stocks, bonds, currencies, precious metals, or real estate with the requirement that there are bounds on the number of investments that can be made from each class. Formally, let the different investment possibilities consist of the set $N = \{1, ..., n\}$ which can be divided into the classes $N_i \subseteq N$, $i = 1, ..., p$, such that $N_i \cap N_k = \emptyset \ \forall \ i \neq k$ and $\bigcup_{i=1}^{p} N_i = N$. With upper and lower bounds U_i and L_i on the number of investment possibilities that can be selected from each class, the *multiple choice knapsack problem* becomes

$$\text{P:} \qquad \text{Max } z = \sum_{j=1}^{n} c_j y_j$$

$$\sum_{j=1}^{n} a_j y_j = b$$

$$L_i \leq \sum_{j \in N_i} y_j \leq U_i, \ \forall \ i = 1,..., p$$

$$y_i = 0 \vee 1 \ \forall j \in N = \{1,..., n\}$$

A special case of the multiple choice knapsack problem arises if $L_i = U_i = 1$, $\forall \ i = 1, ..., p$, so that exactly one investment from each class must be selected. These are *SOS*1 constraints as specified in Definition I.17, where the classes N_i are *SOS*1 sets since $\sum_{j \in N_i} y_j = 1 \ \forall \ i = 1, ..., p$.

In solving knapsack problems with *SOS* constraints, it is beneficial from a computational point of view to use some simple dominance rules in a preprocessing phase.

Dominance Rule 1*:* If in some class N_i there exist k and ℓ such that $c_k \leq c_\ell$ and $a_k \geq a_\ell$, then there is an optimal solution with $y_k = 0$. The reason for this rule is easy. Consider the parameters c_j as unit profits and the parameters a_j as the use of the given resource if one unit of the j-th item is put in the knapsack. Then $c_k \leq c_\ell$ and $a_k \geq a_\ell$ imply that the k-th item provides less benefit per unit at a higher resource consumption than the ℓ-th item, and thus y_k is dominated by y_ℓ.

Dominance Rule 2: If in some class N_i there exists j, k, ℓ such that $a_j < a_k < a_\ell$ and $\frac{c_k - c_j}{a_k - a_j} \le \frac{c_\ell - c_k}{a_\ell - a_k}$, then there is an optimal solution with $y_k = 0$, so that the variable y_k is again dominated. For a proof of this rule, see Sinha and Zoltners (1979).

Example: In order to illustrate the above two dominance rules, consider the following formulation that is part of some larger problem.

$$\begin{aligned} \text{P: Max: } z = \; & 3y_1 + 5y_2 + 7y_3 + y_4 + 5y_5 + \ldots \\ \text{s.t.} \quad & 18y_1 + 20y_2 + 29y_3 + 6y_4 + 25y_5 + \ldots \le 60 \\ & y_1 + y_2 + y_3 + y_4 + y_5 = 1 \\ & y_j = 0 \vee 1 \;\forall j \end{aligned}$$

Since $c_2 = c_5 = 5$ and $a_5 = 25 > 20 = a_2$, y_5 is dominated by y_2 according to Dominance Rule 1; therefore, at least one optimal solution with $y_5 = 0$ will exist. Furthermore, choosing $j = 4$, $k = 1$, and $\ell = 2$, we obtain $a_4 = 6 < 18 = a_1 < 20 = a_2$ and $\frac{3-1}{18-6} = \frac{1}{6} < 1 = \frac{5-3}{20-18}$, so that according to Dominance Rule 2, at least one optimal solution with $y_1 = 0$ exists.

A recent survey of knapsack problems is provided by Lin (1998). A plethora of knapsack problems and algorithms is found in Martello and Toth (1990).

3.2.2 Set Covering, Set Packing, and Set Partitioning Problems

This subsection describes three closely related concepts. Formally, we can define

Definition I.19: Let M be a set and $\{M_j \subseteq M, j \in J\}$ a collection of subsets of M. This collection is a *set covering* of M if $\bigcup_{j \in J} M_j = M$, a *set packing* of M if $M_j \cap M_k = \varnothing \;\forall j, k \in J$ and $j \ne k$, and a *set partitioning* of M if it is both a set covering and set packing of M.

For example, if $M = \{1, 2, 3, 4, 5\}$, then $\{1, 2, 3\}$, $\{2\}$, $\{3, 4, 5\}$ is a set covering but not a set packing of M; {1, 4}, {2, 5} is a set packing but not a set covering of M; and {1, 2, 5}, {3, 4} is a set partitioning of M. If we define the logical variables y_j by $y_j = 1$ if $j \in J$, and 0 otherwise, we then refer to the expression $\sum_{j \in J} y_j \geq 1$ as a *set covering constraint*, $\sum_{j \in J} y_j \leq 1$ as a *set packing constraint*, and to $\sum_{j \in J} y_j = 1$ as a *set partitioning constraint*.

Example: Five daily newpapers are published in a certain country, each paper covering some of the nine regions of the country as shown in Table I.11.

Table I.11

Newspaper #	Region Covered	Cost of Advertisement	Benefit of Advertisement
1	1, 2, 4	3	12
2	2, 3, 6	4	10
3	4, 5, 6	3	14
4	5, 7, 8	7	19
5	6, 8, 9	5	16

Formally, $M = \{1, 2, ..., 9\}$ is the set of all regions, and M_j, $j = 1, ..., 5$, is the set of regions covered by the j-th newspaper. Let c_j denote the cost of an advertisement in the j-th newspaper and define zero-one variables y_j that assume a value of 1 if an advertisement is placed in newspaper j, and 0 otherwise. The *set covering problem* then solves the problem of finding a minimum total cost selection of newspapers such that the advertisement covers the whole country. In general, it can be formulated as

$$P_{SC}: \text{Min } z = \sum_j c_j y_j$$

$$\text{s.t.} \quad \sum_{j:\, i \in M_j} y_j \geq 1 \;\forall\, i \in M$$

$$y_j = 0 \vee 1 \;\forall\, j$$

Given the numerical information in Table I.11, the problem is

$$P: \text{Min } z = 3y_1 + 4y_2 + 3y_3 + 7y_4 + 5y_5$$

$$\text{s.t.} \quad y_1 \geq 1 \quad \text{(Region 1)}$$

$$y_1 + y_2 \geq 1 \quad \text{(Region 2)}$$

$$
\begin{array}{lllllll}
 & y_2 & & & & \geq 1 & \text{(Region 3)} \\
y_1 & + & y_3 & & & \geq 1 & \text{(Region 4)} \\
 & & y_3 & + y_4 & & \geq 1 & \text{(Region 5)} \\
 & y_2 & + y_3 & + & y_5 & \geq 1 & \text{(Region 6)} \\
 & & & y_4 & & \geq 1 & \text{(Region 7)} \\
 & & & y_4 & + y_5 & \geq 1 & \text{(Region 8)} \\
 & & & & y_5 & \geq 1 & \text{(Region 9)}
\end{array}
$$

$$y_j = 0 \vee 1, j = 1, \ldots, 5$$

The structure of the *set packing problem* is very similar to that of the set covering problem. Define d_j as the benefit from placing an advertisement in newspaper j, and require that none of the advertisements should be available in more than one region, then the set packing problem can be formulated as

$$\text{P}_{\text{SP}}\text{: Max } z = \sum_j d_j y_j$$

$$\text{s.t.} \quad \sum_{j:\, i \in M_j} y_j \leq 1 \;\forall\, i \in M$$

$$y_j = 0 \vee 1 \;\forall\, j$$

In our numerical example, the objective function is to Max $z = 12y_1 + 10y_2 + 14y_3 + 19y_4 + 16y_5$ and the constraints are the same as those in the set covering problem except that the inequality signs are reversed.

Finally, consider the *set partitioning problem*. Its formulation is identical to that of the set packing problem, except that the inequality constraints are replaced by equations. In our newspaper example, the constraints require that the advertisement should be seen in each region of the country in exactly one newspaper. The general formulation is then

$$\text{P}_{\text{SPP}}\text{: Max } z = \sum_i d_j y_j$$

$$\text{s.t.} \quad \sum_{j:\, i \in M_j} y_j = 1 \;\forall\, i \in M$$

$$y_j = 0 \vee 1 \;\forall\, j.$$

In our numerical example, it is apparent that the first three set partitioning constraints are inconsistent, as are the last three.

A much celebrated application of set covering and set partitioning comes from *airline crew scheduling*. As an illustration, consider an airline that operates a specialized scheduled flight service between Frankfurt, London and Paris. This service could, for example, transport dangerous chemicals, sensitive medical supplies, or other items that require special attention. The daily scheduled flights are represented by the arrows in Figure I.14.

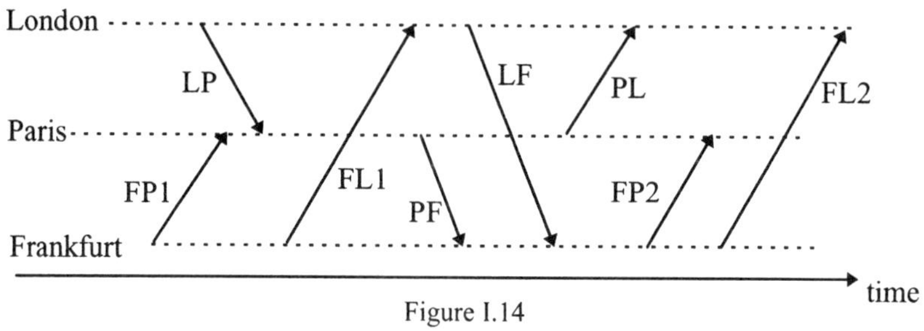

Figure I.14

The tail of an arrow indicates the departure city and time of a flight; the head of an arrow the arrival city and time. The flights are named to indicate departure and arrival city; for instance, LF indicates the flight from London to Frankfurt. Since there are two daily flights from Frankfurt to Paris, they are denoted by FP1 and FP2, respectively.

The airline has flight crews that need to be assigned to operate these flights, so that they return to the same city where they were at the beginning of the day. For instance, a crew could start in Paris, operate flight PF from Paris to Frankfurt, and then operate flight FP2 back to Paris. This process is called *crew pairing*. In general, a pairing is a sequence of flights beginning and ending in the same city. The cost associated with each pairing depends, among other things, on the total time elapsed between the first departure time and the last arrival time of the pairing. It follows from Figure I.14 that there are six possible pairings $j, j = 1, ..., 6$. They are displayed with their associated cost c_j in Table I.12.

Table I.12

Pairing	Flight Sequence	Cost ($)
1	FP1 – PF	6,000
2	LP – PL	8,000
3	LP – PF – FL2	12,500
4	FL1 – FL	6,800
5	PF – FP2	5,200
6	LF – FL2	7,500

The problem is now to select a set of pairings such that all flights are operated and the total cost is minimized. First, define M as the set of all individual flights. For simplicity, we number the flights FP1, LP, FL1, PF, LF, PL, FP2, and FL2 as 1, 2, ..., 8. We then define sets M_j, $j = 1, ..., 6$, as the set of all individual flights in the j-th flight sequence. In our example, $M_1 = \{1, 4\}$, $M_2 = \{2, 6\}$, $M_3 = \{2, 4, 8\}$, $M_4 = \{3, 5\}$, $M_5 = \{4, 7\}$, and $M_6 = \{5, 8\}$. Furthermore, define c_j as the cost if the j-th flight sequence.

We may first try to model this situation as a set partitioning problem. A solution to this problem (if one exists) chooses a set of flight sequences, so that each flight is operated exactly once. Our problem can then be formulated as follows.

$$
\begin{array}{llll}
\text{P: Min } z = 6y_1 + 8y_2 + 12.5y_3 + 6.8y_4 + 5.2y_5 + 7.5y_6 & & & \\
\text{s.t.} & y_1 & = 1 & \text{(Flight FP1)} \\
& y_2 + y_3 & = 1 & \text{(Flight LP)} \\
& y_4 & = 1 & \text{(Flight FL1)} \\
& y_1 + y_3 + y_5 & = 1 & \text{(Flight PF)} \\
& y_4 + y_6 & = 1 & \text{(Flight LF)} \\
& y_2 & = 1 & \text{(Flight PL)} \\
& y_5 & = 1 & \text{(Flight FP2)} \\
& y_3 + y_6 & = 1 & \text{(Flight FL2)} \\
& & y_j = 0 \vee 1, j = 1,..., 6 &
\end{array}
$$

Note that in this formulation $\sum_{j=1}^{6} y_j$ indicates the number of crews needed to operate all the flights and the i-th left-hand side $\sum_{j:\, i \in M_j} y_j$ specifies the number of crews on the i-th flight. It is easy to see that the above model has no feasible solution. Since $y_2 = 1$ by the PL constraint, $y_3 = 0$ by the LP constraint, so that $y_6 = 1$ by the FL2 constraint. However, since $y_4 = 1$ by the FL1 constraint, $y_6 = 0$ by the LF constraint, which is a contradiction.

We may therefore relax the equations in the above problem P and replace them by "≥" constraints, thus requiring only that each flight is operated in at least one flight sequence. Optimizing the resulting set covering model results in the solution $\bar{y}_1 = \bar{y}_2 = \bar{y}_4 = \bar{y}_5 = \bar{y}_6 = 1$, $\bar{y}_3 = 0$ and $\bar{z} = \$33{,}500$. Analysis of the constraints reveals that flights PF and LF are operated by two crews each. This means that one crew needs to travel as passengers on each of these flights, while the other crew operates the flight. This does, of course, add to the cost since a crew is being paid the same for flying as passengers as for operating the flight. In the above solution $\sum_{j=1}^{6} y_j = 5$, i.e., five crews are needed for the minimal cost solution. It turns out that the problem has no feasible solution with four or fewer crews, regardless of the cost.

The above illustrative model is, of course, only a toy example. In practice, most schedules last three days. Within that time frame, each crew has maybe a dozen takeoffs and landings, and the total number of departures for the airline is in the hundreds. As a result, the number of possible pairings could then be astronomically large. So-called *column generating* techniques are commonly used whereby only a subset of all pairings are considered at any point in time with potentially desirable pairings (columns) being generated as and when needed. Successful application of such large-scale techniques enables large airlines to save many millions of dollars per year by finding efficient crew schedules. It should also be pointed out that a number of additional restrictions apply, due to a complex array of aviation authority and labor union rules. For further accounts of this problem, the reader is referred to Gershkoff (1989) and Anbil *et al.* (1991).

CHAPTER 4 REFORMULATION OF PROBLEMS

4.1 Strong and Weak Formulations

In this section we reconsider the task of modeling dealt with in previous chapters. As in linear programming, the basic task is that of defining appropriate variables and expressing the limitations or constraints on these variables in terms of linear relationships. However, straightforward modeling is not always adequate in integer and mixed integer programming. The practical user will soon find that much care needs to be expended in modeling, so that the final solution effort is acceptable. First we consider some common sense measures. Then we outline more complex steps, some of which are conceptual and may be useful at the very outset of creating a model. Finally we describe computational procedures that may be employed to transform a correctly, but poorly formulated model into a well formulated model. In other words, we will attempt to formulate models in such a way that they require less computational effort when solved with the standard procedures described in Chapters 5 and 6 of this part.

There are certain common sense rules that should always be heeded. A typical rule is to use integer variables only when really needed. In some of the examples of the previous chapters, the designation of the variables as integer rather than continuous is not wrong, but it would lead to substantially increased solution times. As in the generalized assignment problem of Section 1.1, the difficulty of a problem is linked to the number of integer variables it includes. As their number increases, the difficulty of the model increases exponentially.

For large problems it may be necessary to go a step further. Sometimes, modelers will relax the integrality conditions for those integer variables that are considered least important to be integer and then solve the resulting problem. Clearly the solution to such a problem is only an approximation of the original problem. Still, such an approximation may be acceptable, particularly if the options are to either solve a relaxed problem or to have a properly formulated model that cannot be solved.

While the difficulty of a linear programming problem is often measured by the number of constraints it includes, this is much less the case in mixed integer programming. As a matter of fact, it is sometimes beneficial to *add* constraints to the problem in order to improve the tractability of the model. Similarly, it may also be useful to increase the number of integer variables, either in an approach called *disaggregation* (for details see Section 4.4), or to use Lagrangean techniques via Lagrangean decomposition (see Section 6.6).

Another common sense rule has to do with the order of variables. The modeler should always bear in mind that some zero-one variables may represent decisions that need to be made first, either in time, or by some other practical priority. If so, they should be positioned in a physical order that places the most important variables first in the data check or in some other manner as described by the solver. Any systematic solution scheme is likely to be more efficient if it deals with the important variables first; there is much evidence that this simple procedure can produce improvements in tractability by orders of magnitude.

In mixed integer programming problems it is useful with a notation that distinguishes the continuous variables $\mathbf{x}$ from the integer variables $\mathbf{y}$. In general, we formulate a mixed integer programming problem as

$$
\begin{aligned}
\text{P:} \quad & \text{Max } z = \mathbf{cx} + \mathbf{dy} \\
& \text{s.t.} \quad \mathbf{Ax} + \mathbf{By} \leq \mathbf{b} \\
& \qquad\;\; \mathbf{x} \geq \mathbf{0} \\
& \qquad\;\; \mathbf{y} \in \mathbb{N}_0
\end{aligned}
$$

where $\mathbf{c}$, $\mathbf{d}$, $\mathbf{A}$, $\mathbf{B}$, and $\mathbf{b}$ are matrices and vectors of appropriate dimensions.

To facilitate our discussion of strong and weak formulations, we provide a formulation for the *Capacitated Plant Location Problem* (*CPLP*). The basis of the capacitated plant location problem is the well-known transportation problem, see e.g., Chapter a. Consider a set of m plants (or origins or sources) indexed by i, and n destinations (or sinks) indexed by j. A given commodity is then to be shipped from the sources, where supplies s_i are available, to the destinations so as to satisfy their demands d_j , $j = 1, \ldots, n$. There are nonnegative unit costs of shipment c_{ij} on the route (arc) from the i-th origin to the j-th destination. There may be variable costs e_i, $i = 1, \ldots, m$ at the plants as well, but they can always be included in the unit transportation costs by replacing c_{ij} by $c_{ij} + e_i \; \forall \; i, j$.

Assume now that there are fixed charges f_i specified for operating the sources i and that we also wish to decide which sources are to be used so that the overall cost of shipping and operating the sources is minimized. There are also shipping capacities κ_{ij} on the arcs a_{ij}. All the parameters, s_i, d_j, κ_{ij}, f_i and c_{ij} are usually assumed to be nonnegative and rational. The formulation is accomplished by introducing, in addition to the x_{ij} , binary decision variables y_i which equal 1 if a plant at site n_i is opened, and 0 otherwise. The resulting Capacitated Plant Location Problem can then be written as

$$\text{P}_{\text{CPLP}}: \quad \text{Min } z = \sum_{i=1}^{m}\sum_{j=1}^{n} c_{ij}\, x_{ij} + \sum_{i=1}^{m} f_i y_i$$

$$\text{s.t.} \quad \sum_{j=1}^{n} x_{ij} \le s_i\, y_i \ , i = 1, ..., m$$

$$\sum_{i=1}^{m} x_{ij} \ge d_j \ , j = 1, ..., n$$

$$0 \le x_{ij} \le \kappa_{ij} \quad \forall\, i = 1, ..., m; j = 1, ..., n$$

$$y_i = 0 \vee 1 \quad \forall\, i = 1, ..., m.$$

The first constraint set requires that for a given plant i, the sum of shipments from the plant must not exceed the supply s_i if the plant is open, and that not opening a plant at n_i (i.e., if $y_i = 0$) prohibits all shipments from this plant. The second constraint set ensures that demands are met. From the above formulation it is apparent that the problem P_{CPLP} is a mixed integer programming problem. The primary computational task is the determination of the combinatorial elements represented by y_i; for a fixed $\mathbf{y}$ the problem reduces to a capacitated transportation problem. While the formulation provided here is correct, it is weak in a sense to be clarified below, and cannot be recommended for computational work. Much of mixed integer programming deals with how one can change from weak to strong formulations, or design computational steps which amount to the same thing within solution procedures.

Example: Consider a capacitated plant location problem with two potential plants and two destinations. The fixed charges for opening plants 1 and 2 are equal to \$20 and \$10, respectively, while the transportation costs to destination 1 from plant 1 and plant 2 are \$10 and \$8, and for destination 2, the transportation costs are \$7 and \$11, respectively. Availabilities at the potential plants are 9 and 8 units, and demands at the destinations are 5 and 6 units respectively. Finally, there are no capacity limits on the arcs that connect the plants to destinations, i.e., $\kappa_{ij} = M >> 0$. The demands must be met exactly, so that no overshipments are permitted. The situation is shown in Figure I.15.

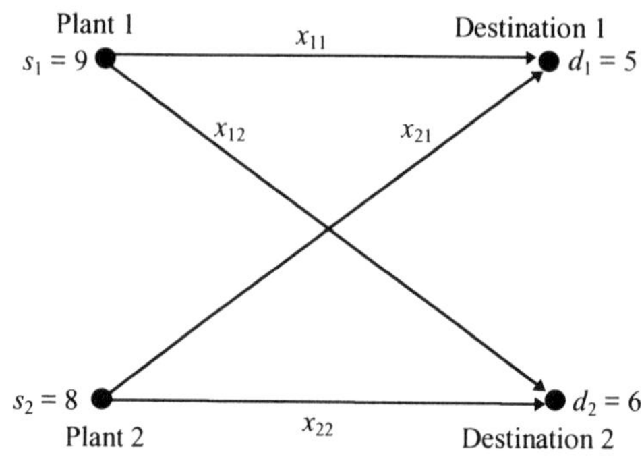

Figure I.15

A straightforward formulation of the model is

$$
\begin{array}{llr}
\text{P}_{\text{CPLP}}\text{: Min } z = 10x_{11} + 7x_{12} + 8x_{21} + 11x_{22} + 20y_1 + 10y_2 & & \\
x_{11} + x_{12} - 9y_1 & \leq 0 & \text{(Plant 1 availability)} \\
x_{21} + x_{22} - 8y_2 & \leq 0 & \text{(Plant 2 availability)} \\
x_{11} + x_{21} & = 5 & \text{(Destination 1 demand)} \\
x_{12} + x_{22} & = 6 & \text{(Destination 2 demand)} \\
x_{11},\ x_{12},\ x_{21},\ x_{22} & \geq 0 & \\
y_1,\ y_2 & = 0 \vee 1 &
\end{array}
$$

The linear programming relaxation of the model has the optimal solution $\bar{y}_1 = {}^2/_3$, $\bar{y}_2 = {}^5/_8$; $\bar{x}_{11} = 0$, $\bar{x}_{12} = 6$, $\bar{x}_{21} = 5$, and $\bar{x}_{22} = 0$ with value of the objective function $\bar{z}_{LP} = 101.58330$. While the above formulation that leads to this solution is undoubtedly correct, it is possible to add constraints that are derived from our knowledge of the model. In particular, consider shipments from the i-th origin to the j-th destination. It is apparent that an upper bound on the shipment x_{ij} is the smaller of s_i and d_j provided that the plant at the i-th origin is actually opened (and that reshipments are not permitted). This can be formulated as the constraint $x_{ij} \leq y_i \min\{s_i, d_j\}$. In our example, these constraints are

$$
\begin{aligned}
x_{11} &\leq 5y_1 \\
x_{12} &\leq 6y_1 \\
x_{21} &\leq 5y_2 \\
x_{22} &\leq 6y_2.
\end{aligned}
$$

With these constraints added, the solution of the linear programming relaxation changes to $\bar{y}_1 = 1$, $\bar{y}_2 = 1$; $\bar{x}_{11} = 0$, $\bar{x}_{12} = 6$, $\bar{x}_{21} = 5$, and $\bar{x}_{22} = 0$ with objective function value $\bar{z}_{LP} = 112$. Therefore, the revised model with the additional constraints contains information that is redundant for the integer problem, but not for its continuous relaxation. The additional constraints that force the integer variable y_i to be 1 if there is any amount of flow from the i-th plant to a destination may also force the continuous variable y_i of the corresponding linear programming relaxation to assume a larger value (probably 1, as in this example) and thus include a larger fraction of the fixed charge in the objective function in the linear programming relaxation. The first formulation above is *weak* in the sense that its integrality gap, as defined in Section 1.2 of this part, is large, whereas the second formulation with the additional constraints has a small (here actually vanishing) integrality gap and is thus referred to as *strong*. It is, of course, not generally the case that the strong formulation yields an integer solution. In fact, one can show that for this problem there is an alternate optimal solution to the linear programming relaxation of the problem which is not integral. In any case, the objective function value of the linear programming relaxation has gone up (i.e., become tighter) by 10%, and this value could be used as a more effective lower bound on the integer optimal value in a systematic mixed integer programming solution procedure. A procedure that adds constraints similar to the process above is called disaggregation; because of its wide applicability, we will describe this procedure later in this chapter. A consequence of the above discussion is the following

Definition I.20: Let LP1 and LP2 be linear programming relaxations of the integer programming problem IP with maximization objective. Denoting by $\bar{z}_{\text{IP}}$, $\bar{z}_{\text{LP1}}$ and $\bar{z}_{\text{LP2}}$ the optimal objective function values of IP, LP1 and LP2, respectively, LP1 is said to be a *stronger* formulation (or relaxation) of IP than LP2 if $\bar{z}_{\text{LP1}} < \bar{z}_{\text{LP2}}$. Conversely, LP2 is called a *weaker* formulation (or relaxation) of IP than LP1.

Since $\bar{z}_{\text{IP}} \le \bar{z}_{\text{LP1}}$ and $\bar{z}_{\text{IP}} \le \bar{z}_{\text{LP2}}$, i.e., both $\bar{z}_{\text{LP1}}$ and $\bar{z}_{\text{LP2}}$ are upper bounds on $\bar{z}_{\text{IP}}$, the stronger formulation LP1 yields a *lower* (i.e., *tighter* or *better*) bound on $\bar{z}_{\text{IP}}$ than the weaker formulation LP2. In other words, strong formulations have smaller integrality gaps than weak formulations. In the literature, the concepts of strong and weak formulations or relaxations are sometimes defined in relationship to the size of their respective feasible regions rather than our definition, which is based on the values of the objective function.

Most solution methods depend critically on the use of relaxed problems that are strong in the sense defined above. In this context, it is necessary to study the problem difficulty of a given application area carefully. Users of mixed integer programming need to find formulations of their models that have strong relaxations. As an example, crew scheduling problems in the airline industry (see, e.g., Section 3.2.3) are so large in size that not only do they require special solution techniques, but analysts will also attempt to aid the solution process by employing the strongest possible formulation of the problem. One way to do so is by way of a technique called logical processing, which is discussed in Section 4.2.

Fortunately, available mixed integer programming codes and modeling systems today make the task of model generation and testing easier than it used to be. They do so by permitting the user to conduct experiments that can often establish whether the problems at hand are easy or hard. The use of such systems, however, requires careful evaluation of the first stages of the solution process. Usually, users run a difficult problem only for a few minutes, and then analyze the behavior of the solution algorithm. There are many instances of integer programming users having used 20 CPU hours without any tangible result. No problem should ever be submitted to a computer, especially for a new model, without the setting of a cutoff parameter such as 5 minutes of CPU time, or 1000 iterations, and saving intermediate results for potential restarts.

The tightening of underlying relaxations is often achieved by adding constraints that involve the integer variables. We have seen in this chapter how the capacitated plant location problem, quite weak in its naive aggregated formulation, can be rendered much better behaved by the addition of constraints. This illustrates the fact that even well thought out models may not convey enough information about the integer variables. There frequently exist additional constraints that, though fully implied by and redundant in the integer programming model, must be spelled out for the linear programming relaxation to make it a better approximation to the integer programming problem.Their addition by the user will tend to help the solution process. Almost all practical problems permit these types of logical add-on constraints. It is important that the application specialist add such constraints to the initially conceived and logically correct, but not practically useful, formulation, in order to aid the solution process. The remainder of this chapter will provide examples of how these ideas can be provided as automatic devices within industrial solution systems. For a reference to the literature in this area, the reader is referred to Wolsey (1989).

4.2 Model Strengthening and Logical Processing

One way of *strengthening* a formulation is to add simple *logical constraints*. These are constraints with zero-one variables that can be derived from given constraints of the problem without requiring further insight into the problem. Some logical constraints are the so-called *cover inequalities* which state that a solution is feasible only if at most a certain number out of a subset of the zero-one variables has a value of one. This type of model strengthening is referred to as *logical processing*, which—since it can be done without any knowledge of the background to the model—is amenable to automatic procedures that could be built into model solvers. By *preprocessing* we will understand logical processing executed after the end of the modeling and before the beginning of the solution procedure, whereas *node processing* refers to logical processing executed at the nodes of a branch and bound procedure as described in Chapter 6.

4.2.1 Single Constraint Procedures

In this section we will demonstrate some simple and efficient automatic procedures for generating and using logical constraints. Formally, let the i-th inequality of a system in zero-one variables be $\sum_{j=1}^{n} a_{ij}\, y_j \le b_i$, where $y_j = 0$ or $1\ \forall\ j$. It is computationally convenient and efficient to rewrite the constraint, so that all coefficients are nonnegative, and the variables are arranged in nonincreasing order of magnitude of their coefficients. The former task is achieved by replacing every variable y_j with a negative coefficient a_{ij} by its logical complement $\bar{y}_j = (1 - y_j)$.

Example: Consider the following linear inequality

$$11y_1 + 0y_2 + 1y_3 - 7y_4 + 0y_5 - 1y_6 + 2y_7 + 1y_8 - 5y_9 + 0y_{10} + 9y_{11} + 0y_{12} \le -4$$

We first apply the transformation $\bar{y}_j = 1 - y_j$ for $j = 4, 6, 9$, so that now all coefficients are nonnegative, and obtain

$$11y_1 + 0y_2 + 1y_3 + 7\bar{y}_4 + 0y_5 + 1\bar{y}_6 + 2y_7 + 1y_8 + 5\bar{y}_9 + 0y_{10} + 9y_{11} + 0y_{12} \le 9.$$

Reordering the variables in order of nonincreasing magnitude of their coefficients, we obtain the constraint

$$11x_1 + 9x_2 + 7x_3 + 5x_4 + 2x_5 + 1x_6 + 1x_7 + 1x_8 + 0x_9 + 0x_{10} + 0x_{11} + 0x_{12} \le 9,$$

where $x_1 = y_1$, $x_2 = y_{11}$, $x_3 = \bar{y}_4 = (1 - y_4)$, and so on. This equivalent presentation highlights the restrictiveness of the constraint in terms of the zero-one nature of the problem. Here, it is immediately obvious that x_1 can not possibly equal 1, and can therefore be fixed at 0. We call this an *implication of degree 1* following from the logical inequality $x_1 \le 0$. In general, a constraint $\sum_{j=1}^{n} a_j y_j \le b$ involving the zero-one variables y_j, $j = 1,\dots, n$ with $a_j \ge a_k\ \ \forall\ j < k$ will produce the degree 1 implications $y_j \le 0$ (i.e., $y_j = 0$) for all j with $a_j > b \ge 0$. In case $b < 0$, the constraint (and with it the entire model) has no feasible solution.

Definition I.21: A set $J \subseteq \{1, 2, \dots, n\}$ is called a *cover* of the constraint $\sum_{j=1}^{n} a_j y_j \le b$, if $\sum_{j\in J} a_j > b$. The inequality $\sum_{j\in J} y_j \le |J| - 1$ is then called a *cover inequality* or *degree* $|J|$ *implication* corresponding to the cover J.

It is easy to demonstrate that any cover inequality is a valid inequality. If this were not true, then $\sum_{j\in J} y_j \geq |J|$ which implies that $y_j = 1 \;\forall\, j \in J$, so that $\sum_{j=1}^{n} a_j y_j \geq \sum_{j\in J} a_j y_j = \sum_{j\in J} a_j > b$, so that the given constraint is violated. In the above example, as $a_1 = 11 > 9 = b$, the degree 1 implication $y_1 \leq 0$ is a cover inequality corresponding to the cover $J = \{1\} \subseteq \{1, 2, ..., 12\}$ of the transformed constraint.

It is also possible to generate other logical constraints in a similar fashion. For instance, implications of *degree 2* involve two logical variables in a dichotomy. They are derived from the constraint $\sum_{j=1}^{n} a_j y_j \leq b$ as $y_i + y_j \leq 1$ for all $a_i + a_j > b$. In the above example, the following degree 2 implications can be generated: $x_2 + x_3 \leq 1$, $x_2 + x_4 \leq 1$, $x_2 + x_5 \leq 1$, $x_2 + x_6 \leq 1$, $x_2 + x_7 \leq 1$, $x_2 + x_8 \leq 1$, and $x_3 + x_4 \leq 1$. In terms of the original variables these conditions are $y_{11} - y_4 \leq 0$, $y_{11} - y_9 \leq 0$, $y_{11} + y_7 \leq 1$, $y_{11} + y_3 \leq 1$, $y_{11} - y_6 \leq 0$, $y_{11} + y_8 \leq 1$, and $-y_4 - y_9 \leq -1$, which contain useful logical information about the problem. Similarly, degree 3 implications are $y_i + y_j + y_k \leq 2$ for all $a_i + a_j + a_k > b$. Our numerical example does not include nondominated degree 3 implications, i.e., those that do not follow from implications of a lower degree. In the above example, the degree 3 implication $y_2 + y_3 + y_4 \leq 2$ follows from the degree 2 implication $x_2 + x_3 \leq 1$ and is therefore dominated. The generalization to implications of a degree higher than 3 are straightforward. Our numerical example produces only one more logical constraint of this kind: the degree 5 implication $x_4 + x_5 + x_6 + x_7 + x_8 \leq 4$. Computer programs such as MPSX-MIP/370 and OSL-MIP use logical constraints within a preprocessing module, i.e., a logical processing procedure at the beginning of the solution process.

Before deriving implications from a single constraint, we may wish to modify it. There are various ways of changing the coefficients of a constraint to make it stronger, i.e., more restrictive. Here, we provide some well-known simple results, which are so efficient that they can (and should) be implemented automatically in every solution method. Consider the constraint $\sum_{j=1}^{n} a_j y_j \leq b$ where b and all a_j are nonnegative, and $y_j = 0 \vee 1 \;\forall\, j$. Clearly, if $\sum_{j=1}^{n} a_j \leq b$, then every binary vector y_j, $j = 1, ..., n$ satisfies the constraint, which is therefore redundant and can be deleted; if $\min_{1\leq j\leq n} \{a_j\} > b$, then the constraint is satisfied only by $y_j = 0 \;\forall\, j$. Consider now the case in which $\min_{1 \le j \le n} \{a_j\} \le b < \sum_{j=1}^{n} a_j$. Assume that $\Delta_k = b - \sum_{j \ne k} a_j > 0$ for some k, and consider the modified constraint $\sum_j a_j y_j \le b + (y_k - 1)\Delta_k$. Clearly, if $y_k = 1$, then this modified constraint equals the original constraint. However, if $y_k = 0$, the right-hand side value of the modified constraint is smaller than that of the original constraint making the modified constraint stronger. The validity of the modified constraint for $y_k = 0$ follows since $\sum_j a_j y_j \le b + (y_k - 1)\Delta_k = b - \Delta_k = b - b + \sum_{j \ne k} a_j = \sum_{j \ne k} a_j$, or simply $\sum_j a_j y_j \le \sum_{j \ne k} a_j$. As $y_k = 0$, $\sum_j a_j y_j = \sum_{j \ne k} a_j y_j$ which yields $\sum_{j \ne k} a_j y_j \le \sum_{j \ne k} a_j$, which is trivially satisfied by all binary solutions y_j, $j = 1,..., n$.

Example: Consider the constraint $3y_1 + 8y_2 + 11y_3 \le 21$. For $k = 3$, we find that $\Delta_3 = b - \sum_{j \ne 3} a_j = 21 - 11 = 10 > 0$, so that the modified constraint becomes $3y_1 + 8y_2 + 11y_3 \le 21 + (y_3 - 1)10$ or simply $3y_1 + 8y_2 + y_3 \le 11$. It is apparent that each binary solution except $y_1 = y_2 = y_3 = 1$ satisfies the modified constraint and the same is true for the original constraint. The original and modified constraints are therefore equivalent from a binary point of view. However, over the set of continuous variables bounded by 0 and 1, we see that any solution with $y_1 = y_2 = 1$ satisfies the original constraint as long as $y_3 \in \left[0; \frac{10}{11}\right]$, whereas in the modified constraint $y_1 = y_2 = 1$ implies that $y_3 = 0$, making the modified constraint more restrictive.

Proceeding with the constraint $3y_1 + 8y_2 + y_3 \le 11$, we can repeat the reduction procedure for $k = 2$. We obtain $\Delta_2 = 11 - (3 + 1) = 7 > 0$, so that another reduction yields the constraint $3y_1 + y_2 + y_3 \le 4$ and finally, since $\Delta_1 = 4 - (1 + 1) = 2 > 0$, we obtain $y_1 + y_2 + y_3 \le 2$ and no further reduction is possible. Since all binary vectors except $\mathbf{y} = [1, 1, 1]$ satisfy the last reduced constraint, it is equivalent to the original one in a binary sense. However, for the continuous relaxation, $[1, 1, y_3]$ satisfies the original constraint for $0 \le y_3 \le \frac{10}{11}$ but the last reduced constraint only for $y_3 = 0$; analogous results hold for $[1, y_2, 1]$ and for $[y_1, 1, 1]$. We conclude that in a linear programming sense, the constraint has been considerably strengthened.

Applying this technique, the constraint $8y_1 + 2y_2 + 14y_3 + 3y_4 + 11y_5 \le 29$ can be reduced to $8y_1 + 2y_2 + 9y_3 + 3y_4 + 9y_5 \le 22$ but no further; the constraint $y_1 + y_2 + 2y_3 \le 2$ cannot be reduced at all. We conclude that the above reduction procedure, which may be considered myopic as it only involves a single constraint, could strengthen the continuous relaxation of an integer programming problem and therefore hopefully reduce the integrality gap.

In general, the above reduction technique is useful if $\sum_{j=1}^{n} a_j$ is not much larger than the right-hand side b since $\sum_{j \neq k} a_j < b$ must hold for at least some k. If this condition does not hold, one might still be able to strengthen the constraint by a procedure called lifting. Unlike the above procedure, where reductions of the right-hand side were matched by reductions in the various coefficients on the left-hand side, in constraint lifting the left-side coefficients are sequentially increased or lifted with no resulting change in the right-hand side. The idea of constraint lifting goes back to Gomory (1969); for later references see Crowder *et al.* (1983), Balas and Zemel (1984), and Nemhauser and Wolsey (1988).

Formally, we can state

Definition I.22: Given a constraint $\sum_{j=1}^{n} a_j y_j \leq b$ with b, $a_j \geq 0\ \forall\, j$ and $y_j = 0 \vee 1\ \forall\, j$. If then for some $\Delta_k > 0$ the inequality

$$\sum_{\substack{j=1 \\ j \neq k}}^{n} a_j y_j + (a_k + \Delta_k) y_k \leq b \tag{1}$$

is valid, then the constraint (1) is said to have been *lifted* from the original constraint with respect to y_k.

The term "lifting" derives from the fact that the hyperplane of the lifted constraint (1) is above that of the original constraint. For a formal description of lifting, consider the constraint

$$\sum_{j=1}^{n} a_j y_j \leq b \tag{2}$$

where, as before, b, $a_j \geq 0\ \forall\, j$ and $y_j = 0 \vee 1\ \forall\, j$. Arbitrarily selecting some a_k, we will investigate for what values of $\Delta_k > 0$ the constraint (1) is a valid inequality. Clearly, any binary vector with $y_k = 0$ that satisfies (2) will also satisfy (1). Let now $y_k = 1$ and denote by α_k the maximum value of $\sum_{j=1}^{n} a_j y_j \leq b$, given that $y_k = 1$. For any $\Delta_k \leq b - \alpha_k$, we then have $a_k + \Delta_k + \sum_{j \neq k} a_j y_j \leq a_k + b - \alpha_k + \sum_{j \neq k} a_j y_j \leq b$.

We conclude that (1) is a valid inequality as long as $\Delta_k \le b - \alpha_k$ holds. In replacing a_k by $a'_k = a_k + \Delta_k$ we say that we have lifted the constraint (2) with respect to the variable y_k. For $\Delta_k = b - \alpha_k$ we say that the lifting is maximum. This lifting procedure is sequentially applied to all variables y_j in constraint (2).

Example: Consider the constraint $4y_1 + 5y_2 + 6y_3 + 8y_4 \le 13$ where $y_j = 0 \vee 1, j = 1, 2, 3, 4$. First we try lifting with respect to y_1. Setting y_1 equal to 1 we obtain $5y_2 + 6y_3 + 8y_4 \le 9$. The largest value of the left hand side not exceeding 9 equals 8, obtained for $y_2 = y_3 = 0, y_4 = 1$. Therefore $\alpha_1 = 4 + 8 = 12$, implying that $\Delta_1 = 13 - 12 = 1$ so that $a_1 = 4$ can be lifted to $a'_1 = 4 + 1 = 5$.

Next, we try to lift the constraint $5y_1 + 5y_2 + 6y_3 + 8y_4 \le 13$ with respect to y_2. The largest possible value of the left-hand side of the constraint without the term involving y_2, *viz.*, $5y_1 + 6y_3 + 8y_4$ is 8 which is achieved for $y_1 = y_3 = 0$, and $y_4 = 1$. This implies $\Delta_2 = 13 - 13 = 0$ so that no lifting is possible with respect to y_2. For lifting with respect to y_3, we find the largest value of the left-hand side of $5y_1 + 5y_2 + 8y_4 \le 7$ is 5. Now $\Delta_3 = 13 - 5 = 8$, so that $a'_3 = 6 + 2 = 8$ and the new constraint is $5y_1 + 5y_2 + 8y_3 + 8y_4 \le 13$. Repeating the process with respect to y_4 reveals that no more lifting is possible. As a result, the constraint $4y_1 + 5y_2 + 6y_3 + 8y_4 \le 13$ has been replaced by the stronger constraint $5y_1 + 5y_2 + 8y_3 + 8y_4 \le 13$.

So far, the lifting procedure has been described as a single constraint procedure. However, in computing the maximum value α_k, it may well be that additional information derived from other constraints could be used, leading to a smaller α_k and consequently more room for lifting. It can also be useful to consider the order in which the coefficients are lifted in order to obtain the strongest possible constraint that results from the lifting process.

4.2.2 Multiple Constraint Procedures

A natural extension of single-constraint procedures is an attempt to involve more than just one constraint in the process. However, most attempts of this nature have failed in practice as the complexity of appropriate procedures approaches that of the original problem. In order to provide some ideas about the complexity of the case of multiple constraints, consider the following two constraints.

$$\begin{aligned} 5y_1 + 6y_2 + 3y_3 &\le 8 \\ 6y_1 + y_2 + 7y_3 &\le 7 \end{aligned}$$

where $y_j = 0 \vee 1, j = 1, 2, 3$.

From the first constraint we obtain the degree 2 implications $y_1 + y_2 \le 1$ and $y_2 + y_3 \le 1$, and from the second constraint the degree 2 implication $y_1 + y_3 \le 1$. Summing the three implications yields $2y_1 + 2y_2 + 2y_3 \le 3$ or $y_1 + y_2 + y_3 \le 1\frac{1}{2}$ which is immediately reduced to $y_1 + y_2 + y_3 \le 1$, which is more restrictive than (i.e., implies) each of the original constraints. The two constraints can therefore be replaced by the single, stronger constraint $y_1 + y_2 + y_3 \le 1$. Although useful, this technique of combining implications from different constraints would require finding such rather particular coincidences, limiting its practical value. Generalization to implications of higher degrees are straightforward. As an example, the four degree 3 implications $y_1 + y_2 + y_3 \le 2$, $y_2 + y_3 + y_4 \le 2$, $y_3 + y_4 + y_1 \le 2$, $y_4 + y_1 + y_2 \le 2$ give rise to the single constraint $y_1 + y_2 + y_3 + y_4 \le 2$ that replaces all four degree 3 implications.

Another preprocessing technique is called *variable fixing* whereby logical constraints are combined in such a way that variables can be fixed. An example will serve to illustrate this technique. Consider the constraint set:

$$2y_1 + 2y_3 + 2y_4 \le 5 \quad (3)$$

$$y_2 + y_3 + 3y_4 = 4 \quad (4)$$

$$y_1 + 2y_2 \le 2 \quad (5)$$

$$3y_1 - 2y_2 + 2y_3 - y_4 \ge -1 \quad (6)$$

$$y_j = 0 \vee 1, j = 1, 2, 3, 4$$

From relation (4) we conclude that $y_4 = 0$ is impossible; therefore $y_4 = 1$ and (4) reduces to $y_2 + y_3 = 1$. Relation (3) is then $2y_1 + 2y_3 \le 3$ which can be tightened to $2y_1 + 2y_3 \le 2$ or simply $y_1 + y_3 \le 1$ since the left-hand side is an even integer. From (5) we conclude that $y_1 + y_2 \le 1$, summing $y_1 + y_3 \le 1$ and $y_1 + y_2 \le 1$ yields $2y_1 + y_2 + y_3 \le 2$. However, since $y_2 + y_3 = 1$ as demonstrated above, this simplifies to $2y_1 \le 1$ which implies $y_1 = 0$. Inserting $y_1 = 0$ and $y_4 = 1$ into (6) leads to $-2y_2 + 2y_3 \ge 0$, i.e., $y_2 \le y_3$. Since $y_2 + y_3 = 1$ we finally conclude $y_2 = 0$, $y_3 = 1$ and our constraint set has the unique feasible solution $y_1 = y_2 = 0$, $y_3 = y_4 = 1$.

This variable fixing process can be quite useful but is difficult to automate and becomes impractical for larger problems. An important use of the procedure is to find inconsistencies that make the feasible region empty. Even if the procedure does not end with a unique solution, it is helpful if at least some variables could be fixed in this way. For example, if inequality (6) were omitted in the above example, we would still have been able to conclude from relationships (3), (4) and (5) the result $y_1 = 0$, $y_2 + y_3 = 1$, and $y_4 = 1$.

Related to variable fixing is the technique of *probing* where variables are tentatively fixed in order to explore the constraint set for inconsistencies. For example, if in the constraint set (3) - (6) above, we probe by tentatively fixing $y_1 = 1$, we obtain $y_2 = 0$ from constraint (5) and then $y_3 = y_4 = 1$ from constraint (4). However, $y_1 = y_3 = y_4 = 1$ violates constraint (3), so that $y_1 = 1$ leads to a contradiction; consequently we must have $y_1 = 0$. Continuing the probing, we would then tentatively set $y_2 = 1$, which via (4) yields $y_3 = 0$, $y_4 = 1$. Relation (6) then implies that $-2-1 \geq -1$ which is a contradiction; therefore $y_2 = 0$, and so on. For further details on variable fixing and probing, an early reference is Brearley *et al.* (1975); see also Guignard and Spielberg (1981).

Finally, we will demonstrate a simple technique called a *bound reduction procedure*. It iteratively tightens the constraints of a model. Consider a system in which all variables and structural constraints have known lower and upper bounds. Formally, the set of constraints is written as

$$L_i \leq \sum_j a_{ij} x_j \leq U_i \quad \forall\, i = 1, \ldots, m$$

$$\ell_j \leq x_j \leq u_j \qquad \forall\, j = 1, \ldots, n.$$

It may now be possible to use the bounds on the variables to tighten the bounds on the constraints, i.e., to find L_i' and U_i' so that $L_i' > L_i$ and/or $U_i' < U_i$. Similarly, it may be possible to use the bounds on the constraints to tighten the bounds on the variables to $\ell_j' > \ell_j$ and/or $u_j' < u_j$. Such a procedure can be applied repeatedly, resulting in a tighter problem formulation that may be considerably easier to solve than the original formulation.

Consider first the tightening of the bounds on the variables. The current upper bound of the i-th constraint is written as $\sum_j a_{ij} x_j \leq U_i$. Suppose now that $a_{ik} > 0$. The i-th upper bound can then be written as $a_{ik}x_k + \sum_{j \neq k} a_{ij} x_j \leq U_i$ or $x_k \leq \frac{1}{a_{ik}}\left(U_i - \sum_{j \neq k} a_{ij} x_j\right)$. Given the existing lower and upper bounds ℓ_j and u_j, we can write

$$x_k \leq u_k' = \frac{1}{a_{ik}}\left(U_i - \sum_{\substack{j \neq k \\ a_{ij} > 0}} a_{ij}\ell_j - \sum_{\substack{j \neq k \\ a_{ij} < 0}} a_{ij} u_j\right) \text{ for } a_{ik} > 0. \tag{7}$$

Similarly, if $a_{ik} < 0$, the i-th upper bound constraint $\sum_j a_{ij} x_j \leq U_i$ results in

$$x_k \geq \ell_k' = \frac{1}{a_{ik}}\left(U_i - \sum_{\substack{j \neq k \\ a_{ij} > 0}} a_{ij}\ell_j - \sum_{\substack{j \neq k \\ a_{ij} < 0}} a_{ij} u_j\right) \text{ for } a_{ik} < 0. \tag{8}$$

Consider now the lower bound of the i-th constraint, *viz.*, $\sum_j a_{ij}x_j \geq L_i$. First, let $a_{ik} > 0$ for some k. Rewriting the constraint results in $a_{ik}x_k + \sum_{j \neq k} a_{ij}x_j \geq L_i$ or $x_k \geq \frac{1}{a_{ik}}\left(L_i - \sum_{j \neq k} a_{ij}x_j\right)$. Using the lower and upper bounds ℓ_j and u_j we can write

$$x_k \geq \ell'_k = \frac{1}{a_{ik}}\left(L_i - \sum_{\substack{j \neq k \\ a_{ij}>0}} a_{ij}u_j - \sum_{\substack{j \neq k \\ a_{ij}<0}} a_{ij}\ell_j\right) \text{ for } a_{ik} > 0. \tag{9}$$

Similarly, if $a_{ik} < 0$, the i-th lower bound constraint $\sum_j a_{ij}x_j \geq L_i$ results in

$$x_k \leq u'_k = \frac{1}{a_{ik}}\left(L_i - \sum_{\substack{j \neq k \\ a_{ij}>0}} a_{ij}u_j - \sum_{\substack{j \neq k \\ a_{ij}<0}} a_{ij}\ell_j\right) \text{ for } a_{ik} < 0. \tag{10}$$

Having tightened the bounds on the variables, we can now attempt to tighten the bounds associated with the constraints. Using again the best known bounds ℓ_j and u_j we obtain

$$U'_i = \sum_{j:a_{ij}>0} a_{ij}u_j + \sum_{j:a_{ij}<0} a_{ij}\ell_j \tag{11}$$

and

$$L'_i = \sum_{j:a_{ij}>0} a_{ij}\ell_j + \sum_{j:a_{ij}<0} a_{ij}u_j \tag{12}$$

The relations (7) - (12) can now be applied repeatedly, until no further tightening is possible.

Example: Consider the set of constraints:

$$7 \leq 2x_1 + 4x_2 - 6x_3 \leq 10 \tag{13}$$
$$0 \leq x_1 - 2x_2 + x_3 \leq 10 \tag{14}$$
$$0 \leq x_1 \leq 1000 \tag{15}$$
$$0 \leq x_2 \leq 2 \tag{16}$$
$$3 \leq x_3 \leq 10 \tag{17}$$

The constraints (13) and (14) have lower bounds $L_1 = 7$ and $L_2 = 0$, respectively, and upper bounds $U_1 = U_2 = 10$, respectively, whereas the variables have lower bounds $\ell_1 = \ell_2 = 0$, and $\ell_3 = 3$, respectively, and upper bounds $u_1 = 1000$, $u_2 = 2$, and $u_3 = 10$, respectively.

The first step is to attempt to tighten the bounds ℓ_j and u_j, $j = 1, 2, 3$. Consider the variable x_1. In constraint (13), $a_{11} = 2 > 0$, so that relation (7) yields $u_1' = ½(10 - 4(0) - (-6)10) = 35$ which replaces the old upper bound $u_1 = 1{,}000$. Similarly, relation (9) yields $\ell_1' = ½(7 - 4(2) - (-6)3) = 8½$ which is considerably tighter than the previous lower bound of 0. The procedure is repeated with the second constraint. As $a_{12} = 1 > 0$, we use again relation (7) which is $u_1' = {}^1/_1(10 - 1(3) - (-2)2) = 11$ which is considerably tighter than the current bound of 35 which it now replaces. Applying (9) yields $\ell_1' = {}^1/_1(0 - 1(10) - (-2)0) = -10$ which is not as tight as the current lower bound of 8½. This completes the process for x_1, with constraint (15) tightened to $8½ \le x_1 \le 11$.

Repeating the process for x_2, we first consider the first constraint and with $a_{12} = 4 > 0$ apply relations (7) and (9) which yield bounds $u_2' = ¼(10 - 2(8½) - (-6)10) = 13¼ > 2 = u_2$, so that no tightening is possible, and $\ell_2' = ¼(7 - 2(11) -(-6)3) = ¾ > 0 = \ell_2$, so that ℓ_2 can be updated. Similarly, we use the second constraint (14). As $a_{22} = -2 < 0$, relations (8) and (10) apply. They result in the bounds $\ell_2' = {}^1/_{-2}(10 - 1(8½) - 1(3)) = ¾$ which equals the current bound ℓ_2, and $u_2' = {}^1/_{-2}(0 - 1(11) - 1(10)) = 10½ > u_2$, so that again, updating is not possible. Constraint (16) has now been tightened to $¾ \le x_2 \le 2$.

The last step in tightening the bounds on the variables uses the variable x_3. In the first constraint (13), $a_{13} = -6$, so that relations (8) and (10) apply. They result in the bounds $\ell_3' = {}^1/_{-6}(10 - 2(8½) - 4(¾)) = 1^2/_3 < 3 = \ell_3$, so that no tightening is possible, and $u_3' = {}^1/_{-6}(7 - 2(11) - 4(2)) = 3^5/_6 < 10 = u_3$, so that u_3 can be tightened significantly. The second constraint (14) has $a_{23} = 1 > 0$, so that relations (7) and (9) apply. They yield the bounds $u_3' = {}^1/_1(10 - 1(8½) - (-2)2) = 5½ > u_3$ which does not allow any tightening, and $\ell_3' = {}^1/_1(0 - 1(11) - (-2)¾) = -9½$ which again allows no tightening. Constraint (17) has been tightened to $3 \le x_3 \le 3\,^5/_6$.

In summary, the bounds on the variables have been tightened to

$$\ell_1 = 8½ \le x_1 \le 11 = u_1$$
$$\ell_2 = ¾ \le x_2 \le 2 = u_2$$
$$\ell_3 = 3 \le x_3 \le 3^5/_6 = u_3.$$

In the second phase of the bound reduction procedure, we attempt to tighten the bounds on the contraints. Invoking relations (11) and (12) on the first contraint, we obtain $U_1' = 2(11) + 4(2) + (-6)3 = 12 > 10 = U_1$ (no tightening) and $L_1' = 2(8\frac{1}{2}) + 4(\frac{3}{4}) + (-6)\,3^5/_6 = -3 < L_1$ (no tightening). Repeating the procedure with the second constraints yields $U_2' = 1(11) + 1(3^5/_6) - 2(\frac{3}{4}) = 13^1/_3 > U_2$ (no tightening), and $L_2' = 1(8\frac{1}{2}) + 1(3) + (-2)2) = 7\frac{1}{2} > 0 = L_1$, so that the lower bound of the second constraint can be tightened.

The two structural constraints now have the following bounds:

$$7 \le 2x_1 + 4x_2 - 6x_3 \le 10$$
$$7\tfrac{1}{2} \le x_1 - 2x_2 + x_3 \le 10$$

The entire process can now be repeated, but will not result in any further tightening of the bounds.

If all coefficients a_{ij} are integer and integrality is imposed on the variables x_j, the procedure can be improved by suitable rounding of fractions as soon as they occur. In the above example, $3 \le x_3 \le 3\frac{5}{6}$ could be strengthened to $3 \le x_3 \le 3$, i.e., $x_3 = 3$. In general, the above bound tightening procedure is easily programmed and not computationally expensive. It can be used to advantage as a preprocessing routine each time a linear or an integer programming problem is to be solved.

All of the procedures described in this section have ignored the objective function. If it is also taken into consideration, some additional variable fixing rules are immediately obvious. For example, consider the problem

$$\text{P: Min } z = \sum_{j=1}^{n} c_j y_j$$

$$\text{s.t.} \quad \sum_{j=1}^{n} a_{ij} y_j \le b_i, \; i = 1, \ldots, m$$

$$y_j = 0 \vee 1, \; j = 1, \ldots, n$$

If there exists a column k such that $c_k > 0$ and $a_{ik} \ge 0 \; \forall \; i = 1, \ldots, m$, then $\bar{y}_k = 0$ must be satisfied at optimum. This is easily proved by considering the constraints as resources. Any activity x_k with $a_{ik} \ge 0$ consumes nonnegative amounts of resources but at the same time increases, i.e., worsens the objective value since $c_k > 0$. If $c_k > 0$ is weakened to $c_k \ge 0$, this result is weakened to the assertion that at least one optimal solution must exist with $\bar{y}_k = 0$, provided, of course, that P has any feasible solution at all. Similarly, we may fix $\bar{y}_k = 1$ if $c_k < 0$ and $a_{ik} \le 0$, $i =$

1, ..., m; if $c_k \leq 0$ there exists an optimal solution with $\bar{y}_k = 1$. Analogous results can be stated and proved for maximization problems.

4.3 Aggregation

The purpose of this section is the discussion of the process of aggregation. By *constraint aggregation* we will understand a procedure whereby two or more constraints are combined into one *aggregated* constraint. In contrast, *variable aggregation* involves replacing two or more variables by an aggregated variable. Variable aggregation is extensively discussed in economic sciences and will not be dealt with here. The purpose of this section is to outline ways in which two constraints in integers can be combined into one. If this were possible, then any integer programming problem could be reduced to a knapsack problem by repeatedly applying such a procedure. The earliest reference to the aggregation of constraints is the classical paper by Mathews (1897) whose procedure was improved upon by Glover (1975). These and other procedures are also described in several classic texts on integer programming, e.g., Garfinkel and Nemhauser (1972), Taha (1975), or Salkin (1975).

Note that the aggregation proposed here is only meant for (in-)equalities in *integer* variables. Combining two constraints for which $\mathbf{x} \in \mathbb{R}^n$, the new constraint will have a different set of solutions than the two old ones, which is easily seen in the graph in Figure I.16 where the original two constraints are $2x_1 + 5x_2 \leq 10$, $x_1 \leq 3$, respectively, and the new aggregated constraint, shown as the dotted line, is $3x_1 + 5x_2 \leq 13$.

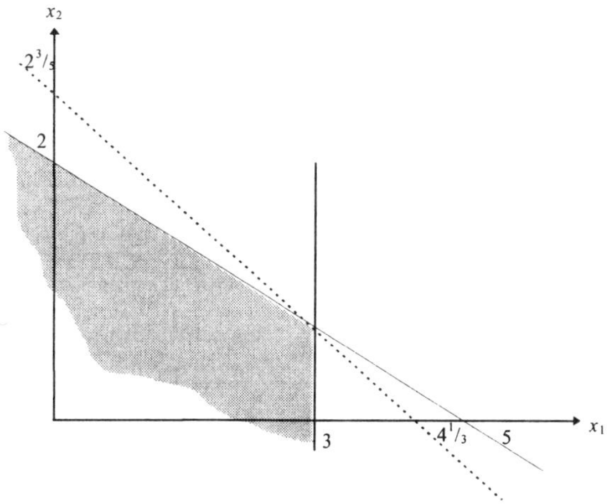

Figure I.16

To formalize, let the original system consist of the two constraints

$$\text{I:} \qquad \mathbf{a}_{1\bullet}\mathbf{x} = b_1 \tag{18}$$
$$\mathbf{a}_{2\bullet}\mathbf{x} = b_2 \tag{19}$$
$$\text{with } \mathbf{x} \in \mathbb{N}_0^n$$

and the goal is to find a constraint

$$\text{II:} \qquad \mathbf{a}_{3\bullet}\mathbf{x} = b_3 \tag{20}$$

which has the same solutions $\mathbf{x} \in \mathbb{N}_0^n$.

Some of the results outlined below have special requirements with respect to the coefficients of the constraints, i.e., $a_{ij} > 0 \ \forall \ i = 1, 2; j = 1,..., n$. Under certain conditions, these requirements can always be satisfied. Ways to achieve this are described below. Throughout the description, it is assumed that all coefficients and parameters are integers.

Case 1: $\exists \ a_{ij} < 0$. Let u_j denote a known upper bound of x_j, then if $a_{ij} < 0$ for some j, let $\bar{x}_j = u_j - x_j$ in the i-th constraint. Thus

$\ldots a_{ij}x_j \ldots \qquad = b_i$ is transformed into

$\ldots a_{ij}(u_j - \bar{x}_j) \ldots = b_i$, or simply

$\ldots -a_{ij}\bar{x}_j \ldots \qquad = b_i - a_{ij}u_j,$

where, by assumption, $-a_{ij} > 0$. This temporarily eliminates all negative coefficients. At the end of the procedure, all variables $\bar{x}_j$ are replaced again by $u_j - x_j$.

Case 2: $\exists \ a_{ij} = 0$. In such a case, Mathews (1897) proposed to replace the system I with constraints (18) and (19) by the system I′ with relations (21), (22), where (21) = (18) + (19) and (22) = (18) + 2(19). In other words, the original system I is replaced by the new system

$$\text{I}': \qquad (\mathbf{a}_{1\bullet} + \mathbf{a}_{2\bullet})\mathbf{x} = b_1 + b_2 \tag{21}$$
$$(\mathbf{a}_{1\bullet} + 2\mathbf{a}_{2\bullet})\mathbf{x} = b_1 + 2b_2 \tag{22}$$

Given that $a_{ij} \geq 0 \ \forall \ i = 1, 2; j = 1, ..., n$, at least one of a_{1j} and a_{2j} is positive for all j, otherwise the variable could be discarded. This property ensures that all coefficients in I′ are positive.

We now show that I and I′ have the same set of solutions. First suppose that some solution $\tilde{\mathbf{x}}$ solves the equations of the system I. It is then easy to show that $\tilde{\mathbf{x}}$ also solves I′. Now suppose that some solution $\tilde{\mathbf{x}}$ solves I′. This implies that $\tilde{\mathbf{x}}$ also satisfies (22) – (21) or $\mathbf{a}_{2\bullet}\tilde{\mathbf{x}} = b_2$, which is (19). Inserting (19) into (21) yields (18), which proves the claim.

An aggregation procedure due to Bradley (1971) considers the system

$$\text{I:} \qquad \mathbf{a}_{1\bullet}\mathbf{x} = b_1 \qquad (23)$$
$$\mathbf{a}_{2\bullet}\mathbf{x} = b_2 \qquad (24)$$

with $\mathbf{x} \in \mathbb{N}_0^n$, $\mathbf{a}_{1\bullet}$, $\mathbf{a}_{2\bullet} \in \mathbf{Z}^n$, b_1, $b_2 \in \mathbf{Z}$, $\mathbf{x} \leq \mathbf{u}$. He then proves

Lemma I.23: The system I is equivalent to the system

$$\text{II: } (\mathbf{a}_{1\bullet} + w\mathbf{a}_{2\bullet})\mathbf{x} = b_1 + wb_2 \qquad (25)$$

with $w > \lambda = \max_{\mathbf{x}} \{|\mathbf{a}_{1\bullet}\mathbf{x} - b_1|\} \geq 0$, where the maximum is taken over $\mathbf{x} \in \mathbb{N}_0^n$, $\mathbf{x} \leq \mathbf{u}$.

Proof: System I implies system II since (24) can be written as $w\mathbf{a}_{2\bullet}\mathbf{x} = \mathrm{w}b_2$ for any $w \in \mathbb{R}$; adding (23) $b_1 = \mathbf{a}_{1\bullet}\mathbf{x}$ to both sides yields system II. Conversely, let $\tilde{\mathbf{x}}$ solve II, i.e., $(\mathbf{a}_{1\bullet} + w\mathbf{a}_{2\bullet})\tilde{\mathbf{x}} = b_1 + wb_2$, with $w > \lambda$. Suppose that $\mathbf{a}_{2\bullet}\tilde{\mathbf{x}} = b_2 + \alpha$ with some $\alpha \in \mathbf{Z}$. Clearly, if $\alpha \neq 0$, then (24) would be violated and the lemma invalid. Here we will demonstrate that $\alpha = 0$. Multiplying $\mathbf{a}_{2\bullet}\tilde{\mathbf{x}} = b_2 + \alpha$ by w yields $w\mathbf{a}_{2\bullet}\tilde{\mathbf{x}} = wb_2 + w\alpha$. Subtracting this from (25) yields $\mathbf{a}_{1\bullet}\tilde{\mathbf{x}} = b_1 - w\alpha$ or $w\alpha = b_1 - \mathbf{a}_{1\bullet}\tilde{\mathbf{x}}$ which, since $w > 0$, implies that $w|\alpha| = |\mathbf{a}_{1\bullet}\tilde{\mathbf{x}} - b|$. But according to the definition $|\mathbf{a}_{1\bullet}\tilde{\mathbf{x}} - b_1| \leq \lambda < w$, so that $w|\alpha| < w$ or $|\alpha| < 1$, and since α is integer it follows that $\alpha = 0$. This proves the lemma which then forms the basis for the aggregation procedure.□

Note that $\lambda = \max_{\mathbf{x}} \{|\mathbf{a}_{1\bullet}\mathbf{x} - b_1|\} = \max\{[\max_{\mathbf{x}:\mathbf{a}_{1\bullet}\mathbf{x}-b_1 \geq 0} \{\mathbf{a}_{1\bullet}\mathbf{x} - b_1\}];[\max_{\mathbf{x}:\mathbf{a}_{1\bullet}\mathbf{x}-b_1 < 0} \{-\mathbf{a}_{1\bullet}\mathbf{x} + b_1\}]\} = \max\{\lambda^+; \lambda^-\}$, so that if we define $\tilde{x}_j^+ = u_j$ if $a_{1j} > 0$, and 0 otherwise, and $\tilde{x}_j^- = u_j$ if $a_{1j} < 0$, and 0 otherwise, then $\lambda^+ = \mathbf{a}_{1\bullet}\tilde{\mathbf{x}}^+ - b_1$ and $\lambda^- = -\mathbf{a}_{1\bullet}\tilde{\mathbf{x}}^- + b_1$.

Example: In order to illustrate Bradley's aggregation procedures, consider the following set of three inequalities (26), (27), and (28):

$$x_1 \leq 3 \qquad (26)$$
$$-x_1 + x_2 \leq 1 \qquad (27)$$
$$2x_1 + 5x_2 \leq 10 \qquad (28)$$
$$x_1, x_2 \in \mathbb{N}_0$$

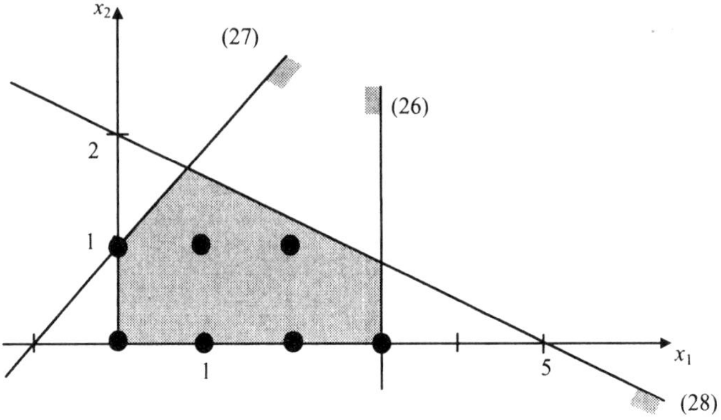

Figure I.17

We first transform the three inequalities to equations by adding slack variables S_1, S_2, and S_3. This results in (29), (30), and (31) as follows:

$$x_1 + S_1 = 3 \quad (29)$$
$$-x_1 + x_2 + S_2 = 1 \quad (30)$$
$$2x_1 + 5x_2 + S_3 = 10 \quad (31)$$

where the implicit bounds are: $x_1 \leq 3$, $S_1 \leq 3$, $x_2 \leq 4$, $S_2 \leq 4$, and $S_3 \leq 10$. Furthermore, in the vector $\mathbf{x}$ we include the two decision and three slack variables. Combine (29) and (30) first. Then $\tilde{\mathbf{x}}^+ = [3, 0, 3, 0, 0]^T$ and $\tilde{\mathbf{x}}^- = [0, 0, 0, 0, 0]^T$ and thus $\lambda^+ = \mathbf{a}_1 \bullet \tilde{\mathbf{x}}^+ - b_1 = [1, 0, 1, 0, 0]\,[3, 0, 3, 0, 0]^T - 3 = 3$ and $\lambda^- = -\mathbf{a}_1 \bullet \tilde{\mathbf{x}}^- + b_1 = -[1, 0, 1, 0, 0]\,[0, 0, 0, 0, 0]^T + 3 = 3$ and hence $\lambda = 3$. Selecting $w = 4$, the new aggregated equation is

$$-3x_1 + 4x_2 + S_1 + 4S_2 = 7 \quad (32)$$

whereas the unchanged constraint (31) is still

$$2x_1 + 5x_2 + S_3 = 10. \quad (31)$$

Now we combine (32) and (31) using (32) as the new first and (31) as the second equation. From (32) we get $\mathbf{x}^+ = [0, 4, 3, 4, 0]^T$ and $\mathbf{x}^- = [3, 0, 0, 0, 0]^T$, so that $\lambda^+ = \mathbf{a}_{4\bullet}\,\tilde{\mathbf{x}}^+ - b_4 = [-3, 4, 1, 4, 0]\,[0, 4, 3, 4, 0]^T - 7 = 28$ and $\lambda^- = -\mathbf{a}_4 \bullet \tilde{\mathbf{x}}^- + b_4 = -[-3, 4, 1, 4, 0]\,[3, 0, 0, 0, 0]^T + 7 = 16$, so that $\lambda = 28$. With $w = 29$, we obtain

$$55x_1 + 149x_2 + S_1 + 4S_2 + 29S_3 = 297 \quad (33)$$

Then (33) has the same solutions $x_1, x_2 \in \mathbb{N}_0$ and $\mathbf{x} \le \mathbf{u}$ as the system comprising (26), (27), and (28) combined, *viz.*, $\mathbf{x}$ = [0, 0], [1,0], [2, 0], [3, 0], [0, 1], [1, 1], and [2, 1] as shown in Figure I.17.

Mathew's (1897) aggregation is based on the following lemma and theorem:

Lemma I.24: If the system I: $\mathbf{a}_{1\bullet}\mathbf{x} = b_1$ (34)

$$\mathbf{a}_{2\bullet}\mathbf{x} = b_2 \quad (35)$$

has a feasible solution $x \in \mathbb{N}_0^n$, then there exists at least one j, such that $b_1 a_{2j} \le b_2 a_{1j}$.

Proof: If $\mathbf{x} = \mathbf{0}$ solves system I, then we must have $b_1 = b_2 = 0$ and $b_1 a_{2j} = 0 = b_2 a_{1j}$ $\forall\, j$. Assume now that $\mathbf{x} \ge \mathbf{0}$ (semipositive) satisfies system I and suppose that the implication is false. Then $b_1 a_{2j} > b_2 a_{1j}$ $\forall\, j$ and thus by $\mathbf{x} \ge \mathbf{0}$ (semipositive) we obtain $b_1 \sum_j a_{2j} x_j > b_2 \sum_j a_{1j} x_j$ or $b_1 \mathbf{a}_{2\bullet}\mathbf{x} > b_2 \mathbf{a}_{1\bullet}\mathbf{x}$ or $b_1 b_2 > b_1 b_2$, which is a contradiction. □

Theorem I.25: Let $\mathbf{x} \in \mathbb{N}_0^n$ and $\mathbf{a}_{i\bullet} \in \mathbb{N}^n$, $i = 1, 2$. If $w \in \mathbb{N}$ and $w > b_2 \max_j \{a_{1j}/a_{2j}\} > b_1 \ge 0$, then the equation

$$\text{II: } (\mathbf{a}_{1\bullet} + w\mathbf{a}_{2\bullet})\mathbf{x} = b_1 + wb_2 \quad (36)$$

has the same set of solutions as system I.

Proof: The proof that system I implies equation (36) in system II is the same as in the proof of Lemma I.24 above. Conversely, we let $\tilde{\mathbf{x}}$ solve system II, i.e. $(\mathbf{a}_{1\bullet} + w\mathbf{a}_{2\bullet})\tilde{\mathbf{x}} = b_1 + wb_2$. As before, supposing that $\mathbf{a}_{2\bullet}\tilde{\mathbf{x}} = b_2 + \alpha$, multiplying by w and subtracting from (36) yields

$$\alpha w = b_1 - \mathbf{a}_{1\bullet}\tilde{\mathbf{x}},\ \alpha \in \mathbf{Z}. \quad (37)$$

We first show that $\alpha > 0$ leads to a contradiction. Since α is integer, $\alpha \ge 1$. By assumption $w > b_1$, so that relation (37) implies that $b_1 - \mathbf{a}_{1\bullet}\tilde{\mathbf{x}} = \alpha w > \alpha b_1 \ge b_1$, i.e., $\mathbf{a}_{1\bullet}\tilde{\mathbf{x}} < 0$, which is impossible since $a_{ij} > 0$ and $x_j \ge 0$ $\forall\, j$. If on the other hand $\alpha < 0$, i.e., $\alpha \le -1$, then $\mathbf{a}_{2\bullet}\tilde{\mathbf{x}} = b_2 + \alpha$ implies that

$$\mathbf{a}_{2\bullet}\mathbf{x} < b_2 \quad (38)$$

Also, $\alpha \le -1$ implies that $b_1 - \mathbf{a}_{1\bullet}\tilde{\mathbf{x}} = \alpha w \le -w$ so that

$$\mathbf{a}_{1\bullet}\tilde{\mathbf{x}} \ge b_1 + w \quad (39)$$

From $w > b_2 \dfrac{a_{1j}}{a_{2j}}$ we obtain $wa_{2j} > b_2a_{1j}$ $\forall\, j$ and then $w\mathbf{a}_{2\bullet}\tilde{\mathbf{x}} > b_2\mathbf{a}_{1\bullet}\mathbf{x}$. Since by (38) and (39) $b_2\mathbf{a}_{1\bullet}\tilde{\mathbf{x}} > (b_1 + w)\,\mathbf{a}_{2\bullet}\tilde{\mathbf{x}}$ we find $w\,\mathbf{a}_{2\bullet}\tilde{\mathbf{x}} > b_1\,\mathbf{a}_{2\bullet}\tilde{\mathbf{x}} + w\mathbf{a}_{2\bullet}\tilde{\mathbf{x}}$ or $b_1\mathbf{a}_{2\bullet}\tilde{\mathbf{x}} < 0$, which is a contradiction. Thus $\alpha = 0$ and $\tilde{\mathbf{x}}$ solves the system I. This completes the proof. □

Example: We illustrate Mathews' procedure by the same problem as above. For convenience, we restate the system with slack variables introduced.

$$
\begin{aligned}
x_1 \qquad + S_1 \qquad\qquad &= 3 &\quad (29)\\
-x_1 + x_2 \qquad + S_2 \qquad &= 1 &\quad (30)\\
2x_1 + 5x_2 \qquad\qquad + S_3 &= 10 &\quad (31)
\end{aligned}
$$

with the implicit upper bounds $x_1 \le 3$, $x_2 \le 4$, $S_1 \le 3$, $S_2 \le 4$, and $S_3 \le 10$. We first perform an upper bound transformation to get rid of the negative coefficient in (30). In particular, use $x_1 = u_1 - \bar{x}_1 = 3 - \bar{x}_1$, so that (30) can be written as $-(3 - \bar{x}_1) + x_2 + S_2 = 1$ or

$$\bar{x}_1 + x_2 + S_2 = 4 \qquad (40).$$

Now combine (26′) and (27″). Although $a_{ij} \ge 0$ $\forall\, i, j$ now holds, $a_{ij} > 0$ $\forall\, i, j$ is not satisfied, so we consider the equivalent system

$$
\begin{aligned}
x_1 + \bar{x}_1 + x_2 + S_1 + S_2 &= 7 &\quad (29) + (40) = (41)\\
x_1 + 2\bar{x}_1 + 2x_2 + S_1 + 2S_2 &= 11 &\quad (29) + 2(40) = (42).
\end{aligned}
$$

In the system (41), (42) we may temporarily treat x_1 and $\bar{x}_1$ as independent variables, although $\bar{x}_1$ will be replaced by $3 - x_1$ by the end of the procedure. We have $w > b_2 \max\limits_j\{a_{1j}/a_{2j}\} = 11 \max\,\{^1/_1, {}^1/_2, {}^1/_2, {}^1/_1, {}^1/_2\} = 11$. Set $w = 12$, so that (41) and (42) are combined to form (43) = (41) + 12(42):

$$
\begin{aligned}
13x_1 + 25\bar{x}_1 + 25x_2 + 13S_1 + 25S_2 \qquad &= 139 &\quad (43)\\
2x_1 \qquad + 5x_2 + \qquad\qquad + S_3 &= 10 \text{ (restated)} &\quad (31).
\end{aligned}
$$

Since again $a_{1j} > 0$ is not satisfied, consider

$$
\begin{aligned}
15x_1 + 25\bar{x}_1 + 30x_2 + 13S_1 + 25S_2 + S_3 &= 149 &\quad (43) + (31) = (44)\\
17x_1 + 25\bar{x}_1 + 35x_2 + 13S_1 + 25S_2 + 2S_3 &= 159. &\quad (43) + 2(31) = (45).
\end{aligned}
$$

Now $w > 159 \max_j \{a_{1j}/a_{2j}\} = 159 \max \{^{15}/_{17}, ^{25}/_{25}, ^{30}/_{35}, ^{13}/_{13}, ^{25}/_{25}, ^{1}/_{2}\} = 159$. Set now $w = 160$, so that (44) and (45) are combined to form (46) = (44) + 160(31):

$$2{,}735x_1 + 4{,}025\,\bar{x}_1 + 5{,}630x_2 + 2{,}093S_1 + 4{,}025S_2 + 321S_3 = 25{,}589. \quad (46)$$

Finally, replacing $\bar{x}_1 = 3 - x_1$, we rewrite (46) as

$$-1{,}290x_1 + 5{,}630x_2 + 2{,}093S_1 + 4{,}025S_2 + 321S_3 = 13{,}514. \quad (47)$$

Again, the seven feasible solutions $[x_1, x_2]$ = [0, 0], [1, 0], [2, 0], [3, 0], [0, 1], [1, 1], and [2, 1] satisfy equation (47), given the upper bounds u_j.

Glover (1975) has suggested an improvement of Mathews' procedure. While in his aggregation the requirement $a_{ij} \geq 0$ is sufficient but not necessary, we now assume that $a_{ij} > 0\ \forall\ i, j$. Glover then uses weights w_1 and w_2, so that the new aggregated constraint is $(w_1\mathbf{a}_{1\bullet} + w_2\mathbf{a}_{2\bullet})\mathbf{x} = w_1b_1 + w_2b_2$. Setting $w_1 = 1$, we obtain the condition (using a proof similar to the one above):

$$w_1a_{1j} + w_2a_{2j} \geq |b_2a_{1j} - b_1a_{2j}|$$

which becomes $w_2 \geq |b_2a_{1j} - b_1a_{2j}| / a_{2j} - a_{1j} / a_{2j}$. Eliminating the absolute value sign we then obtain

$$w_2 > [(b_2 - 1)a_{1j} / a_{2j}] - b_1 \quad \text{or} \quad w_2 > (b_2 - 1) \max \{a_{1j} / a_{2j}\} - b_1, \text{ as well as}$$
$$w_2 > b_1 - (b_2 + 1)a_{1j} / a_{2j} \quad \text{or} \quad w_2 > b_1 - (b_2 + 1) \max \{a_{1j} / a_{2j}\}.$$

We note that w_2 here has the same function as w in the Bradley and Mathews procedures above.

Example: Again, consider the same example as above. We first eliminate the negative coefficient, resulting in the equations

$$x_1 + \bar{x}_1 + x_2 + S_1 + S_2 = 7 \quad (41)$$
$$x_1 + 2\bar{x}_1 + 2x_2 + S_1 + 2S_2 = 11 \quad (42)$$

From the above conditions we then obtain $w_2 > (11 - 1) \max \{^1/_1, ^1/_2, ^1/_2, ^1/_1, ^1/_2\} - 7 = 10\,(1) - 7 = 3$ and $w_2 > 7 - (11 + 1) \max \{^1/_1, ^1/_2, ^1/_2, ^1/_1, ^1/_2\} = 7 - 12(1) = -5$. Therefore we must have $w_2 > 3$. Setting $w_2 = 4$, the equations (41) and (42) can be aggregated to (48) = (41) + 4(42):

$$5x_1 + 9\bar{x}_1 + 9x_2 + 5S_1 + 9S_2 = 51 \quad (48)$$
$$2x_1 + 5x_2 + S_3 = 10 \text{ (restated)} \quad (31).$$

Transforming the equations, so that all coefficients are strictly positive, we obtain

$$7x_1 + 9\bar{x}_1 + 14x_2 + 5S_1 + 9S_2 + S_3 = 61 \qquad (48) + (31) = (49)$$
$$9x_1 + 9\bar{x}_1 + 19x_2 + 5S_1 + 9S_2 + 2S_3 = 71 \qquad (48) + 2(31) = (50).$$

The conditions require that $w_2 > (71 - 1) \max \{\frac{7}{9}, \frac{9}{9}, \frac{14}{9}, \frac{5}{5}, \frac{9}{9}, \frac{1}{2}\} - 61 = 70(1) - 61 = 9$ and also that $w_2 > 61 - (71 + 1) \max \{\frac{7}{9}, \frac{9}{9}, \frac{14}{9}, \frac{5}{5}, \frac{9}{9}, \frac{1}{2}\} = 61 - 72(1) = -11$, i.e., $w_2 > 9$. Setting $w_2 = 10$, the aggregated constraint becomes

$$97x_1 + 99\bar{x}_1 + 204x_2 + 55s_1 + 99s_2 + 21s_3 = 771$$

Finally, with $\bar{x}_1 = 3 - x_1$, we obtain

$$-2x_1 + 204x_2 + 55s_1 + 99s_2 + 21s_3 = 474. \qquad (51)$$

The seven feasible solutions of the original problem are again satisfied by equation (51).

The principal advantage of Glover's improved aggregation method is that the coefficients in the aggregated constraint are smaller than those obtained with Mathews' original aggregation. Given many constraints, they will still be much too large for practical purposes. In that context, Glover and Woolsey (1970) have shown that recursive aggregation of m constraints with common right-hand side values b will result in an equation with right-hand side exceeding $2^{3(2^{m-2})} b^{2^{m-1}}$. With $b = 1$, the right-hand side exceeds 16 million for $m = 5$, 10^{28} for $m = 7$, and 10^{99} for $m = 9$. In other words, aggregating nine constraints with all right-hand side values equal to one will result in a single constraint whose right-hand side value exceeds 10^{99}, an obviously unacceptable result from a computational point of view. Other aggregation procedures similar to those above can be found in Kendall and Zionts (1977). The procedure suggested by Ram *et al.* (1988) aggregates by multiplying the constraint coefficients by the natural logarithms of distinct prime numbers. The result is an aggregated constraint with irrational coefficients, where roundoff errors are of concern.

In summary, aggregation might seem appealing since several constraints are reduced to one, transferring the problem into a knapsack problem, thus allowing specialized solution algorithms to be used. However, since the aggregate constraint coefficients tend to become large or irrational, the advantages of aggregation become dubious.

4.4 Disaggregation

The concept of disaggregation for integer linear programming can be subdivided into *constraint* and *variable disaggregation. Constraint disaggregation* is a procedure that replaces a single constraint by two or more constraints. In general, this leads to a stronger formulation of the problem and is desirable from a numerical solution point of view. It is the opposite of constraint aggregation, which was discussed in Section 4.3 above. As an example of constraint disaggregation consider the constraint $y_1 + y_2 + y_3 - 3y_4 \leq 0$, where $y_j = 0 \vee 1 \; \forall \, j = 1, 2, 3, 4$. Clearly this single constraint can be replaced by the three simpler constraints $y_j - y_4 \leq 0, j = 1, 2, 3$. This disaggregation would give rise to a linear programming relaxation with a smaller feasible region than if the original constraint were used. This and related issues were discussed in Sections 4.1 and 4.2 and will not be repeated here.

Variable disaggregation refers to the process of replacing single variables by two or more new variables. An example of this process was given in Section 2.9 where a semicontinuous variable was replaced by one continuous and one zero-one variable. Another example of variable disaggregation comes from the modeling of production planning. Assume that an amount x_{jt} of a commodity j is produced at time t and that the production capacity is c_{jt}. Define a binary variable y_{jt} that equals 1 if the production process is set up, and 0 if it is not. The production constraint can then be written as $x_{jt} \leq c_{jt}y_{jt}$. The linear programming relaxation of a production planning model involving such constraints often tends to generate fractional values of y_{jt} and large integrality gaps, making it difficult to find good integer solutions. A remedy is to introduce disaggregated variables x_{jts} denoting the production of the commodity j at time t for use in the period s when the demand might be d_{js}. The production constraint can then be written as $x_{jts} \leq (\min \{c_{jt}, d_{js}\})y_{jt}$. Since often $d_{js} << c_{jt}$, the model tends to be much stronger, with the variables y_{jt} being closer to 0 or 1. Other examples of variable disaggregation fall into the area of Lagrangean decomposition, a topic to be discussed in Section 6.6.

CHAPTER 5 CUTTING PLANE METHODS

In this chapter, we introduce a class of methods that were among the first to be designed for the solution of integer programming problems. Throughout the past decades, however, computational evidence has revealed that cutting planes, while appealing from a theoretical point of view, do not appear to work very well if applied to general integer programming problems. They do find use, however, in some specialized problems and in conjunction with other methods.

The basic ideas of cutting plane methods can be presented as follows. As revealed in our discussion of total unimodularity, if all extreme points of the polytope of a linear programming relaxation are integer, then—by virtue of Dantzig's corner point theorem—any solution method that solves the relaxed problem will automatically result in an integer solution. Clearly, very few (if any) general integer programming problems satisfy this property. One way to proceed could be to preprocess the problem by trying to reformulate the problem, so that the new formulation will automatically have integer solutions. Formally, let the original polytope described by an integer programming problem be denoted by $\mathcal{P}$, and let $\mathcal{P}_I$ be the convex hull of $\mathcal{P}$. By definition, all extreme points of $\mathcal{P}_I$ are integer. If there were a simple way to determine $\mathcal{P}_I$ from $\mathcal{P}$, then optimizing the given objective on $\mathcal{P}_I$, e.g., by means of any linear programming code, would easily solve the original linear integer programming problem. Unfortunately, the determination of such a polytope $\wp_I$ is not a simple matter. In fact, we do not even have to find the entire polytope; it would be sufficient to have part of the original polytope modified, so that its extreme points are integer (although it is important to keep in mind that the optimal points of the original problem and its relaxation may be far apart, as discussed in Section 1.1). This is the approach taken by cutting plane methods.

While different cutting plane methods follow different procedures, the general idea is this. First, we solve the linear programming relaxation. If the optimal solution is integer, the problem is solved. If not, then a new constraint, a so-called *cutting plane* (also referred to as a *valid inequality*), is introduced. The problem is then

reoptimized and the procedure is repeated until the optimal solution is integer. Each cutting plane must satisfy two conditions:
(1) it must render the current optimal solution infeasible, and
(2) it must not make infeasible ("cut off") any feasible integer solution point.

In order to visualize valid inequalities, consider the following example:

$$3x_1 + 2x_2 \leq 6$$
$$4x_1 + 8x_2 \leq 12$$
$$x_1, x_2 \in \mathbb{N}_0.$$

The feasible region of this set of constraints with the integrality requirements dropped is shown by the shaded area in Figure I.18.

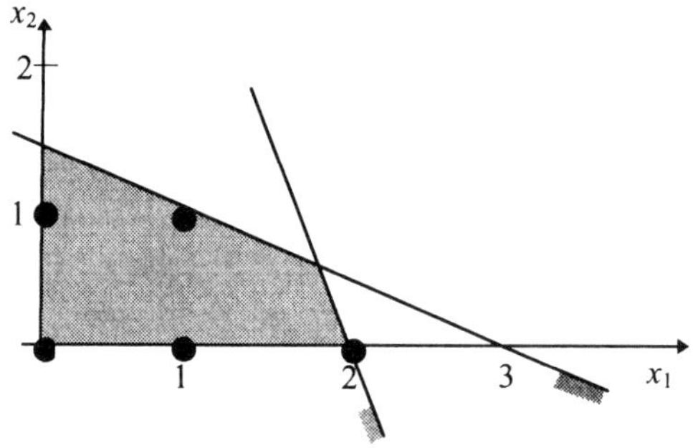

Figure I.18

Figure I.18 also reveals that the following constraints are indeed valid inequalities to the system: $x_2 \leq 1$, $x_1 + x_2 \leq 2$, and $7x_1 + 10x_2 \leq 18$. As a matter of fact, adding the first two of these inequalities to the constraints above results in a polytope with all extreme points integer (i.e., the convex hull of the problem), so that a linear programming algorithm would automatically find an integer solution to the problem.

5.1 Dantzig's Cutting Plane Method

Gomory (1958) and Dantzig (1959) were among the earliest proponents of cutting plane methods, and the first method presented here is generally known as a *Dantzig cut*. Suppose that the linear programming relaxation of an all-integer programming problem was solved and in the optimal solution of the relaxed problem, there exists at least one variable x_k with a noninteger value b_i. As $x_j \in \mathbb{N}_0$ $\forall j$ is required, at least one variable that is currently nonbasic must have a positive value, and since its value must be integer, at least one of the current nonbasic variables must equal a value of one or larger. This implies that the sum of variables that are currently nonbasic, must equal at least one. Defining *nbv* as the set of variables that are presently nonbasic, this constraint is called a Dantzig cut and can be written as

$$\sum_{j \in nbv} x_j \geq 1 \tag{1}$$

Note that relation (1) does indeed satisfy the conditions of a cut: in the current solution all nonbasic variables equal zero, so at present the left-hand side of the relation equals zero, clearly violating the constraint, so that (1) cuts off the current solution. On the other hand, the derivation of the relation has ensured that none of the integer feasible points were made infeasible.

Example: Consider the all-integer programming problem

$$\begin{aligned} \text{P: Max } z &= 3x_1 + 4x_2 \\ \text{s.t.} \quad & 3x_1 + 2x_2 \leq 7 \\ & 4x_1 + 6x_2 \leq 11 \\ & x_1, \quad x_2 \in \mathbb{N}_0. \end{aligned}$$

The final (optimal) simplex tableau for the linear programming relaxation of this problem is shown below as $\mathbf{T}_0$.

$\mathbf{T}_0$:

x_1	x_2	S_1	S_2	1
1	0	3/5	– 1/5	2
0	1	– 2/5	3/10	1/2
0	0	1/5	3/5	8

Since x_2 is noninteger, the process continues. The set of nonbasic variables includes the two slack variables, so that relation (1) leads to the cutting plane $S_1 + S_2 \geq 1$. Multiplying the constraint by (– 1) and adding a new slack variable S_1^* to the left-hand side yields the new tableau

$\mathbf{T}_0'$:

x_1	x_2	S_1	S_2	S_1^*	1
1	0	3/5	– 1/5	0	2
0	1	– 2/5	3/10	0	1/2
0	0	(–1)	– 1	1	– 1
0	0	1/5	3/5	0	8

Using the element in the third column under the slack variable S_1 and the dual simplex method to optimize, we obtain the new optimal solution displayed in tableau

T$_1$:

x_1	x_2	S_1	S_2	S_1^*	1
1	0	0	–4/5	3/5	$1\frac{2}{5}$
0	1	0	7/10	–2/5	9/10
0	0	1	1	–1	1
0	0	0	2/5	1/5	$7\frac{4}{5}$

Notice that the value of the objective function has decreased, but the new solution is not yet integer. The set of nonbasic variables consists of the slack variables S_2 and S_1^*, so that the next Dantzig cut is $S_2 + S_1^* \geq 1$. Again, the constraint is multiplied by (– 1), a new slack variable S_2^* is added to the left-hand side, and a dual simplex step is performed on the pivot in the fourth row under the entering variable S_1^*. The result is the new optimal tableau

T$_2$:

x_1	x_2	S_1	S_2	S_1^*	S_2^*	1
1	0	0	– 7/5	0	3/5	4/5
0	1	0	11/10	0	– 2/5	$1\frac{3}{10}$
0	0	1	2	0	– 1	2
0	0	0	1	1	– 1	1
0	0	0	1/5	0	1/5	$7\frac{3}{5}$

In **T**$_2$, the solution is still not integer, so that we construct the Dantzig cut $S_2 + S_2^* \geq 1$. We terminate at this point, as the introduction of cuts will continue for a long time. Also note that each time a new cut is introduced, the size of the tableau increases by one row and one column. Expressing the slack variables in terms of the decision variables, the first seven Dantzig cuts are as follows:

$$7x_1 + 8x_2 \leq 17 \quad (\text{or } 0.4118x_1 + 0.4706x_2 \leq 1) \quad \text{(I)}$$
$$11x_1 + 14x_2 \leq 27 \quad (\text{or } 0.4074x_1 + 0.5185x_2 \leq 1) \quad \text{(II)}$$
$$15x_1 + 20x_2 \leq 37 \quad (\text{or } 0.4054x_1 + 0.5405x_2 \leq 1) \quad \text{(III)}$$
$$26x_1 + 34x_2 \leq 63 \quad (\text{or } 0.4127x_1 + 0.5397x_2 \leq 1) \quad \text{(IV)}$$
$$41x_1 + 54x_2 \leq 99 \quad (\text{or } 0.4141x_1 + 0.5454x_2 \leq 1) \quad \text{(V)}$$
$$45x_1 + 60x_2 \leq 109 \quad (\text{or } 0.4128x_1 + 0.5505x_2 \leq 1) \quad \text{(VI)}$$
$$86x_1 + 114x_2 \leq 207 \quad (\text{or } 0.4155x_1 + 0.5507x_2 \leq 1) \quad \text{(VII)}$$

The original constraints are shown in Figure I.19. To avoid cluttering the figure, only the cuts (I), (II), and (VII) are displayed.

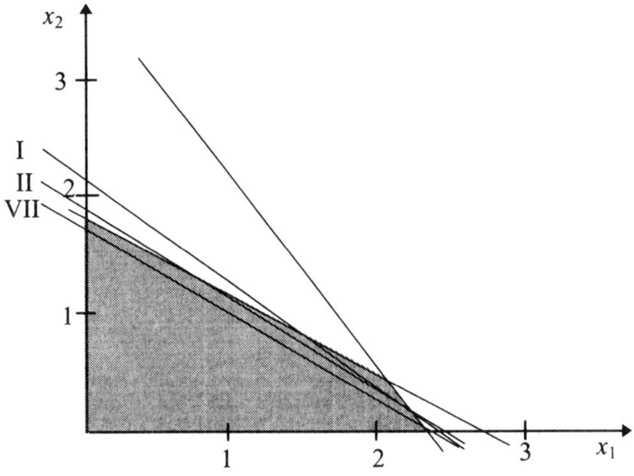

Figure I.19

Note the similarities of the cuts in the expressions in parentheses. Also remarkable are the many redundancies in the constraints, e.g., (IV) dominates (I), (V) dominates (IV), (VI) dominates (IV), and (VII) dominates all other cuts generated here. It can be seen that while each cut does cut off a bit of the current polytope, the cut off piece is very small, resulting in many cuts and very large tableaus. Even this tiny example demonstrates that this type of cut may be very inefficient. In general, cutting planes that cut away a large portion of the feasible region for the relaxed problem are called *deep cuts*, whereas *shallow cuts* only cut away a small portion of the feasible region. Unfortunately, Dantzig cuts are not only quite similar to each other, they are also usually rather shallow, thus severely limiting their usefulness.

A minor modification of Dantzig's cuts is credited by Taha (1975) to Bowman and Nemhauser (1970). The authors observe that the sum of nonbasic variables on the left-hand side of a Dantzig cut can be replaced by the sum of nonbasic variables with noninteger coefficients in the current tableau. Apparently, such a cut is stronger than Dantzig's original cut, but there is not much difference in practical terms as normally only a few of the coefficients in the tableau are integer anyway.

5.2 Gomory's Cutting Plane Methods

Another type of cut was introduced by Gomory (1958). To facilitate the discussion, let a_{ij} and b_i denote the left- and right-hand side values in the current tableau. By assumption, there exists at least one variable $x_k = b_i \notin \mathbf{Z}$ and we say that the i-th row is a *source row*. Define then fractional parameters $f_{ij} = a_{ij} - \lfloor a_{ij} \rfloor$ and $f_i = b_i - \lfloor b_i \rfloor$ with $f_{ij} \in [0; 1[$ and $f_i \in]0; 1[$, so that

$$x_k + \sum_{j \in nbv} \lfloor a_{ij} \rfloor x_j + \sum_{j \in nbv} f_{ij} x_j = \lfloor b_i \rfloor + f_i . \tag{2}$$

As $x_j \in \mathbf{Z}\ \forall j$ is required as well as $\sum_{j \in nbv} \lfloor a_{ij} \rfloor x_j$ and $\lfloor b_i \rfloor \in \mathbf{Z}$, we obtain

$$f_i - \sum_{j \in nbv} f_{ij} x_j \in \mathbf{Z}. \tag{3}$$

With $f_i < 1$ and the sum in (3) nonnegative, the expression in relation (3) must be less than or equal to zero, so that

$$-\sum_{j \in nbv} f_{ij} x_j \leq -f_i . \tag{4}$$

Relation (4) is the *fractional Gomory cut*. As in the case of the Dantzig cut, the current solution violates this new relation (as at present, the left-hand side equals zero and the right-hand side is strictly negative) and the derivation assures that no integer solutions were cut off. At least one dual simplex iteration is needed to find a primal feasible solution again.

Example: Consider again the problem

$$\begin{aligned} \text{P: Max } z &= 3x_1 + 4x_2 \\ \text{s.t. } \quad & 3x_1 + 2x_2 \leq 7 \\ & 4x_1 + 6x_2 \leq 11 \\ & x_1, \quad x_2 \in \mathbb{N}_0. \end{aligned}$$

The final tableau is, as before,

$\mathbf{T}_0$:

x_1	x_2	S_1	S_2	1
1	0	3/5	– 1/5	2
0	1	– 2/5	3/10	1/2
0	0	1/5	3/5	8

At present, $x = [2, ½]$, so that the only potential source row is the second row. Here, we obtain $f_{23} = 3/5$, $f_{24} = 3/10$, and $f_2 = ½$. A Gomory fractional cutting plane is then

$$-3/5S_1 - 3/10S_2 \leq -½.$$

Adding a slack variable S_1^* to the left-hand side, we obtain $-3/5S_1 - 3/10S_2 + S_1^* = -½$. The new tableau is then

$\mathbf{T}_0'$:

x_1	x_2	S_1	S_2	S_1^*	1
1	0	3/5	– 1/5	0	2
0	1	– 2/5	3/10	0	1/2
0	0	– 3/5	– 3/10	1	– 1/2
0	0	1/5	3/5	0	8

and one dual simplex step yields the new optimal solution

$\mathbf{T}_1$:

x_1	x_2	S_1	S_2	S_1^*	1
1	0	0	–1/2	1	3/2
0	1	0	1/2	–2/3	5/6
0	0	1	1/2	–5/3	5/6
0	0	0	1/2	1/3	7⅚

The current solution $x = [1½, 5/6]$ is not integer and at least one more step is required. We arbitrarily choose the second row as source row, although the first row could have also been chosen. The resulting Gomory cut is then $-½S_2 - 1/3\,S_1^* \leq -5/6$. Creating another slack variable S_2^*, adding it to the tableau and performing one dual simplex iteration in the S_2 column (noting the tie for pivot row based on the dual degeneracy), we obtain the next tableau

$\mathbf{T}_2$:

x_1	x_2	S_1	S_2	S_1^*	S_2^*	1
1	0	0	0	4/3	– 1	7/3
0	1	0	0	– 1	1	0
0	0	1	0	–2	1	0
0	0	0	1	2/3	–2	5/3
0	0	0	0	0	1	7

The tableau $\mathbf{T}_2$ exhibits both primal and dual degeneracy. The solution $x = [7/3, 0]$ is still not integer, and an additional step is needed. Using the first row as source row yields the Gomory cut $-1/3\, S_1^* \leq -1/3$. Creating an additional slack variable S_3^*, adding it to the tableau $\mathbf{T}_2$ and performing one dual simplex iteration leads to the final tableau

$\mathbf{T}_3$:

x_1	x_2	S_1	S_2	S_1^*	S_2^*	S_3^*	1
1	0	0	0	0	-1	4	1
0	1	0	0	0	1	-3	1
0	0	1	0	0	1	-6	2
0	0	0	1	0	-2	2	13
0	0	0	0	1	0	-3	1
0	0	0	0	0	1	0	7

Finally, tableau $\mathbf{T}_3$ includes the optimal integer solution $\mathbf{x} = [1, 1]$. However, again, progress was quite slow. Considering the cuts that were generated in the process, and expressing them in terms of the problem variables x_1 and x_2, we find that the cutting planes were:

$3x_1 + 3x_2 \leq 7$, $3x_1 + 4x_2 \leq 7$, and $x_1 + x_2 \leq 2$, or, normalizing them, we obtain

$0.4286x_1 + 0.4286x_2 \leq 1$, $0.4286x_1 + 0.5714x_2 \leq 1$, and $0.5x_1 + 0.5x_2 \leq 1$.

Note again the redundancy of the first cutting plane with respect to either of the two others. In Figure I.20 the original constraints as well as the three cuts are shown.

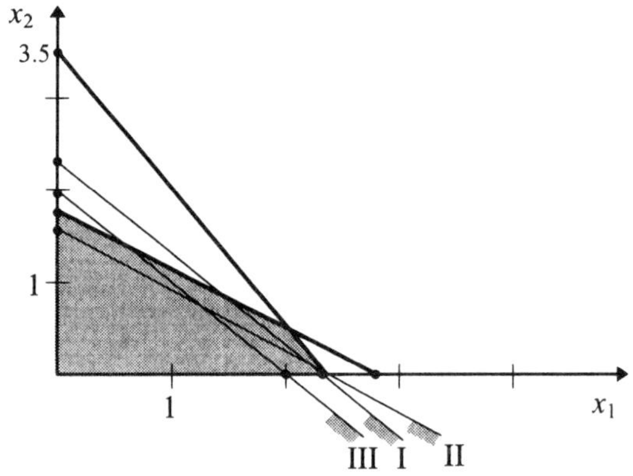

Figure I.20

A very different approach was taken by Gomory (1963) in designing all-integer cuts. The basic idea is to start with a tableau in which all parameters are integers, and retain that property while optimizing. If such a tableau is not initially available, we multiply all constraints by their respective denominators' least common multiples, thereby achieving integrality of all parameters in the tableau. We then add the redundant constraint $\sum_j x_j \le M$ with $M >> 0$ to the tableau and perform a primal simplex iteration on the element in the redundant constraint that belongs to the "most negative" indicator. This step will guarantee that the resulting tableau is dual feasible and, as the pivot equaled one and all parameters in the tableau were integer before the step, the resulting tableau is still all-integer. Find a source row in this tableau, i.e., a row i with $b_i < 0$ (if such a row does not exist, then the current solution is already primal feasible and therefore optimal). First, we determine the pivot in the i-th row according to the usual dual simplex method. In case this pivot equals (–1), we perform a regular dual simplex iteration as integrality of all parameters will be retained. Suppose now that the pivot does not equal (–1). Starting with the constraint $\sum_j a_{ij} x_j = b_i$ where all parameters denote their values in the current tableau, we divide all terms of this constraint by some new parameter $|\lambda| \neq 0$ and decompose $a_{ij}/|\lambda|$ and $b_i/|\lambda|$ into their integer and fractional parts, respectively, i.e., $\frac{a_{ij}}{|\lambda|} = \left\lfloor \frac{a_{ij}}{|\lambda|} \right\rfloor + \frac{f_{ij}}{|\lambda|}$ and $\frac{b_i}{|\lambda|} = \left\lfloor \frac{b_i}{|\lambda|} \right\rfloor + \frac{f_i}{|\lambda|}$. The constraint can then be written as

$$\sum_j \left\lfloor \frac{a_{ij}}{|\lambda|} \right\rfloor x_j - \left\lfloor \frac{b_i}{|\lambda|} \right\rfloor = \frac{f_i}{|\lambda|} - \sum_j \frac{f_{ij}}{|\lambda|} x_j . \tag{5}$$

Since all terms on the left-hand side are integer, the right-hand side must also be integer, i.e.,

$$\frac{f_i}{|\lambda|} - \sum_j \frac{f_{ij}}{|\lambda|} x_j \in \mathbf{Z}. \tag{6}$$

As by definition f_i and $f_{ij} \in [0; 1[$, we also have $f_i/|\lambda|$ and $f_{ij}/|\lambda| \in [0; 1[\ \forall\ \lambda >$ 1. Hence, for all nonnegative x_j, expression (6) must be nonpositive. Consequently, the left-hand side of (5) must also be nonpositive, leading to the Gomory *all-integer cut*

$$\sum_j \left\lfloor \frac{a_{ij}}{|\lambda|} \right\rfloor x_j \le \left\lfloor \frac{b_i}{|\lambda|} \right\rfloor . \tag{7}$$

Clearly, the current solution does not satisfy relation (7), as by assumption $b_i < 0$, so that the right-hand side is less than or equal to –1, whereas the left-hand side is currently zero.

A cut of the type shown in (7) is added to the tableau whenever the pivot found by the usual dual simplex method does not equal (–1). We then choose a value of $\lambda = \min_j \left\{a_{ij} : a_{ij} < 0\right\}$, i.e., the smallest negative coefficient in the source row. (Note that at least one negative coefficient must exist, otherwise the original problem cannot be satisfied by any nonnegative solution). This choice of λ ensures that all negative coefficients of the cut equal (– 1). One of these entries will be chosen as pivot, which, in turn, ensures that all coefficients in the next tableau are again integer.

Example: In order to illustrate the procedure, consider the following problem.

$$\begin{aligned} \text{P: Min } z &= 4x_1 + 5x_2 \\ \text{s.t.} \quad & 2x_1 + 6x_2 \geq 7 \\ & 3x_1 + 2x_2 \geq 5 \\ & x_1, \quad x_2 \in \mathbb{N}_0. \end{aligned}$$

The associated first tableau is

T$_0$:

x_1	x_2	S_1	S_2	1
– 2	– 6	1	0	– 7
– 3	– 2	0	1	– 5
4	5	0	0	0

Selecting the first row as source row, the pivot would be (– 6) in the x_2-column. Setting λ = – 6, we obtain the cutting plane $\left\lfloor \frac{-2}{6} \right\rfloor x_1 + \left\lfloor \frac{-6}{6} \right\rfloor x_2 + \left\lfloor \frac{1}{6} \right\rfloor S_1 \leq \left\lfloor \frac{-7}{6} \right\rfloor$, or simply $-x_1 - x_2 \leq -2$. Adding a slack variable S_1^* to the left-hand side, adding the constraint to the tableau, and performing a dual simplex iteration with the x_1 column as pivot column, we obtain the next simplex tableau

T$_1$:

x_1	x_2	S_1	S_2	S_1^*	1
0	– 4	1	0	– 2	– 3
0	1	0	1	– 3	1
1	1	0	0	– 1	2
0	1	0	0	4	– 2

Since the tableau is not yet primal feasible, at least one further step is needed. The first row is the unique source row, and with $\lambda = -4$, the cutting plane is $\left\lfloor \frac{-4}{4} \right\rfloor x_2 + \left\lfloor \frac{1}{4} \right\rfloor S_1 + \left\lfloor \frac{-2}{4} \right\rfloor S_1^* \leq \left\lfloor \frac{-3}{4} \right\rfloor$, or simply $-x_2 - S_1^* \leq -1$. Adding the constraint to the tableau and performing a simplex iteration with the x_2 column as pivot column, we obtain the tableau

T_2:

x_1	x_2	S_1	S_2	S_1^*	S_2^*	1
0	0	1	0	2	−4	1
0	0	0	1	−4	1	0
1	0	0	0	−2	1	1
0	1	0	0	1	−1	1
0	0	0	0	3	1	−3

Clearly, tableau T_2 is primal and dual feasible, and hence optimal with $\bar{\mathbf{x}} = [1; 1]^T$. We wish to point out that while in both steps above, where λ equaled the value of the pivot, this is not necessarily the case. To summarize, the two cuts were $x_1 + x_2 \geq 2$ and $x_1 + 2x_2 \geq 3$ with no redundancy. Figure I.21 shows the original constraints and the two all-integer cuts.

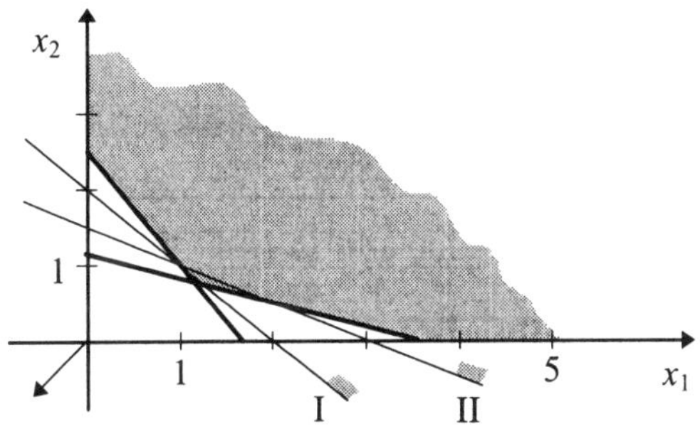

Figure I.21

Young (1965, 1968), Glover (1968), Ben-Israel and Charnes (1962) have used the Gomory cuts described above. However, whereas Gomory works with dual feasible tableaus that are not yet primal feasible, Young *et al.*'s method is primal. In other words, their method commences with a primal feasible all-integer tableau. First, the pivot element is determined in the usual way. In case the pivot equals +1, integrality will be retained in a tableau transformation. Suppose now that a pivot was found in row i, but the pivot does not equal 1. Then the i-th row is considered the source row and the usual Gomory cut (7), *viz.*,

$$\sum_j \left\lfloor \frac{a_{ij}}{|\lambda|} \right\rfloor x_j \leq \left\lfloor \frac{b_i}{|\lambda|} \right\rfloor$$

is introduced with an appropriate value of $|\lambda| > 1$, e.g., $|\lambda|$ equal to the pivot element. The pivot in the cutting plane row is then +1, and a single tableau transformation with this pivot element is performed.

Example: Consider the following all-integer programming problem.

$$\begin{aligned} \text{P: Max } z &= 3x_1 + 4x_2 \\ \text{s.t.} \quad 5x_1 - 4x_2 &\le 11 \\ 3x_1 + 5x_2 &\le 17 \\ x_1, \quad x_2 &\in \mathbb{N}_0. \end{aligned}$$

The initial tableau is

$\mathbf{T}_0$:

x_1	x_2	S_1	S_2	1
5	−4	1	0	11
3	5	0	1	17
−3	−4	0	0	0

Choosing the first column as pivot column, the "5" in the first row is the tentative pivot element. As $5 \ne 1$, we define a cutting plane with $|\lambda| = 5$ and obtain the cut $\left\lfloor \frac{5}{5} \right\rfloor x_1 + \left\lfloor -\frac{4}{5} \right\rfloor x_2 \le \left\lfloor \frac{11}{5} \right\rfloor$, or simply $x_1 - x_2 \le 2$. Adding this row to the tableau, choosing the element in the x_1 column in the cut as pivot element, and performing a regular simplex transformation with this pivot, we obtain the next tableau $\mathbf{T}_1$.

$\mathbf{T}_1$:

x_1	x_2	S_1	S_2	S_1^*	1
0	1	1	0	−5	1
0	8	0	1	−3	11
1	−1	0	0	1	2
0	−7	0	0	3	6

The tableau $\mathbf{T}_1$ is not yet optimal and at least one further iteration is required. Choosing the x_2 column as pivot column, the "1" in the first row is the pivot element, so that integrality will be retained without defining another cut. The pivoting continues for a number of steps until we obtain the final solution $\mathbf{x} = [2, 2]$ with $z = 14$. Here, we summarize the cuts that are defined in the process.

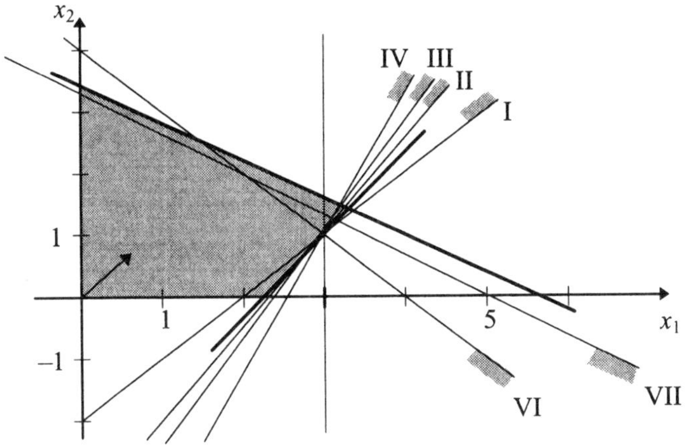

Figure I.22

$x_1 - x_2 \le 2$ (or $0.5x_1 - 0.5x_2 \le 1$)

$-S_1 + S_1^* \le 0$ (or $0.4444x_1 - 0.3333x_2 \le 1$)

$S_1 - 2S_2^* \le 0$ (or $0.4286x_1 - 0.2857x_2 \le 1$)

$S_2^* - 2S_3^* \le 0$ (or $0.4x_1 - 0.2x_2 \le 1$)

$S_3^* - 2S_4^* \le 0$ (or $0.3333x_1 \le 1$)

$S_4^* - 3S_5^* \le 0$ (or $0.25x_1 + 0.25x_2 \le 1$), and

$S_5^* - 3S_6^* \le 1$ (or $0.2x_1 + 0.3x_2 \le 1$).

Note that except for the last cut, all cutting planes lead through the point $[x_1, x_2] = [3, 1]$. Again we notice that the process is painfully slow even for very small examples. Figure I.22 shows the original constraints and the cuts generated in this procedure.

5.3 Cutting Plane Methods for Mixed Integer Programming

Consider now a mixed integer programming problem. Again, we assume that the linear programming relaxation has been solved and that we have at our disposal the optimal tableau whose values are denoted by b_i and a_{ij}. Suppose now that there exists at least one variable $x_k \in \mathbb{N}_0$ that currently assumes a value $x_k = b_i \notin \mathbb{N}_0$. Clearly, if no such variable exists, the problem is solved. The i-th row in the current tableau is then

$$x_k + \sum_{j \ne k} a_{ij} x_j = b_i \qquad (8)$$

with $b_i \notin \mathbb{N}_0$ and $x_k \in \mathbb{N}_0$.

Separating the integer from the noninteger part of the right-hand side by setting $b_i = \lfloor b_i \rfloor + f_i$ with $f_i \in\ ]0; 1[$, relation (8) can be written as $x_k + \sum_{j \neq k} a_{ij} x_j = \lfloor b_i \rfloor + f_i$.

Defining

$$\beta_i = \lfloor b_i \rfloor - x_k \in \mathbf{Z}, \tag{9}$$

allows us to write

$$\sum_{j \neq k} a_{ij} x_j = \beta_i + f_i. \tag{10}$$

We now separate the left-hand side into two parts, one with positive and the other with negative coefficients. Doing so, we obtain

$$\sum_{\substack{j \neq k \\ a_{ij} > 0}} a_{ij} x_j + \sum_{\substack{j \neq k \\ a_{ij} < 0}} a_{ij} x_j = \beta_i + f_i. \tag{11}$$

As $x_j \geq 0\ \forall\ j$ is required, the first sum on the left-hand side of relation (11) is nonnegative, whereas the second sum is nonpositive. Also note that as $\beta_i \in \mathbf{Z}$ and $f_i \in\ ]0; 1[$, the right-hand side of relation (11) cannot equal zero. Hence the left-hand side is either positive or negative; we consider each case separately.

Case 1: The left-hand side of relation (11) is positive.
In this case, the right-hand side must be positive as well, i.e., $\beta_i + f_i > 0$. Then $f_i \in\ ]0; 1[$ implies that

$$\beta_i \in \mathbb{N}_0. \tag{12}$$

Then we can write

$$\begin{aligned} \sum_{\substack{j \neq k \\ a_{ij} > 0}} a_{ij} x_j &\geq \sum_{\substack{j \neq k \\ a_{ij} > 0}} a_{ij} x_j + \sum_{\substack{j \neq k \\ a_{ij} < 0}} a_{ij} x_j && \text{(as } \textstyle\sum_{\substack{j \neq k \\ a_{ij} < 0}} a_{ij} x_j \leq 0\text{)} \\ &= \beta_i + f_i && \text{(by virtue of relation (11))} \\ &\geq f_i && \text{(by virtue of relation (12))} \end{aligned}$$

or simply

$$\sum_{\substack{j \neq k \\ a_{ij} > 0}} a_{ij} x_j \geq f_i. \tag{13}$$

Case 2: The left-hand side of relation (11) is negative.
Then the right-hand side of (11) must also be negative, i.e., $\beta_i + f_i < 0$. Now, $f_i \in\ ]0; 1[$ implies that

$$\beta_i \in \{-1, -2, -3, \ldots\}. \tag{14}$$

Then we can write

$$\begin{aligned}
\sum_{\substack{j \neq k \\ a_{ij} < 0}} a_{ij} x_j &\leq \sum_{\substack{j \neq k \\ a_{ij} > 0}} a_{ij} x_j + \sum_{\substack{j \neq k \\ a_{ij} < 0}} a_{ij} x_j && \Big(\text{as } \sum_{\substack{j \neq k \\ a_{ij} > 0}} a_{ij} x_j \geq 0\Big) \\
&= \beta_i + f_i && \text{(by virtue of relation (11))} \\
&\leq f_i - 1 && \text{(by virtue of relation (14))}
\end{aligned}$$

or simply

$$\sum_{\substack{j \neq k \\ a_{ij} < 0}} a_{ij} x_j \leq f_i - 1. \tag{15}$$

Note that since *either* case 1 *or* case 2 applies, *either* relation (13) *or* relation (15) is satisfied. Multiplying (15) by $f_i/(f_i - 1) < 0$ implies

$$\sum_{\substack{j \neq k \\ a_{ij} < 0}} \frac{f_i}{f_i - 1} a_{ij} x_j \geq f_i. \tag{16}$$

Note that in (13) as well as in (16), each term on the respective left-hand side is positive. As $\sum_{\substack{j \neq k \\ a_{ij} < 0}} \frac{f_i}{f_i - 1} a_{ij} x_j \geq 0$, relation (13) can be written as

$$\sum_{\substack{j \neq k \\ a_{ij} > 0}} a_{ij} x_j + \sum_{\substack{j \neq k \\ a_{ij} < 0}} \frac{f_i}{f_i - 1} a_{ij} x_j \geq \sum_{\substack{j \neq k \\ a_{ij} > 0}} a_{ij} x_j \geq f_i, \text{ or simply}$$

$$\sum_{\substack{j \neq k \\ a_{ij} > 0}} a_{ij} x_j + \sum_{\substack{j \neq k \\ a_{ij} < 0}} \frac{f_i}{f_i - 1} a_{ij} x_j \geq f_i. \tag{17}$$

Using the fact that $\sum_{\substack{j \neq k \\ a_{ij} > 0}} a_{ij} x_j \geq 0$, relation (16) can also be transformed to relation (17). Thus both cases generate the constraint (17), which is referred to as a *Gomory mixed-integer cutting plane.*

By construction, we have ensured that no feasible integer solution was cut off. Clearly, as the current solution has all variables with nonzero coefficients in the i-th row equal to zero, the left-hand side of relation (17) is currently zero, which is not greater than or equal to f_i on the right-hand side that by definition, is positive. This demonstrates that (17) is indeed a cutting plane.

Example: In order to illustrate the use of Gomory's mixed-integer cutting plane algorithm, consider the following problem.

$$\begin{aligned} \text{P: Max } z &= -2x_1 + 3x_2 \\ \text{s.t.} \quad & x_1 + 2x_2 \leq 3 \\ & -3x_1 + 5x_2 \leq 4 \\ & x_1 \geq 0 \\ & x_2 \in \mathbb{N}_0. \end{aligned}$$

The final tableau of the linear programming relaxation is

$\mathbf{T}_1$:

x_1	x_2	S_1	S_2	1
11/5	0	1	−2/5	7/5
−3/5	1	0	1/5	4/5
1/5	0	0	3/5	$2\frac{2}{5}$

As the current solution is $\mathbf{x} = [0, 4/5]^T$ and only x_2 is required to be integer, only the second row is eligible as source row. As $f_2 = 4/5$, Gomory's mixed-integer cutting plane is

$$\frac{\frac{4}{5}}{\frac{4}{5}-1}\left(-\tfrac{3}{5}\right)x_1 + \tfrac{1}{5}S_2 \geq \tfrac{4}{5},$$

or, after adding a new slack variable S_1^*,

$$-\tfrac{12}{5}x_1 - \tfrac{1}{5}S_2 + S_1^* = -\tfrac{4}{5}.$$

After one tableau transformation we obtain the optimal tableau

T_2:

x_1	x_2	S_1	S_2	S_1^*	1
0	0	1	– 7/12	11/12	2/3
0	1	0	1/4	– 1/4	1
1	0	0	1/12	– 5/12	1/3
0	0	0	7/12	1/12	2 ⅓

The current solution is **x** = [1/3, 1] which satisfies all required integrality conditions and is hence optimal. The graphical representation of the problem and the cutting plane are shown in Figure I.23.

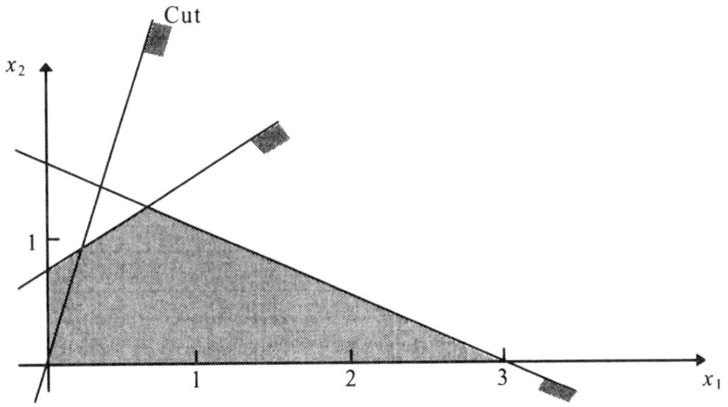

Figure I.23

In summary, all methods face three requirements for an optimal solution: primal feasibility, dual feasibility, and integrality. All methods in this section have in common that they satisfy two of the three requirements at any point during the algorithm (or, at least, after each suboptimization round), so that an optimal solution is reached once the third requirement is satisfied as well. The methods are distinguished on the basis of the criteria that are satisfied throughout the procedure. In particular, the first group consists of methods that satisfy integrality only at optimum; this group includes Dantzig cuts, Gomory's all-integer and mixed-integer cutting plane methods, as well as convexity cuts. The second group comprises methods that satisfy primal feasibility only at optimum; that group includes Gomory's dual cutting plane method. Finally, group 3 consists of primal cutting plane methods (such as that by Young *et al.*) that are characterized by achieving dual feasibility only at optimum. Table I.13 summarizes the main features of these groups.

Table I.13

	Rounding errors	Primal feasibility	Dual feasibility	Integrality
Group I	yes	after each suboptimization round	at all times	only at optimum
Group II	no	only at optimum	at all times	at all times
Group III	no	at all times	only at optimum	at all times

Considering only the two factors of the availability of an (implementable) primal solution at any point during the iterations and the lack of rounding errors, the primal cutting plane methods of group 3 appear superior among cutting plane methods.

CHAPTER 6 BRANCH AND BOUND METHODS

The purpose of this chapter is to introduce and illustrate the most popular and successful computational approach to integer programming problems today. Rather than being a specific algorithm, branch and bound is a general principle that allows the user to finetune the procedure and adjust it to the problem under consideration. The first section introduces the general idea, the second section discusses some specific strategies, and Section 3 then describes a general algorithm. Section 4 demonstrates that while branch and bound procedures have been applied successfully to even rather large problems in practice, the algorithm may fail to converge within a reasonable time even if applied to some very simple problems. The fifth section introduces integer programming duality and general relaxation, and Section 6 explains the concepts of Lagrangean decomposition.

6.1 Basic principles

In this section we provide a description of branch and bound methods which are the foundation of most current commercial codes for integer programming problems. Branch and bound programming is a quite general principle that uses two techniques discussed in Section 1.2. The branch and bound principle was first suggested by Land and Doig (1960), followed by Dakin (1965) and Balas's (1965) additive algorithm for zero-one problems. Some more recent and comprehensive books are those by Schrijver (1986), Nemhauser and Wolsey (1988), and Parker and Rardin (1988). One of these techniques is to relax a mathematical optimization problem by dropping some of its constraints, so that the resulting problem has a larger feasible region and a solution that cannot be worse than that of the original problem. An other technique is to add constraints to a problem, a process that reduces the size of the feasible region, so that the optimal objective value deteriorates. Branch and bound methods can also be viewed as cutting plane methods with an added feature that divides the feasible region into smaller parts which are then dealt with separately. Each successive cut results in a linear programming problem that is represented by a node in a graph, where the arcs correspond to the cuts. The graph is developed and explored as the cuts are successively applied, with the initial node n_1 corresponding to the relaxation of the problem without any additional constraints. By construction, the graph will be a tree, and if we regard its arcs as directed corresponding to the order in which the cuts are applied, we have an arborescence with the initial node n_1 as its root. As the process is directed from n_1 to nodes representing problems with additional cuts, the proper term would be "branch and bound arborescence." However, we will conform to common practice and use the term "branch and bound tree." In a branch and bound tree, the initial node n_1 is usually referred to as the *top node*.

Example 1: As a brief explanation of the basic principle, consider the following problem

$$\begin{aligned} P_{IP}:\ \text{Max } z &= 2x_1 + 2x_2 \\ \text{s.t.} \quad x_1 + 3x_2 &\le 3 \\ 3x_1 + x_2 &\le 6 \\ x_1, \quad x_2 &\in \mathbb{N}_0 \end{aligned}$$

The initial step is to solve the linear programming relaxation P_{LP}, obtained from P_{IP} by relaxing the integrality conditions, and replacing them by the requirements $x_1, x_2 \ge 0$. The feasible set of P_{LP} is shown in Figure I.24.

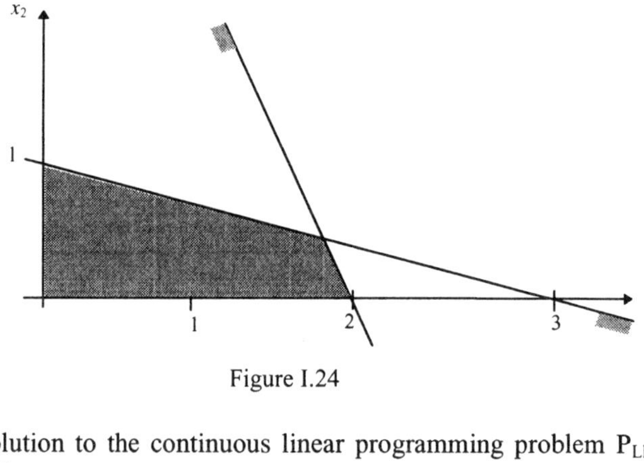

Figure I.24

The optimal solution to the continuous linear programming problem P_{LP} is $x_1 = 1\frac{7}{8}$ and $x_2 = \frac{3}{8}$ with objective value $z_{LP} = 4\frac{1}{2}$. Since the solution does not satisfy the integrality requirements, we choose a variable whose current value is not integer, but which is required to be integer in the original problem; this is the variable we will branch on. In our example, we arbitrarily choose x_2. The process of branching is then as follows. Since $0 < x_2 = \frac{3}{8} < 1$, the optimal (and indeed any) integer solution satisfies either $x_2 \le 0$ or $x_2 \ge 1$. This means that we can partition the original problem into two new problems, a process referred to as "divide and conquer" or, as some cynics say, "double your trouble". One of the two new problems contains all original constraints plus the new constraint $x_2 \le 0$ while the other problem comprises all original constraints plus the new constraint $x_2 \ge 1$. Both new linear programming problems are then solved. This process is called *branching* on the variable x_2. The result of the branching is shown in the *tree* in Figure I.25.

In our example, both branches immediately yield integer solutions, so that the process terminates. The optimum is the solution with the highest objective value among the nodes we did not branch from, *viz.*, $x_1 = 2$ and $x_2 = 0$ with $z = 4$.

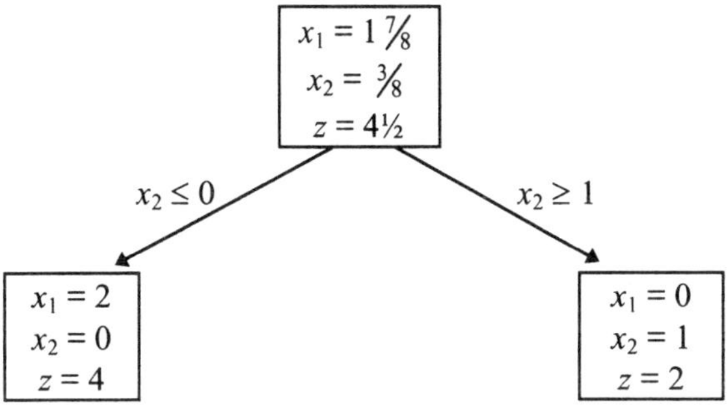

Figure I.25

Example 2: In order to demonstrate how the branch and bound process continues if integer solutions are not found in the first step, consider the following all-integer linear programming problem.

$$\begin{aligned} \text{P}_{\text{IP}}\text{: Max } z &= 4x_1 + 6x_2 + 10x_3 \\ \text{s.t.} \quad & 2x_1 + x_2 + 2x_3 \le 13 \\ & 3x_1 - 2x_2 - 4x_3 \ge 7 \\ & x_1, \; x_2, \; x_3 \in \mathbb{N}_0. \end{aligned}$$

Suppose that the problem is solved by a branch and bound method that at each node branches on the noninteger variable with the smallest subscript and then pursues the left branch. The tree in Figure I.26 shows the results obtained so far. Only four problems have been solved since solutions for nodes or branches that lead to the right have not been calculated. Node n_4 represents the only integer solution known at this point. Note that as we proceed down the tree, more constraints are added and the problems become more constrained. As an example, the problem at node n_4 includes the constraints $x_1 \leq 4$, $x_2 \leq 2$, and $x_3 \leq 0$ in addition to the original two constraints of the relaxed problem P_{LP}. For ease of notation, we denote by z_i the objective value associated with the node n_i, so that $z_{LP} = z_1$.

A few observations are in order. Since the objective function has only integer coefficients, its value must be integral for any integer solution. Using the notation from Section 1.2, $40\frac{2}{7} = \bar{z}_{LP} \geq \bar{z}_{IP}$, which implies that $\bar{z}_{IP} \leq 40$. Furthermore, each node n_i in the tree can be reached from the top node n_1 on a unique path.

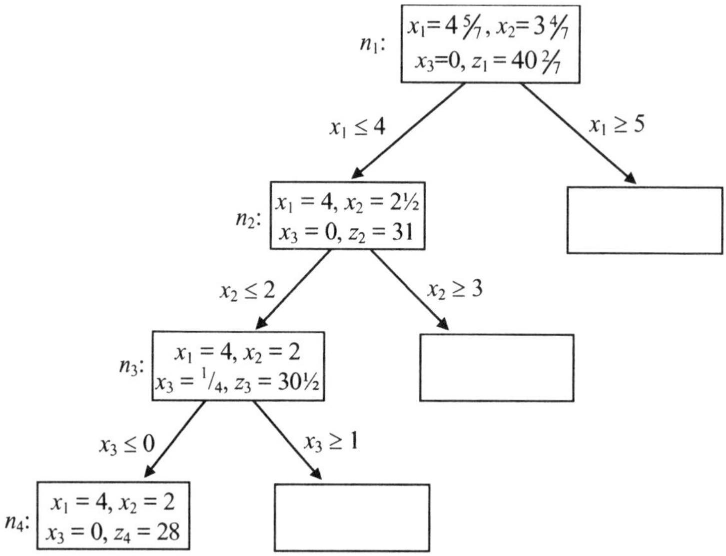

Figure I.26

The graph structure of the solution tree imposes a partial ordering among all nodes; in particular, it uniquely defines *predecessors* and *successors* of any given node. The top node is a predecessor of all other nodes and successor of no other node. Sometimes the immediate successor of a node is called a *child* of the node; similarly, the immediate predecessor or a node would be called the *parent*. For each node, all the constraints of its predecessors apply, plus an additional constraint generated in the branching process. In a maximization problem, all successor nodes have objective function values that are no larger than those of their predecessors. Suppose now that the nodes are numbered in the order in which they are considered, and let z_k denote the objective function value corresponding to node n_k. In our example, $\bar{z}_{\text{LP}} = z_1 = 40\frac{2}{7} \geq z_2 = 31 \geq z_3 = 30\frac{1}{2} \geq z_4 = 28$. Since the solution at node n_4 is integer, we conclude that a lower bound on the objective value of the optimal integer solution is provided by the best known z-value, here $\bar{z}_{\text{IP}} \geq z_4 = 28$. Combining this with the best known upper bound $\bar{z}_{\text{IP}} \leq 40$, obtained above, we find $28 \leq \bar{z}_{\text{IP}} \leq 40$. This is the *bounding* in a branch and bound technique.

As there is still a gap between the lower and upper bound at n_4, the process continues by *backtracking* until a currently unexplored node is found. There are various strategies to choose the node at which the process continues; *depth-first search* as applied in the above example and *breadth-first search* are typical examples. We will discuss various search strategies in the next section. In our example, if we continue the process by pursuing the left branch at each node first, always branching on the resulting noninteger variable with the smallest subscript and backtracking whenever a node is reached from which we cannot proceed, we obtain the full tree displayed in Figure I.27. The nodes n_k are numbered in the order in which they are pursued with the solution. If a node n_k has an integer solution, or corresponds to an infeasible solution, the tree cannot grow further from such a node; this is indicated by a bar under the node. A third reason for stopping the process at a node is explained below.

The unique optimal integer solution at node n_{11} is $\bar{x}_1 = 5$, $\bar{x}_2 = 3$, and $\bar{x}_3 = 0$ with $z_{11} = \bar{z}_{\text{IP}} = \bar{z} = 38$. Additional considerations could have simplified the procedure. For example, after solving the linear programming problem at n_6 with a noninteger solution $z_6 = 28\frac{1}{2}$, it was apparent that the z-value of any integer solution at all its successor nodes is no larger than 28, which is no better than the best integer solution found so far at n_4 where $z_4 = 28$. If we are only interested in one optimal solution, work on the tree would therefore not have to be pursued beyond node n_6. And if z_5 had been, say, 27.9, no successor node of n_5 could have had an integer solution with objective function value better than 27; we could then have started the backtracking already at node n_5. In this way, the tree can be *pruned*. In general, a node at which the tree can be pruned because its potential successors could have only poorer integer solutions than have already been found elsewhere in the tree is said to be *fathomed* or *value dominated*. This is another reason for limiting the growth of a tree. Fathoming is the key reason why many codes attempt to find good integer solutions as early as possible in the process. In that way, large parts of the solution tree need not be explored. Finally, we should mention that in modern branch and bound techniques, no more than two branches emanate from each of the nodes, as distinct from the original approach of Land and Doig (1960).

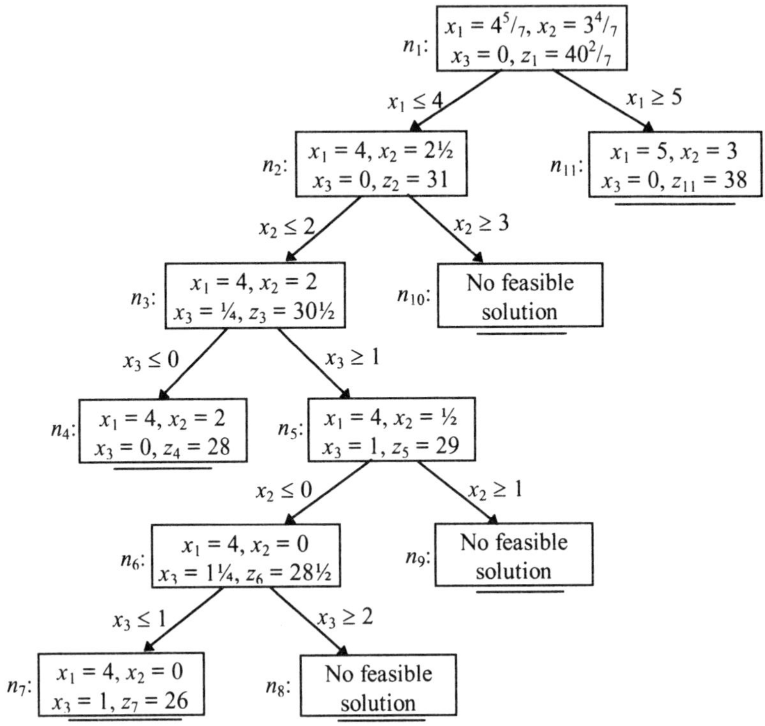

Figure I.27

6.2 Search strategies

Since each node of the solution tree represents a linear programming problem that must be solved in the process, it is obviously important to keep the size of the tree as small as possible. As demonstrated in the previous section, this is accomplished by stopping at a node whose associated problem has an integral solution, is infeasible, or is fathomed (value dominated). A good branch and bound method will bring about such favorable situations as early as possible in the search. In example 2 in the previous section we pursued each path until integrality or infeasibility was reached. This is called a *depth-first search* strategy. The *depth* (or *level*) of a node is the length of the path leading to it from the top node; in our example, node n_6 is at depth 4 whereas node n_{10} is at depth 2. Similarly, the solution tree has grown to a depth of 5. In general, a depth first strategy will pursue each node, using some branching strategy, until integrality, infeasibility, or fathoming has occurred, and then backtrack until the first node is encountered, which can be further pursued. This strategy will bring us to high levels (deep in the tree) early in the search. By contrast, a *breadth-first search* strategy explores all nodes at one level before proceeding to the next. If in our example a breadth first search strategy had been used with the same branching strategy (branch on the noninteger variable with the smallest index and pursue the left branch first), we would have obtained the branch and bound tree displayed in Figure I.28.

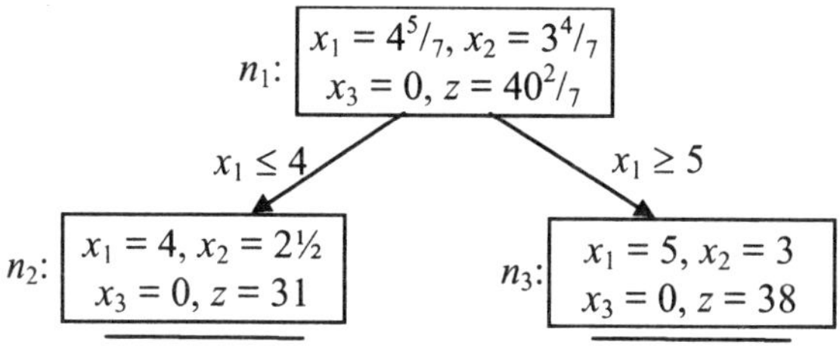

Figure I.28

In Figure I.41, node n_2 is fathomed due to the z-value of 38 obtained at node n_3. We conclude that in this particular example, the breadth-first search strategy was much better than the depth-first search strategy in the sense that the tree is much smaller. In practice, however, it appears that depth-first strategies are more advantageous than breadth-first. One reason for this is that the number of nodes grows exponentially as the level increases, and since feasible (integer) solutions are typically found deeper into the tree, depth-first search tends to find integer solutions sooner than breadth-first search. Another appealing feature of depth-first search is related to the way in which the linear programming problems at each node are solved. Assuming that the simplex method is used for solving the problems, an immediate successor of a node has the same set of constraints as its predecessor, plus the one additional constraint generated in the branching. The dual simplex method is therefore ideally suited for this situation and we will now illustrate how the solution at node n_2 in our example is obtained once the solution at the top node n_1 has been found. The linear programming relaxation of the problem in Example 2 is

$$\begin{aligned}
\text{P}_{\text{LP}}:\ \text{Max } z = z_1 &= 4x_1 + 6x_2 + 10x_3 \\
\text{s.t.} \quad & 2x_1 + x_2 + 2x_3 \le 13 \\
& 3x_1 - 2x_2 - 4x_3 \ge 7 \\
& x_1,\ x_2,\ x_3 \ge 0
\end{aligned}$$

With the slack variable S_1 added to the left-hand side of the first constraint and the excess variable E_2 subtracted from the left-hand side of the second constraint, the primal simplex method is used to solve P_{LP}. The final tableau is then

$\mathbf{T}_0$:

x_1	x_2	x_3	S_1	E_2	1
0	1	2	$^3/_7$	$^2/_7$	$3^4/_7$
1	0	0	$^2/_7$	$-^1/_7$	$4^5/_7$
0	0	2	$^{26}/_7$	$^8/_7$	$40^2/_7$

Tableau $\mathbf{T}_0$ corresponds to the solution obtained at node n_1. Since we decided to branch on x_1, we add the constraint $x_1 \leq 4$. Expressing this constraint in terms of nonbasic variables, we obtain $4^5/_7 - {}^2/_7 S_1 + {}^1/_7 E_2 \leq 4$ or $-{}^2/_7 S_1 + {}^1/_7 E_2 \leq -{}^5/_7$. Introducing the slack variable S_1^* for this constraint, the new tableau is

$\mathbf{T}_0'$:

x_1	x_2	x_3	S_1	E_2	S_1^*	1
0	1	2	$^3/_7$	$^2/_7$	0	$3^4/_7$
1	0	0	$^2/_7$	$^{-1}/_7$	0	$4^5/_7$
0	0	0	($^{-2}/_7$)	$^1/_7$	1	$^{-5}/_7$
0	0	2	$^{26}/_7$	$^8/_7$	0	$40^2/_7$

Performing one tableau transformation step of the dual simplex method using the the circled element in tableau $\mathbf{T}_0'$ as pivot, we obtain the next tableau

$\mathbf{T}_1$:

x_1	x_2	x_3	S_1	E_2	S_1^*	1
0	1	2	0	½	$^3/_2$	2½
1	0	0	0	0	1	4
0	0	0	1	−½	$-^7/_2$	2½
0	0	2	0	3	13	31

This tableau is optimal and corresponds to the solution at node n_2. Pursuing the branch $x_2 \leq 2$ amounts to adding the constraint $2½ - 2x_3 - ½E_2 - {}^3/_2 S_1^* \leq 2$ or $-2x_3 - ½E_2 - {}^3/_2 S_1^* \leq -½$. Adding a slack variable S_2^* to the left hand side of this constraint, we obtain the tableau

$\mathbf{T}_1'$:

x_1	x_2	x_3	S_1	E_2	S_1^*	S_2^*	1
0	1	2	0	½	$^3/_2$	0	2½
1	0	0	0	0	1	0	4
0	0	0	1	–½	$-^7/_2$	0	2½
0	0	(–2)	0	–½	$-^3/_2$	1	–½
0	0	2	0	3	13	0	31

Pivoting on the circled element in Tableau $\mathbf{T}_1'$, the new tableau after one dual simplex step is

$\mathbf{T}_2$:

x_1	x_2	x_3	S_1	E_2	S_1^*	S_2^*	1
0	1	0	0	0	0	1	2
1	0	0	0	0	1	0	4
0	0	0	1	–½	$-^7/_2$	0	2½
0	0	1	0	¼	¾	–½	¼
0	0	0	0	$^5/_2$	$^{23}/_2$	1	30½

This optimal tableau corresponds to the solution at node n_3. Finally, the branch $x_3 \leq 0$ corresponds to adding the constraint $¼ - ¼E_2 - ¾\,S_1^* + ½\,S_2^* \leq 0$ or $-¼E_2 - ¾\,S_1^* + ½\,S_2^* \leq -¼$. With the slack S_3^* for this constraint, we obtain tableau

$\mathbf{T}_2'$:

x_1	x_2	x_3	S_1	E_2	S_1^*	S_2^*	S_3^*	1
0	1	0	0	0	0	1	0	2
1	0	0	0	0	1	0	0	4
0	0	0	1	–½	$-^7/_2$	0	0	2½
0	0	1	0	¼	¾	–½	0	¼
0	0	0	0	(–¼)	–¾	½	1	–¼
0	0	0	0	$^5/_2$	$^{23}/_2$	1	0	30½

After one dual simplex step we obtain the tableau

$\mathbf{T}_3$:

x_1	x_2	x_3	S_1	E_2	S_1^*	S_2^*	S_3^*	1
0	1	0	0	0	0	1	0	2
1	0	0	0	0	1	0	0	4
0	0	0	1	0	−2	−1	−2	3
0	0	1	0	0	0	0	1	0
0	0	0	0	1	3	−2	−4	1
0	0	0	0	0	4	6	10	28

Tableau $\mathbf{T}_3$ corresponds to the solution at node n_4, which is all-integer so that no further branching is necessary. To illustrate the branching leading to a node with no feasible solution, consider again the tableau $\mathbf{T}_1$ and pursue the branch $x_2 \geq 3$ leading to node n_{10} of Figure I.40. Expressed in terms of nonbasic variables, $x_2 \geq 3$ is rewritten as $2\frac{1}{2} - 2x_3 - \frac{1}{2}E_2 - {}^3/_2\,S_1^* \geq 3$ or $2x_3 + \frac{1}{2}E_2 + {}^3/_2\,S_1^* \leq -\frac{1}{2}$. Adding a slack variable S_2^{**} to the left-hand side of this constraint, the tableau is

$\mathbf{T}_1''$:

x_1	x_2	x_3	S_1	E_2	S_1^*	S_2^{**}	1
0	1	2	0	½	$^3/_2$	0	2½
1	0	0	0	0	1	0	4
0	0	0	1	−½	$-^7/_2$	0	2½
0	0	2	0	½	$^3/_2$	1	−½
0	0	2	0	3	13	0	31

The fourth row in $\mathbf{T}_1''$ with a negative right-hand side of $-\frac{1}{2}$ is the only eligible pivot row. However, all coefficients on the left-hand side in this row are nonnegative, so that no dual simplex pivot step is possible. We conclude that no feasible solution exists for node n_{10}.

In general, with the optimal simplex tableau available for a node, the solutions for its immediate successor nodes are obtained by first adding a row to the tableau that corresponds to the *branching cut* (the additional constraint), and then employing the dual simplex method to determine an optimal solution for the next node. In our example, only one dual simplex step was needed for each successor node, and in general only a few iterations are required to find the optimal solution of the next node. It is worth noting that although developing a branch and bound tree involves solving a linear programming problem at each node of the tree, each problem except for the top node can start the optimization from the optimal solution of the immediate predecessor node, whose solution was made infeasible by the addition of the new constraint (a so-called "hot start"). We observe that every branching cut will be binding immediately after it has been applied, if the next solution is feasible. The reason is that if the branching cut were not binding, it could be deleted, resulting in the same solution as previously, which contradicts the fact that both branching cuts cut off the previous solution. However, further down the tree, branching cuts may become again non-binding. As an example, consider the node n_3 in Figure I.40. The constraint $x_3 \geq 1$ on the right branch out of n_3 is no longer binding at node n_6, where $x_3 = 1\frac{1}{4}$. In general, suppose that at a node we branch on the same variable that has already been branched on at some predecessor node. The new branching constraint must then be tighter than the previous one on the same variable, which could therefore be deleted. In the dual simplex tableau, the new row would therefore simply replace the row that corresponds to the previous branching on the same variable. Since branching cuts are $\leq$ or $\geq$ constraints, it follows that the number of additional rows in the tableau never exceeds twice the number of integer restricted variables in the problem.

Back to the construction of a branch and bound tree, there are two decisions to be made each time a branching is to be done, *viz.*:

- which node to pursue next (*node selection strategy*), and
- which variable to branch on (*branch selection strategy*).

We will discuss these two types of strategies in the next two sections.

6.2.1 Node Selection

A number of strategies exist for selecting the next node to branch from. For simplicity of the exposition, assume again that the original problem has a maximization objective. If so far in the search one or more nodes have been encountered for which the solution is integer, we choose the one with the highest objective function value (ties are broken arbitrarily) and call it the *incumbent node* or solution. The collection of nodes that have no branches leading out of them forms the set S. Initially, only the top node is in S.

The nodes in the set S fall into four categories.

(1) The solution at a node $n_i \in S$ is integer, in which case we will not branch further from this node. If the solution is better than that of the existing incumbent(s) n_k, then n_i replaces n_k; if it is as good as n_k it becomes another incumbent.

(2) The node $n_i \in S$ represents a problem that has no feasible solution, making branching on it pointless, since any successor problem will also lack feasible

(3) solutions.

(4) The node n_i is fathomed, i.e., its noninteger solution is no better than the incumbent solution. Again, its potential successors have poorer solutions than the incumbent and are therefore not of interest.

(5) The node n_i has a noninteger solution with an objective function value that is better than that of the incumbent, so that branching from this node might lead to an integer optimal solution. Such a node is called *live* or *active*.

In summary, out of the four possible node types—integer, infeasible, fathomed, and live—we only branch from live nodes. Defining $L \subset S$ as the set of live nodes, the node selection problem addresses the choice of a node $n_i \in L$.

We have already described two node selection strategies: the breadth-first strategy where all live nodes at a given level are considered before nodes on lower levels are examined, and the depth-first strategy where the next node to be considered is a live successor of the latest node that was explored. Since backtracking is needed if the node that was explored last is not live, the strategy is better described as "depth-first with backtracking"; or *last in, first out* (LIFO), borrowing a concept from inventory management. Although a depth-first strategy can be expected to perform better than a breadth-first strategy on average, both of these strategies would be outperformed, again on average, by strategies that make better use of the information gathered during the construction of the tree. One such strategy is the *best-bound-first* strategy that selects the live node with the largest z-value. Formally, the best-bound-first strategy selects the node n_k such that $z_k = \max\{z_\ell: n_\ell \in L\}$.

Another node selection strategy involves the estimate $\hat{z}_\ell$ of the integer optimal solution corresponding to node n_ℓ. Such an estimate can be computed based on the expected *degradation* expressing the deterioration of z_ℓ by requiring the solution point at node n_ℓ to be integral. More specifically, let the solution at node n_i include $\tilde{x}_j = \lfloor \tilde{x}_j \rfloor + f_j$ with $f_j \neq 0$. Using some user-selected coefficients p_j^- and p_j^+, we estimate the decrease in the objective function of $p_j^- f_j$ for branching left at node n_i and of $p_j^+(1 - f_j)$ if branching right. The coefficients p_j^- and p_j^+ can either be user-specified or estimated, e.g., by using dual information at node n_i or information from previous branchings on x_j. A *best-estimate* search strategy selects the node n_k such that $\hat{z}_k = \max\{\hat{z}_\ell : n_\ell \in L\}$. Denoting by $\underline{z}_{\text{IP}}$ the objective function value of the incumbent solution, the *quick-improvement* strategy attempts to quickly improve on the incumbent solution by selecting the node n_k such that $k = \arg\max\{(z_\ell - \underline{z}_{\mathrm{IP}}) / (z_\ell - \hat{z}_\ell)\}$. For further discussion and details on node selection, the reader is referred to Nemhauser and Wolsey (1988).

6.2.2 Branch Selection

Once a node $n_k \in L$ has been selected for further exploration, the next decision concerns the choice of variable to be branched on. Clearly, this variable, while required to be integer, must currently have a noninteger value. In the example above we used the simple strategy of branching on the noninteger variable with lowest index, an obviously arbitrary rule. Another strategy is to branch on the noninteger variable x_k with the "most fractional" value. Formally, let $\tilde{\mathbf{x}} = [\tilde{x}_1, \tilde{x}_2, \ldots, \tilde{x}_n]$ be the solution at the chosen live node, and let $\tilde{x}_j = \lfloor \tilde{x}_j \rfloor + f_j$, so that f_j denotes the *fractional* part of $\tilde{x}_j$. The *most fractional* strategy would branch on the variable whose present value is farthest from the nearest integer or, equivalently, has the value closest to ½, i.e., a variable x_k with $k = \underset{j=1,\ldots,n}{\arg\min}\,\{|f_j - ½|\}$. Unfortunately, experience has failed to identify robust methods for branch selection, and in practice user-specified priorities are employed. Other methods involve degradation measures similar to those in Subsection 6.2.1. More involved methods that employ penalties and use more elaborate computations regarding the penalty coefficients p_j^- and p_j^+ have not turned out to improve the overall efficiency of the search if the additional computational effort it takes to apply them is taken into account.

6.3 A General Branch and Bound Procedure

We are now able to formulate a general branch and bound algorithm. The problem to be solved is an all-integer or mixed-integer programming problem P_{IP} with a maximization objective; its linear programming relaxation is called P_{LP}. It is assumed that specific node and branch selection strategies have been chosen. The algorithm is initialized with node n_1 that includes the optimal linear programming relaxation $\bar{\mathbf{x}}_{\mathrm{LP}}$ with objective value $\bar{z}_{\mathrm{LP}}$. Given that $\bar{\mathbf{x}}_{\mathrm{LP}}$ does not satisfy all of the integrality conditions (otherwise $\bar{\mathbf{x}}_{\mathrm{LP}} = \bar{\mathbf{x}}_{\mathrm{IP}}$ is optimal for the (mixed) integer programming problem as well), set $S := L := \{n_1\}$, $t := 1$, and $\underline{z} := -\infty$.

A General Branch and Bound Algorithm

Step 1: Is $L = \varnothing$? If yes: If $\underline{z} > -\infty$, then the solution $\bar{\mathbf{x}}_{\text{IP}} = \underline{\mathbf{x}}$ with objective value $\bar{z}_{\text{LP}} = \underline{z}$ is optimal, otherwise the integer programming problem has no feasible solution.

If no: Go to Step 2.

Step 2: Choose some node $n_i \in L$ (according to some prescribed criterion) with solution $\mathbf{x}^i$ and objective value z_i.

Step 3: Branch on some variable x_j (according to some prescribed criterion) to the nodes n_{t+1} (with $x_j \le \lfloor x_j \rfloor$) and n_{t+2} (with $x_j \ge \lceil x_j \rceil$). set $L := L \setminus \{n_i\}$, and $S := S \cup \{n_{t+1}, n_{t+2}\} \setminus \{n_i\}$. Solve the linear programming problems at n_{t+1} and n_{t+2}. The results are solutions $\mathbf{x}^{t+1}$ and $\mathbf{x}^{t+2}$ with objective values z_{t+1} and z_{t+2}, respectively.

Step 4: Is $\mathbf{x}^{t+1}$ feasible? If yes: Go to Step 5.

If no: Go to Step 7.

Step 5: Does $\mathbf{x}^{t+1}$ satisfy the integrality requirements of the original problem?

If yes: Go to Step 6.

If no: Set $L := L \cup \{n_{t+1}\}$ and go to Step 7.

Step 6: Is $z_{t+1} > \underline{z}$? If yes: Set $\underline{\mathbf{x}} := \mathbf{x}^{t+1}$, $\underline{z} := z_{t+1}$, and go to Step 7.

If no: Go to Step 7.

Step 7: Repeat Steps 4-6 with $t + 2$ instead of $t + 1$; then set $t := t + 2$ and go to Step 1.

The key to the algorithm is the updating procedure of the sets S and L. In each step when the procedure branches from some node n_i to two nodes n_{t+1} and n_{t+2}, the set of end nodes S is updated to include the new nodes n_{t+1} and n_{t+2} and to exclude n_i. From the set of live nodes L, the node from which the branching takes place is deleted in Step 3, and node n_{t+1} (or n_{t+2}) is added to the set in Step 5 only if its solution is feasible (Step 4) but not yet integer (Step 5).

Example: As an illustration of the above algorithm, consider again the problem of Example 2 of Section 6.1. Here, we choose the node with the best objective value to be branched on next, and we select the "most fractional" variable for the branching. The resulting branch and bound tree is shown in Figure I.29.

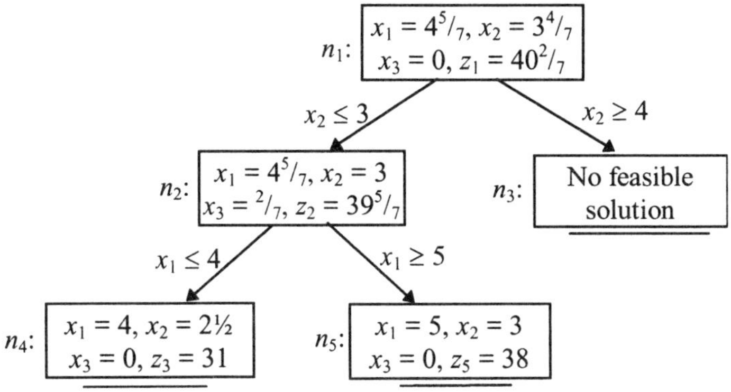

Figure I.29

No further branching is possible due to infeasibility at node n_3, fathoming for node n_4 and integrality for node n_5. The set of L of live nodes during the algorithm is $\{n_1\}$, $\{n_2\}$, $\{n_4\}$, and $\emptyset$. The solution tree contains only five nodes, whereas eleven nodes were required for the same problem with a depth-first node strategy. If a best-bound-first node and a lowest-index variable strategy had been selected, only three nodes would have been required, obtaining the same tree as in Figure I.28. This may demonstrate the difficulty in finding the best strategies to use when a particular problem is being solved.

There is an interesting analogy between cutting plane algorithms and branch and bound methods. We may view the branching constraints as vertical or horizontal cutting planes designed to cut off areas of the feasible region that do not contain any integer points. It is also possible to mix in regular cutting planes with the branching at the nodes of a branch and bound tree. This approach is called *branch and cut*, and is typically used for zero-one problems as well as problems with special structures. The idea is to find valid inequalities for the original problem, which are violated at some nodes in a branch and bound tree. These valid inequalities are then added at these nodes in a cutting plane fashion, thus generating new nodes from which branching can be done as usual. This could be accomplished by introducing an additional step between the existing Steps 2 and 3 in the general branch and bound algorithm above. For details, the reader is referred to Hoffman and Padberg (1985) and Rardin (1998).

6.4 Difficult Problems

When addressing the issue of computational complexity, it is well known that algorithms which perform poorly in the worst case may work well in practice under average conditions. The simplex method for linear programming is a typical example. While the method solves thousands of large-scale practical problems each day, Klee and Minty's (1972) simple examples force the method with the steepest unit ascent pivot column selection rule to go through an exponential number of iterations in the solution process. A similar argument applies to branch and bound procedures. In most cases, branch and bound methods work quite well in practice, considering the fact that integer programming problems are inherently more difficult than linear programming problems. It was Jeroslow (1974) who provided very simple integer linear programming problems, for which branch and bound methods require solution trees with an exponential number of nodes.

In particular, consider the zero-one integer programming problem

$$\begin{aligned} \mathrm{P}_1(n):\ \ & \operatorname{Max} z = y_n \\ \text{s.t.}\ \ & 2y_1 + 2y_2 + \ldots + 2y_n = n \\ & y_j = 0 \vee 1 \ \forall j = 1, \ldots, n \end{aligned}$$

where n is an odd number. Clearly, the left-hand side of the knapsack constraint is even, whereas the right-hand side is odd; hence $\mathrm{P}_1(n)$ has no feasible solution. Applying the branch and bound method of the previous section to the problem $\mathrm{P}_1(3)$ results in the solution tree of Figure I.30.

Solving the problem's linear programming relaxation, the solution at node n_1 results. Branching on y_1 yields the solutions at nodes n_2 and n_3. The process continues at n_2 due to its higher objective value. Branching on y_2 results in n_4 and n_5. The procedure may now continue with either n_5 or n_3. Either way, in three more branchings it is discovered that the problem has no feasible solutions. Worse yet, this result is not specific to this instance of the problem. As a matter of fact, in his paper, Jeroslow proves that *any* branch and bound procedure must explore $2^{\lceil \frac{1}{2} n \rceil}$ nodes before discovering that $\mathrm{P}_1(n)$ does not possess feasible solutions.

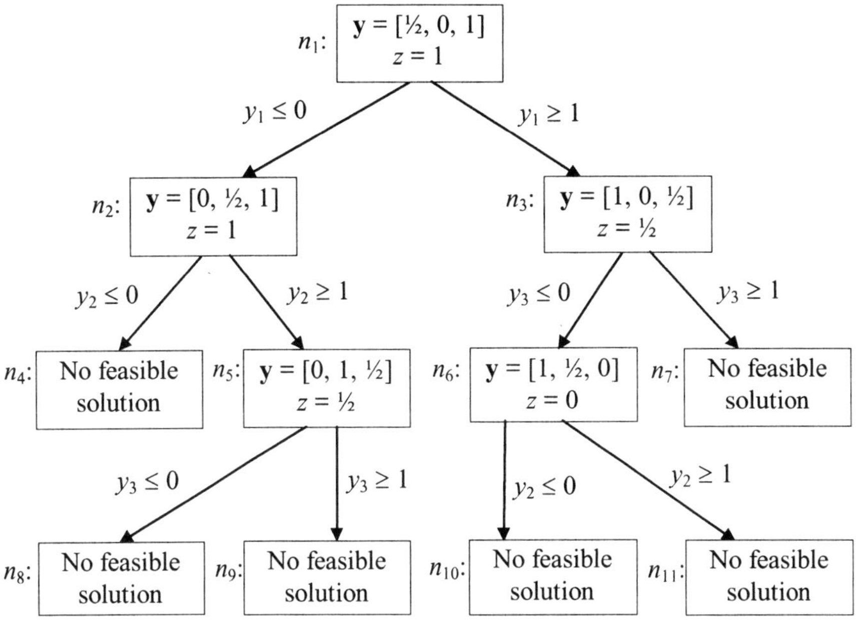

Figure I.30

It may now be conjectured that only problems with no feasible solutions exhibit such convergence problems and that a good phase I method that finds feasible integer solutions will overcome the need for searching an exponential number of nodes. This is, however, not the case. As an illustration, consider the problem

$$\begin{aligned} P_2(n):\ & \text{Max } z = y_n \\ \text{s.t. } & 2y_1 + 2y_2 + \ldots + 2y_{n-1} + y_n = n \\ & y_j = 0 \vee 1 \quad \forall\, j = 1, \ldots, n. \end{aligned}$$

For odd values of n, all feasible solutions have $y_n = 1$, while, for even values of n, all feasible solutions have $y_n = 0$. Since $y_n = 1$ at the top node, for even n we will then obtain a tree identical to the solution tree of $P_1(n-1)$, requiring again the search of an exponential number of nodes before realizing that y_n must take the value zero. This contradicts the conjecture that only problems without feasible solutions may have convergence problems. These examples, albeit contrived, emphasize the need for efficient upper and lower bounding techniques in solution techniques for integer programming problems.

6.5 Integer Programming Duality and Relaxation

We have seen that obtaining tight bounds on the optimal objective function value of an integer programming problem is an important part of many solution algorithms. We will now show how duality concepts can be useful in this regard. Consider therefore the general all-integer programming problem

$$\begin{aligned} \text{P}_{\text{IP}}\text{: Max } z_{\text{IP}} &= \mathbf{cy} \\ \text{s.t. } \quad \mathbf{Ay} &\leq \mathbf{b} \\ \mathbf{y} &\in \mathbb{N}_0^n \end{aligned}$$

where $\mathbf{A}$ is an $[n \times m]$-dimensional matrix, and the linear programming relaxation of P_{IP}, which we will call the *primal* (*linear programming*) *relaxation* of P_{IP}. Following Definition I.5, this primal relaxation can formally be written as

$$\begin{aligned} \text{P}_{\text{LP}}\text{: Max } z_{\text{LP}} &= \mathbf{cx} \\ \text{s.t. } \quad \mathbf{Ax} &\leq \mathbf{b} \\ \mathbf{x} &\in \mathbb{R}_+^n \end{aligned}$$

If $\bar{\mathbf{y}}$ is optimal for P_{IP} and $\bar{\mathbf{x}}$ is optimal for P_{LP}, we know that $\bar{z}_{\text{IP}} = \mathbf{c}\bar{\mathbf{y}} \leq \mathbf{c}\bar{\mathbf{x}} = \bar{z}_{\text{LP}}$, since each $\mathbf{y} \in \mathbb{N}_0^n$ that is feasible for P_{IP} is also feasible for P_{LP}. A simple consequence of this observation is that if $\tilde{\mathbf{y}}$ is feasible for P_{IP} and $\mathbf{c}\tilde{\mathbf{y}} = \bar{z}_{\text{LP}}$, then $\tilde{\mathbf{y}}$ must be optimal for P_{IP}. Since $0 \leq \bar{z}_{\text{LP}} - \bar{z}_{\text{IP}}$ is the integrality gap as formally introduced in Definition I.6, this condition could only occur if the integrality gap is zero, a rather unlikely situation. However, even for problems with a positive integrality gap, the magnitude of $\bar{z}_{\text{LP}} - \mathbf{c}\tilde{\mathbf{y}}$ can be used as an indicator of the quality of the solution $\tilde{\mathbf{y}}$ for P_{IP}. The smaller the value of $\bar{z}_{\text{LP}} - \mathbf{c}\tilde{\mathbf{y}}$, the better the solution although $\bar{z}_{\text{LP}} - \mathbf{c}\tilde{\mathbf{y}}$ is lower bounded by the integrality gap $\bar{z}_{\text{LP}} - \bar{z}_{\text{IP}}$. Given any feasible solution $\tilde{\mathbf{y}}$ to P_{IP}, this quality measure requires finding the solution of P_{LP} in order to obtain the value of $\bar{z}_{\text{LP}}$. Another quality measure which may require less computational effort to obtain is provided by linear programming duality considerations, as follows. Consider the dual problem $\text{P}_{\text{LP}}^{\text{D}}$ of the linear programming relaxation P_{LP} which we will call the *dual* (*linear programming*) *relaxation* of P_{IP}. It is formally written as

$$\begin{aligned} \text{P}_{\text{LP}}^{\text{D}}\text{: Min } z_{\text{LP}}^{\text{D}} &= \mathbf{ub} \\ \text{s.t. } \quad \mathbf{uA} &\geq \mathbf{c} \\ \mathbf{u} &\in \mathbb{R}_+^m \end{aligned}$$

From linear programming duality we know that if $\tilde{\mathbf{u}}$ is a feasible solution of P_{LP}^{D}, then $\tilde{\mathbf{u}}\,\mathbf{b} \geq \bar{z}_{LP} \geq \bar{z}_{IP}$. If, for example, P_{LP}^{D} has unbounded solutions, P_{IP} will then have no feasible solution. The relations between the integer programming problem P_{IP}, its linear programming relaxation P_{LP} and the dual P_{LP}^{D} of P_{LP} are summarized in

Theorem I.25: The following relations hold:

- If P_{LP} has no feasible solution, then neither does P_{IP}.
- If P_{IP} has feasible solutions and P_{LP} has a (finite) optimal solution, then P_{IP} also has a (finite) optimal solution with $\bar{z}_{IP} \leq \bar{z}_{LP}$.
- If P_{IP} and P_{LP}^{D} have feasible solutions $\tilde{\mathbf{y}}$ and $\tilde{\mathbf{u}}$, then both have (finite) optimal solutions $\tilde{\mathbf{y}}$ and $\bar{\mathbf{u}}$, with $\mathbf{c}\tilde{\mathbf{y}} = \bar{z}_{IP} \leq \bar{z}_{LP} = \bar{z}_{LP}^{D} = \bar{\mathbf{u}}\mathbf{b} \leq \tilde{\mathbf{u}}\mathbf{b}$.
- If P_{LP}^{D} has unbounded solutions, then P_{IP} has no feasible solution.

It is possible to extend the notions of duality and relaxation in the context of integer programming. First we introduce the concepts of *weak* and *strong* dual problems of P_{IP}.

Definition I.26: Let P_{IP} be an integer programming problem Max $z_{IP} = \mathbf{cy}$, s.t. $\mathbf{y} \in S_{IP}$. Define P_{IP}^{D} as any problem with Min $z_{IP}^{D} = z(\mathbf{v})$, s.t. $\mathbf{v} \in S_{IP}^{D} \subseteq \mathbb{R}_{+}^{m}$, such that $z(\mathbf{v}) = z_{IP}^{D} \geq z_{IP} = \mathbf{cy}$ $\forall$ $\mathbf{v} \in S_{IP}^{D}$ and $\forall$ $\mathbf{y} \in S_{IP}$. Then P_{IP}^{D} is then called a *weak dual problem* of P_{IP}. Given that P_{IP} and P_{IP}^{D} have finite optimal solutions $\bar{z}_{IP}$ and $\bar{z}_{IP}^{D}$, the *duality gap* is then defined as $\bar{z}_{IP}^{D} - \bar{z}_{IP}$. If the duality gap equals 0, P_{IP}^{D} is a *strong dual* of P_{IP}.

It is now possible to carry the analysis one step further.

Definition I.27: The problem P_{IP}^{Rel} is called a *relaxation* of the original integer programming problem P_{IP} if P_{IP}^{Rel} : Max $z_{IP}^{Rel} = z(\mathbf{w})$, s.t. $\mathbf{w} \in S_{IP}^{Rel} \subseteq \mathbb{R}_{+}^{n}$ is such that $S_{IP}^{Rel} \supseteq S_{IP}$ and $z_{IP} = \mathbf{cy} \leq z(\mathbf{y}) = z_{IP}^{Rel}$ $\forall$ $\mathbf{y} \in S_{IP}$.

It is apparent that the three relaxations defined in Chapter 1 are all special cases of the relaxation in Definition I.27. We can then state

Lemma I.28: Let P_{IP}^{Rel} be any relaxation of P_{IP} and $P_{IP}^{Rel,D}$ be a weak dual of P_{IP}^{Rel}. Then $P_{IP}^{Rel,D}$ is also a weak dual of P_{IP}.

As an illustration of the above discussion, consider again the integer programming problem introduced in Section 1.2. For convenience, the problem is restated here.

$$
\begin{aligned}
P_{IP}\colon \ \text{Max } z_{IP} &= 3x_1 + 5x_2 \\
\text{s.t.} \quad 2x_1 + x_2 &\le 5 \\
3x_1 + 7x_2 &\le 14 \\
x_1 \qquad &\le 2 \\
x_2 &\le 1\tfrac{1}{2} \\
x_1, \quad x_2 &\in \mathbb{N}_0.
\end{aligned}
$$

The linear programming, Lagrangean, and surrogate relaxations discussed in that section have linear programming duals with objective values $11^{7}/_{11}$, 14, and $11^{7}/_{8}$, respectively, so that Lemma I.28 indicates that $\bar{z}_{IP} \le 11^{7}/_{11}$ which can be used as an upper bound in a branch and bound process. (As in this example all coefficients are integer, this upper bound can be strengthened to $\bar{z}_{IP} \le 11$ which happens to be satisfied as equality at optimum). In general, it is possible to define stronger duals than those considered here, that would produce stronger bounds. For details, see the extensive treatment by Nemhauser and Wolsey (1988).

6.6 Lagrangean Decomposition

In Chapter 1, the Lagrangean relaxation of an integer programming problem

$$
\begin{aligned}
P_{IP}\colon \ \text{Max } z_{IP} &= \mathbf{cy} \\
\text{s.t.} \quad \mathbf{Ay} &\le \mathbf{b} \\
\mathbf{Dy} &\le \mathbf{d} \\
\mathbf{y} &\in \mathbb{N}_0^n
\end{aligned}
$$

was defined as the problem

$$\text{LR}(\lambda)\text{: Max } z_{\text{LR}}(\lambda) = \mathbf{cy} + \lambda(\mathbf{b} - \mathbf{Ay})$$
$$\text{s.t.} \quad \mathbf{Dy} \le \mathbf{d}$$
$$\mathbf{y} \in \mathbb{N}_0^n$$

where the vector λ is referred to as the Lagrangean multipliers. In this relaxation we commonly say that the constraints $\mathbf{Ay} \le \mathbf{b}$ have been *relaxed* or *dualized.*

It is well known that for any $\lambda \ge 0$, the optimal objective function value of LR(λ) is an upper bound on the value of the original objective function value, i.e., $\bar{z}_{\text{LR}}(\lambda) \ge \bar{z}_{\text{IP}}$.

A Lagrangean relaxation that is properly set up is an invaluable tool in finding upper bounds that are an essential part of any branch and bound procedure. There are a number of degrees of freedom when setting up and solving Lagrangean relaxations, in particular

- which constraints should be dualized, and
- what values to choose for the Lagrangean multipliers.

Here, we consider the former issue. One idea is to choose constraints to be dualized, so that the remaining constraints are structured and easy to solve; some such solutions can actually be found by inspection. Another possibility is that after dualization, the constraint structure is such that the problem naturally decomposes into two or more smaller subproblems, each of which is again much easier to solve than the original problem.

As an example, consider the standard assignment problem with an additional budget constraint

$$\text{P}_{\text{IP}}\text{: Max } z_{\text{IP}} = \sum_{i=1}^{n}\sum_{j=1}^{n} c_{ij} y_{ij}$$
$$\sum_{j=1}^{n} y_{ij} = 1,\ i = 1, \ldots, n$$
$$\sum_{i=1}^{n} y_{ij} = 1,\ j = 1, \ldots, n$$

$$\sum_{i=1}^{n}\sum_{j=1}^{n} a_{ij}y_{ij} \leq b$$

$$y_{ij} = 0 \vee 1,\ i = 1, \ldots, n;\ j = 1, \ldots, n$$

We will refer to the constraints as the two sets of assignment constraints, and the budget constraint. Without the budget constraint, the problem is a simple assignment problem that is easy to solve; hence we refer to the budget constraint as the *complicating constraint.* Dualizing this constraint, we obtain the problem

$$\text{LR}_1\text{: Max } z_1 = \sum_{i=1}^{n}\sum_{j=1}^{n}(c_{ij} - \lambda a_{ij})y_{ij} + \lambda b$$

$$\text{s.t.} \quad \sum_{j=1}^{n} y_{ij} = 1,\ i = 1, \ldots, n$$

$$\sum_{i=1}^{n} y_{ij} = 1,\ j = 1, \ldots, n$$

$$y_{ij} = 0 \vee 1,\ i = 1, \ldots, n;\ j = 1, \ldots, n$$

which, for any fixed λ, is a standard assignment problem. An intuitive solution approach for P_{IP} might then be to select some value of λ, solve the relaxed assignment problem $\text{LR}_1(\lambda)$ using some assignment problem algorithm and then check if the complicating budget constraint is satisfied. If it is, one would reduce the value of λ and solve the relaxation again. If the complicating constraint is not satisfied, the value of λ would be increased and the relaxation solved again. One can show that the original problem P_{IP} can be iteratively solved in this way.

An alternative to dualizing the budget constraint is to dualize *all* assignment constraints. Using multipliers u_i, $i = 1, \ldots, n$ and v_j, $j = 1, \ldots, n$, this results in the problem

$$\text{LR}_2\text{: Max } z_2 = \sum_{i=1}^{n}\sum_{j=1}^{n}(c_{ij} - u_i - v_j)y_{ij} + \sum_{i=1}^{n} u_i + \sum_{j=1}^{n} v_j$$

$$\text{s.t.} \quad \sum_{i=1}^{n}\sum_{j=1}^{n} a_{ij}y_{ij} \leq b$$

$$y_{ij} = 0 \vee 1,\ i = 1, \ldots, n;\ j = 1, \ldots, n$$

which is a standard zero-one knapsack problem.

Yet another possibility is to relax the first n assignment constraints and the budget constraint. Using multipliers u_i, $i = 1, ..., n$ and λ results in the problem

$$\text{LR}_3\text{: Max } z_3 = \sum_{i=1}^{n}\sum_{j=1}^{n}(c_{ij} - u_i - \lambda a_{ij})y_{ij} + \sum_{i=1}^{n} u_i + \lambda b$$

$$\text{s.t.} \quad \sum_{i=1}^{n} y_{ij} = 1, j = 1, ..., n$$

$$y_{ij} = 0 \vee 1, i = 1, ..., n; j = 1, ..., n$$

The only constraints in problem LR_3 are of the so-called *generalized upper bound* (GUB) type; see e.g., Eiselt *et al.* (1987) or Dantzig and Van Slyke (1967). Again, the problem can be solved efficiently by special purpose algorithms.

In general, the choice of Lagrangean relaxation depends not only on the simplicity of the resulting problem, but also on the quality of the bound that a relaxation produces. So far, we have dualized constraints, so that the resulting constraint structures are simple. Another possibility is to employ *Lagrangean decomposition* which is nothing but a standard Lagrangean relaxation that results in the decomposition of a given problem into two or more separate subproblems. This is, in fact, what happens in the relaxation LR_3 above as for each $j = 1, ..., n$, the subproblem is

$$\text{SUB}_j\text{: Max} \quad \sum_{i=1}^{n}(c_{ij} - u_i - \lambda a_{ij})y_{ij}$$

$$\text{s.t.} \quad \sum_{i=1}^{n} y_{ij} = 1$$

$$y_{ij} = 0 \vee 1, i = 1, ..., n$$

which is completely independent from the subproblems SUB_k with $k \neq j$. As a matter of fact, each subproblem SUB_j can be solved by inspection by setting $y_{i^*j} = 1$ for $i^* = \arg\max_{1 \le i \le n}\{c_{ij} - u_i - \lambda a_{ij}\}$ and $y_{ij} = 0 \ \forall \ i \neq i^*$.

Sometimes a problem does not decompose naturally, regardless of which constraints are dualized. As a remedy, we may introduce so-called *copy constraints*. As an illustration, consider the integer programming problem

$$\text{P}_{\text{IP}}\text{: } \underset{\mathbf{y}}{\text{Max}}\ \{\mathbf{cy} : \mathbf{Ay} \le \mathbf{b}, \mathbf{Dy} \le \mathbf{d}; \mathbf{y} \in \mathbb{N}_0^n\}.$$

Introducing a copy constraint $\mathbf{y} = \mathbf{x}$ with new integer variables $\mathbf{x} \ge \mathbf{0}$, an equivalent formulation of P_{IP} is

$$\text{P}'_{\text{IP}}\text{: } \underset{\mathbf{y},\mathbf{x}}{\text{Max}}\ \{\mathbf{cy} : \mathbf{Ay} \le \mathbf{b}, \mathbf{Dx} \le \mathbf{d}, \mathbf{x} = \mathbf{y}; \mathbf{x}, \mathbf{y} \in \mathbb{N}_0^n\}.$$

Dualizing the copy constraint $\mathbf{y} = \mathbf{x}$, the problem P'_{IP} decomposes into the two independent subproblems:

$$\text{SUB}_1\text{: } \underset{\mathbf{y}}{\text{Max}}\{\mathbf{cy} - \lambda y : \mathbf{Ay} \le \mathbf{b}; \mathbf{y} \in \mathbb{N}_0^n\}, \text{ and}$$

$$\text{SUB}_2\text{: } \underset{\mathbf{x}}{\text{Max}}\{\lambda y : \mathbf{Dx} \le \mathbf{d}; \mathbf{x} \in \mathbb{N}_0^n\}.$$

Depending on the structure of the **A** and **D** matrices, the subproblems may be easily solvable. For example, the capacitated plant location problem, described in section 4.1.1, can be solved using a variety of relaxations with and without copy constraints; see e.g., Guignard (1995) who also discusses efficient subgradient techniques for solving the Lagrangean dual. This paper also considers a generalization of the copy constraint technique called *Lagrangean substitution*; details can be found in the paper.

CHAPTER 7 HEURISTIC ALGORITHMS

In mathematical programming, a *heuristic method*, or *heuristic* for short, is a procedure that determines good or near-optimal solutions to an optimization problem. Early heuristics were distinct methods developed to address specific optimization problems. As an example, consider the *minimum makespan* problem that arises in machine scheduling. In this problem, n given jobs with known processing times must be processed on m machines. The objective is to find a schedule which minimizes the latest finish time of the last job completed. This problem is known to be strongly *NP-complete*. A simple heuristic to solve this problem is the *list heuristic*. This method lists the jobs in some given order, then takes the first m jobs and assigns each to exactly one machine, so that at this point, the completion time of each machine corresponds to the processing time of the job assigned to it. Then, the remaining $(n - m)$ jobs are selected in order of the list and assigned one by one to the machine that has currently the earliest finish time. This heuristic has the advantage of being computationally simple, and it can easily be implemented in real time. Better solutions to the problem, i.e., schedules with a shorter makespan, can be determined by improving on the heuristic in the following way: first sort the jobs on the list in nonincreasing order of processing time, and then apply the same machine selection process as presented above. Both heuristics solve the problem; the first is faster computationally, whereas the second is likely to produce better solutions. Both methods are designed specifically to solve the m machine n job minimum makespan problem, and may not work well if adapted to different problems.

As opposed to exact methods, heuristics carry no guarantee that an optimal solution will be found. The question is then why heuristics should be used at all. There are several reasons. The first case in which heuristics are the solution method of choice includes problems for which no exact solution method or even exact formulation is known. In such a case, a heuristic method is the only choice. The second reason is that even if exact solution methods are known, they may be computationally intensive, rendering them unusable in some situations. Consider the case of a flexible manufacturing system, where jobs must be routed in real time to cells or groups of machines. A good, fast scheduling heuristic may be better in this situation than one which takes so long to solve that the manufacturing system is held waiting for the optimal schedule to be computed. Thirdly, heuristics may also be used to calculate bounds on the optimal solution in branch and bound solution processes.

In contrast to individual heuristic algorithms that are designed to solve a specific problem, *metaheuristics* are strategic problem solving frameworks that can be adapted to solve a wide variety of problems. A metaheuristic framework includes a set of problem-solving principles that are sometimes based on natural or physical processes, and a set of control parameters. Application of a metaheuristic strategy to a specific problem involves choosing, and sometimes fine-tuning, these parameters. Metaheuristic methods are not normally expected to outperform specialized heuristics on a problem. Their main purpose is to deal with difficult problems for which no specialized algorithm is known. However, some studies have demonstrated that certain metaheuristics perform as well or better than speciality heuristics for some well-known difficult problems such as the traveling salesman, graph partitioning and facility layout problems. Such successes have assured their place in the practice of optimization. This chapter discusses four metaheuristic frameworks: *simulated annealing*, *tabu search*, *genetic algorithms*, and *evolutionary algorithms*. All of these are more or less loosely based on the principal of local search, or *neighborhood search*. Due to its importance, the main ideas of neighborhood search are presented in the next section.

Useful references on heuristic methods can be found in the extensive bibliography in Osman and Laporte (1996), the texts by Reeves (1993) and Rayward-Smith (1996), the comparative tutorial by Pirlot (1993), and Michalewicz and Fogel (2000).

7.1 Neighborhood Search

Neighborhood search, also called local search, starts with some initial solution to the problem, then iteratively perturbs that solution in an attempt to improve the objective value. These perturbations, or small transformations of the initial solution, are usually identified by elementary operators on the solution, called *moves*. The set of new solutions that can be obtained by applying moves to a given solution defines its *neighborhood.* The search mechanism is to examine members of the neighborhood of the starting solution, $\mathbf{x}^1$, and then move to a subsequent solution $\mathbf{x}^2$. Then, the neighborhood of $\mathbf{x}^2$ is inspected to determine the next solution $\mathbf{x}^3$. This process is repeated, producing a sequence of solutions $\mathbf{x}^1, \mathbf{x}^2, \ldots$, that is commonly called a *trajectory*, until the search is terminated by a stopping rule.

To formalize matters, let X denote the *solution space* on which the objective function is defined. Furthermore, denote by $S \subseteq X$ the set of feasible solutions. Given a vector $\mathbf{x}$ of decision variables, the objective function can be written as $z(\mathbf{x})$, and the optimization problem is P: Min $z(\mathbf{x})$, $\mathbf{x} \in S$. We can then write

Definition I.29: A move m applied to a solution $\mathbf{x}$ is a rule that perturbs $\mathbf{x} \in X$, so that a new solution $\mathbf{x}' = \mathbf{x} \oplus m \neq \mathbf{x}$ with $\mathbf{x}' \in X$ is generated. Define M as the set of rules that include all moves $\mathbf{x} \oplus m$. If a move m results in a solution $\mathbf{x}' \in X \setminus S$, the move is called *infeasible*, otherwise it is called *feasible*.

Definition I.30: The neighborhood of a solution $\mathbf{x}$ is defined as the set $N(\mathbf{x}) = \{\mathbf{x}': \mathbf{x}' = \mathbf{x} \oplus m, m \in M\}$. A *feasible* neighborhood is the set of solutions produced by feasible moves, i.e., $N(\mathbf{x}) \cap S$.

Definition I.31: The *size* $|N(\mathbf{x})|$ of a neighborhood is the number of elements of $N(\mathbf{x})$ and is equivalent to the number of moves out of $\mathbf{x}$. The average size of all neighborhoods in X is called the *average* neighborhood size.

The average neighborhood size provides an estimate of the number of solutions that can be reached from a given solution $\mathbf{x}$ in one move. It is used in setting parameters that control the patterns of search in heuristic methods. For example, in simulated annealing, a length parameter determines how many solutions are examined at each stage in the process. Good choices for this parameter in a given problem are calculated using the average neighborhood size. Another example is in the Tabu Search method, in which a tabu list length, or tabu tenure, is used to exclude recent moves from consideration. Among other factors, the average neighborhood size is an important factor in choosing values for this parameter. Consider now the following examples of neighborhood search.

Example 1: Let $X = \{0, 1\}^5$, the space of all binary 5-vectors. Define M as the set of moves that flip a single coordinate from 0 to 1 or vice versa. For instance, if $\mathbf{x}^1 = [0, 1, 0, 0, 1]$, then $\mathbf{x}^2 = [1, 1, 0, 0, 1]$ is obtained by flipping the first coordinate. There are five possible moves from any solution $\mathbf{x}$, thus the neighborhood of any solution is of size five, i.e., $|N(\mathbf{x})| = 5$.

Now let M denote the set of moves that swap, i.e., exchange, the values of any two unequal components of $\mathbf{x}$. For instance, if $\mathbf{x}^1 = [0, 1, 0, 0, 1]$, then $\mathbf{x}^2 = [1, 1, 0, 0, 0]$ is obtained by swapping the first and last components. This is a more complex move than the flipping move above. Here, the size of the neighborhood is the product of the number of zeroes and the number of ones in the vector, i.e., $\left(5 - \sum_{i=1}^{5} x_i\right)\left(\sum_{i=1}^{5} x_i\right) \le 6$.

Example 2: Suppose that the problem P is to find an optimal ordering of N jobs in a manufacturing system. A solution to this problem can be described as an N-vector of job numbers, in which the position of each job in the vector defines the order in which the job will be processed. For example, [3, 4, 1, 6, 5, 2] is a possible ordering of 6 jobs, where job 3 is processed first, followed by job 4, then job 1, and so on, finishing with job 2. Define now M as the set of moves that produce new orderings by the swapping of any two jobs. For example, [3, 1, 4, 6, 5, 2] is obtained by swapping the positions of jobs 4 and 1.

Example 3: Consider the traveling salesman problem, whose objective is to find a minimum cost tour of cities to be visited by a salesman; for details, see Chapter 3 in Part II of this volume. A solution to the problem is a listing of cities in the order in which they are visited, similar to the job sequence in Example 2. A type of move that has been used in heuristic methods for traveling salesman problems is to replace the order of a subsection of the sequence by its reverse. Thus, [1, 2, 3, 4, 5, 6] might be replaced by [1, 2, 5, 4, 3, 6], where the subroute [3, 4, 5] is replaced by [5, 4, 3].

It is important that the moves M "cover" the solution space. In other words, it must be possible to reach any point in the solution space X by applying a sequence of moves. In Example 1, it is clear that all possible solutions can be obtained by flipping coordinates. However, the same is not true for interchange moves. As an example, consider the case where the starting solution is [1, 1, 1, 1, 1]. All vectors that result from interchange moves have the same number of zeroes and ones. Hence, it is not possible to cover the solution space $\{0; 1\}^5$ with interchange moves. This observation leads to the following

Definition I.32: A feasible set S is disjoint under a set of moves M if there exist solutions $\mathbf{x}, \boldsymbol{y} \in S$, such that there is no trajectory that leads from $\mathbf{x}$ to $\boldsymbol{y}$.

We are now able to describe a general neighborhood search heuristic. The *evaluation function* $E(\mathbf{x})$ is a function that is used to determine which move is chosen during the search. In simple search applications, the evaluation function is simply the (minimization) objective function $z(\mathbf{x})$. In more complex search heuristics, the evaluation function is related to $z(\mathbf{x})$ but is modified in ways that are specific to the search method being used and the problem under consideration. The search is initialized with an iteration counter $k := 1$, a solution $\mathbf{x}^1$, so that the currently best known solution is $\mathbf{x}^* = \mathbf{x}^1$. Furthermore, some suitable stopping rule has been adopted. The general neighborhood search algorithm can then be described as follows.

Neighborhood Search Algorithm

Step 1: Evaluate $E(\mathbf{x})\ \forall\ \mathbf{x} \in N(\mathbf{x}^k)$.

Step 2: Define $\mathbf{x}^{k+1} := \text{argmin}\ \{E(\mathbf{x}): \mathbf{x} \in N(\mathbf{x}^k)\}$ where ties are broken arbitrarily. Set $k := k + 1$.

Step 3: Is $z(\mathbf{x}^k) < z(\mathbf{x}^*)$? If yes: Set $\mathbf{x}^* := \mathbf{x}^k$ and go to Step 4.
If no: Go to Step 4.

Step 4: Is the stopping rule met? If yes: Stop. The best solution is $\mathbf{x}^*$.
If no: Go to Step 1.

Neighborhood search can be finetuned by choosing different evaluation functions $E(\mathbf{x})$. The most rudimentary form of neighborhood search is the *steepest descent* method. In this basic type of local search, the evaluation function is equal to the cost function $z(\mathbf{x})$ and the process terminates when no more improving moves are found in the neighborhood. Steepest descent (or steepest ascent in case of maximization problems) always terminates at a local optimum, and its trajectory of solutions is nonincreasing in $z(\mathbf{x}^k)$.

Example: Consider a bin packing problem (see Section 3.1.9) with M items, each weighing w_i, to be packed into N bins. One objective of the bin packing problem attempts to place the items so that their weight is evenly distributed across the bins. To formulate this problem exactly, we define binary decision variables $x_{ij} = 1$ if item i is placed in bin j, and 0 otherwise. Under any packing, the weight of the j-th bin is $B_j = \sum_{i=1}^{M} w_i x_{ij}$, whereas the average weight of a bin is $\overline{w} = \frac{1}{N}\sum_{i=1}^{M} w_i$. A possible objective is then to minimize the sum of squared differences between the actual weight of the bins and the average, i.e., $z = \sum_{j=1}^{N} \left[B_j - \overline{w}\right]^2$.

One possible neighborhood structure is defined by moves that swap a pair of items from different bins. Suppose there are $N = 3$ bins into which $M = 10$ items are to be placed, whose weights are $\mathbf{w} = (w_i) = [3, 4, 2, 2, 6, 9, 4, 6, 6, 6]$. The average weight per bin is then $\overline{w} = {}^{48}/_3 = 16$. We can picture a solution to the problem as a vector whose i-th component indicates the bin in which item i is placed. Arbitrarily starting with the solution $\mathbf{x}^1 = [2, 3, 3, 2, 2, 3, 1, 2, 3, 1]$, the three bins have weights 10, 17, and 21, respectively, so that the objective value is $z = 36 + 1 + 25 = 62$. Since the three bins contain two, four and four items, respectively, the neighborhood size is $|N(\mathbf{x}^1)| = 2 \times 4 \times 4 = 32$. Evaluating the objective function value for all solutions in the neighborhood, we find that the unique best move is to swap item 7 in bin 1 with item 6 in bin 3. The resulting solution is $\mathbf{x}^2 = [2, 3, 3, 2, 2, 1, 3, 2, 3, 1]$ with $z(\mathbf{x}^2) = 2$. The procedure is then repeated by exploring all 32 possible moves from $\mathbf{x}^2$, and so on.

Definition I.33: The *attraction basin*, $A(\mathbf{x}')$, of a local optimum $\mathbf{x}'$ is the set of solutions in X for which the search trajectory ends at $\mathbf{x}'$.

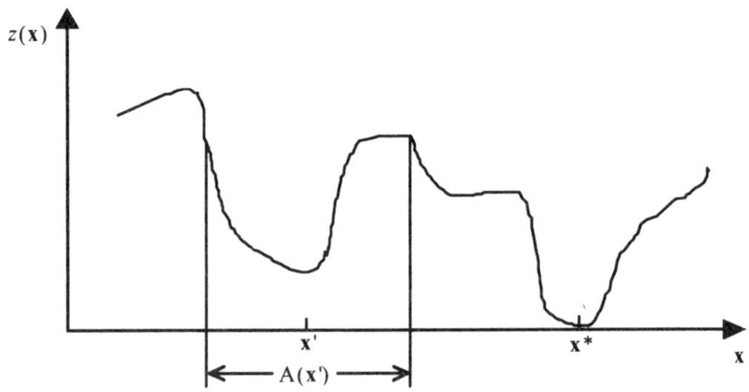

Figure I.31

In other words, given any starting point in $A(\mathbf{x}')$, the search will converge to $\mathbf{x}'$. If $\mathbf{x}^*$ is a global optimum and $A(\mathbf{x}^*)$ equals the feasible set, then the procedure will lead to $\mathbf{x}^*$ from any starting point. Linear programming is a typical example for problems with $A(\mathbf{x}^*) = S$. In general, the concept is shown in Figure I.31 where any starting point in the $A(\mathbf{x}')$ will lead to the local optimum $\mathbf{x}'$, but not to the global optimum $\mathbf{x}^*$.

The difficulty with steepest descent algorithms is that the local optimum which is found may be quite a poor solution to the problem. One way to improve the steepest descent method is to repeat the search from a number of different starting solutions, each time recording the best solution found. This produces a sample of candidate solutions, from which the best is then chosen. If the starting solutions are randomly generated, this will add robustness to the search. This random restart method is called *iterated descent*. The hope is that, in starting the algorithm at many starting points, one of the starting points belong to the attraction basin $A(\mathbf{x}^*)$ of the global optimum $\mathbf{x}^*$, or, at least, to the attraction basin of a high-quality solution.

It is more difficult to define an appropriate neighborhood structure for constrained problems. Let $\mathbf{x}'$ be a solution to a given problem and suppose that the current solution is $\mathbf{x}$. Clearly, if $\mathbf{x}$ and $\mathbf{x}'$ are in disjoint subsets of the feasible region, then neighborhood search will not find the solution. However, even if $\mathbf{x}'$ is reachable from $\mathbf{x}$ on a trajectory, the problem may still be difficult to solve.

Assume that the solution $\mathbf{x}'$ is reachable from $\mathbf{x}$, but only by a limited set of sequences of feasible moves. Feasible regions with this characteristic are unlikely to be adequately explored by neighborhood search. They are, in a sense, probabilistically disjoint, since it is unlikely that the sequence of moves required to traverse from one region of S to another will be generated at any given iteration. This makes it likely that the neighborhood search process will be restricted to one region of S. Problems of this type are usually considered difficult.

The feasible sets of such problems must be "smoothed" so that the search process can freely move across the feasible set. Occasional infeasible moves are usually included in the search trajectory to create more potential trajectories or paths between points in the feasible region. The difficulty arising from infeasible moves is that the search trajectory may wander aimlessly in the infeasible region, and thus become non-productive. To avoid this behavior, the evaluation function is changed so that infeasible solutions are penalized, and so that the search trajectory can be forced to (eventually) produce a feasible solution. These two goals are accomplished by converting the "hard" constraints to "soft" constraints by relaxing them and augmenting the objective function by appropriate penalty terms. In general, if there are K constraints, then K penalty terms must be included in the revised cost function $z_{pen}(\mathbf{x}) = z(\mathbf{x}) + \sum_{k=1}^{K} \alpha_k g_k(\mathbf{x})$, where each term $g_k(\mathbf{x})$ is a nonnegative penalty function that measures the degree to which the solution $\mathbf{x}$ violates the k-th constraint, and α_k is a positive parameter associated with the penalty term g_k. The weighting parameter is employed to emphasize or de-emphasize the relative magnitude of the penalty term $g_k(\mathbf{x})$ in the objective $z_{pen}(\mathbf{x})$. Clearly, if a solution $\mathbf{x}$ is feasible, then $z(\mathbf{x}) = z_{pen}(\mathbf{x})$. Otherwise, there is at least one positive penalty term and hence $z(\mathbf{x}) < z_{pen}(\mathbf{x})$. Increasing all weighting parameters drives the solution process toward the feasible region; reducing a weighting parameter on penalty term g_k tends to allow moves that are infeasible with respect to the k-th constraint.

Pure descent or ascent versions of neighborhood search can be augmented to allow non-improving moves in order to allow the search to escape from a local optimum. Simulated annealing and tabu search strategies are based on this idea. These approaches are described in the next two sections.

7.2 Simulated Annealing

The principle of *simulated annealing* is based on a physical analog, *viz.*, the process of cooling a molten metal. The objective is to produce a solid mass with strong intermolecular bonds by carefully cooling the metal so that its crystalline structure reaches a minimum energy state. If the cooling is done abruptly (so-called "rapid quenching"), energy is trapped in the crystalline structure, resulting in a material with less than optimal strength properties. A gradual cooling schedule, called careful annealing, produces a solid with a lower energy structure, and hence a stronger solid material. This annealing process is done by successively lowering the temperature in stages, waiting long enough at each stage to allow the crystalline structure to reach equilibrium at that temperature. When the temperature has reached the level at which the metal solidifies, the process is complete.

Simulated annealing is a search strategy that mimics the careful annealing of metals. The method was first proposed by Kirkpatrick *et al.* (1982) and Cerny (1985). Kirkpatrick *et al.* first used simulated annealing successfully to solve what is now known as the Ising spin glass problem. In this problem, the cost function equals the amount of magnetic energy in each state. Subsequently, simulated annealing was applied to many combinatorial optimization problems, in which the physical analogy is less apparent.

Simulated annealing is a stochastic neighborhood search process. In each stage, L randomly chosen candidate solutions in the neighborhood of the current solution are evaluated. If a candidate solution improves on the current solution, it is accepted. Otherwise, it is accepted with a probability of $P(T, \Delta) = e^{-\frac{\Delta}{T}}$, which depends on a control parameter T (the temperature in the physical equivalent) and the amount Δ by which the move worsens the current solution. This relation ensures that the probability of accepting a move to a very poor solution is very small. This is accomplished in the algorithm below in Steps 5 and 6. At the completion of each stage, the temperature T is reduced at the cooling rate r. Given a relatively high temperature T at the beginning of the process, the probability of accepting non-improving moves is fairly high. As the process continues, the temperature decreases and non-improving moves become less likely. The search is continued until there is some evidence to suggest that there is a very low probability of improving on the best solution found so far. At this stage, the system is said to be frozen.

Simulated annealing has the advantage that it is simple in concept and implementation. The procedure improves on descent methods by reducing the probability of getting stuck at a poor but locally optimal solution. Simulated annealing has been applied successfully to many hard problems; see, e.g., the bibliography by Osman and Laporte (1996).

Many contributions in the literature that deal with simulated annealing use terms that refer to the original physical process. To simplify the conversion to optimization problems, the most important equivalents in optimization are as follows:

- "achieving a minimum energy state" means finding a global optimal solution,
- "rapid quenching" means steepest descent local search,
- "careful annealing" means simulated annealing process, and
- "the gradual reduction of temperature in cooling stages" means gradual reduction of the probability of accepting a non-improving move.

The simulated annealing search process is initialized with a starting solution $\mathbf{x}^1$. Furthermore, set the iteration counters $\ell := 1$, $k := 1$, the best known solution $\mathbf{x}^* := \mathbf{x}^1$. Furthermore, choose the initial temperature T, a stage length L, a cooling rate r, and a stopping rule. These choices are further discussed below. Furthermore, let

$U[0; 1]$ denote a uniform distribution between 0 and 1 from which random numbers are drawn. The procedure can then be described as follows.

Simulated Annealing Method

Step 1: Is the stopping rule satisfied?

If yes: Stop, $\mathbf{x}^*$ is the best known solution.

If no: Go to Step 2.

Step 2: Is $\ell = L + 1$? If yes: Go to Step 8.

If no: Go to Step 3.

Step 3: Randomly choose $\mathbf{x}' \in N(\mathbf{x}^k)$ and compute the cost change $\Delta = z(\mathbf{x}') - z(\mathbf{x}^k)$.

Step 4: Is $\Delta \geq 0$ (a non-improving move)?

If yes: Generate a random variable $p \in U[0; 1]$ and go to Step 5.

If no: Go to Step 6.

Step 5: Is $p \leq e^{-\frac{\Delta}{T}}$? If yes: Go to Step 6.

If no: Set $\ell := \ell + 1$ and go to Step 2.

Step 6: (Accept the move). Set $\mathbf{x}^{k+1} := \mathbf{x}'$ and $k := k + 1$.

Step 7: Is $z(\mathbf{x}^k) < z(\mathbf{x}^*)$? If yes: Set $\mathbf{x}^* := \mathbf{x}^k$, $\ell := 1$ and go to Step 8.

If no: Set $\ell := \ell + 1$ and go to Step 2.

Step 8: Set $T := rT$ (i.e., reduce the temperature), set $\ell := 1$ and go to Step 1.

Example. The bin packing problem considered in Section 7.1 will now be used as an illustration of the simulated annealing method. We start with temperature $T = 100$. The cooling rate $r = 0.95$ will be used to reduce the temperature in each stage, and $L = 50$ moves evaluated in each stage. Table I.14 shows the first six iterations of the simulated annealing procedure in the first stage where the "swap" and "$p \in U[0; 1]$" columns are randomly generated, and the remaining columns are calculated accordingly. The first move generated is to swap items 2 and 8. The cost of this move is $\Delta = 24$. Since this is not an improving move, a random number is generated, here $p = 0.05 < .7866 = e^{\frac{-24}{100}} = e^{\frac{-\Delta}{T}}$, and the move is accepted. The second and third moves are also uphill (i.e., nonimproving) moves, and are accepted in a similar way. The fourth move is improving, and is automatically accepted. The fifth move, again an uphill move, is rejected since the random

Table I.14

Solution #	Swap		Δ	$p \in U[0;1]$	$e^{-\frac{\Delta}{T}}$	Solution	z	Best solution
0	-	-	-	-	-	[2, 3, 3, 2, 2, 3, 1, 2, 3, 1]	62	62
1	2	8	24	0.05	0.79	[2, 2, 3, 2, 2, 3, 1, 3, 3, 1]	86	62
2	7	4	28	0.41	0.76	[2, 2, 3, 1, 2, 3, 2, 3, 3, 1]	114	62
3	10	8	0	0.98	1.00	[2, 2, 3, 1, 2, 3, 2, 1, 3, 3]	114	62
4	5	6	-18	0.98	1.20	[2, 2, 3, 1, 3, 2, 2, 1, 3, 3]	96	62
5	7	8	56	0.63	0.57	[2, 2, 3, 1, 3, 2, 2, 1, 3, 3]	96	62
6	3	6	98	0.02	0.38	[2, 2, 2, 1, 3, 3, 2, 1, 3, 3]	194	62

number drawn $p = 0.63 > 0.57 = e^{\frac{-56}{100}} = e^{\frac{-\Delta}{T}}$, while the sixth move is downhill and accepted.

It is important to observe the large number of non-improving moves that are accepted in this first stage. In the second stage, the temperature control parameter T is reduced to $T = 0.95(100) = 95$. This will reduce the probability of accepting non-improving moves by five percent. In later stages, T will become near zero, and non-improving moves will be very unlikely to be accepted at all. The graph in Figure I.32 shows the progress of simulated annealing on this problem. The

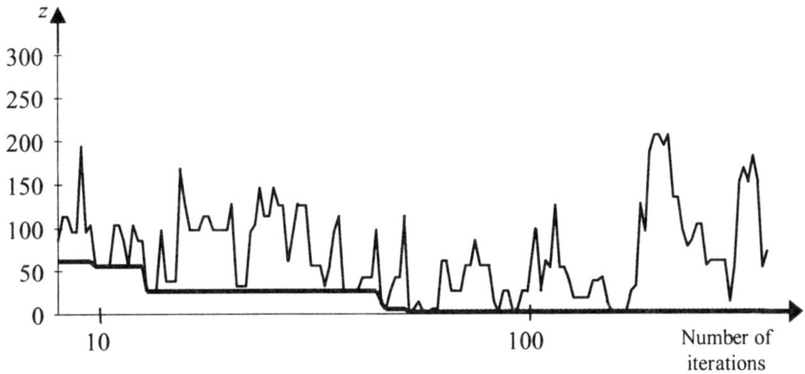

Figure I.32

current solution, shown by the solid line, moves uphill and downhill, while the best solution (the bold line) improves over time. The best solution determined by simulated annealing with this neighborhood structure and parameter settings was $\mathbf{x}^* = [2, 3, 2, 3, 2, 1, 3, 2, 3, 1]$, with a cost of $z^* = 2$. This solution was found at approximately move number 80, and no better solutions were found in later stages.

In comparison, the steepest descent neighborhood search for this problem with the same starting solution terminates after only one iteration, at the same local minimum with objective function value 2. However, this example of steepest descent search is very dependent on the starting solution as shown in Table I.15. In fact, the neighborhood structure based on swap moves is not especially good for this problem, as the number of moves available is limited by the nature of the initial solution. The reason for the poor result shown in row 1 is due to the fact that very few swap moves are possible, since only one element is assigned to bin three. In fact, under this neighborhood structure, there are several regions of the solution space that will never be reached. This is because, using swap moves, we never change the number of items that are assigned to any bin. So, if we start with n_j items assigned to bin j in the initial solution, all subsequent solutions will have this characteristic. The solution space is thus rendered disjoint, since no move will ever reach any solution which has more or less than n_j items assigned to bin j.

Table I.15

	Starting Solution	Best Solution	Cost
1	[2,2,1,1,2,1,2,3,1,1]	[1,2,1,1,2,3,2,2,1,1]	74
2	[2,2,3,2,2,3,1,2,3,1]	[2,3,2,2,3,1,2,2,3,1]	2
3	[1,3,3,2,2,3,1,2,3,1]	[3,2,3,3,2,3,1,2,1,1]	0

This situation can be remedied by changing the moves from *swap* moves to *bin assignment* moves. In other words, instead of swapping two items, a single item is moved from its current bin to another. This is reminiscent of the flipping and swapping moves of Examples 1 and 2, following Definition I.31. Using this neighborhood structure, simulated annealing found an optimal solution [3, 1, 3, 3, 2, 3, 2, 1, 1, 2] to the problem in 310 steps. This simple example emphasizes the importance of defining moves that allow complete and free traversal of the solution space in a neighborhood search. Similar to other neighborhood search heuristics, simulated annealing will be ineffective if unsuitable moves are chosen.

Once an appropriate neighborhood structure has been defined, it is necessary to select stopping rules and parameters that control the search. There are no global or universal rules for setting parameters in simulated annealing, and appropriate settings must be found for each problem. Simulated annealing carries an implicit tradeoff between potential solution quality and computation time. It is generally agreed that, relative to special purpose heuristics and other metaheuristics, simulated annealing requires long runs to reach good solutions. Choices of the control parameters and the stopping rule determine the computational effort.

The rate of cooling, r, has a direct effect on the number of stages which are carried out: the more abruptly temperature is decreased from stage to stage, the sooner the system will be frozen. There are no hard and fast rules for the choice of parameters, but there is some general agreement that, if simulated annealing is to work well, a graph showing the best solution known in iteration k should have roughly the shape shown in Figure I.33.

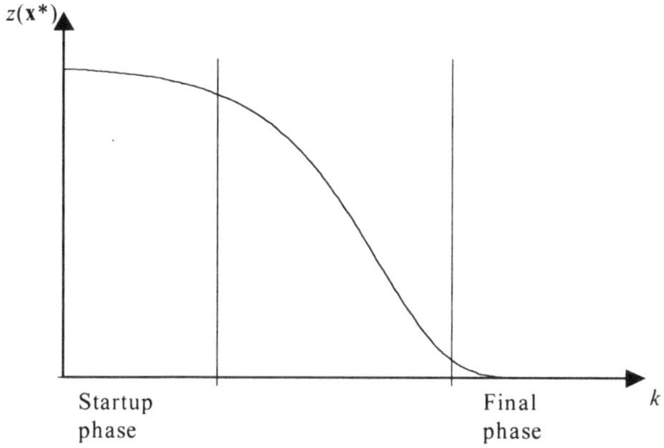

Figure I.33

The startup phase, with its high temperatures, exhibits a gradual but slow improvement in the best solution $z(\mathbf{x}^*)$. In the middle phase, we see rapid improvement in solution quality. Here, solutions should begin to get significantly better than those that could be expected from simple methods such as iterated descent. In the final phase, where temperatures are low, the rate of improvement in solution quality declines. As a general rule, parameter settings should be such that startup and final phases are present, so that the majority of the work occurs in the middle phase.

The final phase length is controlled by the stopping rule, which is used to terminate the search when evidence shows that no further improvement in $z(\mathbf{x}^*)$ is likely. There are two commonly used stopping rules. The first is based on observing solution quality, and the second is based on observing the proportion of moves that are accepted. For that purpose, define a parameter s that specifies the number of non-improving stages to be processed, and a tolerance level ε. As usual, $z(\mathbf{x}^*)$ is the best solution found at any time in the search. If we apply the solution quality stopping rule, the process terminates after s stages have been executed with an improvement less than ε in the best solution found. On the other hand, if we apply the acceptance proportion strategy, we observe the percentage of moves that have been accepted in each stage. When s stages have been executed with the percentage of moves accepted less than ε, the search terminates. Clearly, smaller tolerance settings ε and/or larger values of s result in longer searches at the later stages of simulated annealing. Typical values for ε lie in the range of 1-5%, and a typical number of nonimproving stages is $s = 5$.

The length of the initial phase is determined by the initial temperature. Recall that the probability of accepting a non-improving move is $P(T, \Delta) = e^{-\frac{\Delta}{T}}$, and thus, for a given value of Δ, higher values of T increase the probability of accepting a non-improving move. If computing resources are not restrictive, it makes sense to choose a large initial temperature T, thus allowing more non-improving moves in the early stages of the procedure. Although the time spent at high temperatures may be non-productive, too low a value of T may result in a quenching of the process too early in the search, potentially missing good solutions.

Normally, computing resources and/or time are scarce and we do not have the luxury of arbitrarily setting T to a high value. This implies the need to determine the value of a "sufficiently high" temperature. In analogy with physical annealing, the temperature should be "hot" enough to allow the material to "flow freely". In other words, T should be high enough so that the search process freely moves out of local optima in the early phase of the solution process, by accepting a reasonable number of non-improving moves. Instead of specifying T directly, we prefer to specify p_0, the proportion of non-improving moves that should be accepted at stage 1. A rule of thumb is to set $p_0 \approx ½$. Johnson *et al.* (1989, 1991) demonstrated by experimentation and sensitivity analysis that a value of $p_0 = 0.4$ was appropriate for the graph partitioning problem. Then, the annealing procedure commences with a trial value for T, and a brief experimentation run is carried out where T is adapted until the proportion of moves accepted is approximately equal to p_0.

The most common cooling schedule is *geometric cooling*, which is defined by two parameters, L and r. The number of moves L that are evaluated in one stage is sometimes called the *plateau level*. Once L is chosen, the geometric cooling schedule is given by temperatures equal to T, rT, r^2T, r^3T, ...; i.e., at stage $n + 1$, the temperature is r^nT.

The stage length L determines the number of move evaluations that the random neighborhood search has executed before the temperature is reduced. In analogy with the physical annealing process, the temperature is kept constant until an equilibrium is attained at that level. It is worthwhile to note that the neighborhood size is an important factor in determining a good choice for L, since it is indicative of the total number of possible moves. Thus, problem instances with larger neighborhoods will require a larger value of L to "settle down". Also, rather than specifying stage length, it may be better to set a threshold value for the number of moves which will be accepted in each stage, and use a simple counter to determine the end of the stage.

Many extensions and adaptations of the basic simulated annealing heuristic have been developed. For example, researchers have paid considerable attention to parameter tuning and developing adaptive cooling schedules. A large number of sophisticated variants of simulated annealing are included in the aforementioned bibliography by Osman and Laporte (1996).

7.3 Tabu Search

Tabu search is a memory-based strategy, where information from previous phases of the search is used to direct future phases. The ideas behind tabu search were first developed by Glover (1986), and independently by Hansen (1986). Glover describes the tabu search framework as being based on a set of "principles of intelligent problem solving".

Like simulated annealing, the tabu search method is based on improving descent methods to avoid entrapment at a local optimum. Unlike simulated annealing, however, tabu search is a deterministic neighborhood search method. At each iteration, the neighborhood of a given solution is searched for the best improving move, until a local optimum is reached (similar to the standard descent methods). Then, the least costly non-improving move is chosen. This leads the search away from the current local optimum, hopefully to a solution from which a different neighborhood search may lead to a better solution.

Choosing the best move at each iteration introduces the possibility of endless cycling to the previously discovered best local solution. To illustrate this, consider again the bin-packing example introduced in the previous section. Here, the neighborhood is defined by the bin assignment that simply moves an object from its current location to another bin. We start again with the solution [2, 3, 3, 2, 2, 3, 1, 2, 3, 1], whose cost is $z = 62$. Table I.16 shows the search trajectory that results from choosing the best possible move at each iteration. Here, the notation $[i, j]$ indicates that item i is placed in bin j. In the first iteration the best move is [9, 1], thereby reducing the cost from 62 to 2. This solution is a local optimum; no further improving moves can be found. So, the least increasing move, [4, 1], is taken. This increases the cost to 6. Now, in iteration 3, the best move is an improving move, [4, 2]. This moves item 4 back to bin 2, and the solution trajectory has *cycled* to the same solution that was found in iteration 1. Thereafter, choosing the best move results in continually shifting item 4 between bins 1 and 2.

Table I.16

Move count	Move $[i, j]$	Solution	z	z^*
0		[2, 3, 3, 2, 2, 3, 1, 2, 3, 1]	62	62
1	[9, 1]	[2, 3, 3, 2, 2, 3, 1, 2, 1, 1]	2	2
2	[4, 1]	[2, 3, 3, 1, 2, 3, 1, 2, 1, 1]	6	2
3	[4, 2]	[2, 3, 3, 2, 2, 3, 1, 2, 1, 1]	2	2
4	[4, 1]	[2, 3, 3, 1, 2, 3, 1, 2, 1, 1]	6	2
5	[4, 2]	[2, 3, 3, 2, 2, 3, 1, 2, 1, 1]	2	2

A strategy to prevent such cycling among locally optimal solutions must be a part of any tabu search design. Suppose that k neighborhood searches have been carried out, resulting in a trajectory of solutions $\{\mathbf{x}^1, \mathbf{x}^2, ..., \mathbf{x}^k\}$, and suppose that the moves that have led to these solutions are $\{m^1, m^2, ..., m^{k\text{-}1}\}$, i.e., $\mathbf{x}^\ell = \mathbf{x}^{\ell-1} \oplus m^{\ell-1}$ for $\ell = 1, ..., k$. The idea is then to reject moves that lead back to solutions on this list. The *strict prohibition* approach is to explicitly reject any proposed solution $\mathbf{x}'$ if $\mathbf{x}' = \mathbf{x}^i$ for some $i = 1, ..., k$. However, storing the solution trajectory and explicitly comparing each candidate solution $\mathbf{x}'$ with all k previous solutions would be prohibitively expensive in both computer storage requirements and computing time. Thus, if strict prohibition is to be implemented, a more sophisticated method is required. The *cancellation sequence method* and the *reverse elimination method* by Glover (1990a, 1990b) and Dammeyer and Voss (1993) are two techniques that are used to accomplish the same result more efficiently with respect to space and time requirements. Both work by storing a cancellation sequence of moves, i.e., a sequence of moves that are inverse to the sequence of accepted moves $\{m^1, m^2, ..., m^{k\text{-}1}\}$. By tracing the sequence in reverse order, it is possible to determine whether or not a particular move will lead from the current solution to a solution that was considered previously. This method is feasible with respect to computer storage, and is more efficient in computation than comparing $\mathbf{x}'$ with each $\mathbf{x}^i$ for i = 1, ..., k. Still, these methods are computationally more expensive than the indirect *move prohibition approach.*

The *move prohibition approach* is the most commonly used method of avoiding rapid cycling. It examines the proposed move m', and rejects it if m' reverses any of the last t moves, i.e., one of $\{m^{k-t}, m^{k-t+1}, ..., m^{k-1}\}$. In the example of Table I.49, move [4, 1] in Step 2 assigns item four to bin 1. The Step 3 move [4, 2] assigns item four to bin 2 and in Step 4, move [4, 1] is repeated. The move prohibition approach would forbid this last move as it involves the cycling back to solution $\mathbf{x}^3$.

Rather than actually storing the sequence of previous moves, move prohibition is usually invoked by the *tabu tenure* method. It assigns a temporary *tabu status* to all moves that would reverse the chosen move in any given iteration. This tabu status remains active for a certain number of iterations and it is called the tabu tenure; here denoted by t. When implementing this method, tabu moves may be stored on a list of length t, or a data structure may be set up to store and update a tabu status "flag" for each possible move. Then, when a candidate move is evaluated in the neighborhood search, its tabu status is checked, either by searching through the tabu list, or by checking the move tabu status flag. If the move is tabu, it is rejected.

Most tabu search algorithms are based on move prohibition methods by tabu status, since they are simple in concept and relatively easy to implement. Cycling is still possible, but discouraged by enforcing the tabu status of recent moves.

Given an optimization problem P: Min $z(\mathbf{x})$, s.t. $\mathbf{x} \in F \subseteq X$, we will use the notation $z(\mathbf{x}, H)$ for the move evaluation function. H refers to the recent *history* of the search which is often simply a stored list of *tabu* moves. Updating H is then done by adding a move (or moves) that are inverses of the current move m^*, and dropping the oldest member of the list. Table I.46 shows the search trajectory of the bin-packing example problem when a tabu tenure of length three is applied to an item after it is moved. Here, Δz denotes the decrease of the objective value z. In the basic tabu list scheme that associates a tabu status with moves, H is a list of tabu moves. Furthermore, a candidate solution $\mathbf{x}'$ is found by applying a move m to the current solution $\mathbf{x}^\ell$ (i.e., $\mathbf{x}' = \mathbf{x}^\ell \oplus m$), and the move evaluation function $z(\mathbf{x}', H)$ is formally defined as

$$z(\mathbf{x}', H) = \begin{cases} z(\mathbf{x}') & \text{if } m \notin H \\ \infty & \text{if } m \in H \end{cases}$$

Given a starting solution $\mathbf{x}^1$ and a stopping rule, we initialize the process by setting the iteration counter $\ell := 1$, $\mathbf{x}^* = \mathbf{x}^1$ and $z^* = z(\mathbf{x}^1)$. The tabu search method can then be described as follows.

Tabu Search Method

Step 1: (Neighborhood search): Set $y := \infty$. For each move m_i, evaluate $z(\mathbf{x}', H)$ with $\mathbf{x}' = \mathbf{x}^\ell \oplus m_i \in N(\mathbf{x}^\ell)$. If $z(\mathbf{x}', H) < y$, set $m^* := m_i$ and $y := z(\mathbf{x}', H)$; otherwise continue with the next move m_{i+1}.

Step 2: (Update the current solution and its history): If $y = \infty$ (no move has been found), go to Step 3; otherwise, update the history H by setting $\ell := \ell + 1$ and $\mathbf{x}^\ell := \mathbf{x}^{\ell-1} \oplus m^*$. If $z(\mathbf{x}^\ell) < z^*$, set $z^* := z(\mathbf{x}^\ell)$ and $\mathbf{x}^* := \mathbf{x}^\ell$ and go to Step 3; otherwise leave z^* and $\mathbf{x}^*$ unchanged and go to Step 3.

Step 3: (Termination): Is the stopping condition met?
If yes: Stop. The best solution is $\mathbf{x}^*$ with objective value z^*.
If no: Go to Step 1.

The progress of the tabu search algorithm as applied to the bin-packing example is shown in Table I.17. Examining this table, we see that the first two moves are the same as in Table I.16; item 9 is placed in bin 1, then item 4 in bin 1. In the third iteration, the most improving move would be to move item 4 to bin 2; however

Table I.17

ℓ	Best move	Move chosen	Solution	z	Tabu list H	Δz
0	–	–	[2, 3, 3, 2, 2, 3, 1, 2, 3, 1]	62	∅	–
1	[9, 1]	[9, 1]	[2, 3, 3, 2, 2, 3, 1, 2, 1, 1]	2	∅	-60
2	[4, 1]	[4, 1]	[2, 3, 3, 1, 2, 3, 1, 2, 1, 1]	6	[9]	4
3	[4, 2]	[3, 2]	[2, 3, 2, 1, 2, 3, 1, 2, 1, 1]	14	[9, 4]	-4
4	[4, 3]	[7, 3]	[2, 3, 2, 1, 2, 3, 3, 2, 1, 1]	6	[9, 4, 3]	-12
5	[3, 1]	[2, 1]	[2, 1, 2, 1, 2, 3, 3, 2, 1, 1]	14	[4, 3, 7]	-4
6	[4, 3]	[4, 3]	[2, 1, 2, 3, 2, 3, 3, 2, 1, 1]	2	[3, 7, 2]	-12
7	[3, 1]	[3, 1]	[2, 1, 1, 3, 2, 3, 3, 2, 1, 1]	6	[7, 2, 4]	4
8	[3, 2]	[1, 3]	[3, 1, 1, 3, 2, 3, 3, 2, 1, 1]	24	[2, 4, 3]	-4
9	[1, 2]	[2, 2]	[3, 2, 1, 3, 2, 3, 3, 2, 1, 1]	8	[4, 3, 1]	-18
10	[4, 1]	[4, 1]	[3, 2, 1, 1, 2, 3, 3, 2, 1, 1]	0	[3, 1, 2]	-8

moving item 4 is tabu. The best non-tabu move is to place item 3 in bin 2. In the fourth iteration, moving item 4 is still tabu, so that the best move [4, 3] is not taken and [7, 3] is chosen instead. In iteration five, item 3 is tabu, and the move [3, 1] is forbidden, so instead item 2 is moved to bin 1. Finally, in iteration 6, item 4 is no longer tabu and the best improving move [4, 3] is taken. This process continues until step 10, where an optimal solution with $z^* = 0$ is found and the search terminates.

Notice that the tabu assignment rule used in this example prevents two moves at each iteration, a fairly strong condition. There is another way to enforce tabu status. Suppose that at iteration n, item i is in bin k. If the move $[i, j]$ is chosen, then, for t iterations, the move $[i, k]$ is forbidden. This is a weaker prohibition, banning only a specific move rather than any move of a specific item.

It is possible that a move that is currently tabu leads to a new solution which is better than any of those that have been encountered so far. In such a case, it is clear that the tabu status of this move should be ignored, and the move accepted. Exceptions to tabu status are incorporated in the tabu search method by defining a criterion that indicates whether or not an exception to the rule applies. In order to do this, we define

Definition I.34: An *aspiration criterion* is a condition that overrides the tabu status of a move.

An aspiration that overrides tabu status for moves that improve on the best solution found is called *aspiration by objective*. This aspiration is always in force in tabu search—otherwise, the search might miss the optimal solution. In sophisticated variants of tabu search, more complex evaluation rules are usually formulated by invoking various aspiration criteria. Tabu tenure combined with aspirations forms a flexible structure on which to build successful tabu search methods for hard problems.

The effect of tabu status is to force the search into new regions of the solution space. Note that excluding cycling is not an end in itself; in fact, it will sometimes be desirable to have the search trajectory return to regions of X where high quality solutions have previously been found. The magnitude of the tabu tenure control parameter t controls the number of iterations for which the search will be forced away from the region of the current solution. In fixed tabu search, tabu tenure is specified exogenously. An effective choice for tabu tenure is problem and instance specific. Clearly, it is related to the size of the neighborhood, since setting t too large, e.g., choosing t equal to the number of neighbors of the current solution $\mathbf{x}$

would prohibit all moves. Glover (1990a) suggests that $t = \sqrt{|N(\mathbf{x})|}$ works well in many cases, and that, perhaps surprisingly, values for $t \in [7, 10]$ are good choices for many problems. The magnitude of t also will depend on the nature of the move prohibition strategy. If the tabu status is strong (weak), t should accordingly be adjusted to be smaller (larger). As with simulated annealing, a good value for t is often found by an experimentation phase. Also, dynamic schemes that vary tenure throughout the search process have been found to be more robust and more effective on some problems.

This tabu tenure scheme is based on the *recent* history of the search. A tabu search strategy can also exploit the search history further to recognize the *frequency* with which moves have been taken. In its simplest form, a frequency memory structure is used to penalize moves that have been taken frequently. Suppose that the array $C(m)$ stores the number of times a move m has been taken. One way to take the frequency history into account is to modify the evaluation function to penalize moves according to the magnitude of $C(m)$. With ℓ denoting the iteration number and a penalty parameter $K > 0$, we can define the move evaluation function as the sum of the frequency penalty $\rho = K\left[\dfrac{C(m)}{\ell}\right]$ and the resulting objective function change Δz, *viz*.

$$z(\mathbf{x} \oplus m, H) = K\left[\frac{C(m)}{n}\right] + \Delta z(\mathbf{x})$$

Applied to the bin packing problem with $K = 10$, we obtain the sequence of moves shown in Table I.50. In iteration $\ell = 3$, the move [4, 2] is still the best move, even with its frequency penalty $\rho = 10 \times {}^{1}/_{3} = {}^{10}/_{3}$. However, in iteration 4, it is sufficiently penalized by the frequency penalty, $\rho = 10 \times {}^{2}/_{4} = 5$ resulting in $z(\mathbf{x}, H) = \rho + \Delta z = 5 + 4 = 9$, so that the move [1, 3] is chosen instead. This stops the cycling and moves to a different neighborhood.

Table I.50

ℓ	i, j	Solution	z	Frequency $C[\,]$	Δ	ρ	$\rho + \Delta z$
0		2,3,3,2,2,3,1,2,3,1	62	0,0,0,0,0,0,0,0,0,0			
1	9,1	2,3,3,2,2,3,1,2,1,1	2	0,0,0,0,0,0,0,0,1,0	–60	0	–60
2	4,1	2,3,3,1,2,3,1,2,1,1	6	0,0,0,1,0,0,0,0,1,0	4	0	4
3	4,2	2,3,3,2,2,3,1,2,1,1	2	0,0,0,2,0,0,0,0,1,0	–4	$^{10}/_{3}$	$-^{2}/_{3}$
4	1,3	3,3,3,2,2,3,1,2,1,1	8	1,0,0,2,0,0,0,0,1,0	6	0	6
5	3,2	3,3,2,2,2,3,1,2,1,1	0	1,0,1,2,0,0,0,0,1,0	–8	0	–8

This example uses frequency observations only. More sophisticated tabu search algorithms will use both recency, i.e., the tabu list, and frequency considerations to direct the search trajectory. Extensions to the basic tabu search scheme are well documented in the literature; again, see the bibliography by Osman and Laporte (1996) for details.

7.4 Genetic Algorithms

Genetic algorithms were inspired by the Darwinian theory of evolution, which proposes that populations of species evolve over time to produce organisms that are more complex and fitter for survival on Earth; see Darwin (1859), and Dawkins (1986). Genetic algorithms are designed to mimic this process. They define the population as a collection of solutions to an optimization problem. Each solution has a *fitness value*, which measures how well it solves the problem relative to the other solutions in the population. New generations of solutions are created by selecting "parent" solutions from the existing population and combining them by exchanging elements of the parents to create "child" solutions. These "parents", the mating population, are selected with a probability that is a nondecreasing function of fitness. Good solutions survive to contribute to the next generation, and poor solutions die off. As generations of solutions evolve, the overall fitness of the population improves. The process is continued until the population contains some very good solutions to the problem.

Genetic algorithms were first reported by Holland (1975). They can search relatively large solution spaces quickly, and have been used with some success for hard problems. Genetic algorithms lend themselves naturally to parallel computing and to combination with other search heuristics; a comprehensive reference on genetic algorithms is Goldberg (1989). In addition to developing the problem model, i.e., selecting decision variables $\mathbf{x}$, defining the objective function $z(\mathbf{x})$, and formulating any constraints on the solution space X, a genetic algorithm also requires the definition of an *encoding scheme*, a *fitness function*, as well as *selection*, *crossover*, and *mutation operators*. The encoding scheme transforms a solution $\mathbf{x}$ into a string of finite length. For several reasons, binary strings are the method of choice. The objective function to be optimized will define the fitness function, which is the basis for the selection of the mating populations.

Three standard operators are used in genetic algorithms. They are the selection, crossover and mutation operators. Selection is the operator that selects whether or not an individual solution will be included in the mating population. The crossover operator combines two member solutions of the reproducing population to create new solutions, and the mutation operator makes occasional small, local random changes in a solution.

To implement a genetic algorithm, we first generate an initial population of N solutions, each of which is encoded as a string. These strings are often called *chromosomes*, and their entries *genes*, further emphasizing the genetic analogy. Then, a population of *mating* or reproducing solutions is selected using a probability distribution that favors those solutions with better objective values. Once the mating population has been selected, pairs of strings are subjected to the crossover operator that shuffles the parent genes into two new solutions. This part of genetic algorithms is displayed in Figure I.34. Thus, a child or offspring contains copies of the genetic material of the parents, but it has been rearranged, so that is represents a different solution to the optimization problem. The hope is that combining elements of good solutions will lead to better solutions. Finally, each string in the new population may be mutated. This is usually achieved by changing the value of one character in the string. Mutation is done rarely, i.e., with a very low probability. Mutation preserves diversity in the population and allows for a wider exploration of the search space.

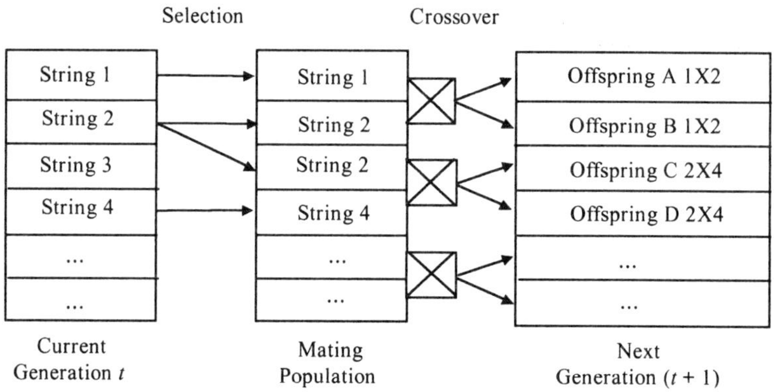

Figure I.34

Suppose now that solutions to a given problem are encoded into binary strings of length 8, and that we have two solutions, *viz*., Parent 1 = [1, 0, 1, 1, 0, 1, 1, 0] and Parent 2 = [0, 0, 1, 1, 1, 0, 0, 1]. To produce children by shuffling the genes in the parents, a crossover point is selected (usually randomly), and the parts of the parent strings before and after the crossover are copied to make two new strings. In particular, the substring of Parent 1 (Parent 2) left of the crossover point is merged with the substring of Parent 2 (Parent 1) to the right of the crossover point. Suppose that crossover is to be done on Parent 1 and Parent 2 after position 5.

Then, the crossover operator produces the two child chromosomes, Child 1 and Child 2 as shown in Table I.19.

Table I.19

Parent 1:	10110	011	Child 1:	10110	001
Parent 2:	00111	001	Child 2:	00111	110

Mutation is usually done by flipping a single element in a string. For example, mutating element 3 in Child 1 = [1, 0, 1, 1, 0, 0, 0, 1] above results in Child 1 (mutated) = [1, 0, 0, 1, 0, 0, 0, 1].

As in other heuristic algorithms, control parameters have to be chosen. In genetic algorithms, these include the population size N, the probability of crossover, α, the probability of mutation, γ, and the number of generations to be produced, T. To illustrate the genetic algorithm operators and the overall procedure, we use a zero-one knapsack problem as an example.

Example: Consider the zero-one knapsack problem

$$\text{P: Max } z = \sum_{i=1}^{n} c_i x_i$$

$$\text{s.t.} \quad \sum_{i=1}^{n} a_i x_i \le b$$

$$x_i = 0 \vee 1 \ \forall \ i = 1, \ldots, n$$

where $n = 6$ and $\mathbf{c} = (c_i) = [9, 2, 15, 7, 12, 14]$, $\mathbf{a} = (a_i) = [9.9, 6.0, 23.0, 3.3, 21.8, 19.6]$, and $b = 40$.

Solutions to this problem can be encoded as binary strings of length 6, where a 1 in position i indicates that item i is included and a 0 indicates that it is not. An initial population of ten randomly generated solutions (chromosomes) is shown in Table I.20.

Table I.20

Solution	Objective Value $z(\mathbf{x}^i)$	Used Weight $\sum_{i=1}^{n} a_i x_i$
$\mathbf{x}^1 = [1, 0, 0, 1, 0, 0]$	16	13.2
$\mathbf{x}^2 = [0, 1, 1, 0, 0, 0]$	17	29.0
$\mathbf{x}^3 = [0, 1, 0, 1, 1, 0]$	21	31.1
$\mathbf{x}^4 = [1, 0, 1, 0, 0, 0]$	24	32.9
$\mathbf{x}^5 = [0, 0, 0, 0, 0, 1]$	14	19.6
$\mathbf{x}^6 = [0, 0, 0, 1, 0, 1]$	21	22.9
$\mathbf{x}^7 = [1, 1, 0, 1, 0, 0]$	18	19.2
$\mathbf{x}^8 = [0, 1, 1, 1, 0, 0]$	24	32.3
$\mathbf{x}^9 = [0, 0, 1, 0, 0, 0]$	15	23.0
$\mathbf{x}^{10} = [0, 0, 0, 0, 0, 1]$	14	19.6

The next important choice to be made in genetic algorithms concerns the selection of the mating population. Typically, selection from the current generation is done in a probabilistic fashion according to the fitness of a solution. Define $f(\mathbf{x})$ to be the fitness function that measures the fitness of a solution $\mathbf{x}$. In many cases, as in this example, the fitness function $f(\mathbf{x})$ is simply the given objective function $z(\mathbf{x})$. It is also possible to form $f(\mathbf{x})$ by augmenting $z(\mathbf{x})$ by penalty terms that express constraint violations.

In order to choose a potentially fit mating population, i.e., particularly fit, good solutions are preferred. In fact, if a solution is fitter than the average fitness of the population, it will automatically be used at least once in the mating population. The recommended way to generate the mating population is a procedure known as *remainder stochastic sampling*. This process involves two steps. For better than average solutions, a number of copies are automatically generated. The remaining positions in the mating population are determined by sampling with replacement from the current generation. Formally, define $f_i = f(\mathbf{x}^i)$ as the fitness of solution $\mathbf{x}^i$, the population fitness $F = \sum_{i=1}^{N} f(\mathbf{x}^i)$, and the *average fitness* of the population $\overline{F} = \frac{1}{N}F$. For each member in the current generation, we calculate the *relative fitness* of $\mathbf{x}^i$ as $q_i = f_i/\overline{F}$. The term q_i is then separated into its integer and fraction parts. The integer part indicates how many copies of $\mathbf{x}^i$ are to be placed directly in the mating population, and the fractional part is proportional to the probability associated with solution $\mathbf{x}^i$ that is used to sample the current population to "fill up" the mating population. After copies have been made, sampling with replacement is performed, where solution $\mathbf{x}^i$ has a probability of $p_i = (q_i - \lfloor q_i \rfloor) \Big/ \sum_{k=1}^{N} (q_k - \lfloor q_k \rfloor)$ being included in the solution (again), until the desired number of solutions is obtained. In our knapsack example, the population fitness in the initial population is F = 184, so that $\overline{F}$ = 18.4 and the relative fitnesses of the ten solutions are $\mathbf{q} = (q_i)$ = [.87, .92, 1.14, 1.30, .76, 1.14, .98, 1.30, .82, .76]. As a result, no copy of solution $\mathbf{x}^1$ is used directly in the mating population, but the probability that the solution is chosen for the mating population is p_1 = 0.87 / 5.99 = .1452. A similar argument applies to solution $\mathbf{x}^2$. As the relative fitness of $\mathbf{x}^3$ is 1.14, the integer part of q_3 indicates that one copy is immediately put in the mating solution, and there is a probability of p_3 = 0.14 / 5.99 = .0234 that it is chosen again. The remaining solutions are dealt with accordingly. As a result, solutions $\mathbf{x}^3$, $\mathbf{x}^4$, $\mathbf{x}^6$, and $\mathbf{x}^8$ are used once in the mating population, and the remaining six positions are filled by sampling from the initial population according to their remainder probabilities p_i. A specific choice of random numbers selects $\mathbf{x}^1$ once, $\mathbf{x}^2$ twice, $\mathbf{x}^7$ twice, and $\mathbf{x}^9$ once, so that the mating population now consists of the ten solutions $\mathbf{x}^1$, $\mathbf{x}^2$, $\mathbf{x}^2$, $\mathbf{x}^3$, $\mathbf{x}^4$, $\mathbf{x}^6$, $\mathbf{x}^7$, $\mathbf{x}^7$, $\mathbf{x}^8$, and $\mathbf{x}^9$.

The next step is to create the next generation from the mating population. This is done by randomly drawing pairs of parents and then applying a crossover procedure. This mimics the exchange of genetic material in the parent chromosomes that occurs when two organisms reproduce. The crossover procedure with randomly chosen crossover points is performed on each pair of strings in the mating population with probability α; here, we choose α = 0.7. With probability $(1 - \alpha)$, the parents are directly copied to the next generation. After applying the crossover procedure, each string is subjected to the mutation operator. A random number in [0, 1] is drawn; if this number is less than the probability of mutation, γ, then a position in the string is selected at random and its value flipped. Mutation probabilities are typically very small. In our example, five pairs of solutions were selected at random, and for each pair, a random number was drawn. If this number was greater than 0.7, then the pairs were copied directly to the next generation. Otherwise, a crossover point was selected at random, and the two offspring resulting from the crossover were placed in the next generation. For the purpose of our illustration, we have chosen to generate a small population of size 10. The results are shown in Table I.21 where the population fitness is F = 178. The best solution found in this next generation is $\mathbf{x}^{13}$, whose fitness $f(\mathbf{x}^{13})$ = 26.0, which is better than any solution in the initial population.

Table I.21

Mating pair	Random number r	Crossover point	Resulting Solution	$z(x^i) = f(\mathbf{x}^i)$	q_i
$\mathbf{x}^6 \times \mathbf{x}^3$	.25	3	$\mathbf{x}^{11} = [0, 0, 0, 1, 1, 0]$ $\mathbf{x}^{12} = [0, 1, 0, 1, 0, 1]$	19 23	0.98 1.19
$\mathbf{x}^4 \times \mathbf{x}^2$	.40	1	$\mathbf{x}^{13} = [1, 1, 1, 0, 0, 0]$ $\mathbf{x}^{14} = [0, 0, 1, 0, 0, 0]$	26 17	1.24 0.88
$\mathbf{x}^7 \times \mathbf{x}^1$	.17	2	$\mathbf{x}^{15} = [1, 1, 0, 1, 0, 0]$ $\mathbf{x}^{16} = [1, 0, 0, 1, 0, 0]$	18 16	0.93 0.82
$\mathbf{x}^7 \times \mathbf{x}^9$	.65	1	$\mathbf{x}^{17} = [1, 0, 1, 0, 0, 0]$ $\mathbf{x}^{18} = [0, 1, 0, 1, 0, 0]$	24 9	1.24 0.46
$\mathbf{x}^2 \times \mathbf{x}^8$	.50	3	$\mathbf{x}^{19} = [0, 1, 0, 1, 0, 0]$ $\mathbf{x}^{20} = [0, 1, 1, 0, 0, 0]$	9 17	0.46 0.88

Various extensions of the basic algorithm are possible. One such extension is the two-point crossover. It is performed by choosing two crossing points (rather than one) at random and the "genes" that lie between them are swapped. For example, if Parent 1 = [0, 0, 0, 0, 1, 0, 0, 0, 1] and Parent 2 = [1, 1, 0, 0, 0, 1, 1, 1, 1], and crossover is to be performed after positions 2 and 6, the resulting children are [0, 0, 0, 0, 1, 0, 0, 1] and [1, 1, 0, 0, 1, 0, 1, 1, 1].

Crossovers and mutations have to be redefined if the original problem is not defined in terms of zero-one strings. As an example, consider a job scheduling problem, where eight jobs must be ordered. Feasible solutions to the problem are all permutations of the job numbers. For example, the solution [3, 2, 4, 1, 5, 6, 7, 8] specifies that job 3 is done first, followed by job 2 and so on. This string of job numbers is a natural encoding of the solution. Neither the one or two point crossover operators will work correctly with these strings. To see this, consider the two solutions, [3, 1, 2, 4, 6, 7, 8, 5] and [8, 2, 4, 3, 1, 5, 6, 7] and a two-point crossover after positions 2 and 5 effected directly on the string representation of the two solutions. The results are the two solutions [3, 1, 4, 3, 1, 7, 8, 5] and [8, 2, 2, 4, 6, 5, 6, 7]. Neither of these are valid orderings, since jobs appear in more than one position. Thus, the crossover method cannot be used without modification on strings encoded in this fashion.

One such modification is as follows. In the two-point crossover after positions 2 and 5, the middle segment [2, 4, 6] of Parent 1 is to be replaced by the middle segment of Parent 2, [4, 3, 1]. Since we cannot have more than one of each job number in the result, we identify those positions in Parent 1 in which the entries in the middle segment of Parent 2 are found. If they are found in anywhere but the middle segment, a *hole H* is created. This way, we "prepare" Parent 1 for crossover by placing a hole for each position which has a gene value of 4 or 3 or 1, *viz.*, positions 1, 2, and 4. Doing so changes the first parent's sequence [3, 1, 2, 4, 6, 7, 8, 5] to [H, H, 2, H, 6, 7, 8, 5]. Moving the holes to the center position results in [2, 6, H, H, H, 7, 8, 5]. At this point, the holes can be filled with the center segment of Parent 2; this process results in [2, 6, 4, 3, 1, 7, 8, 5] which is one of the two children. The other child is created similarly.

A different way of encoding solutions that represent a permutation is the method of random keys. Instead of using the actual permutation values, unique random numbers in the range of [0, 1] are generated. For example, the sequence of random numbers [.112, .011, .314, .781, .532, .976, .245, .034] represents an ordering of jobs, where the smallest random number corresponds to job number 1, the second smallest to job number 2 and the largest to job number 8. In this example, the corresponding job ordering is [3, 1, 5, 7, 6, 8, 4, 2]. A nice property of this random key representation is that a crossover of two vectors of random numbers will almost always represent a feasible ordering, except when two identical random numbers are created in the child. In such a case, the ordering can be resolved by random tie-breaking.

We are now able to summarize the genetic algorithm method to solve an optimization problem P: Min $z(\mathbf{x})$, $\mathbf{x} \in X$. Suppose that the fitness function is $f(\mathbf{x})$, and that a solution $\mathbf{x}^i$ is encoded in a string $\mathbf{s}^i$ of length M. In generation (iteration) t, we distinguish between the set of solutions $\{\mathbf{x}^i\} = X^t$ and the set of strings or chromosomes $\{\mathbf{s}^i\} = S^t$ from which the mating population R^t is selected. Select the (even) population size N, the probability of crossover α, and the probability of mutation γ. As a stop criterion, we will terminate the algorithm after T generations have been performed. The algorithm is initialized by defining the set of N chromosomes S^0, the corresponding N solutions X^0 and setting $t := 0$. Find the best solution $\mathbf{x}^* \in X^0$ that gives the lowest objective function value z^*, and save $\mathbf{x}^*$ and z^*. The genetic algorithm may then be summarized as follows.

A Genetic Algorithm

Step 1: Calculate the fitness values $f_i = f(\mathbf{x}^i)$ for all solutions $\mathbf{x}^i \in X^t$ and determine the population average fitness $\overline{F} = \frac{1}{N}\sum_{i=1}^{N} f_i$.

Step 2: Create R^t, the mating population of encoded solutions. For each string $\mathbf{s}^i \in S^t$, add $\lfloor f_i \rfloor$ copies of $\mathbf{s}^i$ to R^t. Create the remaining members of R^t by sampling with replacement from S^t.

Step 3: (Crossover and Mutation): For ½N steps, randomly choose a pair of parents $\mathbf{s}^i$ and $\mathbf{s}^j$ from R^t. With probability α, perform crossover on $\mathbf{s}^i$ and $\mathbf{s}^j$, producing two new chromosomes $\mathbf{s}^{i'}$ and $\mathbf{s}^{j'}$. If no crossover is done (with a probability 1-α), simply copy $\mathbf{s}^i$ and $\mathbf{s}^j$ to $\mathbf{s}^{i'}$ and $\mathbf{s}^{j'}$. With probability γ, mutate each chromosome to obtain the next generation S^{t+1}. From S^{t+1} determine X^{t+1} and set $t := t + 1$.

Step 4: (Update the best solution): Calculate $z(\mathbf{x}^i)$ for all $\mathbf{x}^i \in X^t$. Set $z^* :=$ $\min\{z^*, \min_{x^i \in X^t} \{z(\mathbf{x}^i)\}\}$, and $\mathbf{x}^* :=$ arg min$\{\mathbf{x}^*; \min_{x^i \in X^t} \{\mathbf{x}^i\}\}$.

Step 5: (Termination): Is $t \geq T$? If no: Go to Step 1.

If yes: Stop and report $\mathbf{x}^*$ and z^*.

The key to the genetic algorithm is that the members of future generations are selected probabilistically for survival and for "reproduction" according to the relative quality of the solution they represent. Thus, in each stage, or generation, good solutions are more likely to be "mated", and the population will hopefully evolve to a superior set of solutions to the problem. As with simulated annealing, there are some obvious parallels between the natural process which inspired the algorithm, and the mechanics of the search process. The main differences between this search methodology and simulated annealing or tabu search were pointed out by Goldberg (1989). They are:

(1) solutions are encoded in strings, often binary strings,

(2) in each generation, a population of solutions is evaluated, and

(3) the search is probabilistic, based on the "fitness" of the members.

7.5 Other Approaches

Genetic algorithms are not the only class of heuristic methods to be based on some aspect of natural evolution. For instance, evolutionary algorithms, which originally arose in the context of optimizing the settings for continuous control parameters in laboratory experiments, emphasize the mutation operator over the crossover operator. Evolutionary algorithms have been applied to discrete optimization problems, but their original focus and that of most of the subsequent applications is on optimizing continuous, real-valued decision variables.

The field of neural networks was inspired by the functioning of neurons and synaptic connections in the human brain. An artificial neural network has a set of nodes (neurons), which can be in different states, weighted connections which are subsequently adapted in feedback loops, and an energy function which is defined over the set of all possible states of the system. As inputs to the system are processed, the system iterates towards a stable or ground state of minimum energy, as defined by the energy function. Neural networks are inherently very adaptable to parallel processing, and have been used very successfully in applications such as pattern recognition and combinatorial optimization. As an example, Hopfield and Tank (1985, 1986) apply neural network to solve the traveling salesman problem. Tutorials on neural networks are provided by Wythoff (1993) and de Werra and Hertz (1989), and a bibiliography is found in Sharda (1994). At this point in time, research has shown that while using neural network structures to solve combinatorial optimization problems is feasible, they have not yet been shown to be as successful as other metaheuristics.

Recently, another heuristic has been suggested by Colorni *et al.* (1991, 1992). Similar to genetic algorithms, it also attempts to mimic a biological phenomenon. In particular, it uses parallel processors (called ants) that independently generate solutions, which are then evaluated against each other. Borrowing from the behavior of ants, a particularly beneficial move, i.e., a move from one point (e.g., a solution) to another or, in terms of networks, from one node to another by way of some arc that significantly improves the quality of the solution and chosen by one ant is likely to be chosen by other ants as well. In biological ant systems, a trail of pheromones is left by each ant, and is subsequently picked up by other ants. This trail is intensified as more and more ants use it, but, at the same time, its intensity diminishes over time.

As an example, consider a standard linear integer programming problem. A move in an "ant system heuristic" could be the simultaneous increase of some variable x_j and decrease of some other variable x_i. If such a change turns out to increase the objective value, it is likely that other processors, regardless of the remaining parts of the solution, will also investigate such a change, increasing the probability that more and more processors make the same change. Bullnheimer *et al.* (1999) have applied ant systems to traveling salesman problems with some success. Due to the novelty of the approach, it is not known at this point what type of problem ant systems work well for, and whether or not they are suitable for general integer programming problems.

As Johnson *et al.* (1991) caution, heuristics are not a panacea, but are potentially useful tools. When the use of exact methods must be ruled out, metaheuristics can provide flexible tools to solve hard problems with the advantage of simple implementation. No one strategy has been shown to dominate any other.

Moreover, we now see a progression to develop hybrid systems that combine selected features from tabu search, simulated annealing and genetic or evolutionary computing. Future research will continue to expand the capabilities of existing heuristics.

PART II NETWORK PATH MODELS

A large variety of problems can be modeled in terms of a network in which the objective is to select a subset of arcs with certain properties. As selecting or not selecting a particular arc for inclusion in a solution is a zero-one decision, all network models that involve choosing arcs can be formulated as integer programming problems with binary variables. In that sense, network path models can be considered special cases of general integer programming problems, and could therefore be modeled and solved using the methods described in Part I. However, special purpose methods exist for solving particular types of network path models, and these are discussed in this part.

Since Chapter c in the Introduction has been devoted to the fundamentals of graphs and networks, in this part we will build on the foundations laid in that chapter and not repeat the material here. Other texts that cover network path models include Bertsekas (1991), Evans and Minieka (1992), Murty (1992), and Ahuja *et al.* (1993).

CHAPTER 1 TREE NETWORKS

One of the most important concepts of graph theory is that of a *tree*. Its importance is due to the fact that a tree structure enables the connection of a set of nodes using a minimal number of edges in such a way that any two nodes of the set are connected by a unique chain. As such, the tree is a fundamental structure in many fields of study: network theory, social science, computer science, transportation, and many others. The simple structure of trees can offer algorithmic advantages for efficiently solving network path problems. Garey and Johnson (1979) have shown this for certain combinatorial optimization problems. Indeed, a number of problems that are *NP-complete* when formulated on a general graph become polynomially solvable when the graph is a tree.

The first section of this chapter presents the most important and simplest of such trees, the minimal spanning tree problem. The second section considers the minimal spanning tree problem with different objective functions and/or additional constraints, the third section extends and applies the notion of connectivity. Finally, the fourth section deals with a problem that involves an embedding of a given tree in the Euclidean plane.

1.1 Minimal Spanning Trees

1.1.1 Definition and Examples

Given a graph $G = (N, E)$ with a set of nodes N and a set of undirected edges E, this section investigates partial graphs $G_T = (N, T)$ of a connected graph G, such that the edge set T forms a tree over the node set N. Such a partial graph of G is called a *spanning tree* of G as all the nodes of G are incident to at least one edge of T. We are particularly interested in weighted graphs, i.e., graphs in which each edge e_{ij} has a numerical value c_{ij} associated with it. These edge weights may symbolize distances, costs, time, or any other feature associated with the edges of a graph. A spanning tree G_T of G then has total cost $c(G_T) = \sum_{e_{ij} \in T} c_{ij}$. A natural problem is then to seek a a *minimal* (*cost*) *spanning tree* (MST) of G, i.e., a spanning tree $G_T^* = (N, T^*)$ such that $c(G_T^*) = \min_{T \subset A}\{c(G_T^*)\}$. Here, we assume that G is connected; in unconnected graphs, minimal spanning trees can be found for each of the connected components of G separately.

Since G is connected, we may remove edges from it until condition (4) of Theorem 39 is satisfied, i.e., the removal of any additional edge would disconnect the graph. For a complete graph with n nodes, there are as many as n^{n-2} spanning trees of G, so that for even moderately dense graphs and $n = |N| > 10$, total enumeration of the spanning trees is out of the question. Fortunately, the minimal spanning tree is one of the few graph-theoretical problems that can easily be solved by efficient low-degree polynomial algorithms.

Before discussing the algorithmic aspects of the problem, we first consider some examples of problems that can be formulated as minimal spanning trees. Applications range from direct use where the graph and costs suggest themselves naturally to indirect use where the minimal spanning tree is a subproblem within a larger and more difficult optimization problem.

Application 1 (*Pipelines*): Consider a petroleum products company that operates a refinery in city n_1, and has oil wells located in oil fields n_2 to n_5. The company has

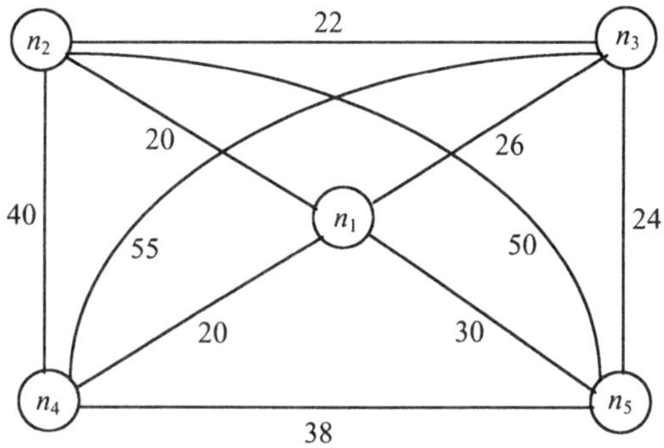

Figure II.1

estimated the costs between the refinery and its oil wells as well as between oil wells. The construction costs (in $100,000) are shown in Figure II.1.

The company's objective is now to construct pipelines so as to connect all oil wells to the refinery at minimal cost. Since all costs are positive, the minimal cost configuration of pipelines forms a minimal spanning tree, as any partial graph connecting all the nodes and having a cycle would cost more than the partial graph resulting from the elimination of one edge from that cycle. The optimal solution to the problem includes the edges e_{12}, e_{14}, e_{23}, and e_{35} with total costs of $86.

Application 2 (*Clustering*): It is sometimes desirable to group together elements of a set S in *clusters*, i.e., disjoint subsets of S, such that elements of the same cluster have more in common with one another than with elements of the other clusters. For example, a set S of patients, all of whom have been diagnosed as having the same disease, can be measured with respect to the same n symptoms. If one can find a meaningful way of clustering these patients based only on the n-vector of the sample, then this may lead to more disease categories to categorize or subdivide the original disease. This technique, called *cluster analysis*, has many applications, among them the taxonomy of biological species and the classification of library material.

An algorithm for the detection of clusters was proposed by Zahn (1971) who suggests calculating a minimal spanning tree based on the Euclidean distances between elements of the sample space and then dropping certain edges so that the resulting tree components correspond to the clusters. He justifies this approach by showing that it agrees with certain principles from gestalt psychology bearing on

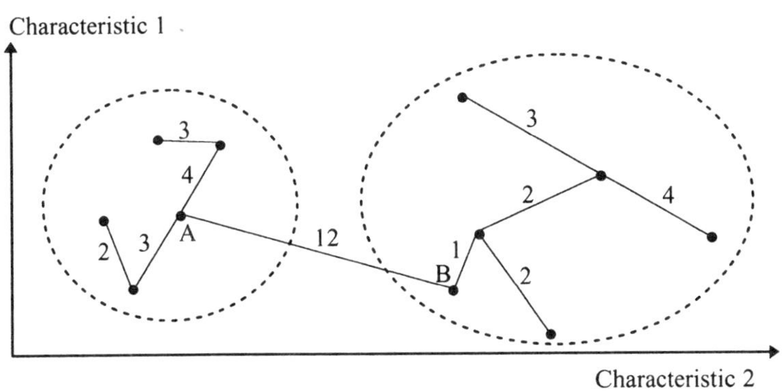

Figure II.2

human perceptual organization of points in $\mathbb{R}^2$. In particular, humans seem to favor groupings of data that represent smaller point-to-point distances. Figure II.2 exhibits a minimal spanning tree on a sample of size 11 based on 2 characteristics.

In order to arrive at the two clusters indicated by the broken contours indicated in Figure II.2, Zahn proposes two criteria: starting with a complete graph on all nodes, (1) drop any edge *e* whose weight is more than a fixed number of sample standard deviations greater than the average of nearby edge weights on both sides of edge *e*, and (2) drop any edge for which the ratio of weight to the two neighboring edge weight averages exceeds a certain factor.

Applying criterion (1) to the edge (A, B) that connects the two clusters A and B, and taking averages on both sides, we obtain average distances of $\frac{1}{4}(2 + 3 + 4 + 3) = 3$, and $\frac{1}{3}(1 + 2 + 2) = 1.67$ within the two respective clusters, including all nodes that are within two edges of A and B, respectively. Thc sample standard deviations are 0.707 and 0.471, respectively, so that edge (A, B) is almost 17 standard deviations larger than the first average and 25 standard deviations larger than the second. This would indicate dropping edge (A, B) since for a normal distribution of edge weights with the same means and standard deviations, an edge whose weight exceeds three standard deviations would occur only about 0.26% of the time. Applying criterion (2) to the same edge, we obtain ratios of 12/3 = 4, and 12/1.67 = 7.19, respectively. Choosing a factor of 2, which usually means the separation is quite apparent, we may again eliminate edge (A, B). For more details on this method of cluster analysis, the reader is referred to Zahn's original work.

Another important class of applications deals with routing problems, while these problems do not involve tree networks *per se*, many of their solution algorithms do. Chapters 3 and 4 of this part are devoted to them.

1.1.2 Solution Techniques

The problem of finding a minimal spanning tree can be formulated as an integer linear program. For that purpose, first define zero-one variables y_{ij} that assume a value of one, if the edge e_{ij} is included in the tree and 0 if it is not. With c_{ij} again denoting the arc value of the edge $e_{ij} \in E$, the problem can be formulated as follows.

$$\text{P: Min } z = \sum_{e_{ij} \in E} c_{ij} y_{ij}$$

$$\text{s.t.} \sum_{e_{ij} \in E} y_{ij} = n - 1$$

$$\text{The set } \{e_{ij} : y_{ij} = 1\} \text{ does not form a cycle}$$

$$y_{ij} = 0 \vee 1 \ \ \forall \ e_{ij} \in E.$$

The first constraint forces the model to choose exactly $(n - 1)$ of the existing edges and the "no cycle" constraints ensure that the resulting subgraph is connected and does not have a cycle, i.e., is a tree. The problem with this formulation is that the set of constraints that ensure that the resulting graph does not include a cycle includes no less than $2^n - n - 1$ constraints. Formulations of the constraints are provided in Section 3. Integer programming formulations of the minimal spanning tree problem with a polynomial number of constraints have been proposed by Magnanti and Wolsey (1995), Dror *et al.* (1990), and these are useful for solving constrained minimal spanning tree problems that are *NP-complete*. Fortunately, the unconstrained minimal spanning tree problem is polynomially solvable. Efficient algorithms have been suggested by Kruskal (1956), Prim (1957), Loberman and Weinberger (1957), Dijkstra (1959), and Yao (1975), all of which in some way exploit the following result.

Lemma II.1: Suppose that a procedure for finding a minimal spanning tree of an undirected graph $G = (N, E)$ has, during the course of constructing the tree, produced the forest $\{G_1, G_2, ..., G_p\}$ of G, where $G_j = (N_j, T_j)$ and the edges of $T = \bigcup_{j=1}^{p} T_j \subset E$ belong to the minimal spanning tree of G. For some subtree (N_i, T_i) with $i \in \{1, 2, ..., p\}$ let $e_{k\ell}$ be the least cost edge with $n_k \in N_i$ and $n_\ell \notin N_i$. Then $e_{k\ell}$ is part of the minimal spanning tree and can be added to the edge set T, thus forming a new forest of G.

Proof: Let $n_k \in N_i$ and $n_\ell \in N_j$ with $i \neq j$. Suppose now that $e_{k\ell}$ is *not* part of the minimal spanning tree T. Let T include instead an edge $e_{\nu\mu}$ with $n_\nu \in N_i$ and $n_\mu \in N_k$ with $k \neq i$ but possibly $k = j$. Furthermore, G_j and G_k are connected by some edge e_{uw}. This situation is shown in Figure II.3. Let the cost of the minimal spanning tree (that, by assumption, does not include the edge $e_{k\ell}$ but the edge $e_{\nu\mu}$) be $c(G)$. Introducing the edge $e_{k\ell}$ generates a cycle that is broken again by the removal of the edge $e_{\nu\mu}$. Note that the removal of $e_{\nu\mu}$ does not change the connectivity of the tree. The total cost of the new spanning tree is then $c(G) + c_{k\ell} - c_{\nu\mu}$. As $c_{k\ell} < c_{\nu\mu}$ by assumption, the cost of the new spanning tree is less than that of the one with edge $e_{\nu\mu}$, violating the assumption that the total cost of the original

tree was minimal.□

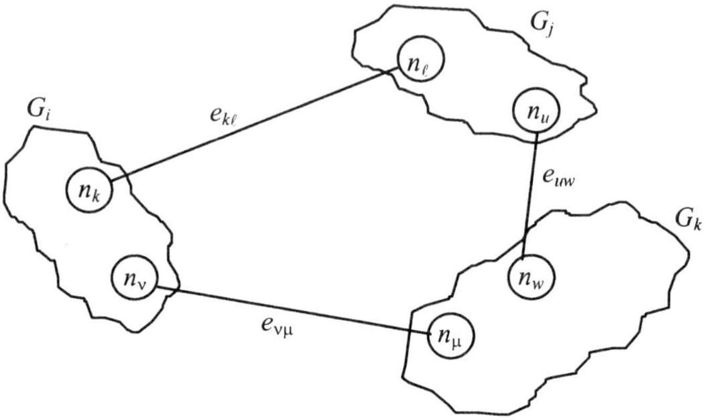

Figure II.3

If the costs of the edges of G are not distinct as assumed above, then several different minimal spanning trees may exist, all of which have the same total cost, but some of whose edges are different. Since any node is in itself a subtree of G, Lemma II.1 implies

Corollary II.2: The smallest cost edge incident to any node of G is part of any minimal spanning tree.

We observe that if we initially take $N_1 = \{n_1\}$, and $T_1 = \emptyset$, then the assumptions of the theorem are trivially satisfied, and $n - 1$ applications of the theorem will produce a minimal spanning tree of G.

The two algorithms presented below differ only in the way they choose the subtrees (N_i, T_i) to which to add the edge e. The first algorithm is due to Prim (1957) and begins with an arbitrarily chosen node n_1. It then adds new edges to the existing connected component (N_1, T_1) until the graph includes $(n - 1)$ edges. We assume that an undirected connected graph $G = (N, E, c)$ is given with $|N| = n$, $|E| = m$, and costs c_{ij} for $e_{ij} \in E$ and $1 \leq i \leq j \leq n$. The algorithm is initialized by setting $N_1 = \{n_1\}$, $T_1 = \emptyset$ and defining labels (α_i, β_i) for all nodes $n_i \in N \setminus \{n_1\}$, such that $\alpha_i = c_{1i}$ denotes the cost of the edge e_{1i} and ∞, if $e_{1i} \notin E$, and $\beta_i = n_1$, which indicates the node that determines α_i. Furthermore, set the iteration counter

k:=1. We can then describe

Prim's Algorithm for Minimal Spanning Trees

Step 1: Find the node $n_{j^*} \in N \setminus N_k$, such that $\alpha_{j^*} = \min_{n_j \in N \setminus N_k} \{\alpha_j\}$.

Set $N_{k+1} := N_k \cup \{n_{j^*}\}$ and $T_{k+1} := T_k \cup \{(\beta_{j^*}, n_{j^*})\}$.

Step 2: Is $k = n - 1$?

If yes: Stop; the graph $G_{k+1} = (N_{k+1}, T_{k+1})$ is a minimal spanning tree.

If no: Go to Step 3.

Step 3: (Update α and β): Set $\alpha_j := \min \{\alpha_j; c_{jj^*}\}\ \forall\ n_j \in N \setminus N_k$. If α_j changes in value, set $\beta_j := n_{j^*}$. Set $k := k + 1$ and go to Step 1.

The algorithm's complexity is easily evaluated. Each iteration requires $O(n)$ operations and $(n - 1) = O(n)$ iterations are required, resulting in an overall complexity of $O(n^2)$.

Example: Consider the graph $G = (N, E, c)$ in Figure II.4.

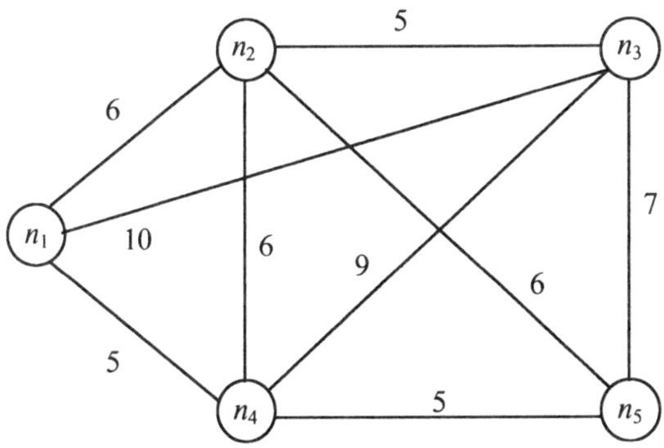

Figure II.4

The algorithm starts with $k = 1$, $N_1 = \{n_1\}$, $T_1 = \emptyset$; $\alpha_2 = 6$, $\alpha_3 = 10$, $\alpha_4 = 5$, $\alpha_5 = \infty$, and $\beta_2 = \beta_3 = \beta_4 = \beta_5 = n_1$. In the first iteration, $\alpha_{j^*} = \min_{n_2, n_3, n_4, n_5} \{\alpha_j\} = 5$, so that j^* = 4. Now $N_2 = \{n_1, n_4\}$, $T_2 = \{e_{14}\}$, and the α- and β-values must be updated. As $c_{24} = 6 = \alpha_2$, α_2 and β_2 remain unchanged. However, $c_{45} = 5 < \infty = \alpha_5$, the label n_5

is revised to $\alpha_5 = 5$ and $\beta_5 = 4$. Similarly, as $c_{34} = 9 < 10 = \alpha_3$, the label of n_3 is revised to $\alpha_3 = 9$, and $\beta_3 = n_4$.

In the second iteration, $\alpha_{j^*} = \min_{n_2,n_3,n_5} \{\alpha_j\} = 5$ with $j^* = 5$. Now $N_2 = \{n_1, n_4, n_5\}$ and $T_3 = \{e_{14}, e_{45}\}$ and the updated α- and β-values are $\alpha_2 = 6$ with $\beta_2 = n_1$ (unchanged), but as $c_{35} = 7 < 9 = \alpha_3$, the label of n_3 is revised to $\alpha_3 = 7$ and $\beta_3 = n_5$. The third iteration finds $\alpha_{j^*} = \min_{n_2,n_3} \{\alpha_j\} = 6$ with $j^* = 2$. The current tree consists of nodes $N_4 = \{n_1, n_2, n_4, n_5\}$ and $T_4 = \{e_{14}, e_{45}, e_{12}\}$, and the updating process reveals that $c_{23} = 5 < 7 = \alpha_3$, and hence the label of n_3 is again revised, this time to $\alpha_3 = 5$ and $\beta_3 = n_2$. In the fourth and last iteration, $\alpha_{j^*} = \alpha_3 = 5$ with $N_5 = N = \{n_1, n_2, n_3, n_4, n_5\}$, and $T_5 = \{e_{14}, e_{45}, e_{12}, e_{23}\}$. As $k = 4 = n - 1$, the process terminates. The graph $G_5 = (N_5, T_5)$ is a minimal spanning tree. Adding the costs of its edges reveals that $c(G_5) = 21$.

The second algorithm considered here is due to Kruskal. Like Prim's algorithm, it uses Lemma II.1 starting with a forest of n subtrees, one for each node of G, and no edges, and adds one edge at a time. The difference to Prim's algorithm is that instead of choosing first a subtree and then finding the shortest edge joining it to another subtree, Kruskal's algorithm chooses first the shortest edge not already added to T, the partially formed minimal spanning forest, and then adds it to T if it does not create a cycle with the edges that already exist in T.

Kruskal's algorithm is initialized by first ordering the edges with respect to their lengths where ties are broken arbitrarily. This resets in an ordered sequence of edges $e_1 = e_{i_1 j_1}$, $e_2 = e_{i_2 j_2}$, ..., $e_m = e_{i_m j_m}$ with costs $c_1, c_2, ..., c_m$, so that $c_1 \le c_2 \le ... \le c_m$. Futhermore, set $T := \emptyset$ and two counters $k := 0$ and $\ell := 1$.

Kruskal's Algorithm for Minimal Spanning Trees

Step 1: Is $k = n - 1$? If yes: Stop, the graph $G = (N, T)$ is a minimal spanning tree.

If no: Go to Step 2.

Step 2: Does the graph $G = (N, T \cup \{e_\ell\})$ include a cycle?

If yes: Set $\ell := \ell + 1$ and go to Step 1.

If no: Set $T := T \cup \{e_\ell\}$, $\ell := \ell + 1$, $k := k + 1$ and go to Step 1.

In the example of Figure II.4, the edges are ordered as $e_1 = e_{14}$, $e_2 = e_{45}$, $e_3 = e_{23}$, $e_4 = e_{25}$, $e_5 = e_{12}$, $e_6 = e_{24}$, $e_7 = e_{35}$, $e_8 = e_{34}$, $e_9 = e_{13}$. Note that many other orderings exist, e.g., any permutation of the first three elements, followed by any permutation of the second three edges is possible. As a consequence, multiple solutions will result. Here, edges e_1, e_2, e_3, and e_4 are introduced, none of them forming a cycle with already existing edges. There are m iterations, each of which, given appropriate data structures, takes O(log m) time. Hence, Kruskal's algorithm takes O(m log m) time and as such is most suitable for relatively sparse graphs, i.e., graphs with $m << \frac{1}{2}n(n-1)$ rather than nearly complete ones that are better treated with Prim's algorithm.

Both Prim's and Kruskal's algorithms are special cases of what are called *greedy* algorithms. The reason is that in each step they both attempt to do what is *locally* best without regard to future consequences (*global* optimality). There is no look-ahead procedure or backtracking. What is so unusual is that these greedy approaches produce globally optimal solutions for this problem. For most problems on graphs, greedy algorithms are heuristics that may produce good feasible, but not necessarily optimal, solutions in reasonable time.

The complexity results stated for the algorithms of Prim and Kruskal may be improved upon for implementation purposes by using appropriate data structures. For details, readers are referred to Tarjan (1983) and Camerini *et al.* (1988).

1.2 Extensions of Minimal Spanning Trees

As we have observed, the minimal spanning tree problem is one of the simplest optimization problems, with a number of efficient algorithms available for its solution. Most applications involving the calculation of minimal weighted spanning trees, however, require that certain constraints be placed on the nodes or edges of the resulting tree, and the majority of these problems are *NP-complete.*

The variants of the minimal spanning tree discussed below are all *NP-complete* unless specifically stated otherwise. Other variants may be found in Tarjan (1983) and Maffioli (1981).

1.2.1 Node-Constrained Minimal Spanning Trees

The objective of a node constrained minimal spanning tree of a given graph $G = (N, E)$ is to find a minimal spanning tree subject to a degree constraint at one or more nodes of N. In other words, certain nodes n_i of N must be adjacent to at most r other nodes of N, i.e., incident to at most r edges of A. Figure II.5 shows a tree with each node n_i satisfying $\delta(n_i) \leq 3$.

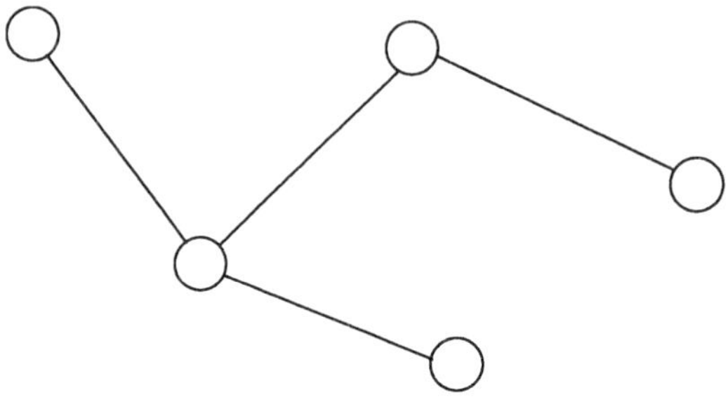

Figure II.5

If the degree constraint $\delta(n_i) \leq r$ applies to only one node $n_i \in N$, then the problem is solvable in polynomial time. This may be the case when the node n_i represents a computer terminal that is linked to terminals represented by the other nodes. In such a case, $\delta(n_i)$ indicates the number of ports available on the computer that can accept cable connections. The minimal spanning tree with this constraint ensures feasible linkage with the terminals at minimal cable connection cost. If, on the other hand, all nodes of N are degree constrained, then the problem is *NP-complete* as Garey and Johnson (1977) have demonstrated.

1.2.2 Edge-Constrained Minimum Spanning Trees

Suppose that a graph $G = (N, E)$ represents a teleprocessing system with node n_1 representing a data processing center and the other nodes n_i, $i = 2,\ldots, n$ representing data terminals that have amounts $d_i \geq 0$ of data to be sent along a unique chain from terminal n_i to the data processing center. The edges $e_{ij} \in E$ are the possible links between any two terminals or between a terminal and the data processing center, each with an associated cost c_{ij} and a capacity $\kappa_{ij} > 0$ that represents the maximum traffic the edge e_{ij} can carry while maintaining an acceptable response time.

The *Constrained Minimum Spanning Tree* problem, as described by Chandy and Lo (1973), seeks the minimum total weight spanning tree of G, such that the traffic flow on each edge e_{ij} of the tree does not exceed κ_{ij}. As an example, consider the graph K_6 with the weight matrix

$$\mathbf{C} = (c_{ij}) = \begin{bmatrix} \infty & 1 & 3 & 4 & 3 & 2 \\ 1 & \infty & 4 & 2 & 3 & 1 \\ 3 & 4 & \infty & 4 & 5 & 6 \\ 4 & 2 & 4 & \infty & 1 & 4 \\ 3 & 3 & 5 & 1 & \infty & 7 \\ 2 & 1 & 6 & 4 & 7 & \infty \end{bmatrix}$$

and let the capacities be $\kappa_{ij} = 8 \ \forall \ e_{ij} \in E$. Furthermore, the amounts of data available at $n_2, n_3, ..., n_6$ ready for shipment to n_1 are 2, 1, 3, 5, and 4. The unconstrained solution is shown in Figure II.6a, where the minimum spanning tree connects n_1 only to nodes n_2 and n_3, so that the entire set of available data except for one unit from n_3 is shipped through the link e_{21}, thus creating a flow of 14. Clearly, this exceeds the given capacity of 8.

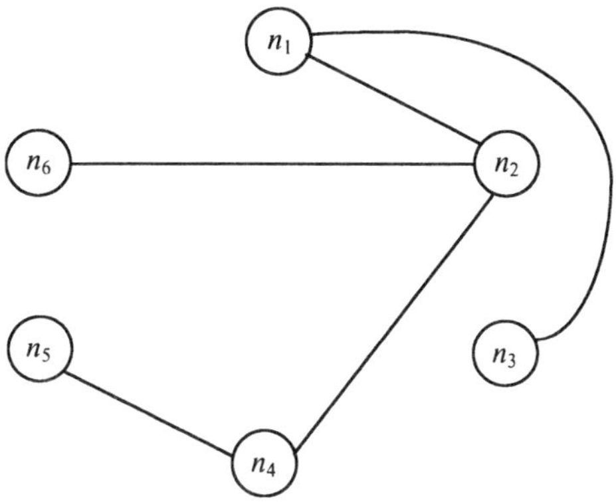

Figure II.6a

The edge-constrained solution for $\kappa_{ij} = 8 \ \forall \ e_{ij} \in E$ is shown in Figure II.6b with the numbers next to the edges indicating the flow towards n_1. Note that in order to reduce the largest required capacity from 14 to 8, it was necessary to spend \$9 rather than \$8 for the spanning tree in Figure II.6b on the next page as opposed to that in Figure II.6a.

Garey and Johnson (1970) have determined that the problem of finding constrained minimum spanning trees is *NP-complete* in general. However, for certain instances, as those described by Edmonds and Karp (1972) in which the capacities are all 1 and all traffic is either 0 or 1, the problem can be solved as a minimum cost network flow problem, and is thus solvable in polynomial time; for details regarding network flow problems, see Part III. Most general solution approaches are either heuristic, such as

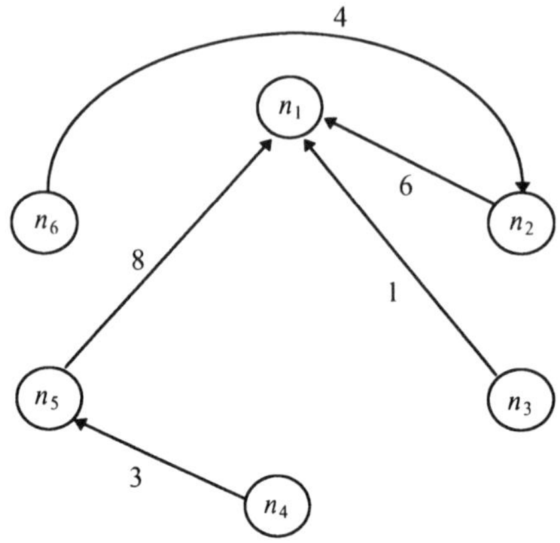

Figure II.6b

those by Martin (1969) and Esau and Williams (1966), or are exact branch and bound algorithms as that by Chandy and Lo (1973).

1.2.3 Alternative Objective Functions

In our development of algorithms for the minimal spanning tree, we have assumed that the edge weights $c_{ij} \in \mathbb{R}$. Consequently, in order to find the *Maximal Spanning Tree* of a connected graph, we need only reverse the sign of the edge weights and apply an algorithm for the minimal spanning tree, so that the maximal spanning tree problem is also solvable in polynomial time.

The problem of finding a path in an undirected graph, such that the longest edge encountered on the path is as short as possible is known as the *bottleneck* or *minimax path* problem. The term "bottleneck" refers to the longest edge of the path; this is the critical edge that determines the objective value. Problems of this type arise in a variety of contexts. Suppose we are taking a biking trip cross-country where food and other supplies are only available at intersections of bicycle paths. Then one criterion we might use is to choose our route in such a way that the maximal distance between these stations is as short as possible. Similarly, an individual confined to a wheelchair might like to choose a route between two locations in a city such that the maximal change in altitude along any given stretch of the route be minimal.

Although problems of this type belong to the class of shortest path problems discussed in Chapter 2, its solution is readily provided by the minimum spanning tree of the graph. In fact, one can prove that the minimax path between any pair of nodes is the unique path in a minimal spanning tree joining these two nodes. Hence the problem can be solved in polynomial time.

Here, we provide a sketch of the proof of the reduction of the bottleneck path problem to a standard minimal spanning tree problem. Let T^* be a minimal spanning tree and suppose that e_{ij} is the maximal weight edge in the unique Path P in T^* connecting some nodes n_r and n_s. If we remove the edge e_{ij}, then the nodes of G are partitioned into two sets N_1 and N_2 as shown in Figure II.7. For any edge $e_{k\ell}$, $n_k \in N_1$, $n_\ell \in N_2$, we must have $c_{k\ell} \geq c_{ij}$ for otherwise we could include the edge $e_{k\ell}$ in T^*, instead of e_{ij}, and obtain a spanning tree of total weight strictly less than that of T^*, which is a contradiction. Since any chain P' in the graph from node n_r to node n_s must include some edge that connects the nodes in N_1 to those in N_2, it follows that the maximum weight edge of P' is at least as large as that of P. Thus, a minimal spanning tree contains the minimax chain between all pairs of nodes of the graph.

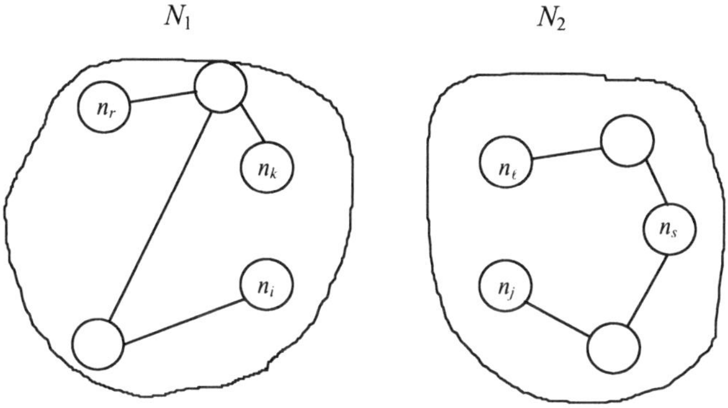

Figure II.7

1.3 Connectivity and Reliability

This section discusses a generalization of the concept of trees and some important applications. The basis of our discussion in this section is the degree of connectivity. To motivate, consider a spanning tree of any given connected graph. Any spanning tree will be minimally connected, as the removal of any of its edges (or interior nodes) will disconnect the tree into disconnected forests. Formally, we can write

Definition II.3: Let $G = (N, E)$ be a connected graph. Two nodes n_i, n_j are k-edge (k-node) connected, if there exist k-edge- (node-) disjoint paths between n_i and n_j. The graph G is called k-edge (k-node) connected, if all pairs of nodes $n_i, n_j \in N$ are at least k-edge (k-node) connected.

Example: Consider the graph $G = (N, E)$ in Figure II.8.

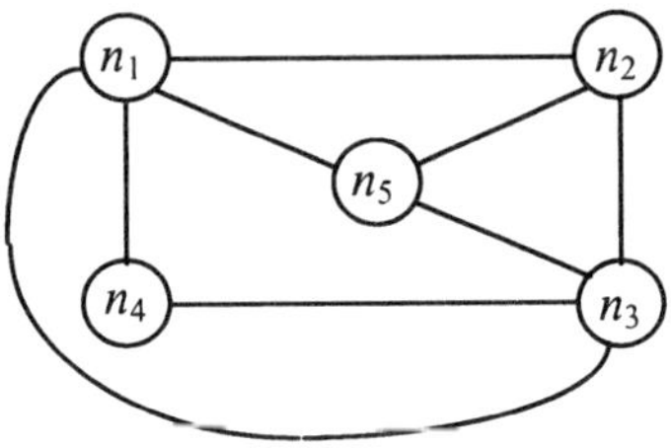

Figure II.8

Nodes n_1 and n_2 are 3 edge-connected, as the paths $p_1 = (n_1, n_2)$, $p_2 = (n_1, n_5, n_2)$, and $p_3 = (n_1, n_4, n_3, n_2)$ are edge-disjoint. The number of edge-disjoint paths between all pairs of nodes (n_i, n_j) is shown in the matrix

$$\begin{bmatrix} \times & 3 & 3 & 2 & 3 \\ 3 & \times & 3 & 2 & 3 \\ 3 & 3 & \times & 2 & 3 \\ 2 & 2 & 2 & \times & 2 \\ 3 & 3 & 3 & 2 & \times \end{bmatrix}.$$

As 2 is the smallest element in this matrix, the graph of Figure II.8 is 2-connected.

The concept of node connectivity is dealt with similarly. Also note that a tree is a 1-connected graph, all of whose pairs of nodes are 1-node *and* 1-edge connected. Similarly, a cycle is a 2-connected graph with all pairs of nodes being 2-node and 2-edge connected. The edge connectivity of any pair of nodes (n_i, n_j) can be determined by defining capacities $\kappa_{ij}\ \forall\ e_{ij} \in E$ and finding the maximal flow from n_i to n_j by any of the pertinent algorithms; see, e.g., Chapter 3 of Part III. A naive approach would solve maximum flow problems between all $\frac{1}{2}n(n-1)$ pairs of nodes. This is, however, unnecessary. As Gomory and Hu (1961) and later Ahuja *et al.*(1993) have demonstrated, only $(n-1)$ such problems need to be solved.

One of the main applications of connectivity is in the field of reliability. Suppose that a given network $G = (N, E)$ includes all possible communication links between users of the network. For example, the nodes may represent users and switching stations, while the edges represent the actual links that may be established. Each link has two features, the cost to establish it and its reliability. The basic idea is then to establish a partial graph $G' = (N, E')$ with $E' \subset E$ that "somehow" balances construction costs and reliability. On the extremes, the graph that includes all possible nodes, i.e., $E' = E$, will provide high reliability at high cost, whereas a minimal spanning tree is vulnerable, i.e., has low reliability at low cost. One way to balance reliability and costs is based on assertions by Boffey (1991) and Sansó *et al.* (1991) who point out that failures of two or more edges or nodes are sufficiently rare to be neglected. This suggests the design of a cost-minimal network that is 2-connected. For a characterization of 2-edge and 2-node connected graphs, it is useful to consider the following

Definition II.4: A *cut node* or *articulation point* (*bridge*) is a node n_i (edge e_{ij}) whose removal increases the number of connected components of the given graph.

Given this definition, we can prove

Lemma II.5: A graph G is 2-edge connected, if and only if G is connected and contains no bridges. G is 2-node connected, if and only if it is 2-edge connected and has no articulation point.

The problem can be formulated by defining zero-one variables $y_{ij}\ \forall\ e_{ij} \in E$, so that $y_{ij} = 1$, if the edge e_{ij} is chosen for the designed graph, i.e., $e_{ij} \in E'$, and 0 otherwise. With construction costs of $c_{ij}\ \forall\ e_{ij} \in E$ and $\overline{S}$ denoting the complement of a set S, we can write

$$\text{P: Min } z = \tfrac{1}{2} \sum_{e_{ij} \in E} c_{ij} y_{ij}$$

$$\text{s.t.} \quad \sum_{\substack{e_{ij} \in E \\ n_i \in S, n_j \in \overline{S}}} y_{ij} \geq 2 \quad \forall\ 0 \subsetneq S \subsetneq N \tag{1}$$

$$y_{ij} = y_{ji} \qquad \forall\ e_{ij} \in E$$

$$y_{ij} = 0 \vee 1 \qquad \forall\ e_{ij} \in E.$$

The objective of this problem minimizes the construction cost, where the factor "½" is required so as to avoid counting the cost of a chosen edge e_{ij} twice, as $y_{ij} = y_{ji} = 1$. The constraints (1) ensure that each subset of nodes is connected to its complement by at least one edge, thus eliminating unconnected components. The constraints (1) appear in many routing problems; the main problem associated with them is that their number is exponential. Typically, they are ignored at first, the relaxed problem without them is solved, and if the resulting solution consists of more that one connected component, the violated constraints of type (1) are added and the problem is resolved.

Monma and Shallcross (1989) list a variety of heuristic algorithms that solve the problem. One such method is the "Greedy Ears" technique. Each iteration commences with a cycle (initially, a randomly chosen node will do). It then finds the node n_i whose shortest distance (based on construction costs c_{ij}) to any node on the current cycle is maximal. This node n_i is then connected to the current cycle by two paths, thus creating an "ear." As the method uses no foresight, it belongs to the class of greedy approaches.

1.4 The Steiner Tree Problem

The *Steiner Problem* in the Euclidean plane has been studied for a long time; see e.g., Melzac (1961), Boyce (1977), and Winter (1985). The problem may be stated as follows. Given $n \geq 3$ points in the Euclidean plane, find the tree with minimum edge costs c_{ij} (= the Euclidean distances) whose nodes include, but are not restricted to, the n given points. Such a tree is called a *Steiner Tree*. In order to construct such a tree, it is usually necessary to introduce additional points, called *Steiner points* or *Steiner nodes*. At the nodes, the edges of the tree meet one another. Figure II.9a depicts 4 points (nodes) in the plane and the edges of K_4 with the Euclidean distance between points next to edges. Figure II.9b shows the usual minimum spanning tree that is constructed from using only the edges from K_4; its total length is 3. Figure II.9c shows a tree spanning the same four nodes but with the addition of one Steiner node, resulting in a tree of a total length of 2.83. Note that the Steiner points may be placed anywhere in the plane.

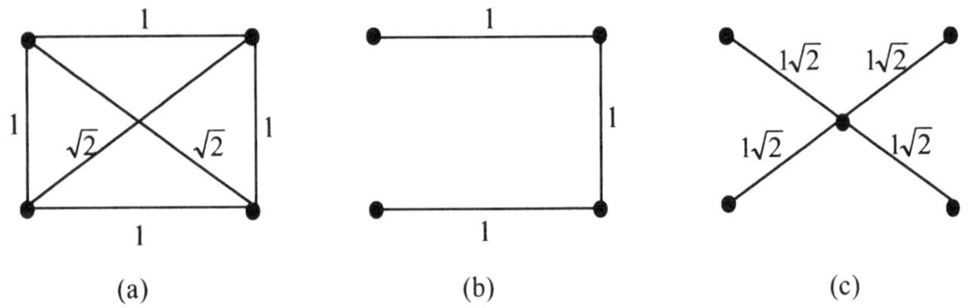

Figure II.9

Some applications call for the minimum weight tree spanning all the nodes of a finite set K in the plane with Steiner points being limited to a finite set S in the plane. Then the Steiner tree on the points of K will be a subgraph of the complete graph on the set of nodes $K \cup S$, where the weight of an edge is equal to the Euclidean distance between the nodes it joins. If we generalize this problem to permit any positive edge weights including ∞, then we have what is called the *Steiner Tree Problem in Graphs* (*STPG*).

The Steiner Tree problem on an undirected, connected graph $G = (N, E)$ with a subset $K \subset N$ of p nodes, is to find a subset $S \subset N \setminus K$ of nodes and a tree T_K over the nodes of $K \cup S$, which is of minimum weight over all such subsets S. The nodes of K are called *K-nodes* and those of S (the Steiner points) S-nodes.

Three simple special cases of the Steiner Tree problem on graphs can easily be solved.

(1) $|K| = 1$. Then T_K consists of the single node of K.

(2) $|K| = 2$. Then T_K is the shortest path between the two nodes of K. Polynomial time algorithms for shortest path problems exist if G contains no cycle of negative length.

(3) $|K| = n$. Then T_K is the minimum spanning tree over the nodes of G.

Considering these easy cases, we will assume that $3 \le |K| \le n - 1$. If $c_{ij} > 0 \; \forall \; a_{ij} \in E$, then there are a number of reduction rules that may yield reductions in the number of nodes and/or edges of the graph, possibly simplifying the task for a subsequent algorithmic treatment. In the six procedures below, d_{ij} represents the length of the shortest length chain between nodes n_i and n_j based on the edge costs, i.e., distances, c_{ij} between the nodes n_i and n_j.

Reduction Rule 1: If $N \setminus K$ contains a node n_i of degree one, then n_i and the unique edge adjacent to it may be removed from G, since the edge does not serve to join K-nodes and its removal can only result in a tree T_K with smaller weight.

Reduction Rule 2: If $N \setminus K$ contains a node n_i of degree two, then the node n_i and its adjacent edges e_{ki} and e_{ij} may be replaced by an edge e_{kj} with weight $c_{kj} = c_{ki} + c_{ij}$ without altering T_K, since any chain in T_K joining K-nodes and containing node n_i will contain nodes n_k and n_j as well. If the addition of the edge e_{kj} creates two parallel edges between the nodes n_k and n_j, then the edge with larger weight may be deleted.

Reduction Rule 3: If E contains an edge e_{ij} such that $c_{ij} > d_{ij}$, then the edge e_{ij} may be removed from E. If $c_{ij} = d_{ij}$, and if there is a chain in G of length d_{ij} not containing e_{ij}, then the edge e_{ij} can also be removed from E. In both cases, the chain from n_i to n_j spans all nodes the edge e_{ij} spans plus possibly additional nodes, and it does so at a weight not exceeding that of the edge e_{ij}.

Reduction Rule 4: If the set K to be spanned contains a node n_k of degree 1, then the unique edge e_{ki} must be in T_K and G may therefore by reduced by deleting n_k. Node n_i must then be declared a K-node in the reduced graph.

Reduction Rule 5: If the set K contains two adjacent nodes n_i and n_j, and there is a K-node n_k, $i \neq k$ and $j \neq k$, such that $c_{ij} \geq d_{ik}$ and $c_{ij} \geq d_{jk}$, then the edge e_{ij} may be removed from E. The reason is that if the chain $(n_k, ..., n_i)$ and the edge e_{ij} are both included in a tree, then the edge e_{ij} may be replaced by the chain $(n_k, ..., n_j)$ that, by assumption, has lesser weight than the edge e_{ij} and connects the same points (plus possibly some additional points).

Reduction Rule 6: If G contains a K-node n_k, such that n_i is the closest and n_j the second-closest neighbor of n_k with respect to weights, and if

$$c_{ki} + d_{i\ell} \leq c_{kj},$$

where $d_{i\ell} = \min\limits_{\mu \in K \setminus \{n_k\}} \{d_{i\mu}\}$, then the edge e_{ki} will be included in T_K. The reasoning behind this rule is somewhat more complex than those of the other rules. If $e_{ik} \in T_K$, then the nodes n_ℓ and n_k must be connected somehow. The cheapest possible way of doing so is by way of the shortest path. Let the path be $(n_\ell, ..., n_u, n_k)$ with length $d_{\ell k}$. The situation is displayed in Figure II.10.

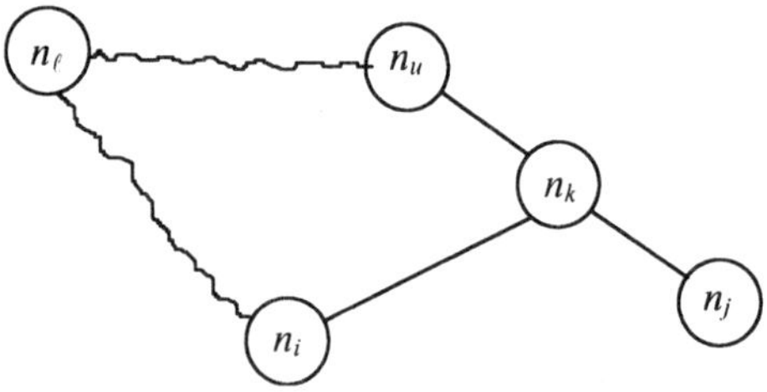

Figure II.10

By assumption, $c_{ki} + d_{i\ell} \leq c_{kj} \leq c_{ku}$. Furthermore, $c_{ku} \leq d_{k\ell}$, so that the path $(n_\ell, ..., n_u,$

n_k) will incur costs $d_{k\ell} \geq c_{ki} + d_{i\ell}$, which cannot be optimal. Hence, $e_{ik} \in T_K$ follows.

Example: In order to illustrate the reduction rules developed in this subsection, consider the graph in Figure II.11, where the numbers next to the edges indicate the edge weights and $K = \{n_1, n_3, n_4, n_6\}$ as indicated by the double circles.

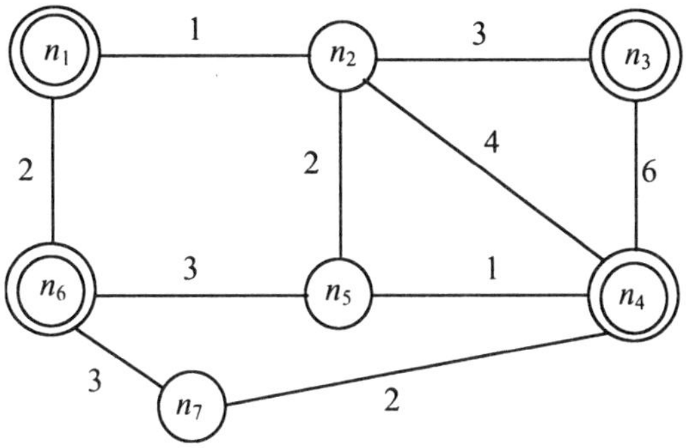

Figure II.11

First, by virtue of Reduction Rule 2, the node n_7 can be eliminated, and its incident edges are replaced by a new edge e_{46} with weight 5. Subsequent application of Reduction Rule 3 on this new edge results in its removal from G. The reduced graph is shown in Figure II.12a.

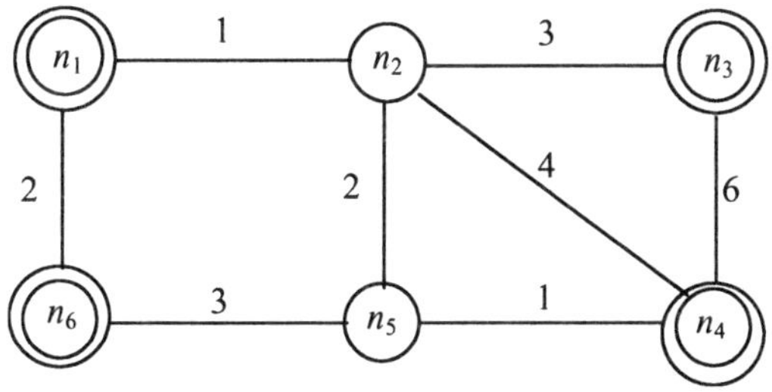

Figure II.12a

At this point, the second part of Reduction Rule 3 can be employed to delete the edge e_{34}, and the first part of the rule may be applied to the edge e_{24}, leading to its deletion. The resulting graph is shown in Figure II.12b on the next page.

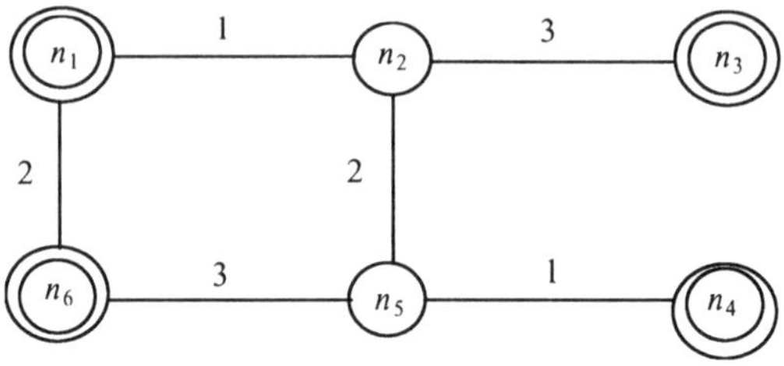

Figure II.12b

At this stage, Reduction Rule 4 may be applied to n_3 and then to n_4. In the first step, n_3 is folded upon n_2 and in the second step, n_4 is folded upon n_5. In those steps, the nodes n_2 and n_5 have become members of K. The resulting graph is shown in Figure II.12c below.

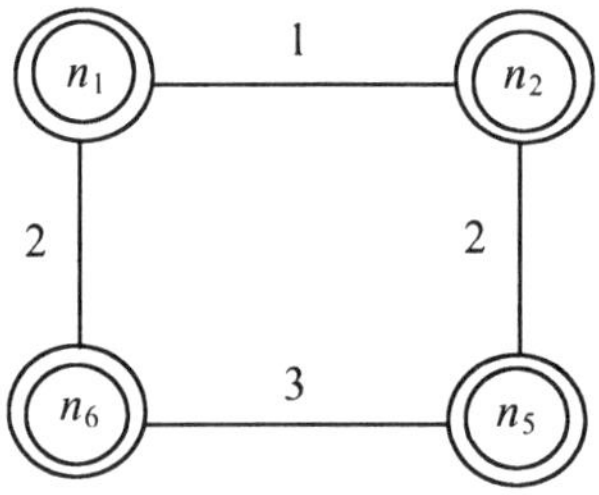

Figure II.12c

We can now apply Reduction Rule 5 to the edge e_{65}, which results in its removal. In the remaining graph, Reduction Rule 4 is applied twice to n_5 and n_6, respectively, resulting in edges e_{16} and e_{25} being included in the tree T_K. The remaining graph consists of only the single edge e_{12}. Another application of Rule 4 includes edge e_{12} in T_K and the process terminates. The original graph G and the resulting spanning tree T_K are shown in Figure II.12d on the next page, where bold edges are included in T_K. The costs of the spanning tree are $c(T_K) = 9$.

While it was possible to solve this particular example by applying a few simple reduction rules, this is generally not possible. However, some simple cases exist. For example, if the graph G is already a tree, then the Steiner Tree over the nodes in $K \subset N$ is trivially solvable in linear time by merely removing successively all pendant

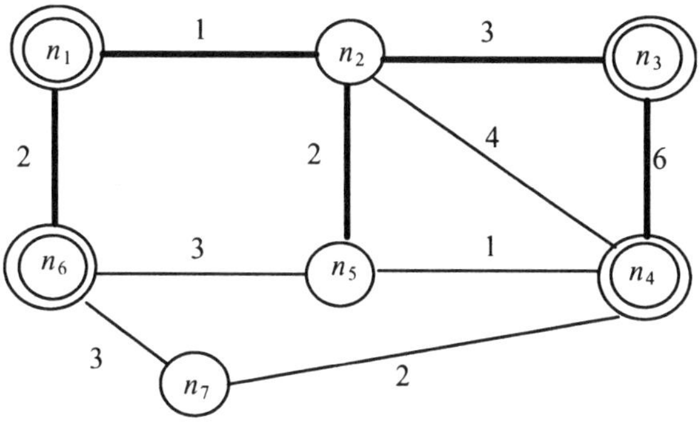

Figure II.12d

nodes in $N \setminus K$ and the single edges incident to them until the only pendant nodes left are those of K. Other examples of classes of graphs on which the Steiner Tree problem on graphs is linearly solvable are given in the survey article of Winter (1987).

The general Steiner problem on graphs has been proven *NP-complete* by Karp (1972). Numerous solution algorithms have been suggested for the problem. Exact methods are due to Hakimi (1971), Aneja (1980), and Beasley (1984), while heuristic methods were proposed by authors such as Rayward-Smith (1983), Plesnik (1981), and Wong (1984). Details of these and other algorithms are provided in the aforementioned article by Winters.

CHAPTER 2 SHORTEST PATH PROBLEMS

2.1 The problem and its formulation

The shortest path problem (*SPP*) constitutes one of the most frequently encountered classes of problems in graph theory. It is certainly the most fundamental of components in the fields of transportation and communication networks. Shortest path problems may be encountered directly, possibly as a result of a clever formulation of a problem not at first sight involving shortest paths, or indirectly as a subproblem in the solution of a more complicated optimization problem. This use of shortest path problems as subroutines motivates the search for algorithms with good theoretical bounds on running time. We also seek computer implementations whose empirical performances are rapid in spite of perhaps weak theoretical bounds for the algorithms they implement. As testimony to the importance of shortest path and related problems, a large number of surveys, annotated bibliographies, and reviews have appeared over the past thirty years. Among them are those by Dreyfus (1969), Pierce (1975), Golden and Magnanti (1977), and Gallo and Pallottino (1988).

Unless stated otherwise, we suppose throughout this chapter that we deal with a directed graph $G = (N, A)$ with $|N| = n$ nodes and $|A| = m$ arcs together with arc lengths c_{ij}. The length of a path p from n_i to n_j is the sum of the lengths of the arcs in p, i.e., $c(p) = \sum_{a_{ij} \in p} c_{ij}$. In case of undirected or mixed graphs, and edge e_{ij} with length c_{ij} can always be represented by two directed arcs a_{ij} and a_{ji} with arc values $c_{ij} = c_{ji}$. The problem is then to find a path p^* in G from an origin or source node n_s to some destination or terminal node n_t such that the length of this path is minimized, i.e., $c(p^*) = \min_p \{c(p)\}$, where the summation is taken over all paths p in G from node n_s to node n_t.

The shortest path problem can be formulated by defining zero-one variables y_{ij} which assume a value of one, if the arc a_{ij} is part of the shortest path, and 0 otherwise.

The shortest path problem can then be written as

$$\text{P}_{\text{SPP}}: \quad \text{Min } z = \sum_{a_{ij} \in A} c_{ij}\, y_{ij}$$

$$\text{s.t.} \quad \sum_{a_{sj} \in A} y_{sj} = 1$$

$$\sum_{j: a_{ij} \in A} y_{ij} - \sum_{j: a_{ji} \in A} y_{ji} = 0 \;\; \forall\, n_i \neq n_s, n_t$$

$$y_{ij} = 0 \vee 1 \;\; \forall\, a_{ij} \in A$$

While the objective function minimizes the sum of the lengths of all arcs on the shortest path, the first constraint ensures that the path originates at the source n_s by forcing the inclusion on the shortest path of one arc that leads out of the source. The remaining constraints guarantee that if the shortest path enters a node, it will also leave it. This type of constraint is discussed further in Part III.

Section 2 considers a representative sample of the long list of formulations of real-world problems which can be cast as shortest path problems. In Section 3 we describe algorithms that find shortest paths from n_s to n_t, from n_s to all other nodes, and shortest paths between all pairs of nodes. Finally, Section 4 will deal with finding not only the shortest, but the k shortest paths (for some given k), a shortest path problem with a fractional objective, and shortest path problems with additional constraints.

2.2 Applications of Shortest Paths

Some shortest path problems arise naturally, such as the problem of finding the shortest traffic route between two locations in a city. This and similar problems are obvious and need not be discussed in detail. A number of problems, however, arise in a context that has apparently nothing to do with shortest paths, nor indeed with networks at all. Some of these cases are discussed in the sections below. Most shortest path problems do, however, occur as subproblems of large real applications, e.g., in multicommodity flow problems as those described in Chapter 4 of Part III. It should be mentioned that all problems in the remainder of this section can also be solved by dynamic programming.

2.2.1 Most Reliable Paths

Communication networks such as those representing telephone, television, computer, carrier pigeon, or even interpersonal communications may be represented by a graph such as that in Figure II.13.

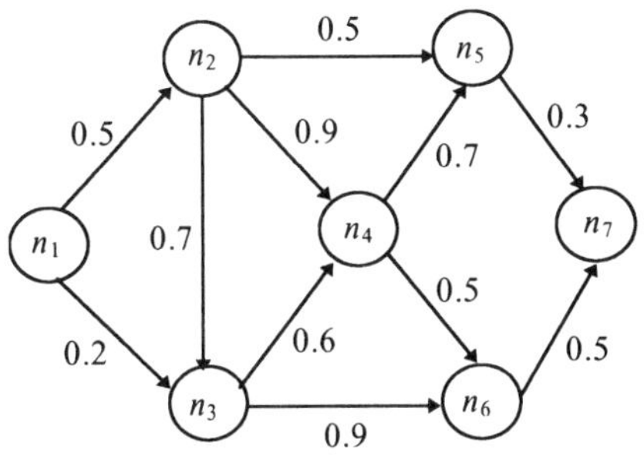

Figure II.13

The nodes represent transmitter and/or receivers of messages and the arcs represent the possibility of transmitting a message between two nodes in the indicated direction. The value r_{ij} next to each arc a_{ij} represents the reliability of the arc, i.e., the known probability that the given message conveyor will actually accomplish its task within a given time frame. We assume further that all message conveyors operate independently of one another.

Suppose now that the objective is to transmit a message in the network of Figure II.13 from node n_1 to node n_7 along a path with maximum total reliability. In other words, the goal is to find a path p from n_1 to n_7 that maximizes

$$\prod_{a_{ij} \in p} r_{ij}$$

The problem may be converted to a shortest path problem by transforming it as follows:

$$\max \prod_{a_{ij} \in p} r_{ij} = \max \sum_{a_{ij} \in p} \ln r_{ij} = -\min \sum_{a_{ij} \in p} \ln r_{ij}$$

where $-\ln r_{ij} \geq 0 \ \forall \ a_{ij} \in A$ as the r_{ij} values denote, so that $r_{ij} \in [0; 1]$. Therefore, solving a shortest path problem on the graph of Figure II.13 with arc lengths $c_{ij} = -\ln r_{ij}$ will identify the most reliable path from node n_1 to node n_7. In this example, the most reliable path is $p = (n_1, n_2, n_4, n_6, n_7)$ that offers a total reliability of $(0.5)(0.9)(0.5)(0.5) = 0.1125$.

2.2.2 Equipment Replacement

It can be shown that any finite horizon, discrete, and finite state dynamic programming problem may be formulated as a shortest path problem in a suitably chosen graph. As an example of such a transformation, we will construct a graph for the example of an equipment replacement problem introduced in Chapter 0b of this volume. In that specific numerical example, a decision to replace a machine could be made at the beginning of each year of a 5-year horizon. The cost of a new machine is $b = 9$, and at the beginning of the first year the current machine is 3 years old. The salvage value of a t year old machine is $s_t = (4 - t)$ for $t \leq 4$, and 0 otherwise. The annual profit derived from a t year old machine is $\pi_t = (16 - t^2)$ for $t \leq 4$, and 0 otherwise. We seek a replacement policy for the machine that minimizes total (non-discounted) costs. Below, we will provide two representations of this problem, both of which are amenable to solution with shortest path algorithms. While the first representation of the problem is more explicit and allows for easier interpretation, the second representation is more compact.

The first representation of the problem creates nodes (i, j), where i denotes the beginning of year i, and j symbolizes the age of the machine at the beginning of year i. As far as arcs are concerned, we have to distinguish between three cases.

Case 1: Keep the machine at the beginning of year i; $i = 1, 2, 3, 4$. In this case, an arc leads from the node (i, j) to node $(i + 1, j + 1)$. The arc value is $\pi_j = \max\{16 - j^2; 0\}$.

Case 2: Replace the machine at the beginning of year i; $i = 1, 2, 3, 4$. This results in an arc leading from a node (i, j) to a node $(i + 1, 1)$. Here, the profit of a new, i.e., 0-year old machine, is 16, the revenue from selling the j-year old machine at the beginning of year i is $\max\{4 - j; 0\}$, and the cost of the new machine is b. In summary, the overall profit of the "replace" decision is $\pi_0 + s_j - b = 16 + \max\{4 - j; 0\} - 9 = 7 + \max\{4 - j; 0\}$.

Case 3: The arcs leading from the nodes $(5, j)$ to the artificial sink "t" are different from those discussed in cases 1 and 2, as they require that the machine is necessarily sold at the end of the fifth year. Also, for simplicity, we will have one arc representing both the "replace" and "keep" decisions. In case we decide to keep the machine, the arc from node $(5, j)$ to t has a value of $\pi_j + \max\{4 - (j+1); 0\} = \pi_j + \max\{3 - j; 0\}$. On the other hand, if we decide to replace the machine, the profit will be $\pi_0 + \max\{4 - 1; 0\} - b = 16 + 3 - 9 = 10$. The arc value will then be $\max\{\pi_j + \max\{3 - j; 0\}; 10\}$. The graph is shown in Figure II.14.

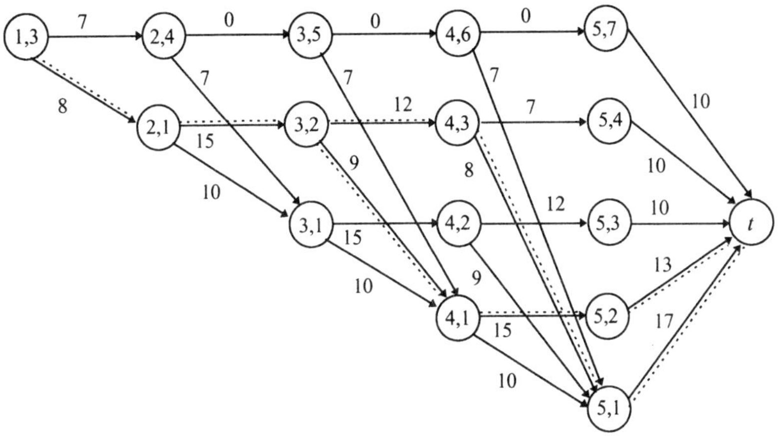

Figure II.14

As the arc values in the network of Figure II.14 represent profits, the objective is to find the profit-maximizing path from the current situation at node (1, 3) to the final situation at node t. Formally, this is the *longest* path in the network. The problem of finding the longest path can easily be converted to one that finds the shortest path by either multiplying all arc values by some negative number, or by subtracting them from an arbitrarily chosen constant. In this example, there are two optimal paths that are shown as broken lines in Figure II.14. Both paths represent sequences of decisions that result in a profit of 60 over the 5-year period. To a decision maker, the two different solutions can be presented as shown in Table II.1.

Table II.1

	Year 1	Year 2	Year 3	Year 4	Year 5
Policy 1	replace	keep	keep	replace	keep
Policy 2	replace	keep	replace	keep	keep

The second possible representation of the problem employs some implicit aggregation, so that it uses considerably less nodes. In particular, this model uses nodes $i = 1, ..., 5$ to represent the beginning of year i, and additional nodes 0 and 6. An arc leading from a node i to a node $(i + k)$ represents a purchase of a new machine at the beginning of year i, and the sale of the machine at the beginning of year $(i + k)$. The node 0 is included to simulate the purchase of the current machine, so that the arc (0, 1) represents the decision to purchase a new machine at the beginning of year 1, whereas the arcs (0, 2), (0, 3), ... indicate that the planners have decided to keep the machine at hand until the beginning of year 2, 3, ... The arc values $c_{i,i+k}$ are then $c_{01} = 1$, and $c_{0j} = 7, j = 2, ..., 6$, while $c_{i,i+k} = s_k - b + \sum_{\nu=0}^{k-1} \pi_\nu = \max\{4 - k; 0\} - 9 + \sum_{\nu=0}^{k-1} \max\{16 - \nu^2; 0\}; i \geq 1; k \in [i + 1; 6]$.

In particular,

$$c_{i,i+1} = \pi_0 + s_1 - b = 10, \qquad i = 1, ..., 5$$

$$c_{i,i+2} = \pi_0 + \pi_1 + s_2 - b = 24, \qquad i = 1, ..., 4$$

$$c_{i,i+3} = \pi_0 + \pi_1 + \pi_2 + s_3 - b = 35, \qquad i = 1, ..., 3$$

$$c_{i,i+4} = \pi_0 + \pi_1 + \pi_2 + \pi_3 + s_4 - b = 41, \qquad i = 1, 2$$

$$c_{i,i+5} = \pi_0 + \pi_1 + \pi_2 + \pi_3 + \pi_4 + s_5 - b = 41, i = 1.$$

For reasons of clarity, the structure of the graph is shown in Figure II.15 without the arc values.

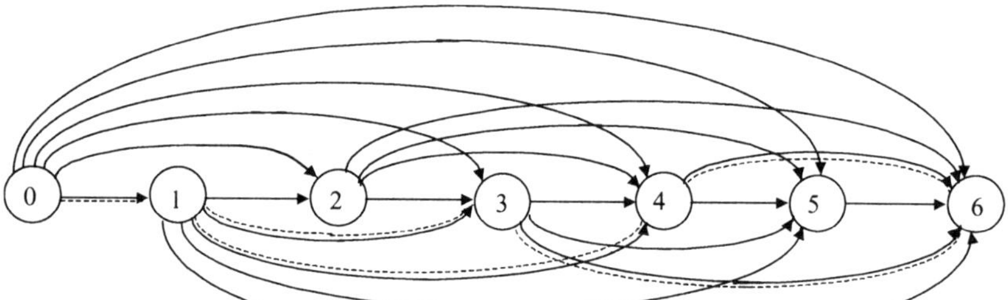

Figure II.15

Again, the problem is to find the longest path from node 1 to node 6. The two solutions are shown by the broken lines in Figure II.15; they result in the two optimal policies in Table II.1.

2.2.3 Functional Approximation

The idea of functional approximation is to approximate a given function, so that the information loss is within given bounds. As an example, consider the piecewise linear function shown in Figure II.16.

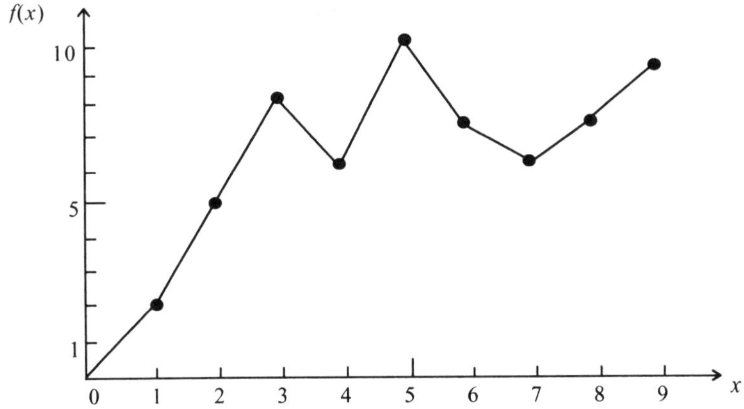

Figure II.16

The original function has ten breakpoints at $x = 0, 1, ..., 9$. For reasons of economy of data requirements, we wish to approximate the function by another piecewise linear function using fewer segments but passing through a subset of the break-points of the original curve. At the same time the approximation is to be within a given error, which is here defined as the maximum vertical distance between the functional values of the original and the new curve. The problem is then to approximate the original curve using a minimal subset of the slope break-points such that each segment of the approximation has an error that does not exceed a prespecified value ε.

This problem can be formulated as a shortest path problem by representing the ten breakpoints of the original function by nodes $n_0, n_1, ..., n_9$. An arc a_{ij} exists if the error introduced in joining break-points at i and j in the original function is at most

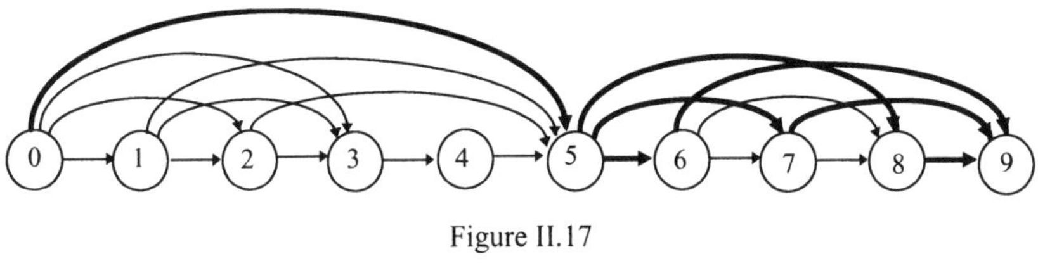

Figure II.17

equal to a perspecified value, e.g., $\varepsilon = 2$. All arcs have length $c_{ij} = 1$. The graph for this example is shown in Figure II.17 on the previous page.

There are three shortest paths, *viz.*, (n_0, n_5, n_8, n_9), (n_0, n_5, n_7, n_9), and (n_0, n_5, n_6, n_9). In Figure II.17, these shortest paths are shown by the bold lines. The three shortest paths correspond to the approximations to the original function obtained by joining the breakpoints at 0, 5, 8, and 9; at 0, 5, 7, and 9; and at 0, 5, 6, and 9 by straight line segments.

2.2.4 Matrix Chain Multiplications

Another application of shortest path problems deals with the multiplication of matrices. In particular, assure that m conformable matrices are to be multiplied to a product of the form $\mathbf{A}_1\mathbf{A}_2...\mathbf{A}_m$, where matrix $\mathbf{A}_i$ is of dimension $[n_{i-1} \times n_i]$, $i = 1,..., m$. It is well known that matrix multiplications can be performed in any order, as long as we keep the required order of the matrices $\mathbf{A}_1\mathbf{A}_2...\mathbf{A}_m$. For instance, the multiplication may be performed from left to right, i.e., $((...((\mathbf{A}_1\mathbf{A}_2)\mathbf{A}_3)...\mathbf{A}_{m-1})\mathbf{A}_m)$, from right to left, i.e., $(\mathbf{A}_1(\mathbf{A}_2...(\mathbf{A}_{m-2}(\mathbf{A}_{m-1}\mathbf{A}_m))...))$, or in any other order.

Depeding on their sizes, different sequences of matrices require different numbers of elementary operations. As a simple example, the matrix product $\mathbf{A}_1\mathbf{A}_2\mathbf{A}_3$ can be computed as $(\mathbf{A}_1\mathbf{A}_2)\mathbf{A}_3$, or as $\mathbf{A}_1(\mathbf{A}_2\mathbf{A}_3)$. In the first instance, $\mathbf{A}_1\mathbf{A}_2$ uses n_1 multiplications and $n_1 - 1$ additions to calculate each element of the resulting product matrix. There are n_0n_2 such elements so that the computation of $\mathbf{A}_1\mathbf{A}_2$ requires $O(n_0n_1n_2)$ operations. $(\mathbf{A}_1\mathbf{A}_2)\mathbf{A}_3$ then uses $O(n_0n_2n_3)$ operations, and the full product requires $O(n_0n_2(n_1 + n_3))$ operations. In the second case, the sequence $\mathbf{A}_1(\mathbf{A}_2\mathbf{A}_3)$ requires $O(n_1n_3(n_0 + n_2))$ operations, and the ratio of their two complexities is $O(n_0n_2(n_1 + n_3)/n_1 + n_3(n_0 + n_2))$ which may be made arbitrarily large by taking, for instance $n_2 = n_0$, $n_3 = n_1$, and n_0 arbitrarily large, in which case $O(n_0n_2(n_1 + n_3)/(n_1 + n_3)) = O(n_0/n_1)$.

If chains of matrices are to be computed many times, then it may well pay to find the order of multiplication of the matrices that minimizes the total number of arithmetical operations. In order to do so, we can model the problem of matrix chain multiplications in an appropriately defined graph. As an example, consider the five matrices $\mathbf{A}[40 \times 60]$, $\mathbf{B}[60 \times 30]$, $\mathbf{C}[30 \times 10]$, $\mathbf{D}[10 \times 40]$, and $\mathbf{E}[40 \times 20]$ to be multiplied to result in **ABCDE**. The first step is to create a graph in which each node represents a certain state of multiplications. An arc exists between nodes n_i and n_j if it is possible to make the transition from state n_i to state n_j by multiplying two matrices in node n_i. The length of the arc between nodes n_i and n_j

is the number of arithmetical operations necessary to make the transition represented by the arc. Figure II.18 illustrates the graph for our five-matrix multiplication problem. Here, the nodes have been arranged with respect to stages. In the first stage (nodes in the first column in Figure II.18), all nodes represent states in which one multiplication is performed. The nodes in the second stage or column all require two matrix multiplications, etc.

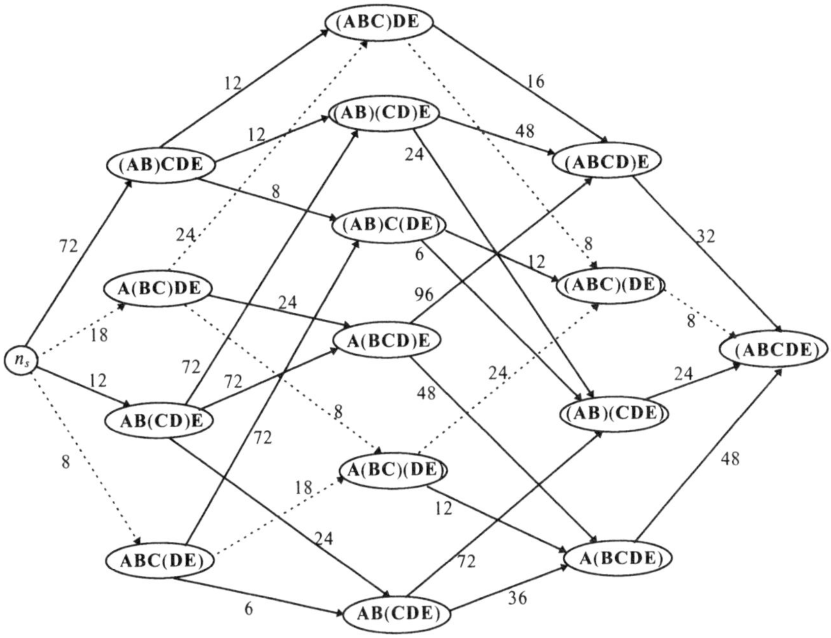

Figure II.18

The artificial source is the point of departure when no multiplications have yet been made, i.e., stage 0. The goal is now to find a shortest path from the source node n_s to the terminal node (**ABCDE**) in this graph. There are three such shortest paths indicated in the graph by broken lines. All are of the type ((**A**(**BC**))(**DE**)) and require 58,000 operations as opposed to the longest path ((**A**(**B**(**CD**)))**E**) for which 212,000 operations are needed, and the regular left-to-right multiplication (**AB**)**C**)**D**)**E**) for which 132,000 operations are needed.

2.3 Solution Methods

In the literature on the shortest path problem and its extensions, many different categories of networks and combinatorial or algebraic solution techniques have been examined. In recent years, researchers have attempted to bring some order to this voluminous literature by classifying a given algorithm according to these and other categories. One such article is that of Deo and Pang (1984). According to their classification scheme, what we will be treating in this chapter are directed graphs with positive and negative deterministic arc lengths. We will also examine the performances of algorithms on dense graphs (i.e., those with $m/n(n-1)$ near 1) and sparse graphs (with $m/n(n-1)$ near 0). The algorithms presented here are either *label setting* or *label correcting techniques*. Label setting algorithms can be applied to graphs with only nonnegative arc lengths, whereas label correcting algorithms apply to graphs whose arc lengths are of arbitrary sign. Both types of algorithms assign a label to each node at each step which is a value equal to the shortest distance so far from a source node to the labeled node. At each step of a label setting algorithm, one or more temporary labels are designated as permanent (the label represents the true shortest distance from the source to the labeled node). Initially only the source node is permanently labeled. The algorithm terminates when the labels of all nodes are permanent. Label correcting algorithms consider all nodes temporarily labeled until the final step when all labels become permanent.

Gallo and Pallottino (1986) have presented a prototype algorithm for the solution of the shortest path problems from which nearly all other algorithms for the problem can be derived. The only difference between all these algorithms is the type of data structure used to implement the choice of candidate nodes from a particular set. Data structures are ways of storing data in computer memory so as to make their retrieval particularly efficient in a given context. A standard reference to data structures is Knuth (1973) and Aho *et al.* (1974).

Some of the algorithms presented in this part only will accept graphs with nonnegative arc lengths, while others only require that the given graph contain no circuit of negative length. In fact, finding shortest paths in graphs that include circuits of negative length is meaningless, as the shortest path will cycle indefinitely on such a circuit, reducing the length of the shortest path towards $-\infty$. Most of the algorithms that accept negative arc lengths have the capability of detecting the presence of negative circuits.

2.3.1 Dijkstra's Method

One of the earliest algorithms proposed for the shortest path problem was proposed by Dijkstra (1959). It is a label setting algorithm and works only on graphs with nonnegative arc lengths. As any label setting algorithm, it first assigns temporary labels to all nodes except the origin which is given a permanent label. Denote now by T the set of nodes with a temporary label, while P is the set of nodes that are permanently labeled. The idea is now to start with $P = \{n_1\}$ and

$T = N \setminus P$, i.e., only the first node is permanently labeled. In each step of the algorithm, one node is chosen and its label is made permanent. The algorithm terminates if all nodes have received a permanent label. In such a case, the shortest paths from n_1 to *all* nodes have been determined and their structure can be retrieved by simple backtracking. If only the shortest path from n_1 to some prespecified node n_t is sought, then the algorithm terminates as soon as $n_t \in P$. The labels of the nodes are instrumental to determine the structure of the shortest path once the algorithm is applied. Dijkstra's algorithm is initialized by setting the labels $L(n_1) = (0, n_1)$, as well as $L(n_j) = (c_{1j}, n_1)$ if $a_{1j} \in A$, and $L(n_j) = (\infty, -)$ otherwise. Once the algorithm terminates, the first part of the label $L(n_j) = (d_j, n_i)$ specifies the length of the shortest path from n_1 to n_j, while the second part shows that n_i is the direct predecessor of n_j on the shortest path to n_j. The algorithm can then be described as follows.

Dijkstra's Algorithm

Step 1: Is $P = N$? If yes: Stop; the longest paths from n_1 to all other nodes have been found.
If no: Go to Step 2.

Step 2: Choose $n_i \in T$, such that $d_i = \min_{k:\, n_k \in T} \{d_k\}$. Set $T := T \setminus \{n_i\}$ and P: $P \cup \{n_i\}$. Update the labels for all nodes n_j: $a_{ij} \in A$ as $L(n_j) := (d_i + c_{ij}; n_i)$ if $d_i + c_{ij} \leq d_j$, and $L(n_j) := L(n_j)$ otherwise. Go to Step 1.

The proof of the validity of Dijkstra's algorithm uses the fact that when the label of node n_j is made permanent via n_i, a shorter path to n_j via some other node n_k could only exist if $d_k + c_{kj} < d_i + c_{ij}$. By contradiction, this could only occur if $c_{kj} < 0$ which, by assumption, is prohibited.

Example: As an illustration of Dijkstra's algorithm, consider the network in Figure II.19.

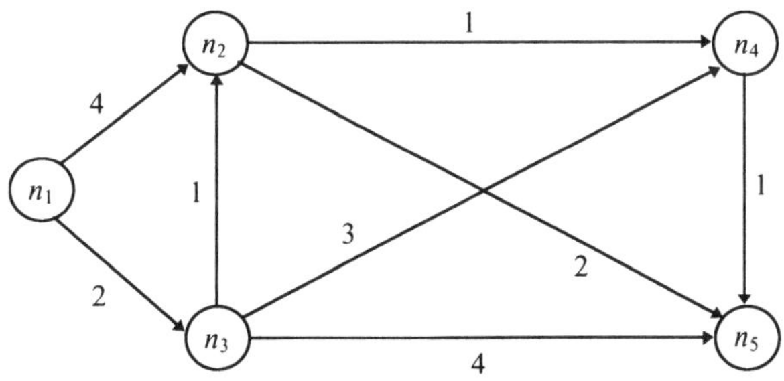

Figure II.19

Initially, the labels are $L(n_1) = (0, n_1)$, $L(n_2) = (4, n_1)$, $L(n_3) = (2, n_1)$, $L(n_4) = (\infty, x)$, and $L(n_5) = (\infty, x)$. At present, only node n_1 is permanently labeled. The node with the smallest first part in its temporary label is n_3, hence the label of n_3 is made permanent. All nodes with temporary labels that are immediate successors of n_3 are now revised. That results in $L(n_2) = (3, n_3)$, $L(n_4) = (5, n_3)$, and $L(n_5) = (6, n_3)$. Now, n_2 is the next node whose label is made permanent, and the labels of its successors are examined. As $d_2 + c_{24} = 3 + 1 = 4 < 5$, which is the current temporary label of n_4, the new temporary label is $L(n_4) = (4, n_2)$. Similarly, $L(n_5) = (5, n_2)$. At this point, only n_4 and n_5 have temporary labels, and n_4 has the smaller (first part of the) label. Thus, the label of n_4 is made permanent and the label of n_5 is examined. As $d_4 + c_{45} = 4 + 1 = 5$, which equals the present label of n_5, we either leave $L(n_5) = (5, n_2)$, or revise it to $L(n_5) = (5, n_4)$. In the next step, the label of n_5 is made permanent and the algorithm terminates. In summary, the permanent labels of the nodes are $L(n_1) = (0, n_1)$, $L(n_2) = (3, n_3)$, $L(n_3) = (2, n_1)$, $L(n_4) = (4, n_2)$, and $L(n_5) = (5, n_2)$ or $(5, n_4)$.

As in dynamic programming, the objective value of the problem is readily available, whereas finding the optimal solution of the problem requires an additional step. Here, we know that the lengths of the shortest paths from n_1 to n_j, $j = 1, \ldots, 5$ are 0, 3, 2, 4, and 5, respectively. In order to find the shortest path from n_1 to, say, n_5, we need to backtrack the path from n_5. The permanent label of n_5 specifies n_2 (or n_4) as the direct predecessor of n_5. Arbitrarily, choose n_4. The label of n_4 indicates that its direct predecessor on the shortest path is n_2 whose label, in

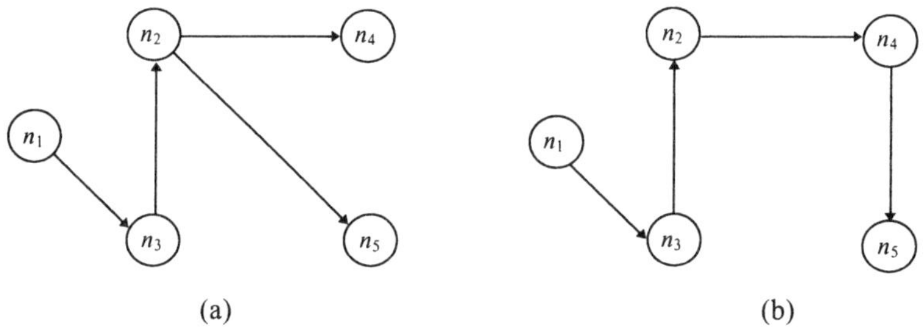

Figure II.20

turn, identifies n_3 as its direct predecessor. Finally, n_3 is preceded directly by the node n_1. Putting the arcs together results in the shortest path $(n_1, n_3, n_2, n_4, n_5)$ whose length is, as determined earlier, $d_5 = 5$. If more than a single shortest path is sought, it is useful to construct the shortest path tree $G^* = (N, A^*)$. The tree is constructed by including $a_{ij} \in A^*$, if $L(n_j) = (d_j, n_i)$. In our example, the two shortest path trees (the tie in $L(n_5)$ results in non-uniqueness) are shown in Figure II.20a and b on the previous page.

The network in Figure II.21 provides an example which demonstrates that Dijkstra's algorithm may fail in case negative arc lengths are present.

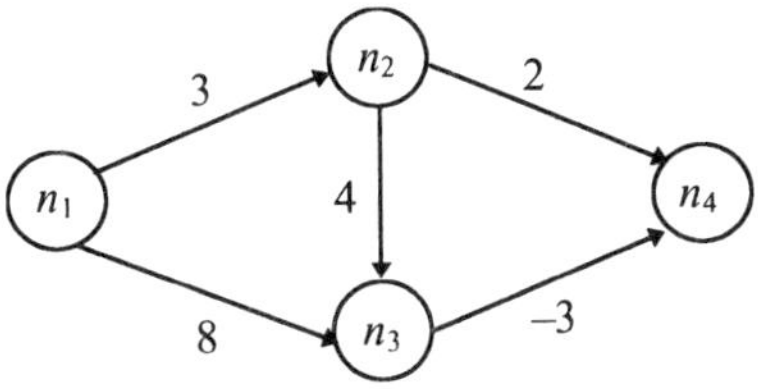

Figure II.21

Applying Dijkstra's algorithm results in the permanent labels $L(n_1) = (0, n_1)$, $L(n_2) = (3, n_1)$, $L(n_3) = (7, n_2)$, and $L(n_4) = (5, n_2)$. Backtracking reveals that the shortest path is (n_1, n_2, n_4) with length 5. However, the path (n_1, n_2, n_3, n_4) has length $3 + 4 - 3 = 4$ and is the shortest path that Dijkstra's algorithm failed to find.

The computational complexity of Dijkstra's algorithm is dominated by finding the minimum and the updating of the labels in step 2. Step 2 is of $O(n)$, and at most $(n - 1) = O(n)$ iterations are necessary, as in each iteration one label becomes permanent. As a result, the complexity of the algorithm is $O(n^2)$.

While there is no possibility to reduce the number of iterations needed to solve the problem, step 2 of Dijkstra's algorithm can be time consuming if not performed in an intelligent fashion. Indeed, a brute force method would compare all of the d_j, $n_j \in T$, in order to pick the minimum value. Alternatively, one could sort the d_j, $n_j \in T$ in nondecreasing order and maintain this sorted order at each iteration. Then, choosing the minimum value d_j amounts to choosing the first element of this sorted list. In addition, instead of updating the labels d_j for all $n_j \in T$, we note that only the labels of nodes n_j need be updated if n_i was made permanent in this step, and $a_{ij} \in A$. Dial (1965) made these observations and created a code for the shortest path problem, which for general problem instances has been shown empirically to be an extremely efficient algorithm for problems with nonnegative arc lengths.

2.3.2 The Bellman-Ford-Moore Algorithm

In this section, we describe an algorithm that is somewhat more general than Dijkstra's method in that although it does allow negative arc lengths, $c_{ij} < 0$ for some $a_{ij} \in A$, it still assumes the absence of negative circuits in G. The following algorithm is of the label correcting type, i.e., each node $n_j \in N$ is assigned a label representing a certain shortest path distance in G, and this label is corrected (i.e., decreased or unchanged) at each iteration until the final iteration, when it is equal to the length of a shortest distance path from node n_1 to node n_j.

First define these labels. Let d_j^t, $n_j \in N$, denote again the length of the shortest path from node n_1 to node n_j using at most t arcs. By this definition, $d_1^t = 0$. Suppose now that we have already calculated d_j^{t-1} for $j = 1, ..., n$. Now, either a shortest path from node n_1 to node n_j using at most t arcs has also at most $t - 1$ arcs, in which case $d_j^t = d_j^{t-1}$, or the last node in the shortest path is node n_i with a shortest path distance from node n_1 equal to d_k^{t-1}, in which case $d_j^{t-1} = d_k^{t-1} + c_{ij}$. Then the label d_j^t must satisfy $d_j^t = \min\{d_j^{t-1}, \min_{i \neq j}\{d_k^{t-1} + c_{ij}\}\}$ with initial conditions $d_j^1 = 0$, $d_j^1 = c_{ij}$, if $a_{ij} \in A$, and $d_j^1 = \infty$ otherwise. These labels satisfy $d_j^1 > d_j^2 > ... > d_j^{n-1} = d_j$, where the d_j are the solutions of Bellman's equations $d_1 = 0$, and $d_j = \min_{i \neq j}\{d_k + c_{ij}\}$, because any elementary path in the n node graph G contains at most $(n - 1)$ arcs. Thus the d_j are the shortest path distances we seek. This iterative procedure may terminate before $t = (n - 1)$ if $d_j^t = d_j^{t-1} \ \forall n_j \in N$.

The Bellman-Ford-Moore algorithm is initialized with a set of labels $d_1^0 = 0$, and $d_j^0 = \infty \ \forall j = 2, ..., n$, the set $J_1 = \{j: a_{1j} \in A\}$, and the iteration counter $t = 1$. The algorithm can then be described as follows.

The Bellman-Ford-Moore Algorithm

Step 1: Compute $d_j^t = \min\{d_j^{t-1}, \min_{i: a_{ij} \in A}\{d_i^{t-1} + c_{ij}\}\}$, and define

$J_t' = J_t \cap \{j : d_j^t < d_j^{t-1}\}$.

Step 2: Is $J_t' = \emptyset$? If yes: Stop; the current labels denote the lengths of the shortest paths from n_1 to all other nodes.

If no: Set $t := t + 1$, define $J_t := \{j: (i \in J_{t-1}') \wedge (a_{ij} \in A)\}$, and go to Step 1.

The above algorithm assumes that G does not contain any negative circuits. If there were such a circuit, then $J_n \neq \varnothing$, and the nodes of J_n would be contained in negative circuits. Thus, if it is not known whether or not the given network has circuits of negative length, then we must add a test for $t = n$ when $J_t = \varnothing$ before stopping. This test serves as a detector of negative circuits in G. As far as computational complexity is concerned, there are $O(n)$ iterations, each of which requires $O(n^2)$ elementary opertions for a total complexity of $O(n^3)$.

Example: Consider the network shown in Figure II.22.

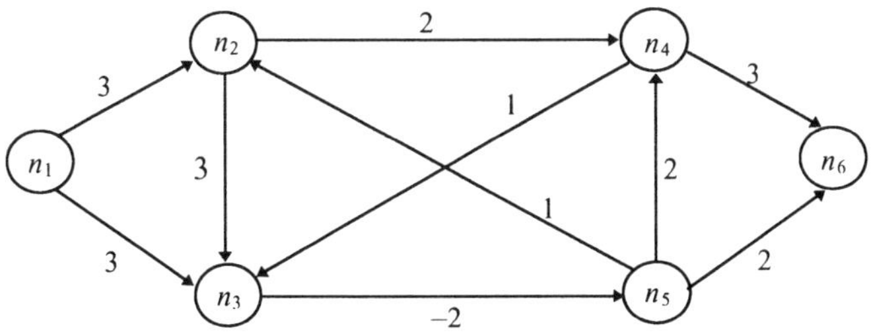

Figure II.22

Initially, the labels of the nodes are $d_1^0 = 0$, and $d_2^0 = d_3^0 = d_4^0 = d_5^0 = d_6^0 = \infty$, the iteration counter $t = 1$, and the set $J_1 = \{2,3\}$. In the first iteration, the labels are then recomputed as $d_2^1 = \min\{\infty; 3\} = 3$, $d_3^1 = \min\{\infty; 3\} = 3$, $d_4^1 = d_5^1 = d_6^1 = \infty$. As the labels of n_2 and n_3 have changed, $J_1' = \{2, 3\}$. The second iteration has $t = 2$, and $J_2 = \{3, 4, 5\}$ as the set of all successors of J_1'. We can then compute the new labels $d_2^2 = 3, d_3^2 = \min\{\infty; 3 + 3\} = 3, d_4^2 = \min\{\infty; 3 + 2\} = 5, d_5^2 = \min\{\infty; 3 - 2\} = 1, d_6^2 = \infty$, so that $J_2' = \{4, 5\}$. In the third iteration $t = 3$, all successors of J_2' are included in $J_3 = \{2, 3, 4, 6\}$. This allows us to compute the revised labels $d_2^3 = \min\{3; 1 + 1\} = 2$, $d_3^3 = \min\{3; 3 + 3, 5 + 1\} = 3$, $d_4^3 = \min\{5; 3 + 2, 1 + 2\} = 3$, $d_5^3 = \min\{1; 3 - 2\} = 1$, and $d_6^3 = \min\{\infty; 5 + 3, 1 + 2\} = 3$, so that $J_3' = \{2, 4, 6\}$. In iteration $t = 4$, the set $J_4 = \{3, 4, 6\}$, and the labels are revised to $d_2^4 = 2$, $d_3^4 = \min\{3; 0 + 3, 2 + 3, 3 + 1\} = 3$, $d_4^4 = \min\{3; 2 + 2, 1 + 2\} = 3$, $d_5^4 = 1$, and $d_6^4 = \min\{3; 3 + 3, 1 + 2\} = 3$, so that $J_4 = \varnothing$. At this point, the algorithm terminates. The shortest path tree is shown in Figure II.23 on the next page.

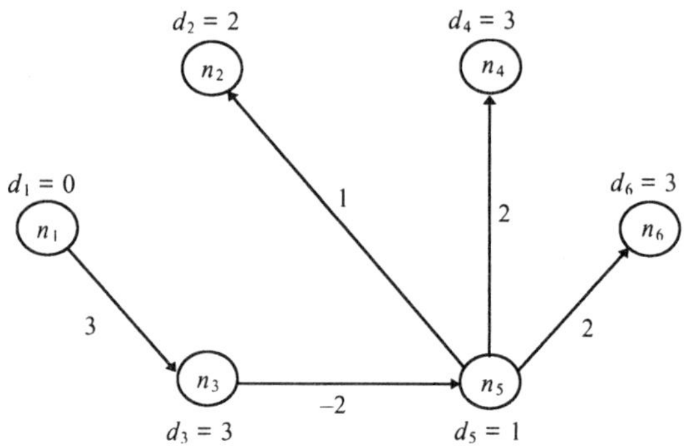

Figure II.23

In some applications, a shortest path must be found in a circuitless graph. One example of such a graph is the equipment replacement problem discussed in Section 2.2.2. For such graphs it can be easily shown that there exists a labeling of the nodes that can be carried out in $O(n^2)$ time such that $a_{ij} \in A$ if and only if $i < j$. This fact renders the solution of Bellman's equations particularly simple. In fact, these equations become

$$d_1 = 0, \text{ and}$$

$$d_j = \min_{i<j}\{d_i + c_{ij}\}, j = 2, \ldots, n$$

and the values d_j can therefore be calculated sequentially for $j = 2, \ldots, n$, because d_2 depends only on d_1, d_3 only on d_2 and d_1, etc., and d_n depends only on $d_1, d_2, \ldots, d_{n-1}$. The calculations involved are $O(n^2)$ comparisons and $O(n^2)$ additions for a total complexity of $O(n^2)$.

2.3.3 The Floyd-Warshall Algorithm

In the previous sections we have considered algorithms solving problems that involve finding shortest paths from one given node to either one other or all other nodes. However, in many real-life applications, it is required to find shortest paths between all pairs of nodes. Obviously, we could apply the Dijkstra or Bellman-Ford-Moore algorithms n times to obtain all shortest paths, depending on whether or not all arc distances are nonnegative. However, algorithms exist that solve the problem directly. The original algorithm of this type is due to Floyd (1962) and Warshall (1962). It is a label correcting algorithm that runs in $O(n^3)$ time. Once again, arc lengths may be of arbitrary sign, but no negative circuits should be present in the graph.

Define now d_{ij}^k as the length of the shortest path from node n_i to node n_j, given that the path does not use any of the nodes n_k, n_{k+1}, ..., n_n (except possibly n_i and n_j). Suppose that we have already calculated d_{ij}^k $\forall$ i and j. Either the shortest path p from node n_i to node n_j that does not use nodes n_{k+1}, ..., n_k does not use n_k as well, in which case $d_{ij}^{k+1} = d_{ij}^j$, or it does. In the latter case, the principle of optimality implies that the portion of p from node n_i to node n_k has length d_{ik}^k and the portion from node n_k to node n_j has length d_{kj}^k, so that $d_{ij}^{k+1} = d_{ik}^k + d_{kj}^k$. The recursive equation is then

$$d_{ij}^{k+1} = \min\{d_{ij}^k, d_{ik}^k + d_{kj}^k\}.$$

Initially, define the matrix $\mathbf{D}^1 = (d_{ij}^1)$ with $d_{ij}^1 = 0$ if $i = j$, $d_{ij}^1 = c_{ij}$ if $a_{ij} \in A$, and $d_{ij}^1 = \infty$ otherwise. Defining the shortest unrestricted distance between n_i and n_j as d_{ij}, we have $d_{ij}^{n+1} = d_{ij}$. In addition to the length of the shortest path, it is also possible to keep track of detour nodes used in the process. This will allow us to determine the actual shortest paths in a backtracking process. For that purpose, define p_{ij} as the predecessor node of n_j in the shortest path from n_i to n_j. Initially, $p_{ij}^1 = i$ if $a_{ij} \in A$, and 0 otherwise. The values p_{ij}^1 are collected in the matrix $\mathbf{P}^1$. The recursive equation is then

$$p_{ij}^{k+1} = \begin{cases} p_{ij}^k \text{ if } d_{ij}^{k+1} = d_{ij}^k \\ p_{kj}^k \text{ otherwise} \end{cases}.$$

Here again, $p_{ij} = p_{ij}^{n+1}$. We can now formally state

The Floyd-Warshall Algorithm

Step 1: Is $k = n + 1$? If yes: Stop, $\mathbf{D} = \mathbf{D}^{n+1}$ is the shortest path matrix and $\mathbf{P} = \mathbf{P}^{n+1}$ is the associated detour matrix.

If no: Go to Step 2.

Step 2: Determine $d_{ij}^{k+1} = \min\{d_{ij}^k; d_{ik}^k + d_{kj}^k\}$ and $p_{ij}^{k+1} = p_{ij}^k$, if $d_{ij}^{k+1} = d_{ij}^k$, and p_{kj}^k otherwise. Set $k := k + 1$ and go to Step 1.

In each iteration, each of the O(n^2) elements is computed in constant time. Given that O(n) iterations are required, the Floyd-Warshall algorithm has computational complexity $O(n^3)$. The major drawback to the Floyd-Warshall algorithm is the storage requirement of the $O(n^2)$ values in the matrices $\mathbf{D}^k$ and $\mathbf{P}^k$. For this reason, large scale models are sometimes solved in practice by an n-fold application of a single source algorithm. In order to reduce redundancy in such a sequential approach, different methods have been proposed that compute the shortest spanning tree of a given source node while exploiting the information which is contained in the shortest spanning tree of the previous source node of the sequence. Two such methods may be found in Gallo and Pallottino (1986), and Florian *et al.* (1981).

Negative circuits in the given graph G may be detected during the execution of the Floyd-Warshall algorithm. In particular, G contains a negative circuit if and only if $d_{ii}^k < 0$ for some i in some iteration k.

Example: Consider the graph of Figure II.24.

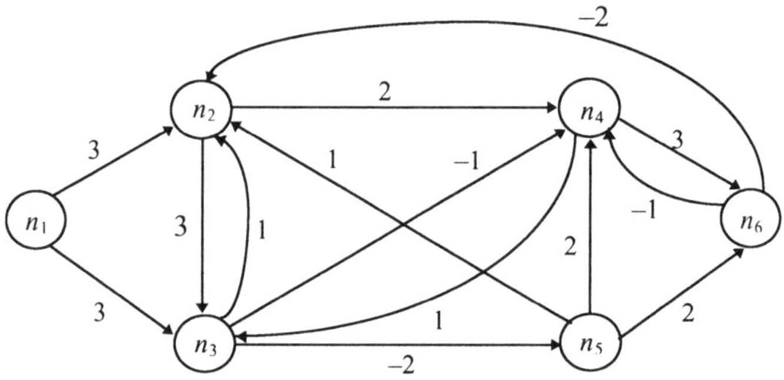

Figure II.24

Below, we show the sequences of matrices $\mathbf{D}^k$ and $\mathbf{P}^k$, $k = 1, ..., 7$. In each matrix $\mathbf{D}^k$, the k-th row and column are outlined to facilitate the calculation of the matrix $\mathbf{D}^{k+1}$. Similarly the k-th row of the matrix $\mathbf{P}^k$ is outlined for ease of calculation of the matrix $\mathbf{P}^{k+1}$.

$$\mathbf{D}^1 = \begin{bmatrix} 0 & 3 & 3 & \infty & \infty & \infty \\ \infty & 0 & 3 & 2 & \infty & \infty \\ \infty & 1 & 0 & -1 & -2 & \infty \\ \infty & \infty & 1 & 0 & \infty & 3 \\ \infty & 1 & \infty & 2 & 0 & 2 \\ \infty & -2 & \infty & -1 & \infty & 0 \end{bmatrix} \qquad \mathbf{P}^1 = \begin{bmatrix} 0 & 1 & 1 & 0 & 0 & 0 \\ 0 & 0 & 2 & 2 & 0 & 0 \\ 0 & 3 & 0 & 3 & 3 & 0 \\ 0 & 0 & 4 & 0 & 0 & 4 \\ 0 & 5 & 0 & 5 & 0 & 5 \\ 0 & 6 & 0 & 6 & 0 & 0 \end{bmatrix}$$

$$\mathbf{D}^2 = \begin{bmatrix} 0 & 3 & 3 & \infty & \infty & \infty \\ \infty & 0 & 3 & 2 & \infty & \infty \\ \infty & 1 & 0 & -1 & -2 & \infty \\ \infty & \infty & 1 & 0 & \infty & 3 \\ \infty & 1 & \infty & 2 & 0 & 2 \\ \infty & -2 & \infty & -1 & \infty & 0 \end{bmatrix} \qquad \mathbf{P}^2 = \begin{bmatrix} 0 & 1 & 1 & 0 & 0 & \infty \\ 0 & 0 & 2 & 2 & 0 & \infty \\ 0 & 3 & 0 & 3 & 3 & \infty \\ 0 & 0 & 4 & 0 & 0 & 4 \\ 0 & 5 & 0 & 5 & 0 & 5 \\ 0 & 6 & 0 & 6 & 0 & 0 \end{bmatrix}$$

Note that $\mathbf{D}^2 = \mathbf{D}^1$, i.e., the node n_1 was never used as a detour node. This is, of course, not surprising as the node n_1 has indegree zero, prohibiting its use as a detour node. In the next iteration, we obtain

$$\mathbf{D}^3 = \begin{bmatrix} 0 & 3 & 3 & 5 & \infty & \infty \\ \infty & 0 & 3 & 2 & \infty & \infty \\ \infty & 1 & 0 & -1 & -2 & \infty \\ \infty & \infty & 1 & 0 & \infty & 3 \\ \infty & 1 & 4 & 2 & 0 & 2 \\ \infty & -2 & 1 & -1 & \infty & 0 \end{bmatrix} \qquad \mathbf{P}^1 = \begin{bmatrix} 0 & 1 & 1 & 2 & 0 & 0 \\ 0 & 0 & 2 & 2 & 0 & 0 \\ 0 & 3 & 0 & 3 & 3 & 0 \\ 0 & 0 & 4 & 0 & 0 & 4 \\ 0 & 5 & 2 & 5 & 0 & 5 \\ 0 & 6 & 2 & 6 & 0 & 0 \end{bmatrix}$$

In the computation of $\mathbf{D}^3$, only the element d_{14}^3 changed, indicating that the node n_2 was used as a detour node.

$$\mathbf{D}^4 = \begin{bmatrix} 0 & 3 & 3 & 2 & 1 & \infty \\ \infty & 0 & 3 & 2 & 1 & \infty \\ \infty & 1 & 0 & -1 & -2 & \infty \\ \infty & 2 & 1 & 0 & -1 & 3 \\ \infty & 1 & 4 & 2 & 0 & 2 \\ \infty & -2 & 1 & -1 & -1 & 0 \end{bmatrix} \qquad \mathbf{P}^4 = \begin{bmatrix} 0 & 1 & 1 & 3 & 3 & 0 \\ 0 & 0 & 2 & 2 & 3 & 0 \\ 0 & 3 & 0 & 3 & 3 & 0 \\ 0 & 3 & 4 & 0 & 3 & 4 \\ 0 & 5 & 2 & 5 & 0 & 5 \\ 0 & 6 & 2 & 6 & 3 & 0 \end{bmatrix}$$

Computing $\mathbf{D}^4$, we notice many changes in the fifth column, showing that the node n_3 was extensively used as detour node in the shortest path to n_5.

$$\mathbf{D}^5 = \begin{bmatrix} 0 & 3 & 3 & 2 & 1 & 5 \\ \infty & 0 & 3 & 2 & 1 & \infty \\ \infty & 1 & 0 & -1 & -2 & 2 \\ \infty & 2 & 1 & 0 & -1 & 3 \\ \infty & 1 & 3 & 2 & 0 & 2 \\ \infty & -2 & 0 & -1 & -1 & 0 \end{bmatrix} \qquad \mathbf{P}^5 = \begin{bmatrix} 0 & 1 & 1 & 3 & 3 & 4 \\ 0 & 0 & 2 & 2 & 3 & 4 \\ 0 & 3 & 0 & 3 & 3 & 4 \\ 0 & 3 & 4 & 0 & 3 & 4 \\ 0 & 5 & 4 & 5 & 0 & 5 \\ 0 & 6 & 4 & 6 & 3 & 0 \end{bmatrix}$$

Computing $\mathbf{D}^5$ shows that d_{53}^5, d_{63}^5, and d_{65}^5 have decreased. In these cases, the node n_4 was used as a detour node.

$$\mathbf{D}^6 = \begin{bmatrix} 0 & 2 & 3 & 2 & 1 & 3 \\ \infty & 0 & 3 & 2 & 1 & 3 \\ \infty & -1 & 0 & -1 & -2 & 0 \\ \infty & 2 & 1 & 0 & -1 & 1 \\ \infty & 1 & 3 & 2 & 0 & 2 \\ \infty & -2 & 0 & 1 & 2 & 0 \end{bmatrix} \qquad \mathbf{P}^6 = \begin{bmatrix} 0 & 5 & 1 & 3 & 3 & 5 \\ 0 & 0 & 2 & 2 & 3 & 5 \\ 0 & 5 & 0 & 3 & 3 & 5 \\ 0 & 5 & 4 & 0 & 3 & 5 \\ 0 & 5 & 4 & 5 & 0 & 5 \\ 0 & 6 & 4 & 6 & 3 & 0 \end{bmatrix}$$

The computation of $\mathbf{D}^6$ indicates various changes in the second and sixth columns, showing the extensive use of the node n_5 on shortest paths to n_2 and n_6.

$$\mathbf{D}^7 = \begin{bmatrix} 0 & 1 & 3 & 2 & 1 & 3 \\ \infty & 0 & 3 & 2 & 1 & 3 \\ \infty & -2 & 0 & -1 & -2 & 0 \\ \infty & -1 & 1 & 0 & -1 & 1 \\ \infty & 0 & 2 & 1 & 0 & 2 \\ \infty & -2 & 0 & -1 & -2 & 0 \end{bmatrix} \qquad \mathbf{P}^7 = \begin{bmatrix} 0 & 6 & 1 & 3 & 3 & 5 \\ 0 & 0 & 2 & 2 & 3 & 5 \\ 0 & 6 & 0 & 3 & 3 & 5 \\ 0 & 6 & 4 & 0 & 3 & 5 \\ 0 & 6 & 4 & 6 & 0 & 5 \\ 0 & 6 & 4 & 6 & 3 & 0 \end{bmatrix}$$

At this point, the algorithm terminates with the shortest path matrix $\mathbf{D} = \mathbf{D}^7$ and its associated detour matrix $\mathbf{P} = \mathbf{P}^7$. To demonstrate the use of the detour matrix, consider the shortest path between n_1 and n_2. The element $d_{12} = d_{12}^7 = 1$ indicates the length of the path. The element $p_{12}^7 = 6$ indicates that n_6 is the direct predecessor of n_2 on the shortest path. The direct predecessor of n_6 is then n_5, as $p_{16}^7 = 5$. Now, the direct predecessor of n_5 on the shortest path is n_3, as $p_{15}^7 = 3$. Finally, the direct predecessor of n_3 is n_1, as shown by $p_{13}^7 = 1$. Putting the pieces together, we obtain the path $p_{12} = (n_1, n_3, n_5, n_6, n_2)$. The k-th row of the matrix **P**, $k = 1, ..., n$, can then be used to construct the shortest path tree with the root n_k.

2.4 Extensions of the Basic Problem

Until now, we have only considered algorithms that determine unrestricted paths between nodes of the given graph $G = (N, A)$. Sometimes, decision makers may be interested not only in the shortest, but k shortest paths. The next section addresses these problems. Alternatively, it may be desired to consider factors other than distance, such as time or resource usage. The last two sections of this chapter deal with problems in this category.

2.4.1 The k-Shortest Paths Problem

Consider again a decision maker whose task is to find a path on which units are to be shipped from a given node (the source) to another node (the sink). Almost by definition, the decision maker will have to consider various objectives, e.g., time, money, risk, or ecological impact, just to name a few criteria. Incorporating these criteria in a model is usually done in one of two ways. We either consider the criteria as objectives and solve an appropriate multiobjective optimization problem, or define upper bounds on all but one of the criteria, formulate them as constraints, and optimize the objective determined by the remaining criterion. Another possibility is to generate not "the optimal" solution, but the k best solutions for some $k \geq 1$, provide the decision maker with them, and give him the possibility to consider the tradeoffs between objectives. For instance, the objective may be flat in the vicinity of the optimal solution, so that the decision maker has to give up very little of one objective to make significant gains in terms of another objective. The approach described in this section falls into that category.

A variety of methods for the solution of k-shortest path path problems have been described in the literature. A good variety of algorithms can be found in the annotated bibliography by Deo and Pang (1984). The algorithm presented in this section is due to Shier (1976). His technique is a label correcting method that computes vectors of k-shortest paths for all nodes $n_i \in N$. The method does allow nonelementary paths as well as negative arc lengths, but circuits of negative length are prohibited.

All standard shortest path methods use basic operations, such as addition and comparison. In order to deal with k-vectors, these operations have to be generalized. In order to do so, we will need the concept of the set of extended real numbers $\mathbb{R}_\infty = \mathbb{R} \cup \{\infty\}$, where we assume that a k-vector $\mathbf{a} \in \mathbb{R}^k_\infty$ has its components in strictly increasing order, i.e., $a_1 < a_2 < \ldots < a_k$. It is convenient to assume that $\infty < \infty + 1 = \infty$. Furthermore let j-min$\{c_1, \ldots, c_n\}$ denote the j-th smallest distinct element of the set or vector that includes $c_1, c_2, \ldots, c_n$. We can then write

Definition II.6: Let **a**, **b**, and $\mathbf{c} \in \mathbb{R}^k_\infty$. Then *generalized minimization* is defined as $\mathbf{a} \oplus \mathbf{b} = \mathbf{c} = (c_j)$ with $c_j = j\text{-min}\{a_1, \ldots, a_k; b_1, \ldots, b_k\}$ $\forall\, j = 1, \ldots, k$. In other words, the k-vector **c** includes the k smallest distinct elements of **a** and **b**.

Similarly, we can write

Definition II.7: Let **a**, **b**, and $\mathbf{c} \in \mathbb{R}^k_\infty$. Then *generalized addition* is defined as $\mathbf{a} \otimes \mathbf{b} = \mathbf{c} = (c_j)$ with $c_j = j\text{-min}\{a_i + b_\ell;\ i, \ell = 1, \ldots, k\}$. In other words, **c** is the k-vector that includes the k smallest distinct sums of elements in **a** and **b**.

As an example, if $\mathbf{a} = [1, 3, 4, 8]$ and $\mathbf{b} = [2, 3, 5, 7]$, then $\mathbf{a} \oplus \mathbf{b} = [1, 2, 3, 4]$, and $\mathbf{a} \otimes \mathbf{b} = [3, 4, 5, 6]$. It is apparent that for $k = 1$, $\mathbf{a} \oplus \mathbf{b} = \min\{a, b\}$ reduces to the usual minimization, and $\mathbf{a} \otimes \mathbf{b} = a + b$ is the usual addition. It is also important to note that in this generalized algebra, the multiplication of a matrix $\mathbf{A} = (a_{ij})$ with a vector $\mathbf{b} = (b_i)$, where a_{ij} and b_i are k-vectors, follows the same rules as the usual matrix-vector multiplication. As an example, let $\mathbf{A} = \begin{bmatrix} (3,4) & (2,0) \\ (\infty,1) & (3,4) \end{bmatrix}$ and $\mathbf{b} = \begin{bmatrix} (1,6) \\ (4,0) \end{bmatrix}$. Then $\mathbf{A} \otimes \mathbf{b} = \begin{bmatrix} (3,4) \otimes (1,6) \oplus (2,0) \otimes (4,0) \\ (\infty,1) \otimes (1,6) \oplus (3,4) \otimes (4,0) \end{bmatrix} = \begin{bmatrix} (4,5) \oplus (0,2) \\ (2,7) \oplus (3,4) \end{bmatrix} = \begin{bmatrix} (0,2) \\ (2,3) \end{bmatrix}$. A multiplication of this type is required by the algorithm described below.

The Double Sweep algorithm described below will use generalized minimization and addition to calculate the lengths of the k shortest paths from the node n_1 to all other nodes in the graph $G = (N, A, \mathbf{c})$. The method is initialized with a guess of the lengths of the k shortest paths from n_1 to nodes n_j, $i = 2, \ldots, n$. This guess vector $\mathbf{d}_j^{\nu}$ in iteration ν will be revised downward during the calculations until an optimum has been reached at $\mathbf{d}_j^{*}$. The revisions take place in two phases, a backward sweep and a forward sweep; hence the name of the method. Each backward sweep starts with a vector $\mathbf{d}_j^{\nu}$ of overestimates and revises the path lengths in a backward fashion, starting with node n_n, and continuing with n_{n-1}, n_{n-2}, ... in that order, until n_1 is reached. This revision is accomplished for each node n_j by comparing the latest estimate of the k-shortest path from n_1 to n_j with the paths from n_1 to n_j that use one detour node. This is followed by a forward sweep that is similar to a backward sweep, except that it scans the nodes in a forward fashion, i.e., first n_1, then n_2, followed by n_3, n_4, etc. Backward and forward sweeps alternate with each other, so that estimates $\mathbf{d}_j^{\nu}$ with odd values of ν are computed in backward sweeps, and estimates $\mathbf{d}_j^{\nu}$ with even values of ν are calculated in forward sweeps.

To formalize matters, let $\mathbf{C} = \mathbf{c}_{ij}$ denote the matrix of arc lengths, so that each $\mathbf{c}_{ij} = (c_{ij}^1, c_{ij}^2, \ldots, c_{ij}^k)$ consists of k components, the ℓ-th of which denotes the ℓ-th shortest arc from n_i to n_j. If $a_{ij} \notin A$, then $\mathbf{c}_{ij} = (\infty, \infty, \ldots, \infty)$ and if $i = j$, then $\mathbf{c}_{ii} = (0, \infty, \infty, \ldots, \infty)$. If there is a single arc from n_i to n_j with length of, say, 5, then $\mathbf{c}_{ij} = (5, \infty, \infty, \ldots, \infty)$. If there are multiple arcs from n_i to n_j, e.g., four arcs of lengths 2, 3, 3, and 6 and we are interested in the 5 shortest paths, then $\mathbf{c}_{ij} = (2, 3, 6, \infty, \infty)$. We can now decompose the matrix $\mathbf{C}$ into two matrices $\mathbf{L} = (\ell_{ij})$ and $\mathbf{U} = (\mathbf{u}_{ij})$ that are lower and upper triangular matrices, respectively. These matrices are defined, so that $\ell_{ij} = \mathbf{c}_{ij}\ \forall\ i > j$ and $(\infty, \infty, \ldots, \infty)$ otherwise, while $\mathbf{u}_{ij} = \mathbf{c}_{ij}\ \forall\ i < j$ and $(\infty, \infty, \ldots, \infty)$ otherwise.

Given some estimates $\mathbf{d}_j^0\ \forall\, j$, we calculate

$$\mathbf{d}_j^{\nu} = [\mathbf{d}^{\nu} \otimes \ell_{.j}] \oplus \mathbf{d}_j^{\nu-1} \quad \text{for } j = n, n-1, n-2, \ldots, 1 \qquad (1)$$

in a backward sweep, and

$$\mathbf{d}_j^{\nu} = [\mathbf{d}^{\nu} \otimes \mathbf{u}_{.j}] \oplus \mathbf{d}_j^{\nu-1} \quad \text{for } j = 1, 2, \ldots, n \qquad (2)$$

in a forward sweep where $\ell_{.j}$ and $\mathbf{u}_{.j}$ denote the j-th column of the matrix $\mathbf{L}$ and $\mathbf{U}$, respectively. As mentioned earlier, the procedure starts with a backward sweep and, after that, backward and forward sweeps alternate. The procedure terminates when the estimates in two successive forward and backward sweeps are equal. As each double sweep requires $O(n^2)$ generalized additions and minimizations, and there are a total of kn values to compute, the algorithm requires $O(kn^3)$ time. For details, see, e.g., Evans and Minieka (1992).

Example: Consider the graph $G = (N, A, \mathbf{c})$ in Figure II.25 and determine the 3 shortest paths from n_1 to all other nodes.

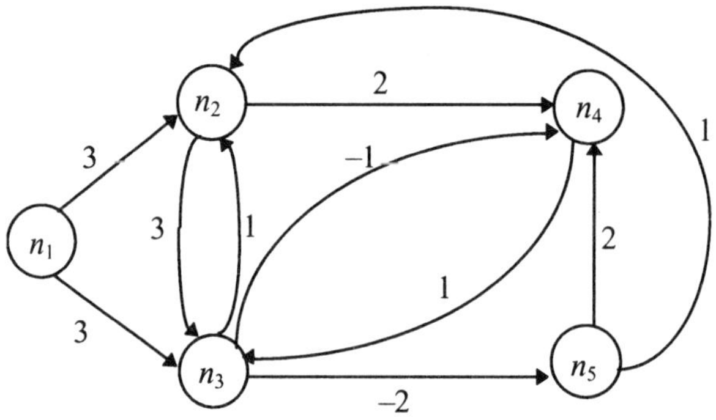

Figure II.25

Given this graph, the arc lengths can be written in the distance matrix

$$\mathbf{C} = \begin{bmatrix} (0,\infty,\infty) & (3,\infty,\infty) & (3,\infty,\infty) & (\infty,\infty,\infty) & (\infty,\infty,\infty) \\ (\infty,\infty,\infty) & (0,\infty,\infty) & (3,\infty,\infty) & (2,\infty,\infty) & (\infty,\infty,\infty) \\ (\infty,\infty,\infty) & (1,\infty,\infty) & (0,\infty,\infty) & (-1,\infty,\infty) & (-2,\infty,\infty) \\ (\infty,\infty,\infty) & (\infty,\infty,\infty) & (1,\infty,\infty) & (0,\infty,\infty) & (\infty,\infty,\infty) \\ (\infty,\infty,\infty) & (1,\infty,\infty) & (\infty,\infty,\infty) & (2,\infty,\infty) & (0,\infty,\infty) \end{bmatrix}.$$

As defined above, we can write the lower and upper triangular matrices $\mathbf{L}$ and $\mathbf{U}$ as

$$\mathbf{L} = \begin{bmatrix} (\infty,\infty,\infty) & (\infty,\infty,\infty) & (\infty,\infty,\infty) & (\infty,\infty,\infty) & (\infty,\infty,\infty) \\ (\infty,\infty,\infty) & (\infty,\infty,\infty) & (\infty,\infty,\infty) & (\infty,\infty,\infty) & (\infty,\infty,\infty) \\ (\infty,\infty,\infty) & (1,\infty,\infty) & (\infty,\infty,\infty) & (\infty,\infty,\infty) & (\infty,\infty,\infty) \\ (\infty,\infty,\infty) & (\infty,\infty,\infty) & (1,\infty,\infty) & (\infty,\infty,\infty) & (\infty,\infty,\infty) \\ (\infty,\infty,\infty) & (1,\infty,\infty) & (\infty,\infty,\infty) & (2,\infty,\infty) & (\infty,\infty,\infty) \end{bmatrix} \text{ and}$$

$$\mathbf{U} = \begin{bmatrix} (\infty,\infty,\infty) & (3,\infty,\infty) & (3,\infty,\infty) & (\infty,\infty,\infty) & (\infty,\infty,\infty) \\ (\infty,\infty,\infty) & (\infty,\infty,\infty) & (3,\infty,\infty) & (2,\infty,\infty) & (\infty,\infty,\infty) \\ (\infty,\infty,\infty) & (\infty,\infty,\infty) & (\infty,\infty,\infty) & (-1,\infty,\infty) & (-2,\infty,\infty) \\ (\infty,\infty,\infty) & (\infty,\infty,\infty) & (\infty,\infty,\infty) & (\infty,\infty,\infty) & (\infty,\infty,\infty) \\ (\infty,\infty,\infty) & (\infty,\infty,\infty) & (\infty,\infty,\infty) & (\infty,\infty,\infty) & (\infty,\infty,\infty) \end{bmatrix}$$

We initialize the Double Sweep algorithm with distance estimates $\mathbf{d}_1^0 = (0, \infty, \infty)$, and $\mathbf{d}_j^0 = (\infty, \infty, \infty)$ for $j = 2, 3, 4, 5$. In iteration number $\nu = 1$, we perform a backward sweep, starting with n_j.

For $j = 5$, we have $\mathbf{d}_5^1 = [[\mathbf{d}_1^1 \otimes (\infty, \infty, \infty)] \oplus [\mathbf{d}_2^1 \otimes (\infty, \infty, \infty)] \oplus [\mathbf{d}_3^1 \otimes (\infty, \infty, \infty)] \oplus [\mathbf{d}_4^1 \otimes (\infty, \infty, \infty)] \oplus [\mathbf{d}_5^1 \otimes (\infty, \infty, \infty)]] \oplus \mathbf{d}_5^0 = [(\infty, \infty, \infty) \oplus (\infty, \infty, \infty) \oplus (\infty, \infty, \infty) \oplus (\infty, \infty, \infty) \oplus (\infty, \infty, \infty)] \oplus (\infty, \infty, \infty) = (\infty, \infty, \infty)$.

For $j = 4$ we obtain $\mathbf{d}_4^1 = [[\mathbf{d}_1^1 \otimes (\infty, \infty, \infty)] \oplus [\mathbf{d}_2^1 \otimes (\infty, \infty, \infty)] \oplus [\mathbf{d}_3^1 \otimes (\infty, \infty, \infty)] \oplus [\mathbf{d}_4^1 \otimes (\infty, \infty, \infty)] \oplus [\mathbf{d}_5^1 \otimes (\infty, \infty, \infty)]] \oplus \mathbf{d}_4^0 = [(\infty, \infty, \infty) \oplus (\infty, \infty, \infty) \oplus (\infty, \infty, \infty) \oplus (\infty, \infty, \infty) \oplus (\infty, \infty, \infty)] \oplus (\infty, \infty, \infty) = (\infty, \infty, \infty)$.

Continuing with $j = 3$, $j = 2$, and $j = 1$ results in $\mathbf{d}_3^1 = \mathbf{d}_2^1 = (\infty, \infty, \infty)$, and $\mathbf{d}_1^1 = (0, \infty, \infty)$, i.e., in the first backward sweep no change has occurred.

The first forward sweep commences with $j = 1$ and $\mathbf{d}_1^2 = [[\mathbf{d}_1^2 \otimes (\infty, \infty, \infty)] \oplus [\mathbf{d}_2^2 \otimes (\infty, \infty, \infty)] \oplus [\mathbf{d}_3^2 \otimes (\infty, \infty, \infty)] \oplus [\mathbf{d}_4^2 \otimes (\infty, \infty, \infty)] \oplus [\mathbf{d}_5^2 \otimes (\infty, \infty, \infty)]] \oplus \mathbf{d}_1^1 = [(\infty, \infty, \infty) \oplus (\infty, \infty, \infty) \oplus (\infty, \infty, \infty) \oplus (\infty, \infty, \infty) \oplus (\infty, \infty, \infty)] \oplus (0, \infty, \infty) = (0, \infty, \infty)$ and again, no change has occurred. Let now $j = 2$. We then have $\mathbf{d}_2^2 = [[\mathbf{d}_1^2 \otimes (3, \infty, \infty)] \oplus [\mathbf{d}_2^2 \otimes (\infty, \infty, \infty)] \oplus [\mathbf{d}_3^2 \otimes (\infty, \infty, \infty)] \oplus [\mathbf{d}_4^2 \otimes (\infty, \infty, \infty)] \oplus [\mathbf{d}_5^2 \otimes (\infty, \infty, \infty)]] \oplus \mathbf{d}_2^1 = [(3, \infty, \infty) \oplus (\infty, \infty, \infty) \oplus (\infty, \infty, \infty) \oplus (\infty, \infty, \infty) \oplus (\infty, \infty, \infty)] \oplus (\infty, \infty, \infty) = (3, \infty, \infty)$ and a downward reduction of the estimation of the length of the shortest path from ∞ to 3 has occurred. For $j = 3$ we obtain $\mathbf{d}_3^2 = [[\mathbf{d}_1^2 \otimes (3, \infty, \infty)] \oplus [\mathbf{d}_2^2 \otimes (3, \infty, \infty)] \oplus [\mathbf{d}_3^2 \otimes (\infty, \infty, \infty)] \oplus [\mathbf{d}_4^2 \otimes (\infty, \infty, \infty)] \oplus [\mathbf{d}_5^2 \otimes (\infty, \infty, \infty)]] \oplus \mathbf{d}_2^2 = [(3, \infty, \infty) \oplus (6, \infty, \infty) \oplus (\infty, \infty, \infty) \oplus (\infty, \infty, \infty) \oplus (\infty, \infty, \infty)] \oplus (3, \infty, \infty) = (3, 6, \infty) \oplus (3, \infty, \infty) = (3, 6, \infty)$.

Continuing with $j = 4$, and 5, these steps are then followed by another backward sweep. The results of the computations are summarized in Table II.2.

Table II.2

$\nu =$	0	1	2	3	4	5
$\mathbf{d}_1^\nu$	$(0, \infty, \infty)$	$(0, \infty, \infty)$	$(0, \infty, \infty)$	$(0, \infty, \infty)$	$(0, \infty, \infty)$	$(0, \infty, \infty)$
$\mathbf{d}_2^\nu$	(∞, ∞, ∞)	(∞, ∞, ∞)	$(3, \infty, \infty)$	$(2, 3, 4)$	$(2, 3, 4)$	$(2, 3, 4)$
$\mathbf{d}_3^\nu$	(∞, ∞, ∞)	(∞, ∞, ∞)	$(3, 6, \infty)$	$(3, 4, 6)$	$(3, 4, 5)$	$(3, 4, 5)$
$\mathbf{d}_4^\nu$	(∞, ∞, ∞)	(∞, ∞, ∞)	$(2, 5, \infty)$	$(2, 3, 5)$	$(2, 3, 4)$	$(2, 3, 4)$
$\mathbf{d}_5^\nu$	(∞, ∞, ∞)	(∞, ∞, ∞)	$(1, 4, \infty)$	$(1, 4, \infty)$	$(1, 2, 3)$	$(1, 2, 3)$

The estimates in iteration 5 are the same as those in iteration 4, so that the algorithm has converged and the k-shortest paths from n_1 to all other nodes n_j are given as $\mathbf{d}_j^* = \mathbf{d}_j^5 = \mathbf{d}_j^4$. Notice the nonincreasing values of the estimate, i.e., $\mathbf{d}_j^{\nu+1} \leq \mathbf{d}_j^\nu \;\forall\, j, \nu$.

It is now possible to retrieve not only the lengths of the k-shortest paths, but also the paths themselves. In order to do so, we need another backward recursion. As an illustration, determine the second-shortest path from n_1 to n_4 in the above example. As the second component of the vector $\mathbf{d}_4^5$ indicates, the length of this path is 3. The original graph structure reveals that the immediate predecessors of n_4 could be n_2, n_3, and n_5 with arc lengths 2, – 1, and 2, respectively. If n_2 were the immediate predecessor of n_4 on the second shortest path from n_1 to n_4, then the shortest or second-shortest path from n_1 to n_2 equals the second component of $\mathbf{d}_4^5$ minus c_{24} which is 3 – 2, = 1. However, as $\mathbf{d}_2^5$ indicates, the two shortest paths from n_1 to n_2 are of length 2 and 3, respectively, so that n_2 cannot be the immediate predecessor of n_4. Similar arguments indicate that n_3 and n_5 may both be predecessors of n_4. Repeating this procedure results in the second shortest path from n_1 to n_4 as $[n_1, n_3, (n_4, n_3)^r, n_5, n_4]$ with $r \geq 0$ indicating that the circuit (n_3, n_4, n_3) may be included arbitrarily often, as its length is zero. The results in an infinate number of second-shortest paths.

Another k-shortest path algorithm has been developed by Dreyfus (1969). As Evans and Minieka (1992) demonstrate, the Dantzig, Floyd, and Warshall algorithms can also be suitably modified to determine the k-shortest paths.

2.4.2 The Minimum Cost-to-Time Ratio Problem

Consider a merchant marine vessel, a so-called tramp steamer, that sails from port to port in order to pick up and deliver goods. Each route from a port n_i to another port n_j along a direct link a_{ij} nets a cetain profit p_{ij} and takes t_{ij} time units. The vessel is not required to visit any (or all) part(s) but follows the objective to maximize its profit per time unit. Formally speaking, this means that the planner will want to identify a cycle C that optimizes the objective

$$\text{Max} - z = \sum_{a_{ij} \in C} p_{ij} \Big/ \sum_{a_{ij} \in C} t_{ij} ,$$

or, equivalently, defining costs $c_{ij} = -p_{ij} \ \forall \ a_{ij} \in A$, as

$$\text{Min } z = \sum_{a_{ij} \in C} c_{ij} \Big/ \sum_{a_{ij} \in C} t_{ij} .$$

Or discussion below will assume that the latter formulation with minimization objective is used.

Suppose now that an arbitrary circuit $\tilde{C}$ has been chosen. The value of the objective function is $\tilde{z}$ which represents an upper bound on the objective value $\overline{z}$ of the optimal solution; i.e., $\tilde{z} \geq \overline{z}$. Define now arc values $d_{ij} = c_{ij} - \tilde{z}\, t_{ij} \ \forall \ a_{ij} \in A$. Given these values, we now determine the circuit of minimum length $\hat{C}$, so that $\hat{C}$ includes at least two nodes. This can be accomplished by using the Floyd-Warshall algorithm. If any of the diagonal elements in the final distance matrix is negative, a circuit of negative length exists. Assuming that the length of the circuit $\hat{C}$ with respect to the arc lengths d_{ij} is $\hat{d}$, either $\hat{d} < 0$ or $\hat{d} = 0$. Suppose that $\hat{d} < 0$. As $\hat{d} = \sum_{a_{ij} \in \hat{C}} d_{ij} = \sum_{a_{ij} \in \hat{C}} c_{ij} - \tilde{z} \sum_{a_{ij} \in \hat{C}} t_{ij} < 0$ or, equivalently, $\tilde{z} > \sum_{a_{ij} \in \hat{C}} c_{ij} \Big/ \sum_{a_{ij} \in \hat{C}} t_{ij}$, and as the right-hand side value of this strict inequality equals $\hat{z}$, $\tilde{z} > \hat{z}$ and thus the new solution is a strict improvement over the previous one. In such a case, we would redefine $\tilde{C} := \hat{C}$, $\tilde{z} := \hat{z}$, determine arc value d_{ij} on the basis of the new value of $\tilde{z}$, and find a circuit of minimum length.

In case $\hat{d} = 0$, a similar argument reveals that $\hat{z} = \tilde{z}$, i.e., the best solution that could be found equals the one that is already known. This is an indication of optimality. The above two cases imply an iterative procedure that is illustrated by the following

Example: Consider the undirected graph $G = (N, A, t_{ij}, p_{ij})$ shown in Figure II.26.

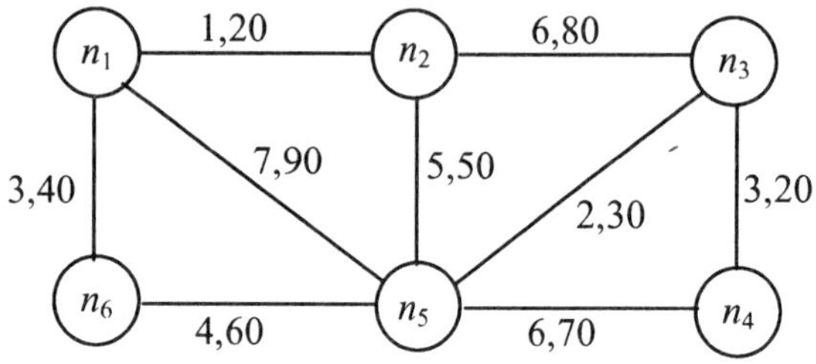

Figure II.26

An initial solution is provided by the circuit $\tilde{C} = (a_{12}, a_{25}, a_{51})$ with $\tilde{z} = \frac{-190}{14} = -12.3077$. Defining arc values $d_{ij} = -p_{ij} - \tilde{z}\, t_{ij} = 12.3077 t_{ij} - p_{ij}$, we obtain the graph in Figure II.27.

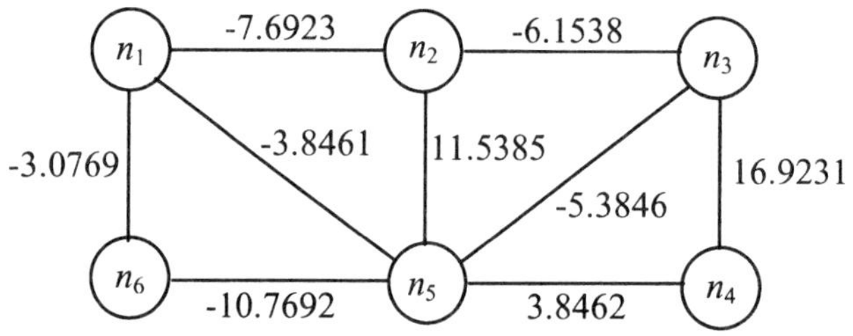

Figure II.27

The circuit of minimal length is $\hat{C} = (a_{12}, a_{23}, a_{35}, a_{56}, a_{61})$ with $\hat{d} = -33.0768$. In this example, all arcs on $\hat{C}$ happen to have negative length which, in general, is not required. As $\hat{d} < 0$, a new solution has been found, and we redefine $\tilde{C} := \hat{C} = (a_{12}, a_{23}, a_{35}, a_{56}, a_{61})$ which has an objective value $\tilde{z} = \frac{-230}{16} = -14.375$. Defining new arc lengths $d_{ij}\ \forall\ a_{ij} \in A$, we obtain the graph shown in Figure II.28.

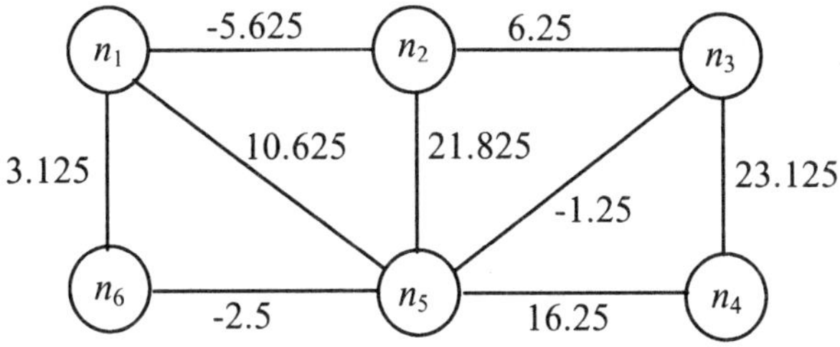

Figure II.28

The circuit with the shortest length is now equal to $\hat{C} = (a_{12}, a_{23}, a_{35}, a_{56}, a_{61})$ with $\hat{d} = 0$, indicating that the current solution is optimal.

A few notes are in order with respect to the above procedure. First, the procedure is finite, as each iteration provides a strict improvement and the number of different solutions cannot exceed the number of circuits which, in turn, is finite. Secondly, it is not necessary to find the circuit with the shortest length; any circuit with negative length will do. To prove optimality of a solution, it is, of course, required that all circuits have nonnegative length, which is achieved if the shortest circuit has length $\hat{d} = 0$. Finally, starting with an upper *and* a lower bound on the optimal objective value $\bar{z}$ allows the use of binary search. Lawler (1976) has shown that such a procedure requires $O(n^3 \log n)$ steps in the worst case.

2.4.3 Resource-Constrained Shortest Path Problems

The cost-to-time ratio problem in the previous section introduced a resource, *viz.*, time, in addition to the cost that is to be minimized. In that problem, cost and time were combined in a fractional objective. In this section, we consider p resources that are consumed if an arc is used. However, rather than including the resource consumptions in the objective, we will consider them in individual constraints. In order to formalize matters, let r_i^k denote the consumption of the k-th resource at node n_i and define r_{ij}^k as the consumption of the k-th resource on the arc a_{ij}. Furthermore, let ℓ^k and u^k denote the lower and upper bounds on the availabilities of the k-th resource, $k = 1, ..., p$. The objective is now to determine a cost-minimal (i.e., shortest) path that uses between ℓ^k and u^k resource units, for all $k = 1, ..., p$. For simplicity, consider the case in which $\ell^k = 0 \ \forall \ k$. The resource-constrained shortest path problem can then be formulated as

$$\text{P: Min } z = \sum_{a_{ij} \in A} c_{ij} x_{ij}$$

$$\text{s.t.} \quad \sum_{a_{ij} \in A} r_{ij}^k x_{ij} + \sum_{i \in N} r_i^k \tfrac{1}{2} \left(\sum_{j: a_{ij} \in A} x_{ij} + \sum_{j: x_{ji} \in A} x_{ji} \right) \le u^k \quad \forall\, k = 1, \ldots, p$$

$$\sum_{j: a_{ij} \in A} x_{ij} - \sum_{j: a_{ji} \in A} x_{ji} = 0 \quad \forall\, n_i \in N \backslash \{n_s, n_t\}$$

$$\sum_{j: a_{sj} \in A} x_{sj} = 1$$

$$x_{ij} = 0 \vee 1 \quad \forall\, a_{ij} \in A,$$

where $x_{ij} = 1$, if the arc a_{ij} is on the resource-constrained shortest path, and 0 if it is not. Here, the term in parentheses in the first set of constraints either equals 0 in case n_i is not on the shortest path, or equals 2, if n_i is on the shortest path. A naive approach to solving the problem would enumerate the paths, starting with the shortest, and continue until, for the first time, a feasible solution with respect to the resource constraints is found. Such a solution would necessarily be optimal. Unfortunately, the computational burden of this type of procedure is prohibitive. Instead, Beasley and Christofides (1989) suggest an approach based on branch and bound. Their method uses lower bounds obtained by Lagrangean relaxation. In particular, defining Lagrangean multipliers λ_k for $k = 1, \ldots, p$ and dualizing the resource constraints results in the Lagrangean function

$$L(\mathbf{x}, \lambda) = \sum_{a_{ij} \in A} c_{ij} x_{ij} + \sum_{k=1}^{p} \lambda_k \left[\sum_{a_{ij} \in A} r_{ij}^k x_{ij} + \tfrac{1}{2} \sum_{n_i \in N} r_i^k \left(\sum_{j: a_{ij} \in A} x_{ij} + \sum_{j: x_{ji} \in A} x_{ji} \right) - u^k \right]$$

that is to be minimized subject to the usual flow conservation, flow requirement, and zero-one constraints. Define now parameters $d_{ij} = c_{ij} + \sum_{k=1}^{p} \lambda_k \left[r_{ij}^k + \tfrac{1}{2} \left(r_i^k + r_j^k \right) \right]$, the Lagrangean function can be written as

$$L(\mathbf{x}, \lambda) = \sum_{a_{ij} \in A} d_{ij} x_{ij} - \sum_{k=1}^{p} \lambda_k u^k .$$

For given Lagrangean multipliers λ_k, $k = 1, \ldots, p$, the parameters d_{ij} are known and the problem with the Lagrangean objective is nothing but a standard shortest path problem with arc values d_{ij} and a constant term $\sum_{k=1}^{p} \lambda_k u^k$ that does, of course, not affect the optimization. Beasley and Christofides (1989) report encouraging results obtained with lower bounding by the function.

More general formulations include nonnegative lower bounds ℓ^k for all resources $k = 1, \ldots, p$. Also, we may define ℓ_i^k and u_i^k as lower and upper bounds of the availability of resource k at node n_i. If then $\ell_i^k \geq 1 \; \forall \; n_i \in S \subseteqq N$ for any nonempty subset S of N, and $\ell_i^k = 0 \; \forall \; n_i \in N \setminus S$, then the model requires that we find the shortest path from n_s to n_t, so that each node in the set S is visited at least once. This problem is commonly referred to as the node-constrained shortest path problem. If $S = N$, then each node must be visited at least once, which is the traveling salesman problem discussed in the next chapter.

CHAPTER 3 TRAVELING SALESMAN PROBLEMS AND EXTENSIONS

The Traveling Salesman Problem (*TSP*) is one of the most widely studied combinatorial optimization problems. While its statement is deceptively simple, it remains one of the most challenging problems in operations research. Hundreds of articles have been written on the problem. The book edited by Lawler *et al.* (1985) provides an insightful and comprehensive survey of major research results until that date. The purpose of this chapter is present some exact and approximate algorithms for the traveling salesman problem.

3.1 The Problem and its Applications

The traveling salesman problem is defined on a graph $G = (N, A, c)$, where c_{ij} denote the costs (lengths, distances) associated with the arc a_{ij}. Throughout this chapter, we assume that G is complete. If this is not originally the case, then G is constructed from the original graph by determining the shortest paths between all pairs of nodes by one of the pertinent methods. The problem is then to determine a circuit of minimal distance that "covers" all nodes. Formally, we can write

Definition II.8: A *Hamiltonian tour* (or Hamiltonian circuit) is a circuit in a given graph $G = (N, A, c)$ that includes each node $n_i \in N$ exactly once. A *traveling salesman tour* is a Hamiltonian tour in G whose length is minimal with respect to c_{ij}.

A problem is called symmetrical if $c_{ij} = c_{ji}$ $\forall$ $n_i, n_j \in N$; otherwise it is asymmetrical. Also, the costs c_{ij} are said to satisfy the *triangle inequality* if and only if $c_{ik} + c_{kj} \geq c_{ij}$ $\forall$ $n_i, n_j, n_k \in N$. Typical scenarios that satisfy the triangle inequality are Euclidean problems, in which N is a set of points in $\mathbb{R}^2$ and c_{ij} are the straight line distances between n_i and n_j. We can prove

Lemma II.9: Let G be a complete graph for which the triangle inequality is satisfied. Then among the minimal length circuits that include each node at least once, there exists at least one traveling salesman tour.

As an illustration, consider the two graphs in Figure II.29.

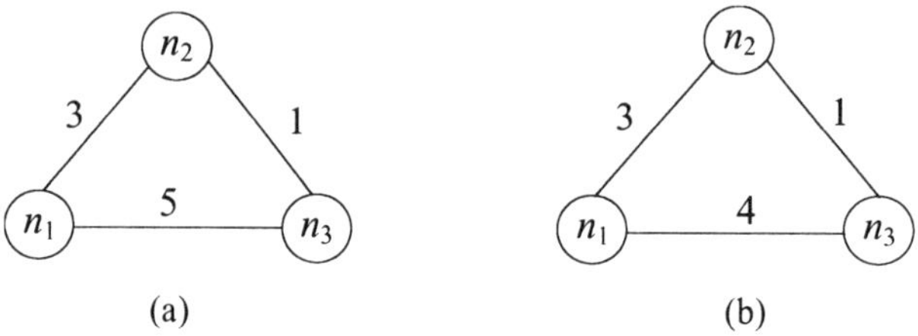

Figure II.29

The graph in Figure II.29a violates the triangle inequality as $c_{12} + c_{23} = 3 + 1 = 4 < 5 = c_{13}$ whereas the graph in Figure II.29b satisfies the triangle inequality. All shortest length circuits in II.29a have length 8, but none of them, e.g., n_1, n_2, n_3, n_2, n_1, is a traveling salesman tour. On the other hand, among the shortest length circuits in Figure II.29b, n_1, n_2, n_3, n_1 is a traveling salesman tour.

3.1.1. Applications

The most common interpretation of the traveling salesman problem is that of a salesman seeking the shortest tour to visit n clients or cities. This basic structure is part of several vehicle routing applications which add a number of side constraints to these problems, and are discussed further in section 4. Several interesting problems not associated with routing are presented below. While applications correspond to asymmetric problems, the other applications have a symmetric cost structure.

Application 1: *Computer wiring*; Lenstra and Rinnooy Kan (1975): Some computer systems can be described as modules with pins attached to them. It is often desired to link these pins by means of wires, so that exactly two wires are attached to each pin and total wire length is minimized.

Application 2: *Wallpaper cutting*; Garfinkel, (1977): Suppose that n sheets must be cut from a roll of wallpaper on which a pattern of length ℓ is repeated. For sheet i, denote by a_i and b_i the starting and ending points on the pattern, where $0 \le a_i \le b_i\, \ell$. Then cutting sheet j immediately after sheet i results in a waste between the end of sheet i and the beginning of sheet j of the amount

$$c_{ij} = \begin{cases} a_j - b_i & \text{if } b_i \le a_j \\ \ell + a_j - b_i & \text{if } b_i \ge a_j \end{cases}. \tag{1}$$

A simple example may illustrate the two cases. First, let the pattern be of length $\ell = 10$, and suppose that the i-th sheet extends from points 1 to 4, while the j-th sheet covers the area between points 7 and 9. Clearly, $b_i = 4 < 7 = a_j$, so that the waste covers the area between points 4 and 7, i.e., $a_j - b_i = 3$ units. This situation is shown in Figure II.30a. Suppose now that the i-th sheet covers the area between 2 to 8, while the j-th pattern extends from points 4 to 7. Now $b_i = 8 > 4 = a_j$. As Figure II.30b demonstrates, the waste consists of the area between the end of the i-th sheet and the end of the pattern, plus the area between the beginning of the pattern and the top of the j-th pattern at a_j, for a total waste of $\ell + a_j - b_i$.

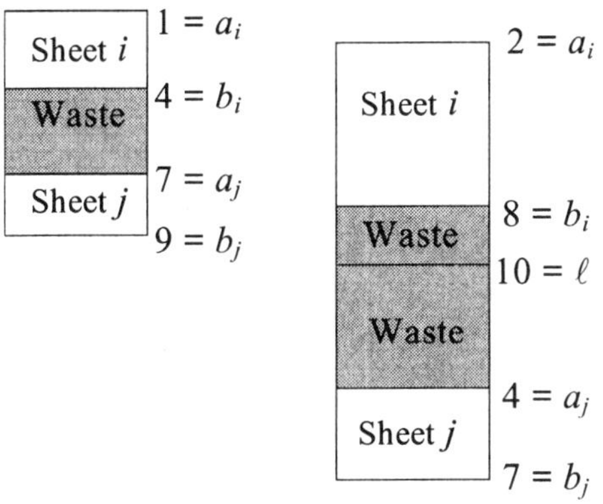

(a) (b)

Figure II.30

The objective is to order the n sheets so as to minimize total waste. In order to formulate the problem as a traveling salesman problem, introduce a dummy sheet $n + 1$ with $c_{i,\, n+1} = 0$ and $c_{n+1,\, i} = 0\ \forall\ i, j = 1, \ldots, n$. Alternatively, define a point b_{n+1} as the end of the roll position at the start of cutting, and assume that after the last sheet a final cut must be made to restore the roll to its original position. Then $a_{n+1} = b_{n+1}$, and c_{ij} can be defined as in (1) for all $i, j = 1, \ldots, n + 1$.

Application 3: *Hole punching*; Reinelt (1992): In several manufacturing contexts, it is necessary to punch holes in boards or sheets of metal so as to minimize the time required to drill the holes and move the drill to the positions at which holes are to be drilled. Such problems frequently occur in the manufacturing of metal sheets and in the construction of circuit boards. These problems are often of large scale and must be solved in real time.

Application 4: *Job sequencing*: Suppose that n jobs are to be performed sequentially on a single machine. Let c_{ij} denote the changeover time if job j is executed immediately after job i; then the task is to determine a sequence that minimizes the total changeover time. The problem can be modeled as a traveling salesman problem by defining a dummy node n^* which is connected to all other nodes of the given graph by arcs (n^*, n_j) and (n_j, n^*) with zero costs. As the changeover costs naturally satisfy the triangle inequality, a traveling salesman tour will solve the problem.

Application 5: *Dartboard design*; Eiselt and Laporte (1991): Dartboards can be visualized as circular targets with concentric circles, and 20 sectors identified by the numbers 1 to 20. Players throw darts at target points on the board. In the most common version of the game, the objective is to reduce an initial value of 301 to zero by subtracting scores. The game rewards accuracy in the throw and it is often more important to hit one's target than to merely register a large score. A natural objective for designing a dartboard is therefore to position the 20 numbers around the board so as to maximize players' risk. For fairly accurate players, it is reasonable to assume that the sector that is hit is always the targeted sector or its neighbor. Let $[\pi(1), ..., \pi(20)]$ be any permutation of the numbers 1, ..., 20. In what follows, $\pi(k)$ must be interpreted as $\pi(k \bmod 20)$ whenever $k < 1$ or $k > 20$. Consider a player aiming at $\pi(k)$ and hitting $\pi(k \pm 1)$ with probability p, and $\pi(k)$ with probability $1 - 2p$. A possible objective is to design a dartboard that maximizes the expected sum of square deviations, i.e., $\sum_{k=1}^{20} \{p[\pi(k-1) - \pi(k)]^2 + p[\pi(k+1) - \pi(k)^2]\}$. Since p is a constant, this problem is equivalent to solving a traveling salesman problem with $c_{ij} = (i-j)^2$.

Application 6: *Crystallography*; Bland and Shallcross (1989): In crystallography, some experiments consist of taking a large number of X-ray intensity measurements on crystals by means of a detector. Each measurement requires that a sample of the crystal be mounted on an apparatus and that the detector be positioned appropriately. The order in which the various measurements on a given crystal are made can be seen as the solution of a traveling salesman problem. In practice, these problems are of large scale and obtaining good solutions can considerably reduce the time needed to carry out all measurements.

Application 7: *Job sequencing in kilns*; Gilmore and Gomory (1964): Assume that $(n-1)$ jobs are to be processed sequentially in a kiln. Job i requires a starting temperature of a_i and must be finished at temperature b_i. Further assume that the initial kiln temperature is a_n and that the final temperature must be b_n. Then the problem can be formulated as a traveling salesman problem with $c_{ij} = \int_{b_i}^{a_i} f(x)dx$, if $b_i \leq a_i$, and $c_{ij} = \int_{a_i}^{b_i} g(x)dx$, if $a_i \leq b_i$ where f and g are cost density functions and $f(x) + g(x) \geq 0$ for all x, since otherwise it would be profitable to keep changing the kiln temperature.

In general, one can prove that the traveling salesman problem is *NP-hard*. However, a few cases exist which can be solved in polynomial time. They include

- Problems with $c_{ij} = 0 \; \forall \; i \geq j$;
- The wallpaper cutting problem in Application 2, and
- The job sequencing problem on kilns of Application 7.

3.1.2 Integer Linear Programming Formulations

A large number of exact and heuristic algorithms have been proposed for the traveling salesman problem. These can be best understood and explained in the context of integer linear programming. In this section, we examine a number of such formulations and algorithms derived from them.

One of the earliest formulations is due to Dantzig, Fulkerson and Johnson (1954). It associates a binary variable x_{ij} to each arc a_{ij}, with $x_{ij} = 1$ if a_{ij} is used on the tour and zero otherwise. The problem in this formulation can then be written as

$$\text{P}_{\text{DFJ}}: \quad \text{Min } z = \sum_{i \neq j} c_{ij} x_{ij}$$

$$\text{s.t.} \quad \sum_{\substack{j=1 \\ j \neq i}}^{n} x_{ij} = 1, \qquad i = 1, \ldots, n \tag{2}$$

$$\sum_{\substack{i=1 \\ i \neq j}}^{n} x_{ij} = 1, \qquad j = 1, \ldots, n \tag{3}$$

$$\sum_{\substack{n_i, n_j \in S \\ i \neq j}} x_{ij} \leq |S| - 1 \quad \forall\, S \subset N;\ 2 \leq |S| \leq n-2 \tag{4}$$

$$x_{ij} = 0 \vee 1 \ \forall\, i, j = 1, \ldots, n;\ i \neq j. \tag{5}$$

In this formulation, the objective function clearly describes the cost of an optimal tour. Constraints (2) and (3) are *degree constraints*: they specify that each node of G is left exactly once (2) and entered exactly once (3). Constraints (4) are *subtour elimination constraints*: they prohibit the formation of *subtours*, i.e., tours on subsets of less than n nodes. If there were such a subtour on a subset S of node indexes, it would contain $|S|$ arcs and as many nodes, thus violating constraint (4) for this subset since its left-hand side would equal $|S|$ and its right-hand side would equal $|S| - 1$. Because of degree constraints, subtours over one node (and hence, over $n - 1$ nodes) cannot occur. Therefore it is valid to define constraints (4) for $2 \leq |S| \leq n - 2$ only. Finally, constraints (5) impose binary conditions on the variables. This formulation of the traveling salesman problem is nothing but an assignment problem with additional subtour elimination constraints.

Defining $\overline{S} = N \setminus S$, and noting that each node n_i of S is the origin of one arc to another node of S or to a node of $\overline{S}$, $|S| = \sum_{\substack{n_i, n_j \in S \\ i \neq j}} x_{ij} + \sum_{n_i \in S} \sum_{n_j \in \overline{S}} x_{ij}$, and an alternative equivalent form of constraints (4) is

$$\sum_{n_i \in S} \sum_{n_j \in \overline{S}} x_{ij} \geq 1 \quad \forall\, S \subset N;\ 2 \leq |S| \leq n-2 \tag{6}$$

The geometric interpretation of the *connectivity constraints* (6) is that in each solution of the traveling salesman problem, there must be at least one arc pointing from S to its complement, so that S cannot be disconnected. This formulation contains n (n – 1) binary variables, $2n$ degree constraints and $2^n - 2n - 2$ subtour elimination constraints. Even for moderate values of n, it would be unrealistic to expect to be able to solve problems with this formulation by means of an integer programming code. The model is usually relaxed and solved by means of specialized algorithms.

Miller, Tucker and Zemlin (1960) have proposed an alternative formulation that reduces the number of subtour elimination constraints at the expense of extra variables u_i, $i = 2, \ldots, n$. Their subtour elimination constraints can be expressed as

$$u_i - u_j + (n-1)x_{ij} \le n-2 \qquad \forall\, i, j = 2, \ldots, n; i \ne j \tag{7}$$

$$1 \le u_i \le n-1 \qquad \forall\, i = 2, \ldots, n. \tag{8}$$

Constraints (7) ensure that the solution contains no subtour on a set of nodes $S \subseteq N \setminus \{n_1\}$ and hence, no subtour involving less than n nodes. Constraints (8) ensure that the u_i variables are uniquely defined for any feasible tour. In order to see how constraints (7) operate, suppose there were a subtour $(n_{i_1}, n_{i_2}, \ldots, n_{i_k}, n_{i_1})$ with $k < n$. Writing constraints (7) for each arc of that subtour and summing up these constraints yields $k\,(n - 1) \le k\,(n - 2)$, which is a contradiction. A proper tour involving all n nodes will, however, be feasible as it will involve node n_1, and hence not be prohibited by the constraints (7) that are only defined for nodes $n_2, \ldots, n_k$. Also note that if an arc a_{ij} is included in the solution, then $x_{ij} = 1$ and constraint (7) becomes $u_i - u_j + n - 1 \le n - 2$ or $u_i \le u_j - 1$, i.e., the two variables assigned to the end points of a_{ij} differ by at least 1. Given that any traveling salesman tour includes n arcs, the difference between the smallest and largest node variables is at least n. Combining this with constraint (8) implies that $u_i = u_j - 1$ for all a_{ij} on the tour.

Desrochers and Laporte (1991) have demonstrated that constraints (7) can be strengthened by introducing an extra term in their left hand-side to yield

$$u_i - u_j + (n-1)\,x_{ij} + (n-3)\,x_{ji} \le n-2 \quad \forall\, i, j = 2, \ldots, n;\ i \ne j. \tag{9}$$

In spite of its relative compactness, Wong (1980) has shown that the formulation with subtour elimination constraints (7) and (8) is weaker than the formulation with the constraints (5) or (6) in the sense of providing a smaller lower bound. Finally, Langevin *et al.* (1990) and Padberg and Sung (1991) have proposed and compared a number of alternative formulations, but none of these seems to have a stronger linear relaxation than the Dantzig *et al.* formulation.

3.2 Exact Algorithms

Many traveling salesman problems are solved by branch and bound algorithms. In the context of mathematical programming, they can best be viewed as initially relaxing some of the problem constraints, and then regaining feasibility through an enumerative process. As usual, the quality of a branch and bound algorithm is directly related to the quality of the bound provided by the relaxation. For the traveling salesman problem, an initial lower bound can be obtained from the Dantzig *et al.* formulation with constraints (2) – (5) by relaxing the subtour elimination constraints (4). The resulting problem is a standard assignment problem which Carpaneto *et al.* (1988) show to be solvable in $O(n^3)$ time. Thus, a valid lower bound on the value of an optimal solution is the bound defined by the constraints of an assignment problem (2) and (3), and the nonnegativity requirements on the variables.

Several authors have proposed branch and bound algorithms for the traveling salesman problem, based on the relaxation of the assignment problem. Early contributions are those by Eastman (1958), Little *et al.* (1963), Shapiro (1966), Murty (1968), Bellmore and Malone (1971), Garfinkel (1973), and Smith *et al.* (1977), followed by Carpaneto and Toth (1980), and Balas and Christofides (1981). A survey by Balas and Toth (1985) indicates that the Balas and Christofides algorithm was then probably the best available, followed by that of Carpaneto and Toth. We have chosen to describe the latter algorithm because of its relative simplicity.

In the Carpaneto and Toth algorithm, the problem solved at each node of the search tree is a modified assignment problem in which some x_{ij} variables are fixed at 0 or at 1. If the solution to this problem consists of a unique tour that covers all nodes, it is feasible for the traveling salesman problem. Otherwise, it consists of a number of subtours. One of these subtours is selected and broken by creating subproblems in which all arcs of the subtour are one at a time prohibited. Denote by z^* the cost of the best solution known so far, and let $\bar{z}_k$ symbolize the value of the objective function at node k of the solution tree. Here, we denote the nodes of the solution tree by their subscripts k = 1, 2,... in order not to confuse nodes k in the solution tree and nodes n_i in the given graph G. The algorithm by Carpaneto and Toth (1980) is initialized with a value of z^* that may be obtained with a heuristic method. Choosing some node selection rule in the branch and bound process and setting and iteration counter k:=1 and s:=1 as the number of subproblems investigated so far. We first solve the relaxed problem with constraints (2), (3), and (5) that provide an objective value $\bar{z}_1$. If $\bar{z}_1 \geq z^*$, the heuristic solution is optimal and the process terminates. Otherwise, node 1 of the solution tree represents the present problem and is the only live node. A simplified version of the algorithm then proceeds as follows.

Carpaneto and Toth Algorithm

Step 1: Does any live node exist in the solution tree?

If yes: According to the chosen node selection rule, choose node k and go to Step 2.

If no: The node representing the best feasible tour is optimal. If no such node exists, then there exists no feasible tour.

Step 2: Does the solution represented by node k include subtours?

If yes: Go to Step 3.

If no: Stop, the solution in node k is optimal.

Step 3 (Branching): Let the subtour with the smallest number of arcs not already fixed at 1 (ties are broken arbitrarily) be $(n_1, n_2, ..., n_r, n_1)$. Branch from node k to nodes $s + 1$, $s + 2$, ..., $s + r$, so that, in addition to the variables fixed earlier, at the node $s + 1$ we set x_{12}:= 0, and at each node $s + \ell$, $\ell = 2, ..., r$, set $x_{\ell,\ell+1}$:= 0, where $x_{r,r+1} = x_{r1}$ (the arc is excluded) as well as $x_{\nu,\nu+1}$:= 1 $\forall$ $\nu = 1, ..., \ell$ (all arcs before the excluded arc will be included).

Step 4: For each node $s + \ell$, solve an assignment problem including the variables preset in Step 3. Let the objective value be denoted by $\bar{z}_{s+\ell}$. If $\bar{z}_{s+\ell} \geq z^*$, disregard the node. If $\bar{z}_{s+\ell} < z^*$ and the solution includes no subtours, set z^*:= $\bar{z}_{s+\ell}$, if it does includes subtours, the node $(s + \ell)$ is live. Set k:= $s + r$ and go to Step 1.

Example: To illustrate the Carpaneto and Toth algorithm, consider the following traveling salesman problem with $n = 10$ and the distance matrix $\mathbf{C} = (c_{ij})$.

$$\mathbf{C} = \begin{bmatrix} - & 9 & 8 & 3 & 5 & 1 & 3 & 3 & 7 & 4 \\ 2 & - & 6 & 1 & 3 & 6 & 1 & 9 & 9 & 7 \\ 9 & 2 & - & 8 & 8 & 5 & 1 & 3 & 7 & 9 \\ 3 & 6 & 9 & - & 5 & 4 & 8 & 9 & 6 & 2 \\ 8 & 5 & 6 & 5 & - & 3 & 7 & 6 & 4 & 5 \\ 9 & 2 & 8 & 2 & 8 & - & 9 & 4 & 3 & 3 \\ 4 & 1 & 3 & 6 & 6 & 5 & - & 4 & 2 & 8 \\ 7 & 7 & 6 & 1 & 1 & 9 & 4 & - & 6 & 9 \\ 1 & 6 & 9 & 6 & 2 & 8 & 6 & 3 & - & 3 \\ 6 & 8 & 2 & 5 & 6 & 7 & 7 & 5 & 7 & - \end{bmatrix}$$

One of the solutions of the assignment problem contains the two subtours (n_6, n_8, n_5, n_9, n_1, n_6) and (n_2, n_4, n_{10}, n_3, n_7, n_2), and has a total cost of 11 + 7 = 18. Branching is done on the first of these subtours and $r = 5$ subproblems are thus created from node 1, with the following sets of included and excluded arcs:

Node 2: $x_{68} = 0$,

Node 3: $x_{85} = 0$; $x_{68} = 1$,

Node 4: $x_{59} = 0$; $x_{68} = x_{85} = 1$,

Node 5: $x_{91} = 0$; $x_{68} = x_{85} = x_{59} = 1$,

Node 6: $x_{16} = 0$; $x_{68} = x_{85} = x_{59} = x_{91} = 1$.

The solutions of the assignment problems have objective values 18, 19, 18, 19 and 19, respectively. They are shown in Figure II.31 below. The process continues by

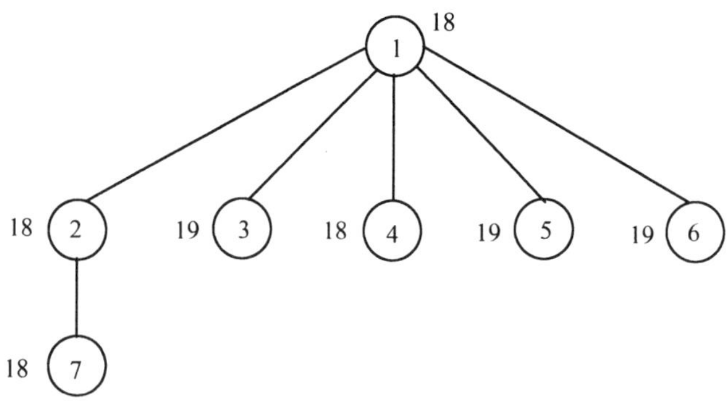

Figure II.31

selecting a live node with the lowest objective value, e.g., node 2. The solution generated at node 2 is (n_7, n_2, n_7) and $(n_1, n_6, n_4, n_{10}, n_3, n_8, n_5, n_9, n_1)$. Branching is therefore donc on the first of these subtours. The first of the two resulting subproblems with the variable fixing of $x_{72} = 0$ (in addition to the fixing of $x_{68} = 0$ at the predecessor node 2) yields the Hamiltonian circuit $(n_1, n_8, n_5, n_6, n_2, n_4, n_{10}, n_3, n_7, n_9, n_1)$ with cost 18. As node 7 is one of the live nodes with the lowest objective value, it is therefore optimal.

Using their algorithm, Carpaneto and Toth have consistently solved randomly generated traveling salesman problems with 240 nodes in less than one minute on a CDC 6600. The main limitation of this algorithm appears to be computer memory rather than CPU time.

Fischetti and Toth (1992) have proposed an alternative which they have embedded in Carpaneto and Toth's (1980) branch and bound algorithm on a variety of randomly generated problems and on some problems described in the literature. The success of the algorithm depends on the type of problem considered. For the easiest problem type, the authors report having solved 2000-node problems in an average time of 8329 seconds on an HP 9000/840 computer.

Balas and Christofides (1981) have proposed an algorithm that uses a stronger relaxation than the relaxation of the assignment problem. Its description and the computational effort required for the lower bound computations are much more involved, but the resulting search trees are smaller and the procedure is overall more powerful. Their procedure solves randomly generated 325-node problems in less than one minute on a CDC 7600 to optimality.

More recently, Miller and Pekny (1991) have developed a branch and bound algorithm based on the relaxation of the assignment problem. It solves a problem on a partial graph that contains only some of the original arcs. A test is then performed so as to whether or not the solution on the partial graph is also optimal for the original problem. If so, the algorithm terminates; otherwise, more arcs are added and the procedure is repeated. The authors have applied this procedure to randomly generated problems. Instances involving up to 5000 nodes were solved within 40 seconds on a Sun 4/330 computer. The largest problem reported solved by this approach contains 500,000 nodes and requires 12,623 seconds of computing time on a Cray 2 supercomputer.

The Carpaneto and Toth algorithm described above was applicable to symmetrical and asymmetrical problems. However, in the case of asymmetrical problems, algorithms based on solutions to assignment subproblems often contain several subtours containing only two nodes and the elimination of these subtours frequently requires excessive computing times. In general, symmetrical problems are better handled by specialized algorithms that operate on an undirected graph. The asymmetrical traveling salesman problem defined on a graph $G = (N, E)$ includes binary variables x_{ij} that equal 1 if and only if edge e_{ij} is used in the optimal solution. These variables are only defined for $i < j$. The formulation of the problem is then

$$\text{P}_{\text{SYM}}: \quad \text{Min } z = \sum_{i<j} c_{ij} x_{ij} \tag{10}$$

$$\text{s.t.} \quad \sum_{i<k} x_{ik} + \sum_{j>k} x_{kj} = 2 \qquad \forall\ k = 1,\ldots, n \tag{11}$$

$$\sum_{\substack{n_i, n_j \in S \\ i<j}} x_{ij} \le |S| - 1 \quad \forall\ S \subset N;\ 3 \le |S| \le n-3 \tag{12}$$

$$x_{ij} = 0 \vee 1 \quad \forall\ i, j = 1,\ldots, n;\ i < j \tag{13}$$

In this formulation, constraints (11) specify that each node has a degree of 2. Constraints (12) are subtour elimination constraints that need not be defined for $|S| = 2$ and $n - 2$ since constraints (11) and (13) combined prevent the formation of subtours involving 2 (and thus $n - 2$) nodes. Finally, constraints (13) impose binary conditions on the variables.

Similar to the Dantzig *et al.* formulation, connectivity constraints equivalent to (12) can be derived. These can be written as

$$\sum_{\substack{n_i \in S, n_j \in \overline{S} \\ \text{or } n_j \in S, n_i \in \overline{S}}} x_{ij} \ge 2 \qquad \forall\ S \subset N;\ 3 \le |S| \le n-3 \tag{14}$$

In order to show the equivalence of (12) and (14), consider $D(S) = \sum_{n_k \in S} \left(\sum_{i<k} x_{ik} + \sum_{j>k} x_{kj} \right)$, the sum of degrees of nodes of S. Each edge e_{ij} for which $n_i, n_j \in S$ makes a contribution of 2 to $D(S)$, whereas edges e_{ij} with $n_i \in S$, $n_j \in \overline{S}$ or $n_j \in S$, $n_i \in \overline{S}$, contribute by only one unit. Hence, $D(S) = 2 \sum_{\substack{n_i, n_j \in S \\ i>j}} x_{ij} + \sum_{\substack{n_i \in S, n_j \in \overline{S} \\ \text{or } n_j \in S, n_i \in \overline{S}}} x_{ij}$, and the equivalence follows.

A solution algorithm based on the lower bounds obtained by 2-matching relaxation, i.e., the problem formulation P_{SYM} with the constraints (12) omitted, can then be devised. The principles behind such a method were first laid out by Dantzig *et al.* (1954, 1959), followed by Martin (1966), Miliotis (1976, 1978) and Land (1979). Based on this body of work, it was Miliotis who developed the first completely automatic procedure for solving symmetrical traveling salesman problems using this relaxation. He proposed several algorithms that differ in the order in which the violated constraints are introduced and in the way integer solutions are reached. One of these is a pure cutting plane algorithm.

Miliotis's Algorithm

Step 1: Solve the relaxed problem of P_{SYM} with the subtour elimination constraints (12) deleted and with the zero-one conditions replaced by $x_{ij} \in [0; 1] \ \forall \ e_{ij} \in E$. Let $\bar{\mathbf{x}}$ denote the solution.

Step 2: Is $\bar{\mathbf{x}} \in \mathbb{N}_0$?
If yes: Go to Step 3.
If no: Go to Step 4.

Step 3: Does $\bar{\mathbf{x}}$ include any subtours?
If yes: Go to Step 5.
If no: Stop; $\bar{\mathbf{x}}$ is an optimal solution of P_{SYM}.

Step 4: Use Gomory cutting planes to find an integer solution to the problem. Let $\bar{\mathbf{x}}$ be the new solution and go to Step 3.

Step 5: Eliminate one or several subtours by introducing the appropriate subtour elimination constraints and reoptimize. Result: a new solution $\bar{\mathbf{x}}$. Go to Step 2.

A variant of this problem consists of reaching integrality in Step 4 by a branch and bound method. Subtour elimination constraints are then introduced in individual subproblems through the branching process. These constraints remain, however, valid for the whole of the branch and bound tree.

As indicated by Miliotis, the two steps that consist of checking for integrality and then for violated subtour elimination constraints can be switched, leading to a stronger algorithm. It must be observed that subtour elimination constraints can be applied to fractional components, i.e., connected subgraphs having fractional x_{ij} variables associated with some of their edges. Efficient procedures for generating violated subtour elimination constraints have been suggested in the literature. The basic idea behind this line of research consists of introducing several types of valid constraints before branching on fractional variables in order. To increase the value of the linear programming relaxation and, thus, to limit the growth of the search tree. These constraints are particularly powerful since they are facets of the polytope of integer solutions of P_{SYM}. Following Padberg and Rinaldi (1990), the most commonly used constraints are

- Subtour elimination constraints

$$\sum_{n_i, n_j \in S} x_{ij} \leq |S| - 1 \quad \forall \quad S \subset N;\ 3 \leq |S| \leq n - 3 \text{ and} \tag{15}$$

- 2-matching inequalities

$$\sum_{n_i, n_j \in H} x_{ij} + \sum_{e_{ij} \in E'} x_{ij} \leq |H| + ½(|E'| - 1) \tag{16}$$

for all $H \subset N$ and all $E' \subset E$ satisfying
(i) $|\{n_i, n_j\} \cap H| = 1\ \forall\ e_{ij} \in E'$;
(ii) $|\{n_i, n_j\} \cap \{n_k, n_\ell\}| = 0,\ \forall\ e_{ij} \neq e_{k\ell} \in E'$;
(iii) $|E'| \geq 3$ and odd.
The set H is called the *handle* and the edges of $E' = \{e_{i_1 j_1}, \ldots, e_{i_s j_s}\}$ are called *teeth*. A handle and teeth are shown in Figure II.34.

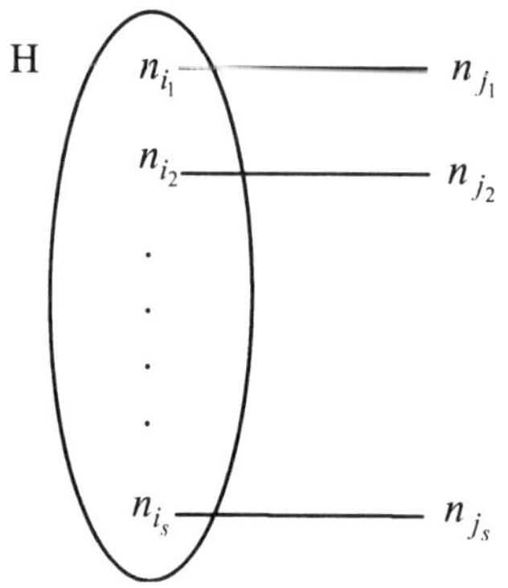

Figure II.34

- Comb inequalities

$$\sum_{n_i, n_j \in H} x_{ij} + \sum_{k=1}^{s} \sum_{n_i, n_j \in T_k} x_{ij} \leq |H| + \sum_{k=1}^{s} (|T_k| - 1) - ½(s - 1) \tag{17}$$

for all $H, T_1, \ldots, T_s \subseteq N$ satisfying
(i) $|T_k \cap H| \geq 1 \quad \forall \quad k = 1, \ldots, s$;
(ii) $|T_k \setminus H| \geq 1 \quad \forall \quad k = 1, \ldots, s$;
(iii) $T_k \cap T_\ell = 0\ \forall \quad 1 \leq k < \ell \leq s$;
(iv) $s \geq 3$ and odd.

Other valid inequalities, such as clique tree inequalities have been developed.

Padberg and Rinaldi (1990, 1991) suggest efficient procedures for identifying violated instances of these three types of constraints and of several other families of inequalities. By incorporating these procedures within an integer programming code, the authors have solved to optimality traveling salesman problems with between 51 and 2392 nodes. Out of the 26 problems that were attempted, 10 required no branching at all.

3.3 Heuristic Algorithms

Since the traveling salesman problem is *NP*-hard, it is natural to approach it by means of heuristic algorithms. One type of heuristic are methods with a guaranteed *worst-case* performance. In the following discussion, we assume that the given problem is symmetric and that the triangle inequality holds. A simple method with a guaranteed worst-case performance was put forward by Christofides (1976). First, a tour is constructed by determining the minimal spanning tree of the given graph, and then a nonelementary circuit is determined that traverses each edge exactly twice. Denote this circuit by "2MST" and its length by c(2MST). Similarly, MST denotes the minimal spanning tree and c(MST) is its length. It is apparent that c(2MST) = 2c(MST). A tour TSP is then constructed form the nonelementary circuit 2MST by repeatedly replacing subpaths (n_i, n_j, n_k) where n_j is visited more than once by shortcuts (n_i, n_k) until a feasible tour is found. As the triangle inequality holds, $c(\text{TSP}) \leq c(\text{MST}) = 2c(\text{MST})$, so that the TSP tour found with this method is no longer than twice the length of the optimal traveling salesman tour, i.e., a 100% error bound. For directed problems, no methods with guaranteed error bounds are known.

Most heuristics for traveling salesman problems have been designed so as to provide good solutions in the average case. Basically, heuristics fall into three categories:

- in case a tour must be constructed from scratch, there are *construction heuristics*,
- in case a tour is known but may be improved upon there are *improvement heuristics*, and
- in those cases in which a solution with (illegal) subtours is known, these subtours will be merged by *merge heuristics*.

Some of the more successful methods merge features from at least two classes of heuristics.

First consider construction heuristics.

The *nearest neighbor algorithm* due to Rosenkrantz *et al.* (1977) constructs a feasible tour by taking at each step the decision that is the most advantageous in the next step. The main advantage of this myopic Greedy-type algorithm lies in its simplicity. The method is initialized with a starting node n_i, a node n_j that is presently the last node on the tour, and a set N^* of nodes that are already connected by the tour. Initially, $N^* = \{n_i\}$ and $n_j = n_i$.

The Nearest Neighbor Algorithm

Step 1: Find a node n_ℓ, so that $c_{j\ell} = \min\limits_{n_\nu \notin N^*} \{c_{j\nu}\}$. Set $N^* := N \cup \{n_\ell\}$ and $n_j := n_\ell$.

Step 2: Is $N^* = N$?
If yes: Add a_{ji} to the existing tour and stop.
If no: Go to Step 1.

The complexity of this procedure is $O(n^2)$. A possible modification is to consider in turn all n nodes as a starting point. The overall algorithm complexity is then $O(n^3)$, but the resulting tour is generally better.

Example: Consider the 10-city problem introduced in Section 3.2 and assume that it was decided to start with node n_1. The closest successor of n_1 is n_6 as $c_{16} = 1$. The successors closest to n_6 are n_2 and n_4 with $c_{62} = c_{64} = 2$. Arbitrarily choosing the former, the method proceeds and finds the tour (n_1, n_6, n_2, n_4, n_{10}, n_3, n_7, n_9, n_5, n_8, n_1) whose total length is $c_{16} + c_{62} + \ldots + c_{81} = 26$. Notice that at n_1 we have $(n - 1) = 9$ choices or degrees of freedom of which node to choose next. Each time a choice is made, the number of degrees of freedom decreases by 1 until, in the very last step (here at n_8), there is no degree of freedom left and we must return to n_1 via the arc a_{81}. As this arc can be arbitrarily long, the performance of this method can be arbitrarily poor. Repeating to apply the nearest neighbor method with different starting nodes can help to find better solutions. In this example, the tour starting at n_2 has length 20, while the tour that starts at n_6 has a length of 18; the latter tour happens to be optimal.

Another type of construction method are the *Insertion algorithms* by Rosenkrantz *et al.* (1977), Stewart (1977), Norback and Love (1977, 1979), Or (1976), and others.

A generic insertion algorithm is initialized with an arbitrary subtour (n_i, n_j, n_i) with N^* denoting the set of nodes already on the tour; initially $N^* = \{n_i, n_j\}$. Also, the current tour includes the set of arcs A^*, which is initialized as $A^* = \{a_{ij}, a_{ji}\}$.

An Insertion Algorithm

Step 1: Does the existing tour include all nodes?
If yes: Stop; a tour has been found.
If no: Go to Step 2.

Step 2: Find n_k, so that $n_k = \arg\min\limits_{\substack{n_\ell \notin N^* \\ a_{ij} \in A^*}} \{ c_{i\ell}, + c_{\ell j} - c_{ij} \}$. Set $N^* := N \cup \{n_k\}$, $A^* := A \cup \{a_{ik}, a_{kj}\}$ and go to Step 1.

As the nearest neighbor algorithm, the insertion method could be initialized with any pair $(n_i, n_j\}$ and could be repeated for all pairs, so that the best resulting tour can be chosen. Criteria other than the smallest net increase of the length of the tour, as in the above Step 2 (again, a Greedy-type criterion), could be employed. Depending on the choice of criterion, the complexity of the Insertion Algorithm varies between $O(n^2)$ and $O(n \log n)$.

Example: Consider again the 10-city problem introduced above. Let (n_1, n_4, n_1) by the initial tour, so that $N^* = \{n_1, n_4\}$ and $A^* = \{a_{14}, a_{41}\}$. All nodes in $N \setminus N^*$ are eligible for insertion into the tour. Inserting n_2 between n_1 and n_4 would result in the tour (n_1, n_2, n_4, n_1) and the net increase in the length of the tour is $c_{12} + c_{24} - c_{14} = 9 + 1 - 3 = 7$; similarly, the increase of n_2 between n_4 and n_1 would result in the tour (n_1, n_4, n_2, n_1) with a net increase of $c_{42} + c_{21} - c_{41} = 6 + 2 - 3 = 5$. Repeating this process for n_3, n_5, n_6, n_7, n_8, n_9, and n_{10}, and choosing the best alternative results in the three-node tour (n_1, n_6, n_4, n_1) with a net increase of zero. This process is now repeated. In the next step, one of the elegible nodes n_2, n_3, n_5, n_7, n_8, n_9, n_{10} will be inserted between n_1 and n_6, n_6 and n_4, or n_4 and n_1. It is apparent that, as the algorithm progresses, the number of nodes that may be inserted decreases by one in each iteration, while the number of positions in which the new node may be inserted increases by one.

A third type of construction heuristic is the *patching algorithm* by Karp (1979) designed for asymmetrical traveling salesman problems. The algorithm starts with a solution that involves at least two subtours and merges them. In particular, it selects an arc a_{ij} on the first subtour and an arc $a_{k\ell}$ on the second subtour such that the cost $c_{i\ell} + c_{kj} - c_{ij} - c_{k\ell}$ of merging the two circuits is minimized. Further details can be found in Balas and Toth (1985).

Consider now tour improvement heuristics. A popular method is the r-Opt algorithm by Lin (1965). It starts with an arbitrary tour and can be described as follows.

Lin's r-Opt Algorithm

Step 1: Remove r non-incident arcs from the tour and tentatively reconnect the resulting r chains in all possible ways.

Step 2: Does any reconnection yield a shorter tour?
If yes: Consider this tour as a new initial solution and go to Step 1.
If no: Stop, a local optimum has been found.

This heuristic was originally devised for symmetrical traveling salesman problems. In this case, the number of candidate solutions in each iteration is of the order of n^r

since there are $\binom{n}{r}$ ways to remove arcs and $r!$ ways to reconnect the resulting undirected chains. However, since the arcs that are removed cannot share a node, not all these reconnections are feasible. In general, r is taken as 2 or 3. The algorithm may be adapted for use with directed graphs as well. To demonstrate what is involved in the application of the method, consider an undirected graph, let $r = 2$, and consider a tour $(n_1, ..., n_i, n_j, ..., n_k, n_\ell, ..., n_1)$. Removing the two edges (n_i, n_j) and (n_k, n_ℓ) results in the two paths $(n_\ell, ..., n_i)$ and $(n_j, ..., n_k)$. These two paths may be reconnected by introducing the edges (n_i, n_k) and (n_j, n_ℓ), in which case the new tour is $(n_\ell, ..., n_i, n_k, ..., n_j, n_\ell, ..., n_i)$ The cost chang is $- c_{ij} - c_{k\ell} + c_{ik} + c_{j\ell}$.

The case of $r = 3$ allows reconnection of the pieces also in the case of directed graphs. Given an original tour $(n_1, ..., n_i, n_j, ..., n_k, n_\ell, ..., n_s, n_t, ..., n_1)$ and deleting the three arcs/edges (n_i, n_j), (n_k, n_ℓ), and (n_s, n_t), the resulting three paths $(n_t, ..., n_i)$, $(n_j, ..., n_k)$, and $(n_\ell, ..., n_s)$ can be reconnected by introducing the arcs/edges (n_i, n_ℓ), (n_s, n_j), and (n_k, n_t), resulting in the new tour $(n_t, ..., n_i, n_\ell, ..., n_s, n_j, ..., n_k, n_t)$ with a cost change of $-c_{ij} - c_{k\ell} + c_{st} + c_{i\ell} + c_{sj} + c_{kt}$. In the 10-city example above, removing the arcs a_{23}, a_{56}, and a_{89} from the tour $(n_1, n_2, ..., n_{10})$ results in the paths (n_3, n_4, n_5), (n_6, n_7, n_8), and (n_9, n_{10}, n_1, n_2). These paths are reconnected to a tour by introducing the arcs a_{26}, a_{59}, and a_{83}, resulting in the tour $(n_3, n_4, n_5, n_9, n_{10}, n_1, n_2, n_6, n_7, n_8, n_3)$ with a cost change of $- 6 - 3 - 6 + 6 + 4 + 6 = +1$, so that the exchange results in a cost increase and will not be made.

Or (1976) has proposed a simplified exchange procedure requiring only $O(n^2)$ operations at each step, but producing tours nearly as good on average as those obtained with a 3-opt algorithm. Or's algorithm starts again with a given tour, removes from the tour a chain of s consecutive nodes, starting with the node in position t, and tentatively insert it between all remaining pairs of consecutive nodes on the tour. Whenever the tentative insertion decreases the cost of the tour, the insertion is implemented.

Other tour improvement methods are based on the general principle of Tabu Search. For details regarding the general method, readers are referred to Chapter 7 of Part I. As applied to the traveling salesman problem, Tabu Search examines successive neighbors $\mathbf{x}'$ of a solution $\mathbf{x}$ where the objective is allowed to deteriorate in a move in order to avoid local minima. To prevent cycling, solutions that have already been examined are forbidden and inserted in a constantly updated "tabu list".

Composite algorithms combine features of construction and improvement methods. Several composite algorithms have been analyzed by Golden and Stewart

(1985). More recently, another composite, named GENIUS, was devised by Gendreau *et al.* (1990). For details, readers are referred to the original contributions.

3.4 Vehicle Routing Problems

The objective of Vehicle Routing Problems (*VRP*) is to design optimal delivery or collection routes from one or several depots to a number of geographically scattered cities or customers, subject to side constraints. Vehicle routing problems exist in a wide variety of contexts and play a central role in the fields of physical distribution and logistics; see, e.g., the surveys of Bodin *et al.* (1983), Christofides (1985a), Laporte and Nobert (1987), Laporte (1990), as well as the classification scheme proposed by Desrochers *et al.* (1990).

Formally let $G = (N, A)$ be a graph in which the set of nodes represents cities with the *depot* located at node n_1. With each arc a_{ij} is associated a nonnegative *value* c_{ij} equal to the shortest distance from n_i to n_j. In some contexts, c_{ij} can be interpreted as a *travel cost* or as a *travel time*. In case of undirected problems, the arcs a_{ij} are replaced by edges e_{ij}. In addition, assume there are m vehicles available at the depot, where $\underline{m} \leq m \leq \overline{m}$. When $\underline{m} = \overline{m}$, m is said to be fixed, and when $\underline{m} = 1$ and $\overline{m} = n - 1$, m is said to be *free*. When m is not fixed, it often makes sense to associate a fixed cost f on the use of a vehicle. For the sake of simplicity, we will ignore these costs and, unless otherwise specified, assume that all vehicles are identical and have the same capacity κ. The vehicle routing problem consists of designing a set of least cost vehicle routes in such a way that

- each city in $N \setminus \{n_1\}$ is visited exactly once by exactly one vehicle,
- all vehicle routes start and end at the depot, and
- some side constraints are satisfied.

Note that since c_{ij} is defined as a shortest distance, the first condition is not restrictive in the sense that each customer will be served exactly once, although several routes may include any given city.

Without any additional side constraints, the problem is identical to an m-TSP, which minimizes the length of m tours, all including the depot n_1, while covering all nodes of G. The most common side conditions include restrictions related to the following five categories.

(1) *Capacity restrictions.* A nonnegative weight (or demand) d_i is attached to each city n_i, $i > 1$ and the sum of weights of any vehicle route may not exceed the vehicle capacity κ. *Capacity constrained vehicle routing problems* are referred to as *CVRPs*.

(2) The number of cities on any route is bounded above by q. Note that this is a special case of capacity constraints in (1) with $d_i = 1 \ \forall\ i > 1$ and $\kappa = q$.
(3) *Total time restrictions.* The length of any route may not exceed a prescribed bound L; this length is made up of intercity travel times c_{ij} and of stopping times δ_i at each city n_i on the route. Time or *distance constrained vehicle routing problems* will be referred to as *DVRPs*.
(4) *Time windows.* City n_i must be visited within the time interval $[a_i, b_i]$.
(5) *Precedence relations* between pairs of cities: city n_i may have to be visited before city n_j.

This list is by no means exhaustive. A number of other interesting and meaningful variants are described by Golden and Assad (1988). Here, we will concentrate on capacity- and distance-constrained vehicle routing problems. To simplify the exposition, we assume that the graph is directed and that arcs a_{ij} exist for all pairs $n_i, n_j \in N$. In order to formulate the problem, define again zero-one variables $x_{ij}\ \forall\ i, j$ which assume a value of 1, if the arc/edge is included in the optimal solution, and 0 otherwise. In case of a capacity-constrained vehicle routing problem, if $d_i + d_j > \kappa$ for any pair of nodes n_i, n_j, then a variable x_{ij} need not be defined. The problem can then be written as

$$P_{\text{VRP1}}: \text{ Min } z = \sum_{i \neq j} c_{ij} x_{ij} \tag{18}$$

$$\text{s.t.} \quad \sum_{\substack{j=1 \\ i \neq j}}^{n} x_{ij} = 1 \qquad \forall\ i = 1, \ldots, n \tag{19}$$

$$\sum_{\substack{i=1 \\ i \neq j}}^{n} x_{ij} = 1 \qquad \forall\ j = 1, \ldots, n \tag{20}$$

$$\sum_{n_i, n_j \in S} x_{ij} \leq |S| - v(S) \qquad \forall\ S \subset N \setminus \{n_1\}; |S| \geq 2 \tag{21}$$

$$x_{ij} = 0 \vee 1 \qquad \forall\ i, j = 1, \ldots, n;\ i \neq j. \tag{22}$$

As for the traveling salesman problem, the objective function of this formulation and constraints (19), (20) and (22) define a modified assignment problem in which assignments on the main diagonal are prohibited. Constraints (21) are subtour elimination constraints in which $v(S)$ is an appropriate lower bound on the number of vehicles required to visit all vertices of S in the optimal solution. These constraints are obtained by observing that for any $S \subset N \setminus \{n_1\}$, $|S| \geq 2$. Defining $\bar{S}$ as $N \setminus S$, we must have

$$\sum_{n_i \in S} \sum_{n_j \in \bar{S}} x_{ij} \geq v(S)$$

Due to constraints (19), each node in S has an outdegree of 1, so that we obtain

$$|S| = \sum_{n_i, n_j \in S} x_{ij} + \sum_{n_i \in S} \sum_{n_j \in \bar{S}} x_{ij}$$

The value of $v(S)$ depends on the type of vehicle routing problem under consideration. In case of capacity constraints, it is valid to take

$$v(S) = \left\lceil \frac{1}{\kappa} \sum_{n_i \in S} d_i \right\rceil$$

The subtour elimination constraint corresponding to a set of nodes $S = \{n_1, n_2, n_3\}$ with demands $d_1 = 3$, $d_2 = 9$, and $d_3 = 4$, and capacity $\kappa = 10$ is then

$$x_{12} + x_{13} + x_{21} + x_{23} + x_{31} + x_{32} \le 3 - \left\lceil \frac{3+9+4}{10} \right\rceil = 1.$$

In case of distance constraints, $v(S)$ is not as easy to determine *a priori*. Usually, however, a lower bound on its value can easily be determined during the course of a branch and bound process. Failing this, it is always valid to choose $v(S) = 1$. It is worth observing that constraints (21) play a dual role: one the one hand, they ensure that all vehicle routes satisfy the capacity or maximum length restrictions, and on the other, they also guarantee that the solution contains no subtour disconnected from the depot, since every subset S of $N \setminus \{n_1\}$ will be linked to its complement.

Following the Laporte and Nobert (1987) survey, exact algorithms for the vehicle routing problems can be classified into three broad categories: direct tree search methods, dynamic programming techniques, and integer linear programming algorithms.

The first algorithm to be discussed here is due to Laporte *et al.* (1986). It exploits the relationship between the vehicle routing problem and one of its relaxations, the m-TSP which itself can be transformed into a 1-TSP. Given the graph $G = (N, A)$ with a depot at node n_1 and m vehicles based at the depot, the m-TSP consists of establishing m least cost vehicle routes starting and ending at the depot in such a way that each remaining node is visited exactly once. Given an upper bound $\overline{m}$ on m, the m-TSP can be transformed into a 1-TSP by introducing $\overline{m} - 1$ artificial depots; letting $n' = n + \overline{m} - 1$, $N' = \{n_1, ..., n_{n'}\}$ and $A' = A \cup \{a_{ij}: n_i, n_j \in N', i \neq j;$ n_i or $n_j \in N' \setminus N\}$. The distances c'_{ij} between nodes in N' are

$$c'_{ij} = \begin{cases} c_{ij} & \text{if } n_i, n_j \in N \\ c_{i1} & \text{if } n_i \in N \setminus \{n_1\}, n_j \in N' \setminus N \\ c_{1j} & \text{if } n_i \in N' \setminus N, n_j \in N \setminus \{n_1\} \\ \gamma & \text{if } n_i, n_j \in (N' \setminus N) \cup \{n_1\} \end{cases} \tag{23}$$

where the value of γ depends on the variant of the problem considered. If the parameter γ is set equal to ∞, the solution includes the minimum distance for $\overline{m}$

vehicles; $\gamma = 0$ yields the minimum distance for at most $\overline{m}$ vehicles, and $\gamma = -\infty$ yields the minimum distance for the minimum number of vehicles.

To illustrate this transformation, consider an n = 7-city problem, with $\overline{m}$ = 3 vehicles and the following distance matrix **C**, where $c_{ii} = \infty$ indicates that arcs a_{ii} do not exist and hence cannot be included in any feasible solution.

$$\mathbf{C} = \begin{bmatrix} \infty & 4 & 6 & 9 & 2 & 3 & 4 \\ 1 & \infty & 5 & 7 & 2 & 6 & 2 \\ 3 & 8 & \infty & 7 & 5 & 1 & 3 \\ 4 & 9 & 6 & \infty & 4 & 5 & 6 \\ 2 & 4 & 1 & 8 & \infty & 7 & 2 \\ 6 & 6 & 1 & 9 & 5 & \infty & 3 \\ 4 & 2 & 6 & 4 & 9 & 8 & \infty \end{bmatrix}$$

The transformation yields the following 9×9 distance matrix

$$\mathbf{C}' = \left[\begin{array}{ccccccc:cc} \infty & 4 & 6 & 9 & 2 & 3 & 4 & \infty & \infty \\ 1 & \infty & 5 & 7 & 2 & 6 & 2 & 1 & 1 \\ 3 & 8 & \infty & 7 & 5 & 1 & 3 & 3 & 3 \\ 4 & 9 & 6 & \infty & 4 & 5 & 6 & 4 & 4 \\ 2 & 4 & 1 & 8 & \infty & 7 & 2 & 2 & 2 \\ 6 & 6 & 1 & 9 & 5 & \infty & 3 & 6 & 6 \\ 4 & 2 & 6 & 4 & 9 & 8 & \infty & 4 & 4 \\ \hdashline \infty & 4 & 6 & 9 & 2 & 3 & 4 & \infty & \infty \\ \infty & 4 & 6 & 9 & 2 & 3 & 4 & \infty & \infty \end{array}\right]$$

A natural algorithm now presents itself. Like the assignment based algorithm for the traveling salesman problem, the capacitated or distance-constrained vehicle routing problem is solved through a branch and bound process in which subproblems are assignment problems. The only difference lies in the definition of illegal subtours. These now include

- subtours over a set S of nodes of $N \setminus \{n_1\}$, and
- vehicle routes violating capacity or maximum length restrictions. These forbidden routes consist of paths $(n_{i_1}, n_{i_2}, \ldots, n_{i_r})$, where $n_{i_\ell}, n_{i_r} \in \{n_1, n_{n+1}, \ldots, n_{n+\overline{m}-1}\}$, $n_{i_2}, \ldots, n_{i_{r-1}} \in N \setminus \{n_1\}$ and $\sum_{t=2}^{r-1} d_{i_t} > \kappa$ or $\sum_{t=1}^{r-1} c_{i_t, i_{t+1}} > L$.

These routes can be eliminated by partitioning the current infeasible subproblem, as in the Carpaneto and Toth algorithm for the traveling salesman problem, taking $v(S) = 1$. More sophisticated partitioning schemes for $v(S) \geq 1$ are detailed in Laporte *et al.* (1986).

Suppose now that a solution to the transformed problem has been found. It is then required to transform this solution back to the original problem. This is done by applying the following rules. Given that $x_{ij} = 1$, i.e., the arc a_{ij} belongs to the solution, then

- if $n_i \in N \setminus \{n_1\}$ and $n_j \in N' \setminus N$, replace a_{ij} by a_{i1}
- if $n_i \in N' \setminus N$ and $n_j \in N \setminus \{n_1\}$, replace a_{ij} by a_{1j}, and
- if $n_i, n_j \in N' \setminus N$, delete a_{ij}.

This process is illustrated in Figure II.33 where $n = 7$ and $\overline{m} = 3$. The transformed solution consists of the single tour $(n_1, n_5, n_8, n_3, n_7, n_4, n_6, n_2, n_9, n_1)$. It can be interpreted as two vehicle routes (n_1, n_5, n_1), and $(n_1, n_3, n_7, n_4, n_6, n_2, n_1)$.

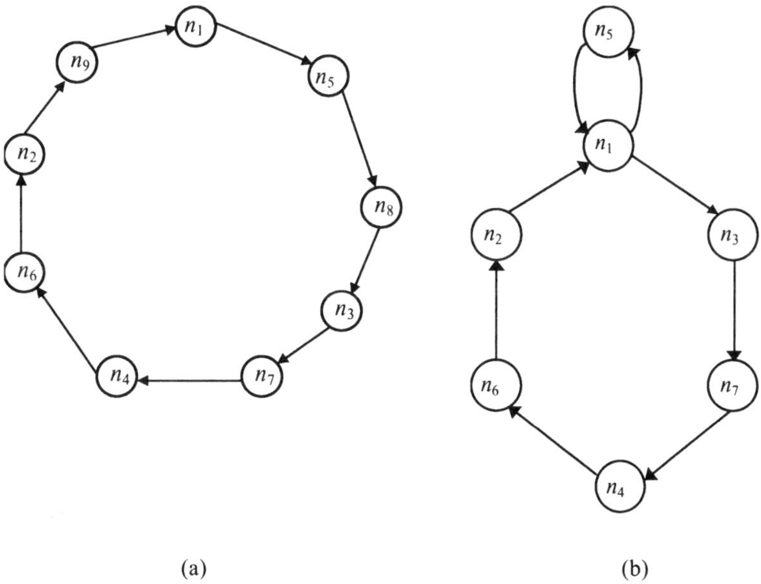

Figure II.33

Using this methodology, Laporte *et al.* (1986) have solved to optimality asymmetrical randomly generated capacity-constrained vehicle routing problems involving up to 260 vertices.

Dynamic programming was first proposed for vehicle routing problems by Eilon *et al.* (1971). With this approach, Christofides *et al.* (1981) obtained lower bounds on optimal solutions for 10 problems containing 10 to 25 nodes. The ratio of lower bound to optimum varied between 93.1% to 100%. Christofides (1985) reported that capacity-constrained vehicle routing problems of up to 50 nodes could be solved systematically with this approach.

An entirely different approach was suggested by Balinski and Quandt (1964) who suggested a set partitioning formulation for vehicle routing problems. The idea is to first identify clusters $S_1, S_2, ..., S_r$ where $\emptyset \subsetneq S_j \subsetneq N$ and one vehicle serves one cluster. Clearly, r is exponential in the number of nodes. Consider now any cluster S_j. While S_j contains many routes—in principle, all permutations of the nodes in S_j are possible—some of these routes may not be feasible due to their length (in case of a distance-constrained problem) or their capacity (in case of a capacity-constrained problem). Let c_j^* denote the length of the shortest feasible route of the nodes in the cluster S_j. Note that the determination of c_j^* requires the solution of a traveling salesman problem.

Define now zero-one variables x_j which are one if a vehicle serves cluster S_j on an optimal route, and let $a_{ij} = 1$ if node n_i is on the optimal route in cluster S_j, and 0 otherwise. The vehicle routing problem can then be formulated as

$$\text{P}_{\text{VRP2}}: \quad \text{Min } z = \sum_{j=1}^{r} c_j^* x_j \tag{24}$$

$$\text{s.t.} \quad \sum_{j=1}^{r} a_{ij} x_j = 1 \quad \forall\, n_i \in N \backslash \{n_1\} \tag{25}$$

$$x_j = 0 \vee 1 \;\; \forall\, j = 1, ..., r \tag{26}$$

Note that there is no need to explicitly include capacity or distance constraints in the formulation; the tour associated with cluster S_j is by assumption feasible with respect to any constraint that may be imposed. Other than the fact that traveling salesman tours must be determined for each of the exponentially many clusters, the main difficulty with this formulation is the large number of binary variables x_j that can run into the millions in most real-life cases. Only in extremely constrained problems i.e., in problems with very few feasible solutions will the number of variables be small enough to enable the problem to be solved directly.

However, Toregas and Revelle (1972) have demonstrated that if the number of variables is relatively small and the objective is to minimize the number of vehicles, i.e., $c_j^* = 1 \; \forall \, j = i, \ldots, r$, the linear relaxation of P_{VRP2} often provides an integer solution. This property is often referred to as "integer friendliness" and occurs in a number of problems related to location models. If the solution $\bar{\mathbf{x}}$ is noninteger and gives a fractional objective value, then the cutting plane $\sum_{j=1}^{r} x_j > \left\lceil \sum_{j=1}^{r} \bar{x}_j \right\rceil$ can be introduced. Orloff (1976) has demonstrated that very few cuts are generally required to reach integrality.

A natural way around the difficulties just mentioned is to use a column generation. This relaxation technique has been applied to the field of vehicle routing by Rao and Zionts (1968), Foster and Ryan (1976), Agarwal *et al.* (1989), and Desrochers *et al.* (1992). In column generation, a reduced problem containing only a restricted subset of all possible columns, i.e., variables, is repeatedly solved. The linear relaxation of the reduced problem provides an optimal dual variable vector λ. Checking for optimality implies computing the column s of least marginal cost, i.e., determining the column s satisfying

$$c_s^* - \lambda \, \mathbf{a}'_{\bullet s} = \min_{j \in J} \{ c_j^* - \lambda \, \mathbf{a}'_{\bullet j} \}, \tag{27}$$

where $\mathbf{a}'_{\bullet j}$ is the column vector of constraint coefficients of variable x_j. If the marginal cost of x_s is nonnegative, the current solution is optimal and the procedure terminates. Otherwise, x_s enters the basis and the problem is reoptimized. Since all feasible solutions of value routing problems must be integer, this procedure must be used in conjunction with a branch and bound algorithm. Solving (27) is done using an algorithm for a shortest path problem constrained in the same manner as the original vehicle routing problem. This procedure has been applied with success by Desrosiers *et al.* (1984) and Desrochers *et al.* (1991) to the solution of vehicle routing problems with time windows containing up to 100 nodes. As expected, the method performs better on tightly constrained problems as the number of feasible columns is then smaller.

Other exact approaches include a three-index formulation by Fisher and Jaikumar (1978, 1981) with binary variables x_{ijk} that equal 1 if the arc a_{ij} is traversed by vehicle k, and 0 otherwise. In their 1981 paper, the authors report computational results for problems up to 99 nodes. Laporte *et al.* (1985) suggest a two-index formulation for undirected problems with capacity and distance constraints. The authors have applied their constraint relaxation algorithm to problems containing up to 60 nodes. Loosely constrained problems are easier than tight problems since they require fewer subtour elimination constraints. In contrast, dynamic programming and set partitioning work better on tightly constrained problems.

Heuristic algorithms for the vehicle routing problems can often be derived from procedures designed for traveling salesman problems. The nearest neighbor algorithm, insertion algorithms and tour improvement procedures can be applied almost without modifications as long as care is taken to ensure that only feasible vehicle routes are created.

The classical merge algorithm proposed by Clarke and Wright (1964) solves capacitated vehicle routing problems in which the number of vehicles is free. The method starts with vehicle routes containing the depot and one other node. The general step merges two existing tours if it is profitable (i.e., cost saving) and feasible to do so. The step considers two tours $(n_1, ..., n_i, n_1)$ and $(n_1, n_j, ..., n_1)$ and examines deleting the arcs or edges a_{i1} and a_{1j}, and replacing them with a_{ij} instead, thus creating one tour $(n_1, ..., n_i, n_j, ..., n_1)$. Merging the two tours results in the cost change $\Delta_{ij} = c_{ij} - c_{i1} - c_{1j}$. The algorithm is then initialized by creating a set of tours T^* that include tours (n_1, n_i, n_1) $\forall$ $i = 2, ..., n$. Furthermore, the cost changes Δ_{ij} are sorted in nondecreasing order. The algorithm can then formally be described as follows.

The Clarke and Wright Algorithm

Step 1: Choose the lowest Δ_{ij}, such that $n_i \in T_k$, $n_j \in T_\ell$, $k \neq \ell$ and $T_k, T_\ell \in T^*$.

Step 2: Is $\Delta_{ij} < 0$? If yes: Go to Step 3.
If no: Stop; a solution has been found.

Step 3: Tentatively merge T_k and T_ℓ by introducing a_{ij} and deleting a_{i1} and a_{1j}. If this merger is not feasible with respect to capacity or distance constraints, set $\Delta_{ij} := M >> 0$ and go to Step 1; otherwise, make the merger permanent, call the resulting tour T_p and set $T^* := T^* \cup T_p \setminus \{T_k, T_\ell\}$. Go to Step 1.

This procedure can be executed in $O(n^2 \log n)$ time. This complexity can be reduced by using appropriate data structures; see Golden *et al.* (1977) and Paessens (1988). Gaskell (1967), Yellow (1970) and Paessens (1988) have also proposed a number of variants of this method. The Clarke and Wright algorithm implicitly ignores vehicle fixed costs and fleet size. Vehicle costs f can easily be taken into account by adding this constant to every c_{1j}, $j = 2, ..., n$. Solutions with a fixed number of vehicles can be obtained by repeating Step 3 until the required number of routes has been reached, even if the savings become negative.

Example: Consider a problem with five nodes and the cost matrix

$$\mathbf{C} = \begin{bmatrix} 0 & 2 & 4 & 5 & 2 \\ 2 & 0 & 5 & 9 & 3 \\ 4 & 5 & 0 & 7 & 7 \\ 5 & 9 & 7 & 0 & 6 \\ 2 & 3 & 7 & 6 & 0 \end{bmatrix}$$

Furthermore, assume that the demands at the nodes are $d_2 = 3$, $d_3 = 4$, $d_4 = 2$, and $d_5 = 6$. The capacity of a vehicle is $\kappa = 10$ in this capacity constrained vehicle routing problem.

The initial tours are shown in Figure II.34a. The costs of the four tours are 26.

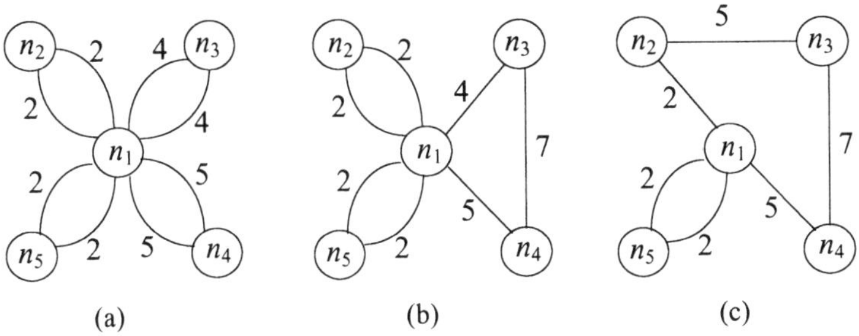

Figure II.34

The cost changes Δ_{ij} are $\Delta_{23} = -1$, $\Delta_{24} = 2$, $\Delta_{25} = -1$, $\Delta_{34} = -2$, $\Delta_{35} = 1$, and $\Delta_{45} = -1$. The largest improvement $\Delta_{34} = -2$ is obtained by merging the tours (n_1, n_3, n_1) and (n_1, n_4, n_1) by introducing the edge e_{34} as shown in Figure II.34b. Doing so is feasible as the combined tour has a total demand of $d_3 + d_4 = 6$ which does not exceed the capacity of the vehicle.

The next tour merger has these candidates as indicated by the cost changes $\Delta_{23} = \Delta_{25} = \Delta_{45} = -1$. However, only the first two are feasible as combining the routes (n_1, n_3, n_4, n_1) and (n_1, n_5, n_1) exceeds the capacity of the vehicle. Arbitrarily choosing $\Delta_{23} = -1$, we combine the tours (n_1, n_3, n_4, n_1) and (n_1, n_2, n_1) which results in the solution shown in Figure II.34c with cost 24. No further combinations of tours can be made.

CHAPTER 4 ARC ROUTING

Arc routing problems are among the earliest problems known in the theory of graphs. The source is the Swiss mathematician Leonhard Euler and his famed Königsberg Bridge Problem posed in 1736. It concerns a walk across the seven bridges of Königsberg that lead across the Pregel River; see Figure II.35a. The question Euler asked was whether or not a walk would exist on which each of the bridges is crossed exactly once. When presenting the problem and its solution to the St. Petersburg Academy, Euler did not use the geographical picture of the city, river, and bridges, but a graph-theoretical representation of them which represents bridges as edges, and the ends of the bridges at an island or the banks of the river as nodes. This representation is shown in Figure II.35b.

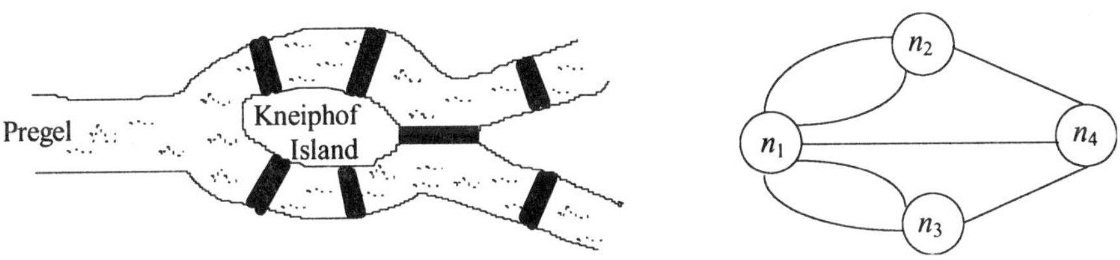

(a) (b)

Figure II.35

He also presented necessary and sufficient conditions for a walk across the bridges to exist. More than a hundered years later, Hierholzer (1873) went beyond Euler's existence problem: given that Euler's existence conditions were satisfied, his concern was to determine one of these existing tours on the edges of the given graph. Historical details about the problem along with a facsimile of Euler's original article can be found in Euler (1953) and Korte (1989). Fleischner (1990) provides translations of the original papers by Euler and Hierholzer.

It took almost a hundred more years before another related problem was posed: what is to be done if Euler's conditions are violated? The Chinese mathematician Meigu Guan (1962) of Shangtun Normal College who, during the "Cultural Revolution" worked for some time as a post office worker, formulated what has since been dubbed the "Chinese Postman Problem". Its objective is to find a tour of minimal length, on which each arc or edge is traversed *at least* once.

Applications of arc routing problems are found in many areas in routing. Typical examples are snow removal, street cleaning, the design of routes for police patrols, and mail delivery. Clearly, some of these problems go much beyond the simple versions of the problem. As an example, snow removal systems will choose a route that removes the snow from interstate highways first, serving regional highways next, followed by smaller roads and neighborhood streets. Such hierarchies are commonplace in a number of problems that involve public services.

4.1 Euler Graphs and Cycles

In this section, we define Euler graphs and give necessary and sufficient conditions for their existence. We then present a simple method that determines an Eulerian cycle, given that one exists. For the remainder of this chapter, consider a given graph $G = (N, A, E)$, where the set A includes all directed arcs, while the set E comprises all undirected edges. Either set may be empty. We can then define an *Euler graph* as a graph G that is *unicursal*, i.e., it includes a circuit that uses each arc and/or edge of G exactly once. In the following, we will distinguish between three cases: undirected, directed, and mixed graphs. First consider undirected graphs, which represent Euler's original problem. One can then prove

Theorem II.10: An undirected graph G is Eulerian, if and only if G is connected and the degrees of all of its nodes are even.

The proof of the theorem is easy and it is sketched here. Suppose that an Eulerian tour in some given graph is $(n_1, n_2, ..., n_n, n_1)$. As we walk on edge e_{ij} from n_i to n_j, we remove e_{ij} from the graph. That removal decreases the degrees of n_i and n_j by 1 each. Hence, in the first step, the degrees of n_1 and n_2 are decreased by 1, in the next step the degrees of n_2 and n_3 are decreased by 1, and so forth. This means that the degree of n_1 is decreased by 1, and the degree of each succeeding node on the path is decreased by 2 as we pass through it, *viz.*, by 1 as we enter the node, and by 1 as we leave it. Upon entering n_1 again, which is the last node on the tour, its degree decreases again by 1, so that now the degrees of all nodes in the graph have decreased by a multiple of 2, i.e., by an even number. Given that the degrees of all nodes were even in the beginning, at the end of the tour all degrees are zero, meaning that all edges of the graph are used. The result of Theorem II.10 also answers Euler's original problem in the negative, as not all of the nodes of the graph in Figure II.37b are of even degree.

Analogous to Theorem II.10, we can specify a condition for the existence of Euler circuits in directed graphs.

Theorem II.11: A directed graph G is Eulerian, if and only if it is strongly connected and its indegrees equal its outdegrees for all nodes.

The proof is similar to that above in the undirected case. The basic idea is that on a Eulerian tour in a directed graph, each node must be entered exactly as often as it is left. Finally, consider the mixed case.

Definition II.12: A mixed graph G is said to be *even*, if each node is incident to an even number of edges and arcs. A graph G is called *balanced*, if for each set $S \subseteq N$, the difference between the number of arcs leading from S to $N \setminus S$ and the number of arcs leading from $N \setminus S$ to S is less than or equal to the number of edges that connect S and $N \setminus S$. A graph G is said to be *symmetric* if $\delta^+(n_i) = \delta^-(n_i)\ \forall\ n_i \in N$.

Given these definitions, Ford and Fulkerson (1962) prove the following

Theorem II.13: A mixed graph G is Eulerian, if and only if it is even and balanced.

It is apparent that, in contrast to the conditions of Theorems II.10 and II.11, the combinatorial nature of the conditions of Theorem II.13 are difficult to ascertain. As an illustration, consider the mixed graphs in Figure II.36 on the next page.

While the graph in Figure II.36a does include an Eulerian path, e.g., (e_{12}, a_{24}, a_{43}, e_{32}, e_{25}, e_{54}, e_{41}), the graph in Figure II.36b does not. The reason for the latter is

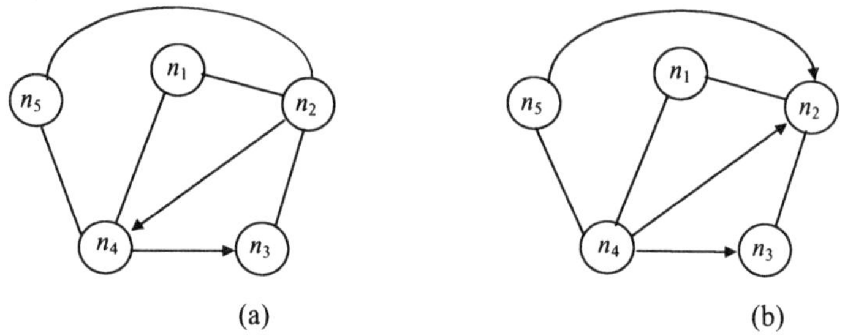

Figure II.36

that while the graph in Figure II.36b is even, it is not balanced. To see that, consider the set $S = \{n_4, n_5\}$. There are three arcs leading from S to $N \setminus S$ (*viz.*, a_{42}, a_{43}, and a_{52}) while no arc is leading from $N \setminus S$ to S. This difference of 3 is larger than the number of edges between S and $N \setminus S$, of which there is only one, *viz.*, e_{14}. Hence, the given graph is not Eulerian.

Lemma II.14: If a mixed graph G is even and symmetric, then it is also balanced.

Lemma II.14 and Theorem II.13 imply that any even and symmetric graph is Eulerian. Note, however, that symmetry is not a necessary condition for unicursality. The following method, due to Ford and Fulkerson (1962), will first derive a symmetric graph from any given graph G (if one exists) and then orient the remaining undirected edges, so that a directed Euler graph results. The algorithm is initialized with a mixed graph $G = (N, A, E)$ and can be described as follows.

Transformation of a Mixed Graph into a Symmetric Graph

Step 1: Define the graph $G' = (N', A')$, such that $a_{ij} \in A$ implies that $a'_{ij} \in A'$, and $e_{ij} \in E$ implies that a'_{ij} and $a'_{ji} \in A'$. The lower bound of an arc a'_{ij} is $\ell_{ij} = 1$, if $a_{ij} \in A$, and $\ell_{ij} = 0$ if $e_{ij} \in E$. The upper bounds are $u_{ij} = 1\ \forall$ $a'_{ij} \in A$.

Step 2: Determine a feasible circulation in G', e.g., with one of the feasible flow algorithms of Chapter 3 of Part III. If no feasible circulation exists, G is not Eulerian. Otherwise, let x_{ij} denote the flow on arc a'_{ij}.

Step 3: Define the graph $G^* = (N, A^*, E^*)$, where $a^*_{ij} \in A^*\ \forall\ a_{ij} \in A$, $a^*_{ij} \in A^*\ \forall$

$e_{ij} \in E$ with $x_{ij} = 1$ and $x_{ji} = 0$, and $e^*_{ij} \in E^*$ for the remaining edges. Note that G^* is symmetric.

Step 4: Is $E^* = \varnothing$?

If yes: Stop; the graph G^* is Eulerian.

If no: Go to Step 5.

Step 5: Apply the end-pairing algorithm or Fleury's algorithm (described below) to assign directions to the remaining edges.

Example: Consider again the graph in Figure II.36a. Applying Step 1 of the above algorithm results in the graph $G' = (N, A', \ell, u)$ shown in Figure II.37a with arcs $a_{ij} \in A$ shown as bold lines.

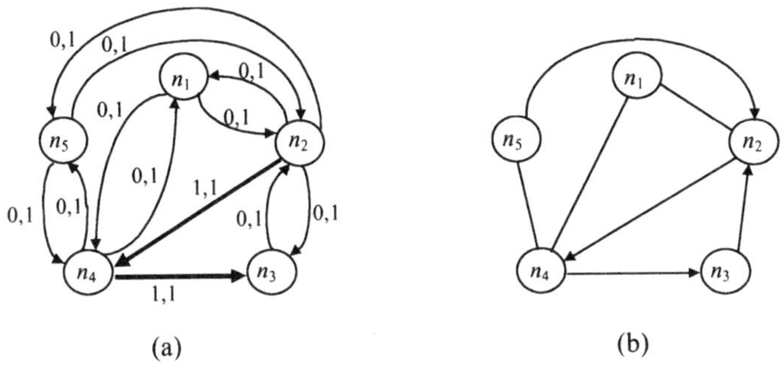

Figure II.37

A feasible circulation has $x_{24} = x_{43} = x_{32} = 1$ and $x_{ij} = 0$ otherwise, and the resulting graph G^* is shown in Figure II.37b. Arbitrarily choosing the node n_4, Step 5 finds the circuit $(n_4, n_1, n_2, n_5, n_4)$, so that the final graph G^* includes arcs a^*_{41}, a^*_{12}, a^*_{25}, and a^*_{54}. Together with the already existing arcs a^*_{24}, a^*_{43}, and a^*_{32}, the directed graph G^* is Eulerian.

Having stated the existence conditions for Eulerian graphs G, we will now present algorithms that find an Eulerian tour, given that the underlying graph is Eulerian. Again, we must distinguish between undirected, directed, and mixed graphs. First, consider undirected graphs. The most time-consuming task of *Fleury's algorithm*, as reported by Kaufmann (1967), is the identification of *bridges* which are edges whose removal from the graph increases the number of connected components each of which includes at least one edge. Defining B as the set of bridges and starting the tour at some node n_i, we can then formally state an algorithm whose idea was first put forward by Fleury (1885).

Fleury's Algorithm

Step 1: From n_i, move to the node n_j along edge e_{ij} only if $e_{ij} \notin B$. Set $E := E \setminus \{e_{ij}\}$, and determine B.

Step 2: Is $\delta(n_i) = 0$? If yes: Set $N := N \setminus \{n_i\}$, $n_i := n_j$ and go to Step 3.

If no: Set $n_i := n_j$ and go to Step 1.

Step 3: Is $N = \emptyset$? If yes: Stop, an Euler tour has been found.

If no: Go to Step 1.

While this algorithm is easy to program, it requires the repeated identification of bridges. This is a rather time-consuming task; Edmonds and Johnson (1973) have proposed a number of alternative algorithms that can be implemented in $O(n)$ time. One of the methods they propose is the *end-pairing algorithm*. The idea of the method is to trace individual circuits that do not necessarily cover all nodes or edges, and then merge them. For a survey of algorithms that find Eulerian cycles in graphs, readers are referred to Fleischner (1991).

Example: Consider the graph in Figure II.38.

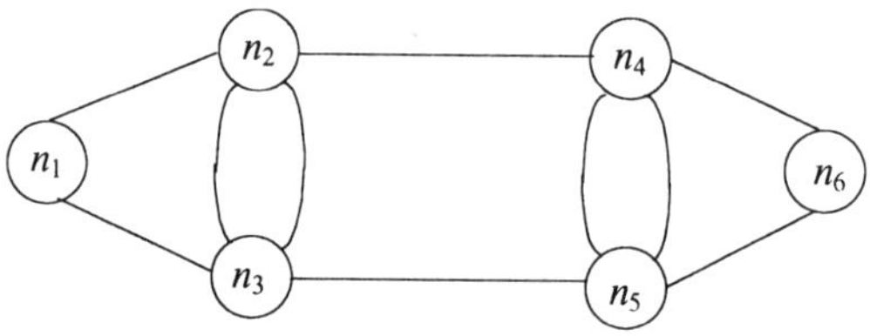

Figure II.38

All nodes have even degrees, so that the given graph is indeed Eulerian. At present, the graph does not contain any bridges. Arbitrarily, the tour commences at the node n_2. Moving to n_4 and removing the edge e_{24} establishes edge e_{35} as a bridge. Moving on from n_4 to n_5 requires the removal of e_{45}. At n_5, we cannot move on to n_3, as such a move would be along a bridge. Continuing instead back to n_4, and further to n_6 and then to n_5 finds the edge e_{35} no longer as a bridge, as its removal still disconnects the graph into two parts, but one of them includes no more edges. The next moves lead us to n_3, n_2, n_3, n_1 and back to n_2, which completes one of the many Eulerian tours in this graph.

In case of directed graphs, an algorithm was described by van Aardenne-Ehrenfest and De Bruijn (1951). For details, readers are referred to the survey by Eiselt *et al.* (1995a,b).

4.2 Constructing Eulerian Graphs

In this section, we will consider graphs G that are *not* Eulerian. In these cases, we will construct a supergraph G* of G, such that G^* is Eulerian and includes an Eulerian tour that is shorter than the Eulerian tour in any other subgraph of G. First consider undirected graphs. Define c_{ij} as the length of the shortest path from node n_i to node n_j, and let $\mathcal{N}(n_i)$ denote the neighbors of the node n_i, i.e., the nodes incident to n_i. Define now variables x_{ij} as the number of copies of the edge e_{ij} that have to be introduced in order to make the graph Eulerian. Given the set T of nodes with odd degree, the problem can then be formulated as

$$\text{P: Min z} = \sum_{e_{ij} \in E} c_{ij} x_{ij}$$

$$\text{s.t.} \quad \sum_{n_{ij} \in \mathcal{N}(n_i)} x_{ij} = \begin{cases} 1 \,(\text{mod } 2), \text{if } n_i \in T \\ 0 \,(\text{mod } 2), \text{if } n_i \in N \setminus T \end{cases}$$

$$x_{ij} = 0 \vee 1 \; \forall \; e_{ij} \in E.$$

This problem can be solved as a perfect matching problem on a graph whose nodes are the nodes in T and where all arcs $e_{ij} \in E$ with $i < j$ are included.

Example: Consider the graph in Figure II.39a.

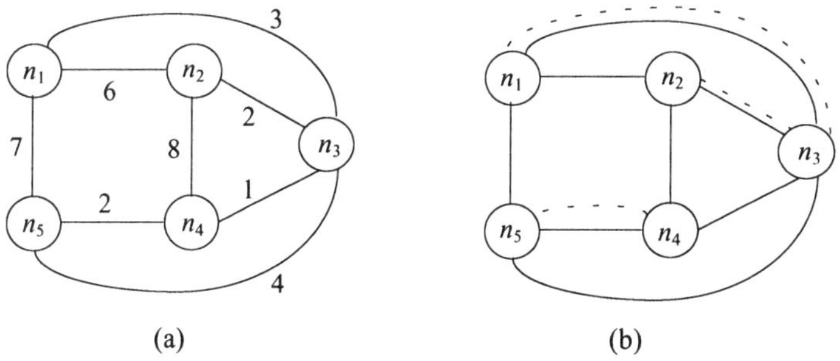

Figure II.39

The nodes n_1, n_2, n_4 and n_5 have odd degrees, requiring their membership in the set T. The shortest path matrix on all nodes is as follows:

$$\mathbf{C} = \begin{bmatrix} 0 & 5 & 3 & 4 & 6 \\ 5 & 0 & 2 & 3 & 5 \\ 3 & 2 & 0 & 1 & 3 \\ 4 & 3 & 1 & 0 & 2 \\ 6 & 5 & 3 & 2 & 0 \end{bmatrix}$$

The problem has a solution $\bar{x}_{12} = \bar{x}_{21} = 1$, and $\bar{x}_{45} = \bar{x}_{54} = 1$. In other words, copies of all edges on the shortest paths between nodes n_1 and n_2 and between nodes n_4 and n_5 have to be introduced. As the two respective shortest paths are (n_1, n_3, n_2) and (n_4, n_5), copies of the edges e_{13}, e_{23}, and e_{45} are introduced as shown in Figure II.39b, where the broken lines show the additional copies of arcs introduced in the process. The resulting graph is Eulerian, and an appropriate tour can then easily be determined by any of the pertinent algorithms.

Consider now the case of directed graphs. Assuming that the given graph G is not Eulerian, then there exists at least one node n_i whose indegree $\delta^+(n_i)$ does not equal its outdegree $\delta^-(n_i)$. Define $S = \{n_i: \delta^+(n_i) > \delta^-(n_i)\}$ and $T = \{n_i: \delta^+(n_i) < \delta^-(n_i)\}$. Clearly, $S \cap T = \emptyset$ and $S \cup T \subseteq N$. It is apparent that copies of arcs have to be added, leading out of nodes in S, so as to make up for the deficit of outgoing arcs at these nodes. Similarly, copies of arcs have to be introduced so as to lead into nodes in the set T to make up for the deficit of incoming arcs. Defining again c_{ij} as the length of the shortest path from n_i to n_j, the problem can be formulated as follows.

$$\text{P: Min } z = \sum_{i:n_i \in S} \sum_{j:n_j \in T} c_{ij} x_{ij}$$

$$\text{s.t. } \sum_{j:n_j \in T} x_{ij} = -\delta^+(n_i) + \delta^-(n_i) \quad \forall\, n_i \in S$$

$$\sum_{j:n_{ij} \in S} x_{ij} = \delta^+(n_j) - \delta^-(n_i) \quad \forall\, n_i \in S$$

$$x_{ij} \in \mathbb{N}_0 \;\forall\, n_i \in S;\, n_j \in T.$$

It is apparent that this formulation is a standard transportation problem as described in Chapter a of the Introduction. A small example may explain its use.

Example: Consider the graph in Figure II.40a.

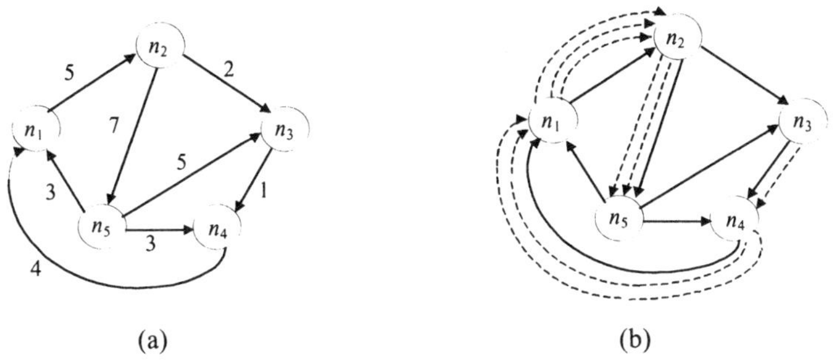

Figure II.40

Clearly, the graph is not Eulerian; its differences $\delta^+(n_i) - \delta^-(n_i) = -1, -1, -1, -1$, and 2 for the nodes $n_1, n_2, \ldots, n_5$, so that the sets $S = \{n_1, n_3, n_4\}$ and $T = \{n_2, n_5\}$. The problem can then be formulated as

$$\text{P: Min } z = 5x_{12} + 12x_{15} + 10x_{32} + 17x_{35} + 9x_{42} + 16x_{45}$$

$$\begin{aligned}
\text{s.t. } x_{12} + x_{15} &= 1\\
x_{32} + x_{35} &= 1\\
x_{42} + x_{45} &= 1\\
x_{12} + x_{32} + x_{42} &= 1\\
x_{15} + x_{35} + x_{45} &= 1\\
x_{ij} \in \mathbb{N}_0 \; \forall \, n_i \in S;\ n_j \in T.&
\end{aligned}$$

The problem is highly degenerate. One of its optimal solutions is $\bar{x}_{15} = \bar{x}_{35} = \bar{x}_{42} = 1$ and $\bar{x}_{ij} = 0$ otherwise. This implies that one copy each of the shortest paths (n_1, n_2, n_5), $(n_3, n_4, n_1, n_2, n_5)$, and (n_4, n_1, n_2) must be introduced, resulting in the graph shown in Figure II.40b, where additional copies of arcs that are introduced are shown as broken lines. Note that although the optimal values of the variables happen to be zero and one, three additional copies of the arc a_{12} are introduced, two additional copies of a_{25}, etc.

Consider now mixed graphs. As in the undirected and directed cases, we must determine a minimal-cost augmentation in order to solve the problem. Note that some graphs cannot be made Eulerian; the graph $G = (N, A, E)$ with $N = \{n_1, n_2, n_3\}$, $A_2 = \{a_{12}, a_{13}\}$ and $E = \{e_{23}\}$ is a pertinent example. Exact algorithms are based on the integer linear programming formulation of the problem; results are reported by Grötschel and Win (1992) and Picard and Queyranne (1980). Early heuristics are due to Edmonds and Johnson (1973), which were subsequently improved by Frederickson (1979). Each of Frederickson's two heuristics has a complexity of $O((\max\{n^3; |A| \max\{|A|; |E|\})^2)$ and a worst-case ratio of 2; i.e., the tour each of these heuristic algorithms finds is never more than twice as long as the optimal tour. If applied in succession, the worst-case error bound reduces to $\frac{5}{3}$. Both heuristics rely on shortest path and minimum cost algorithms; for details about such methods, readers are referred to parts II and III of this book. Christofides *et al.* (1984) propose heuristics that exhibit good average-case behavior: they average 3% above the optimum, with 17% in the worst case.

A close relative to the Chinese Postman Problem on mixed graphs is the *Windy Postman Problem* introduced by Minieka (1979). It is defined on undirected graphs and the cost of traversing depends on the direction an edge is traversed in: in one direction, the postman must walk into the wind resulting in higher costs, whereas in the other direction, he can walk with the wind, which is easier and thus less costly. It is easy to demonstrate that the mixed chinese postman problem is a special case of the windy postman problem: to each edge $e_{ij} \in G$ in the mixed problem with costs c_{ij} we assign an edge e_{ij} in the windy postman problem with equal costs of traversal in both directions, and to each arc $a_{ij} \in G$ with costs c_{ij} in the mixed problem, we assign an edge e_{ij} inthe windy problem with costs of c_{ij} from n_i to n_j and costs of ∞ from n_j to n_i. Brucker (1981) and Guan (1984) independently proved that the Windy Chinese Postman Problem is *NP-hard*. Win (1989) showed that the problem is solvable in polynomial time if the underlying graph G is Eulerian.

4.3 Rural Postman Problems

Rural postman problems (*RPP*) are defined on a graph $G = (N, A \cup E)$, with a set of *required arcs* and *edges* $R \subseteq A \cup E$. The objective in this problem is to find a cost-minimal tour that traverses each arc or edge in R at least once, whereas the use of the remaining arcs and edges is optimal. The visualization is a postman who has to deliver the mail on all streets in a number of villages (the set R), whereas the travel along the roads from one village to another along arcs $(A \cup E) \setminus R$ is not required and merely used to get from one connected set of required arcs to another.

As all routing problems, the rural postman problem comes in directed, undirected, and mixed versions. In this section, we concentrate on the undirected version. Lenstra and Rinnooy Kan (1976) have proved that the general rural postman problem is *NP-hard*. However, if $R = A$ in the directed, or $R = E$ in the undirected problem, then the usual Chinese Postman Problem results, which can be solved in polynomial time. The problem is also easy, if the graph $G_R = (N, R)$ is connected. In that case, we determine shortest paths between all nodes with odd degree (which may or may not include nonessential arcs) and proceed similarly to the standard Chinese Postman Problem.

Consider now the undirected problem with a given graph $G = (N, E)$. In general, the graph $G_R = (N, R)$ will not be connected. In that case, we modify the given graph G to a modified graph $G' = (N', E')$ as follows.

- The set N' includes all nodes $n_i \in N$ that are incident to at least one required edge.
- The set E' includes all required edges.
- Add to E' edges e_{ij} $\forall$ n_i, $n_j \in N'$ whose length c_{ij} denotes the length of a shortest path between n_i and n_j.
- Delete all nonrequired edges e_{ij} for which $c_{ij} = c_{ik} + c_{kj}$ for some k, as well as any edge that is parallel to another edge with the same cost.

The result of the above procedure is a graph G' that includes $p \geq 2$ connected components G_1, G_2, ..., G_p which are spanned by the required edges R. Frederickson (1979) has designed an efficient heuristic algorithm for the problem. It can formally be described as follows.

Frederickson's Heuristic

Step 1: Determine a minimal spanning tree T that connects the components G_1, G_2, G_p.

Step 2: Determine a minimum cost matching for all nodes with odd degrees in the graph with edge set $R \cup T$. Denote by M the set of edges in this matching.

Step 3: Determine an Eulerian cycle on the graph with the edge set $R \cup T \cup M$.

Example: Consider the graph $G = (N, E)$ in Figure II.41, where the bold lines denote the required edges $R = \{e_{12}, e_{13}, e_{23}, e_{45}\}$.

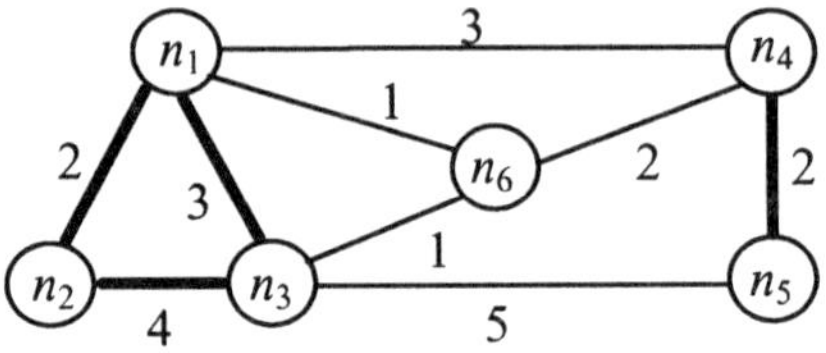

Figure II.41

The graph $G' = (N', E')$ then includes the nodes $N' = \{n_1, n_2, n_3, n_4, n_5\}$, but not n_6 as the latter is not incident to any of the required edges. Adding edges e'_{ij} for all nodes $n_i \in N'$ results in the graph in Figure II.42a.

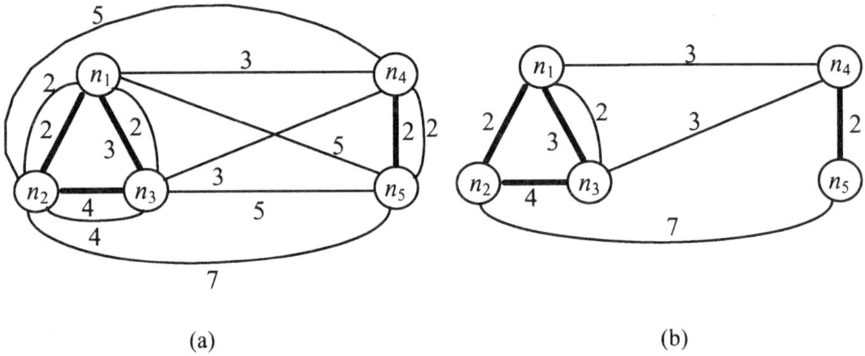

Figure II.42

As $c_{24} = c_{21} + c_{14}$, $c_{15} = c_{14} + c_{45}$, and $c_{35} = c_{34} + c_{45}$, the edges e'_{24}, e'_{15}, and e'_{35} can be deleted. Subsequent deletion of parallel edges of equal length results in the graph $G' = (N', E')$ in Figure II.42b. Frederickson's algorithm commences to work on the graph $G' = (N', E')$. The connected components G_1 and G_2 include the nodes $\{n_1, n_2, n_3\}$ and $\{n_4, n_5\}$, respectively. The minimal spanning tree that connects the two components includes the single edge e_{14} which may either represent the edge e_{14} or the path (e_{16}, e_{64}) in the original graph. The graph at the end of Step 1 of Frederickson's heuristic includes all edges in $R \cup T$ and is shown in Figure II.43a on the next page.

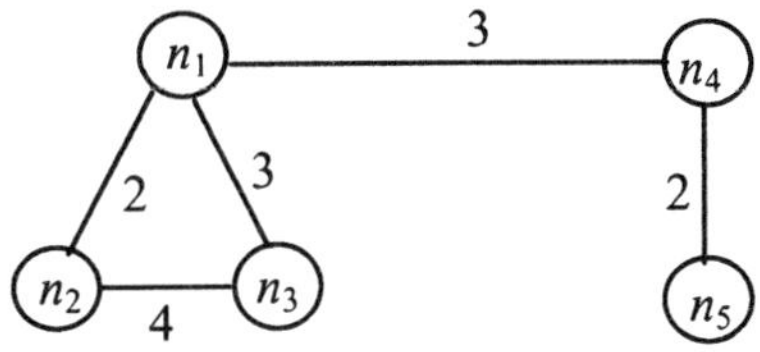

(a)

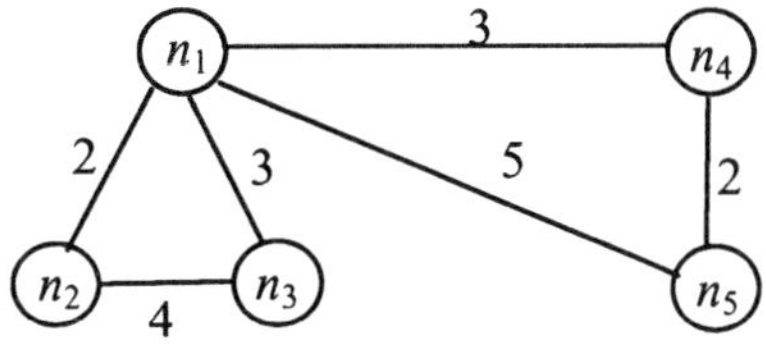

(b)

Figure II.43

This graph includes n_1 and n_5 as nodes with odd degrees. A minimal matching M introduces the edge e'_{15} with length 5. The resulting graph is shown in Figure II.43b. This graph is now Eulerian and any appropriate technique, e.g., the end pairing algorithm or Fleury's method, can be used to find a Eulerian tour. Here, one such tour is $(n_1, n_4, n_5, n_1, n_3, n_2, n_1)$.

Frederickson (1979) proved that if the edge lengths c_{ij} satisfy the triangular inequality, then the solution found by his algorithm is within 50% of the optimal solution. A more recent improvement heuristic for the undirected problem was described by Hertz *et al.* (1999). An exact branch-and-cut algorithm was developed by Ghiani and Laporte (2000).

The directed rural postman problem can be solved by a similar method; here, the minimal spanning tree in Step 1 is replaced by a shortest spanning arborescence, and the matching in Step 2 reduces to a transportation problem. Christofides *et al.* (1986) report computational results that, on average, are within 1.3% of the optimal solution.

A special case of the mixed rural postman problem is the *Stacker crane* problem, with $R = A$, and none of the edges being required. The name derives from the set A whose members symbolize the movements of a crane. A typical application is also found in delivery problems whose objective is to minimize unnecessary travel. Frederickson *et al.* (1978) have proved the problem to be *NP-hard*, as even the special case with $e_{ij} = 0 \; \forall \; a_{ij} \in A$ reduces to a traveling salesman problem.

4.4 The Capacitated Arc Routing Problem

Capacitated arc routing problems (*CARP*) are probably the most applicable and most difficult routing problems. They are defined on a graph $G = (N, A, E, c, w)$ in which weights $w_{ij} \geq 0$ are assigned to arcs and edges. Let now a fleet of m homogeneous vehicles with capacity W be located at some node, say, n_1. The problem is then to determine m tours involving n_1, so that the sum of weights on each tour does not exceed W, and the sum of costs of all tour is minimized. The capacitated arc routing problem is the most general of the arc routing problems. For instance, if $W = |R|$ and $w_{ij} = 1$ if $e_{ij} \in R$ and $w_{ij} = 0$ otherwise, the problem reduces to the rural postman problem. While the CARP comes in undirected, directed, and mixed versions, we only examine the undirected problem in this section. Two different formulations exist for the problem, one by Golden and Wong (1981) and the other by Belenguer and Benavent (1991). Golden and Wong's formulation uses binary variables x_{ijk} and y_{ijk}, both relating to vehicle k traversing edge e_{ij}. The difference between them is that x_{ijk} denotes all edges e_{ij} that are traversed by vehicle k, while $y_{ijk} = 1$, if the edge e_{ij} is not just traversed, but also serviced, by vehicle k. The problem can then be formulated as follows.

$$\text{P}_{\text{CARP}}\text{: Min } z = \sum_{k=1}^{m} \sum_{e_{ij} \in E} c_{ij} x_{ijk}$$

$$\text{s.t.} \quad \sum_{e_{ij} \in E} x_{ijk} - \sum_{e_{ji} \in E} x_{jik} = 0 \quad \forall\, n_i \in N;\ k = 1, \ldots, m. \qquad (1)$$

$$\sum_{k=1}^{m} (y_{ijk} + y_{jik}) = 1, \text{ if } w_{ij} > 0 \text{ and 0 otherwise}, \quad \forall\, e_{ij} \in E \qquad (2)$$

$$x_{ijk} \geq y_{ijk} \quad \forall\, e_{ij} \in E;\ k = 1, \ldots, m \qquad (3)$$

$$\sum_{e_{ij} \in E} w_{ij} y_{ijk} \leq W \quad \forall\, k = 1, \ldots, m \qquad (4)$$

$$\{x_{ijk}\} \text{ does not form subtours} \qquad (5)$$

$$x_{ijk}, y_{ijk} = 0 \vee 1 \quad \forall\, e_{ij} \in E;\ k = 1, \ldots, m \qquad (6)$$

In this formulation, the constraints (1) require that a vehicle that enters a node will also leave it, constraints (2) ensure that each edge e_{ij} with positive weight is served at least once, constraints (3) guarantee that an edge is served by a vehicle only if the vehicle actually traverses the edge. These latter constraints are the only link between the variables x_{ijk} and y_{ijk}; without them, P_{CARP} would be decomposable into two problems, one involving the variables x_{ijk}, and the other comprising the variables y_{ijk}. Constraints (4) require that no tour exceeds the capacity of a vehicle, and constraints (5) are the usual subtour elimination constraints.

The success of this or any other integer programming formulation of the problem are good lower bounds. For details, see the survey article by Eiselt *et al.* (1995). Below, we briefly describe some heuristic methods that have been developed for the capacitated arc routing problem. As in the traveling salesman problem, heuristics fall into two different categories: *construction algorithms* that construct a feasible solution, and *improvement algorithms*, that attempt to improve an already existing feasible solution. Three popular heuristics for the capacitated arc routing problem are:

- The *Construct-Strike Algorithm* was first suggested by Christofides (1973) and later improved by Pearn (1989). Christofides' method constructs feasible tours and then removes them from the existing graph. This step ensures that the removal of a tour from the graph does not disconnect the graph. When it is no longer possible to find tours, a Euler tour is determined on the remaining graph; followed by the repeated search for feasible tours. Pearn's modification of the method does away with the restriction that the graph remains connected after removing all arcs on a tour. Christofides' method is of complexity $O(mn^3)$, while Pearn's modification requires $O(mn^4)$ steps in the worst case.
- The *Path Scanning Algorithm* by Golden *et al.* (1983) constructs feasible tours one at a time in Greedy fashion. The procedure is repeated for a variety of criteria, and the best resulting solution is chosen. Each such run requires $O(n^3)$ steps.
- The *Augment-Merge Algorithm* is also due to Golden *et al.* (1983). It is based on ideas put forward by Clarke and Wright (1964) for vehicle routing problems which is described in Section 3.4. The method starts with all edges belonging to different tours. The procedure then merges tours, resulting in an $O(n^3)$ algorithm.
- A *Tabu Search Algorithm* was described by Hertz *et al.* (2000). It first constructs an initial solution, and then gradually improves it by moving required edges from their current route to an adjacent route. During the search process, intermediate non-improving or infeasible moves are considered.

For a survey of a variety of other heuristics, see, e.g., Eiselt *et al.* (1995).

PART III NETWORK FLOW AND NETWORK DESIGN MODELS

There are many problems in engineering, operations research and other disciplines that involve the flow of some items or substances through a network. Examples include the flow of vehicles through a road network, packets of information through a computer network, people through a "network" of corridors in a building, water through pipes, and electricity through a regional or national grid. Sometimes there is a time associated with any unit of flow along each arc of the network.

The chapters in this part cover a variety of issues associated with network flow and design. The first chapter introduces and formulates the problem and states some fundamental results. The second chapter describes a number of applications of network flow models, some of which are not obviously connected with the concept of flows, but their structure reduces to network flows. The third chapter surveys algorithms for network flow problems, and Chapter 4 introduces and investigates approaches to multicommodity flow problems. Finally, Chapter 5 introduces the concept of flows with congestion and outlines some ways to solve the resulting difficult problems.

CHAPTER 1 BASIC PRINCIPLES OF NETWORK MODELS

This chapter first formulates some standard network flow problems. Subsequently, some transformations are explained that allow the reduction of a variety of problems to standard mathematical structures. We then introduce the dual problems, optimality conditions, and finally some results regarding feasibility and integrality.

1.1 The Problem and its Formulation

Given a network $G = (N, A)$, an *arc flow* x_{ij} is a nonnegative real number associated with an arc $a_{ij} \in A$. The *capacity* κ_{ij} of a_{ij} is an upper bound on the flow that can pass along a_{ij} per time unit when constraints imposed by flows in other arcs are ignored. Similarly, ℓ_{ij} provides a lower bound on the flow in a_{ij}. A node n_s at which flow originates is called an *origin* or a *source*, whereas a node n_t at which the flow terminates is called a *destination* or a *sink*. Nodes that are neither sources nor sinks are *intermediate nodes* or *transshipment nodes*. In this chapter we deal with the case of a single origin-destination (or *O-D*) pair (n_s, n_t). It is assumed that flow is neither created nor destroyed along the way. The constraints of a flow problem can then be written as follows.

$$\sum_{j:a_{ij}\in A} x_{ij} - \sum_{j:a_{ji}\in A} x_{ji} = 0 \qquad \forall\, j \neq s, t \tag{1}$$

$$x_{ij} \geq \ell_{ij} \qquad \forall\, a_{ij} \in A \tag{2}$$

$$x_{ij} \leq \kappa_{ij} \qquad \forall\, a_{ij} \in A \tag{3}$$

$$x_{ij} \geq 0 \qquad \forall a_{ij} \in A \tag{4}$$

The constraints (1) are generally known as *conservation equations*, *(flow) balancing equations*, or *Kirchhoff node equations* in reference to their counterpart in electrical networks. In essence, they specify that the total flow into a node n_j must equal the total flow out of a node. Constraints (2) and (3) are lower and upper bound constraints that restrict the flow in each of the arcs of the network. The usual nonnegativity constraints (4) complete the problem. Any flow $\mathbf{x}$ that satisfies the constraints (1)-(4) is called a *flow pattern*.

There are two main objective functions used in flow problems. The first type maximizes the number of flow units that are sent out of the source n_s through the network to the sink n_t. Such flows are referred to as *maximal flow patterns*. Their objective function value $f_{\mathbf{x}}$ related to a given flow pattern $\mathbf{x}$ equals the number of flow units that leave the origin, i.e.,

$$\text{Max } f_{\mathbf{x}} = \sum_{j:a_{sj}\in A} x_{sj} \qquad (5)$$

By adding the conservation constraints, it is readily verified that $f_{\mathbf{x}}$ also equals the total flow into n_t, i.e., $f_{\mathbf{x}} = \sum_{i:a_{it}\in A} x_{it}$. As it is easily shown that flow through a network can be maximized while keeping $x_{is} = 0 \ \forall \ a_{is} \in A$ and, similarly, $x_{tk} = 0 \ \forall \ a_{tk} \in A$, most maximal flow problems are set up in such a way that the source n_s has no incoming arcs and the sink n_t has no outgoing arcs.

The second objective minimizes the total cost of all shipments through the network. Assuming that unit transport costs are c_{ij} and transport costs are linear in the quantity x_{ij} shipped along an arc a_{ij}, the total transport costs can be written as

$$z = \sum_{a_{ij}\in A} c_{ij} x_{ij}\,. \qquad (6)$$

Flow problems with these two objective functions are discussed in detail in the remainder of this chapter.

Another representation of network flows considers complete paths from source to sink, taken by flow units, rather than arc flows. For that purpose, define a *(directed) path flow* from the source n_s to the sink n_t along a path π as a number $y_\pi \geq 0$, and the collection $\mathbf{y} = (y_\pi)$ will be termed a network flow in "path form". Then, the total flow in any arc a_{ij} is given by

$$x_{ij} = \sum_{\pi} \delta^{\pi}_{ij} y_{\pi} \tag{7}$$

where $\delta^{\pi}_{ij} = 1$ if arc a_{ij} is on path π, and 0 otherwise. An alternative formulation of the maximal flow is now at hand.

$$\text{P: Max } f_y = \sum_{\pi} y_{\pi}$$

$$\text{s.t. } \ell_{ij} \leq \sum_{\pi} \delta^{\pi}_{ij}\, y_{\pi} < \kappa_{ij} \qquad \forall\, a_{ij} \in A$$

$$y_{\pi} \geq 0 \qquad \forall\, \pi.$$

Notice that there is no need for any explicit conservation constraints such as those in relation (1) since the flow is conserved at intermediate nodes on each path.

So far, our discussion has centered on flows in networks. A concept intimately related to flows is that of *cuts*. Let $\bar{S}$ denote the complement of a set S. Then we can write

Definition III.1: A *cut* in a network $G = (N, A)$ is a partition (S, T) of N, such that $\varnothing \underset{\neq}{\subset} S \underset{=}{\subset} N$, $T = \bar{S}$, $n_s \in S$, and $n_t \in T$. The arcs in the cut are $A^{S,T} = \{a_{ij}: n_i \in S, n_j \in T\}$ and the *value* (or *capacity*) *of the cut* is $\kappa(S, T) = \sum_{a_{ij} \in A^{S,T}} \kappa_{ij}$. The cut with the smallest capacity is called a *minimal cut*.

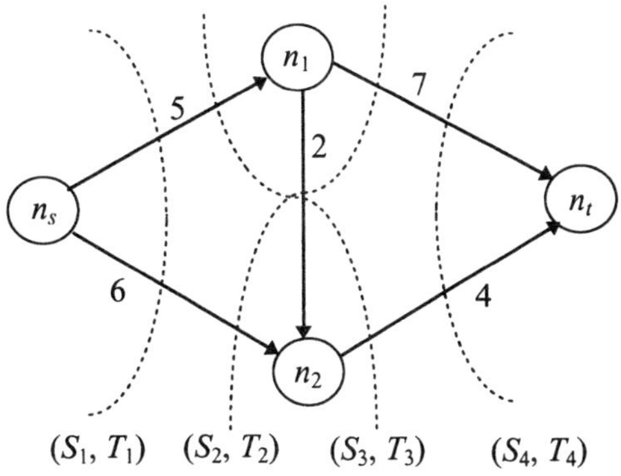

Figure III.1

Example: Consider the network in Figure III.1. As shown in the figure, the network includes four cuts: $S_1 = \{n_s\}$, $S_2 = \{n_s, n_1\}$, $S_3 = \{n_s, n_2\}$, and $S_4 = \{n_s, n_1, n_2\}$, and $T_k = \overline{S}_k$ generate cuts (S_k, T_k) with sets of arcs $A^{S_1,T_1} = \{a_{s1}, a_{s2}\}$, $A^{S_2,T_2} = \{a_{s2}, a_{12}, a_{1t}\}$, $A^{S_3,T_3} = \{a_{s1}, a_{2t}\}$, and $A^{S_4,T_4} = \{a_{1t}, a_{2t}\}$. Note that in Figure III.2, the arcs in the cuts are those arcs which cross the line that divides the subsets of nodes S_k and T_k, provided that the arc is directed from a node in S_k to a node in T_k. As an example, the cut (S_3, T_3) includes the arcs a_{s1} and a_{2t}, but not the arc a_{12}, as it is directed from T_3 to S_3. The cuts in the network have capacities $\kappa(S_1, T_1) = 5 + 6 = 11$, $\kappa(S_2, T_2) = 6 + 2 + 7 = 15$, $\kappa(S_3, T_3) = 5 + 4 = 9$, $\kappa(S_4, T_4) = 7 + 4 = 11$, so that (S_3, T_3) is the cut with the smallest capacity, i.e., the minimal cut. An important relationship between the value of the minimal cut and the value of the maximal flow is explored in Section 1.3.

1.2 Transformations of Flow Problems

This section presents a variety of flow problems that can be transformed to standard minimal cost flow problems as introduced above. The purpose is to demonstrate that minimal cost flow problems are indeed a very general class of models that are able to encompass a wide variety of applications.

Problems with multiple sources and sinks. Let the given flow problem include sources $n_{s^1}, n_{s^2}, \ldots, n_{s^p}$ and sinks $n_{t^1}, n_{t^2}, \ldots, n_{t^r}$. Assuming that the goods that shipped through the network are homgeneous, so that a customer at destination n_{t^ℓ} cannot identify the source the good originates from (if this assumption is violated, we have a case of multicommodity flows dealt with in Chapter 4), then the following transformation can be applied. Create a supersource n_s and a supersink n_t, and connect them with arcs a_{s,s^k} $\forall\, k = 1, \ldots, p$ and $a_{t^\ell,t}$ $\forall\, \ell = 1, \ldots, r$. All flows then originate at n_s and are shipped to n_t.

Flow and Circulatory Networks. So far, flow problems have been presented with three different types of nodes: the source, the sink, and all other nodes that are frequently referred to as intermediate or transshipment nodes. Such flow problems can be transformed to *circulations* by allowing the flow that is "somehow" created at the source, conserved at intermediate nodes, and that "somehow" disappears at the sink, to be returned from the sink to the source. This is accomplished by introducing an additional arc a_{ts} whose function is to return all units that arrive at the sink back to the source. This arc is very useful to control the flow through the network as a whole, as shown in some of the transformations discussed below. If not otherwise stated, we set $\ell_{ts} = 0$, $\kappa_{ts} = M >> 0$, and $c_{ts} = 0$.

Shortest path problems. Any shortest path problem can be transformed into a more general minimal cost flow problem by setting $\ell_{ij}: = 0$ and $\kappa_{ij}: = 1 \; \forall \; a_{ij} \in A$, essentially modeling the shortest path as a path from source to sink with a flow of one. As the formulation so far would allow a zero flow, it is necessary to provide an incentive to ship one unit out of the source or, alternatively, into the sink. This can be accomplished by using the corresponding circulatory network and setting either $\ell_{ts} = \kappa_{ts} = 1$ (thus forcing a single unit from sink to source that, by virtue of the conservation equations, must be returned through the network to the sink) or by providing an incentive by setting $c_{ts} = -M << 0$ (coupled with $\ell_{ts} = 0$ and $\kappa_{ts} = 1$). In both cases, the arc lengths are used as costs $c_{ij} \; \forall \; a_{ij} \in A$.

Maximal flow problems. In case the flow from source to sink through a network is to be maximized, the lower and upper bounds on the arc flows are already known. In order to push as many units as possible through the network, an appropriate incentive must be introduced. In case of a flow network, we would set $c_{sj} = -1$ (or any other negative constant) $\forall \; a_{sj} \in A$, coupled with $c_{ij} = 0$ for all other arcs in the network. Similarly, we could set $c_{it} = -1 \; \forall \; a_{it} \in A$ and $c_{ij} = 0$ for all other arcs. In case of a circulatory network, it is sufficient to set $\ell_{ts}: = 0$, $\kappa_{ts}: = M >> 0$, and $c_{ts} = -1$ or any negative constant. In all three cases, each additional flow unit from source to sink or any route provides a cost reduction by "1" or whatever negative constant is chosen, so that a solution method would attempt to push as many units through the network, which is the desired result.

Maximization of profit. Consider the case in which profits, rather than costs, form the basis of the optimization. In case profits p_{ij} are generated by shipping flow units along an arc $a_{ij} \in A$, defining costs $c_{ij}: = -p_{ij}$ as negative profits will solve the problem. Many times, however, profits are generated not by sending units along an arc, but upon arrival at the sink; e.g., after completion of the shipment at the point of sale. Given unit shipping costs of $c_{ij} \; \forall \; a_{ij} \in A$ and a profit of p per unit, we can either redefine $c_{sj}: = c_{sj} - p \; \forall \; a_{sj} \in A$ or, equivalently $c_{it}: = c_{it} - p \; \forall \; a_{it} \in A$. In both cases, all other arcs have the same costs as before. Another possibility is to employ a circulatory network with $\ell_{ts} = 0$, $\kappa_{ts} = M >> 0$, and $c_{ts} = -p$.

Constraints on flows in nodes. At times, it many be required to impose restrictions on the flow through a node in addition to constraints on arc flows. This can be accomplished by splitting the node in question, say n_i, into two nodes n_i' and n_i'', so that (1) all arcs leading into n_i now lead into n_i', (2) all arcs leading out of n_i now lead out of n_i'', and (3) a new arc $a_{i'i''}$ is created. It is now possible to impose lower and upper bounds on this new arc, as well as introduce costs for flows through the original node n_i, which will now appear as $c_{i'i''}$.

The Capacitated Transshipment Problem. The main feature of the capacitated transshipment problem is that each node n_i has a flow balance b_i associated with it that denotes the difference between the flow into n_i and the flow out of n_i. Hence, $b_i > 0$ indicates that more units flow into n_i than flow out of it, resulting in a net supply, thus the term "supply node". Similarly, all nodes n_i with $b_i < 0$ are "demand nodes", whereas nodes n_i with $b_i = 0$ are usually called "transshipment nodes". Furthermore, arcs in the network have capacity κ_{ij} $\forall$ $a_{ij} \in A$.

It is apparent that any standard network flow problem is a capacitated transshipment problem with the source as the only supply node (with unknown supply), the sink as the only demand node (with unknown demand), and all other nodes as transshipment nodes. The node equations for all nodes in a capacitated transshipment problem are

$$\sum_{j:a_{ij} \in A} x_{ij} - \sum_{j:a_{ji} \in A} x_{ji} = b_i \quad \forall\, n_i \in N \tag{8}$$

A capacitated transshipment problem defined on a given graph $G = (N, A)$ can be remodeled as a standard flow problem by creating a network $G' = (N', A')$ where $N' = N \cup \{n_s, n_t\}$ and $A' = A \cup A_1 \cup A_2$, such that $A_1 = \{a_{si}: n_i \in N, b_i > 0\}$ and $A_2 = \{a_{it}: n_i \in N, b_i < 0\}$. To each arc $a_{si} \in A_1$, a capacity of $\kappa_{si} = b_i$ is assigned while each arc $a_{it} \in A_2$ has a capacity of $\kappa_{it} = -b_i$. A maximal flow is then determined in the network G'. Clearly the value $f_{\mathbf{x}}$ of such a maximal flow cannot exceed the sum of the capacities of arcs leading out of n_s, i.e., $f_{\mathbf{x}} \leq \sum_{j:a_{sj} \in A'} \kappa_{sj} = \sum_{a:n_i \in A'} b_i$. If $f_{\mathbf{x}}$ equals the total supply, then a feasible solution to the capacitated transshipment problem is obtained merely by ignoring flows in arcs $a_{ij} \in A' \setminus A$. If $f_{\mathbf{x}}$ is strictly smaller than the total supply, then the transshipment problem has no feasible solutions.

Flow models with required flow value. In some applications, the flow through a network is exogenously given. If such a flow value $\tilde{f}$ is known, then setting $\ell_{ts} = \kappa_{ts} = \tilde{f}$ in the corresponding circulatory network will ensure that exactly $\tilde{f}$ units are shipped from source to sink (provided a feasible solution exists).

Piecewise linear cost functions. In some applications, the unit transportation costs depend on the quantity that is shipped on an arc. In particular, suppose that the unit transportation costs are constant in each of the q intervals, and that these unit costs are no lower for higher shipments. Formally, we assume that $c_{ij}(x_{ij}) = c_{ij}^k$ $\forall$ $x_{ij} \in [b_{ij}^{k-1}; b_{ij}^k]$ such that $c_{ij}^k \geq c_{ij}^{k-1}$ $\forall$ $k = 1, ..., q$ and $b_{ij}^q = \kappa_{ij}$. Such a cost function is shown in Figure III.2.

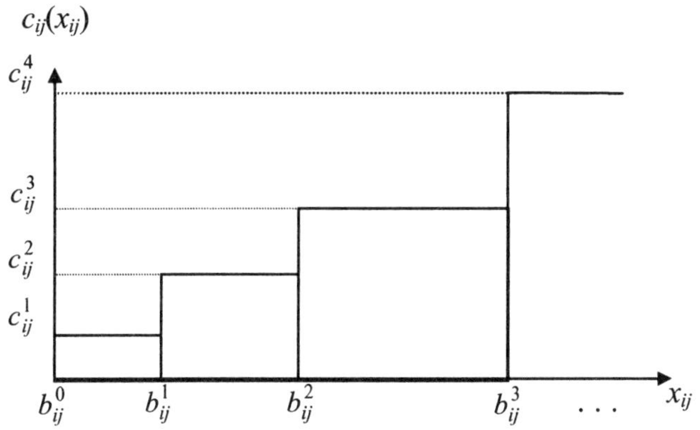

Figure III.2

We can now replace the single arc with $(\ell_{ij}, \kappa_{ij}; c_{ij}) = (0, \kappa_{ij}; c_{ij}(x_{ij}))$ by q parallel arcs with $(0, b_{ij}^1; c_{ij}^1)$, $(0, b_{ij}^2 - b_{ij}^1; c_{ij}^2)$, $(0, b_{ij}^3 - b_{ij}^2; c_{ij}^3), ...,$ $(0, b_{ij}^q - b_{ij}^{q-1}; c_{ij}^q)$. At this point, any standard solution method can be applied. The transformation works because it is cheaper to first use arcs that relate to smaller quantities. If the cost function were, say, $c_{ij}(x_{ij}) = 5$ for $x_{ij} \in [0; 10]$ and 3 for $x_{ij} \in [10; 25]$, then a cost-minimizing flow algorithm would first route flows on the second arc that is, however, only usable if the first arc were used first. A nondecreasing step function such as that in Figure III.2 ensures that flow units are assigned in the proper sequence.

1.3 Duality and Optimality Conditions

This section will develop optimality conditions for general flow problems. For that purpose, consider the primal minimum cost flow problem in a circulatory network as

$$\text{P: Min } z = \sum_{a_{ij} \in A} c_{ij} x_{ij}$$

$$\text{s.t.} \quad -\sum_{j:a_{ij} \in A} x_{ij} + \sum_{j:a_{ji} \in A} x_{ji} = 0 \quad \forall\, i \tag{9}$$

$$x_{ij} \geq \ell_{ij} \quad \forall\, a_{ij} \in A \tag{10}$$

$$x_{ij} \leq \kappa_{ij} \quad \forall\, a_{ij} \in A \tag{11}$$

$$x_{ij} \geq 0 \quad \forall\, a_{ij} \in A$$

with the usual node equations (9), lower bound constraints (10), and upper bound constraints (11). The nonnegativity constraints on x_{ij} are implied by the constraints (10) and the nonnegativity of the lower bounds; they are explicitly included here for the construction of the dual constraints.

In order to set up the dual, assign dual variables $u_i \in \mathbb{R}$ to the constraints (9), and dual variables $v_{ij} \geq 0$ and $w_{ij} \geq 0$ to the constraints (10) and (11), respectively. The dual problem can then be written as

$$\text{P}_\text{D}\text{: Min } z_\text{D} = \sum_{a_{ij} \in A} \kappa_{ij} w_{ij} - \sum_{a_{ij} \in A} \ell_{ij} v_{ij}$$

$$\text{s.t.} \quad u_i - u_j + w_{ij} - v_{ij} \geq -c_{ij} \tag{12}$$

$$u_i \in \mathbb{R} \quad \forall\, i = 1, \ldots, n \tag{13}$$

$$w_{ij} \geq 0 \quad \forall\, a_{ij} \in A \tag{14}$$

$$v_{ij} \geq 0 \quad \forall\, a_{ij} \in A. \tag{15}$$

The dual variables u_i assigned to the nodes of the network are usually referred to as *node potentials*. Defining $\bar{c}_{ij} = c_{ij} + u_i - u_j$, the dual constraints (12) can be rewritten as

$$\bar{c}_{ij} + w_{ij} - v_{ij} \geq 0 \quad \forall\, a_{ij} \in A. \tag{16}$$

The conditions of weak complementary slackness (see, e.g., Eiselt *et al.* (1987)) can then be written as

$$(\kappa_{ij} - x_{ij}) w_{ij} = 0 \quad \forall\, a_{ij} \in A \tag{17}$$

$$(\ell_{ij} - x_{ij}) v_{ij} = 0 \quad \forall\, a_{ij} \in A \tag{18}$$

$$(\bar{c}_{ij} + w_{ij} - v_{ij})x_{ij} = 0 \quad \forall\, a_{ij} \in A. \tag{19}$$

Without loss of generality, let

$$\kappa_{ij} > 0; \tag{20}$$

if $\kappa_{ij} = 0$, the arc a_{ij} can simply be dropped from consideration as it cannot carry any flow. Furthermore, we let

$$\ell_{ij} < \kappa_{ij} \quad \forall\, a_{ij} \in A. \tag{21}$$

All edges with $\ell_{ij} = \kappa_{ij}$ must carry exactly $x_{ij} = \ell_{ij} = \kappa_{ij}$ units of flow, so that such an arc relates to feasibility conditions. We now distinguish between three cases.

Case 1: $x_{ij} = \ell_{ij}$. Then relation (17) can be written as $(\kappa_{ij} - \ell_{ij})w_{ij} = 0$. As $\kappa_{ij} > \ell_{ij}$ by relation (21), $\kappa_{ij} - \ell_{ij} > 0$ and hence $w_{ij} = 0$. Then relation (16) reduces to $\bar{c}_{ij} - v_{ij} \geq 0$ or $\bar{c}_{ij} \geq v_{ij}$ and, with $v_{ij} \geq 0$ as per relation (15), we obtain $\bar{c}_{ij} \geq 0$.

Case 2: $x_{ij} = \kappa_{ij}$. Then relation (18) implies that $(\ell_{ij} - \kappa_{ij})v_{ij} = 0$ and, as by virtue of relation (21), $(\ell_{ij} - \kappa_{ij}) < 0$, $v_{ij} = 0$ results. Relation (19) can then be rewritten as $(\bar{c}_{ij} + w_{ij})\kappa_{ij} = 0$. As $\kappa_{ij} > 0$ by (20), we obtain $\bar{c}_{ij} + w_{ij} = 0$ or $\bar{c}_{ij} = -w_{ij}$. As $w_{ij} \geq 0$ as per relation (14), we obtain the result $\bar{c}_{ij} \leq 0$.

Case 3: $\ell_{ij} < x_{ij} < \kappa_{ij}$. The inequality $\ell_{ij} < x_{ij}$ and relation (18) imply that $v_{ij} = 0$; similarly, the inequality $x_{ij} < \kappa_{ij}$ and relation (17) imply that $w_{ij} = 0$. Replacing v_{ij} and w_{ij} in (19) results in $\bar{c}_{ij}\, x_{ij} = 0$ and as $x_{ij} > \ell_{ij}$ and $\ell_{ij} \geq 0$ imply $x_{ij} > 0$, $\bar{c}_{ij} = 0$ follows.

The findings in the above three cases can be summarized as

$x_{ij} = \ell_{ij}$ implies $\bar{c}_{ij} \geq 0$

$x_{ij} = \kappa_{ij}$ implies $\bar{c}_{ij} \leq 0$, and

$x_{ij} \in \,]\ell_{ij}; \kappa_{ij}[$ implies $\bar{c}_{ij} = 0$.

These implications can be replaced by their equivalents

$$\bar{c}_{ij} < 0 \rightarrow x_{ij} \neq \ell_{ij} \quad (22)$$

$$\bar{c}_{ij} > 0 \rightarrow x_{ij} \neq \kappa_{ij} \quad (23)$$

and $\bar{c}_{ij} \neq 0 \rightarrow x_{ij} \leq \ell_{ij}$ or $x_{ij} \geq \kappa_{ij}$. (24)

Invoking feasibility, relation (22) can be written as $\bar{c}_{ij} < 0 \rightarrow x_{ij} > \ell_{ij}$. As $\bar{c}_{ij} < 0$ also implies that $\bar{c}_{ij} \neq 0$, relation (24) applies, so that $x_{ij} \leq \ell_{ij}$ (a contradiction to $x_{ij} > \ell_{ij}$) or $x_{ij} \geq \kappa_{ij}$. Due to the feasibility of the flows, $x_{ij} = \kappa_{ij}$, so that relation (22) can be rewritten as

$$\bar{c}_{ij} < 0 \rightarrow x_{ij} = \kappa_{ij} \quad (25)$$

Similarly, we obtain

$$\bar{c}_{ij} > 0 \rightarrow x_{ij} = \ell_{ij} \quad (26)$$

and $\bar{c}_{ij} = 0 \rightarrow x_{ij} \in [\ell_{ij}; \kappa_{ij}]$. (27)

The relations (22) – (27) allow us to state

Theorem III.2: A flow pattern $\mathbf{x}$ is optimal for a minimal cost flow problem, if and only if

$\bar{c}_{ij} > 0$ and $x_{ij} = \ell_{ij}$, (28)

$\bar{c}_{ij} < 0$ and $x_{ij} = \kappa_{ij}$, or (29)

$\bar{c}_{ij} = 0$ and $x_{ij} \in [\ell_{ij}; \kappa_{ij}]$. (30)

1.4 Some Fundamental Results

In this section, we will state and prove three important results in network flow theory. For that purpose, consider the formulation

$$\text{Max } f_{\mathbf{x}} \quad \text{or} \quad \text{Min } z = \sum_{a_{ij} \in A} c_{ij} x_{ij}$$

$$\text{s.t.} \qquad \sum_{j} x_{sj} - f_{\mathbf{x}} = 0$$

$$\sum_{j:a_{ij}\in A} x_{ij} - \sum_{j:a_{ji}\in A} x_{ji} = 0 \qquad \forall\, i \neq s, k$$

$$\sum_{i:a_{it}\in A} x_{it} - f_{\mathbf{x}} = 0$$

$$x_{ij} \geq \ell_{ij} \qquad \forall\, a_{ij} \in A$$

$$x_{ij} \leq \kappa_{ij} \qquad \forall\, a_{ij} \in A$$

$$x_{ij} \geq 0 \qquad \forall\, a_{ij} \in A.$$

Defining **A** as the node-arc incidence matrix as in Definition 31 and $\mathbf{e}_i$ as a unit vector with the "1" in the i-th position, the problem can be written in matrix form as follows.

$$\text{Max}\, f_{\mathbf{x}} \;\text{ or }\; \text{Min}\, \mathbf{cx}$$

$$\text{s.t.} \quad \mathbf{Ax} + [-\mathbf{e}_1 + \mathbf{e}_n]f_{\mathbf{x}} = \mathbf{0}$$

$$\mathbf{x} \geq \ell$$

$$\mathbf{x} \leq \kappa$$

$$\mathbf{x} \geq \mathbf{0}$$

Notice that the column of the single variable $f_{\mathbf{x}}$ has—like all other variables x_{ij}—only two nonzero values in the first n rows. In particular, $f_{\mathbf{x}}$ has a "–1" in the first node that corresponds to the source, and a "+1" in row n that corresponds to the sink. If this column were appended to the node-arc incidence matrix **A**, it would be identical to the column of an arc a_{ts} which would transform the original flow problem to a circulation. Hence, the matrix $[\mathbf{A}, -\mathbf{e}_1 + \mathbf{e}_n]$ is still a node-arc incidence matrix. Heller and Thompkins (1956) have proved

Lemma III.3: The node-arc incidence matrix of a graph is totally unimodular.

As demonstrated in Section I.3 of part I, any linear programming problem with a matrix of coefficients that is totally unimodular and with integer right-hand side values will have at least one all-integer optimal solution, regardless of the objective function. The addition of identity matrices (for the lower and upper bounds) does not change that result. This allows us to state

Theorem III.4: A flow problem with integer right-hand side values has all-integer solutions, and the dual of a flow problem with integer objective function coefficients has all-integer solutions.

We now apply this result to the maximum flow problem. The problem can be restated as

$$\begin{aligned} \text{P: } & \text{Max } f_{\mathbf{x}} \\ \text{s.t. } & \mathbf{Ax} + [-\mathbf{e}_1 + \mathbf{e}_n] f_{\mathbf{x}} = \mathbf{0} \\ & \mathbf{x} \leq \boldsymbol{\kappa} \\ & \mathbf{x} \geq \mathbf{0} \\ & f_{\mathbf{x}} \in \mathbb{R} \end{aligned}$$

Assigning dual variables **u** to the conservation equations and dual variables **w** to the upper bound constraints, the dual problem can be written as

$$\begin{aligned} \text{P}_\text{D}\text{: } & \text{Min } \mathbf{w}\boldsymbol{\kappa} \\ \text{s.t. } & \mathbf{uA} + \mathbf{w} \geq \mathbf{0} & (31) \\ & -u_s + u_t = 1 & (32) \\ & \mathbf{u} \in \mathbb{R} \\ & \mathbf{w} \geq \mathbf{0} \end{aligned}$$

The constraints can be rewritten as

$$u_i - u_j + w_{ij} \geq 0 \quad \forall\, a_{ij} \in A$$

$$-u_s + u_t = 1.$$

We can assume that $u_i \geq 0\ \forall\ i$, as any solution with some $u_i < 0$ can be modified by adding a positive constant to all variables, so that $u_i \geq 0\ \forall\ i$ results. Doing so does not affect feasibility (as each column of **A** consists of exactly one "+1" and one "–1") or the value of the objective function. Invoking Theorem III.4, we obtain $u_i \in \mathbb{N}_0\ \forall\ i$.

Lemma III.5: There exists at least one solution (**u**, **w**) with $\mathbf{u} \in \{0; 1\}^n$.

Proof: Given that $u_i \in \mathbb{N}_0 \ \forall\ i$, the set $Q = \{n_i: u_i > 1\}$ has a complement $\overline{Q} = \{n_i: u_i = 0 \vee 1\}$ based on the current solution **u**. Define now a new solution $\tilde{\mathbf{u}}$, so that $\tilde{u}_i = u_i - 1 \ \forall\ n_i \in Q$ and $\tilde{u}_i = u_i \ \forall\ n_i \in \overline{Q}$. Then the left-hand side of each relation in (31) and (32) is unchanged for all arcs a_{ij} with $n_i, n_j \in Q$ or $n_i, n_j \in \overline{Q}$. For arcs a_{ij} with $n_i \in \overline{Q}$ and $n_j \in Q$, the left-hand side of the relation (31) increases by one, so that again feasibility is guaranteed. Consider now an arc a_{ij} with $n_i \in Q$ and $n_j \in \overline{Q}$. Clearly, the left-hand side of the relation (31) has decreased by one in the new solution. This will result in an infeasible solution only if $u_i - u_j + w_{ij} = 0$ or, equivalently, $u_j = u_i + w_{ij} > 1$ (by definition as $n_i \in Q$) and $w_{ij} \geq 0$. However, $u_j > 1$ implies that $u_j \in Q$ which contradicts the assumption that $n_j \in \overline{Q}$. Repeatedly replacing **u** by $\tilde{\mathbf{u}}$ as above ultimately leads to a binary vector **u**. □

Lemma III.5 immediately implies

Corollary III.6: There exists at least one solution with $w_{ij} = 0 \vee 1 \ \forall\ a_{ij} \in A$.

The proof is easy; as $\kappa_{ij} > 0 \ \forall\ a_{ij} \in A$, any $w_{ij} > 1$ could be reduced to one, thus improving the objective value, while not violating feasibility since $u_i, u_j = 0 \vee 1$ as demonstrated above, so that $(\mathbf{u}, \mathbf{A})_i \geq 1$ for each row i in the constraint set (31).

A result of Lemma III.5 is that there exist sets S and T, such that $S = \{n_i: u_i = 0\}$ and $T = \overline{S} = \{n_i: u_i = 1\}$ and, due to constraint (32), $n_s \in S$ and $n_t \in T$. Then (S, T) is a cut with $w_{ij} = 1 \ \forall\ a_{ij}: n_i \in \mathrm{S}, n_j \in \mathrm{T}$ and $w_{ij} = 0$ otherwise. The value of the minimal cut equals $\kappa(S, T)$ which, by virtue of duality theory, equals $f_{\overline{\mathbf{x}}}$, the value of the maximal flow. This is summarized in Ford and Fulkerson's (1956) famed *min cut-max flow theorem*:

Theorem III.7: For any given network with capacities $\kappa_{ij} > 0$, the value of a maximal flow equals the value of a minimal cut.

An alternative, constructive, proof of the theorem is provided in Section 3.1.1. In essence, a minimal cut can be seen as a bottleneck in a network, and the theorem states that the largest possible flow will equal the capacity of a bottleneck.

In the following discussion, it will prove useful to define a set of nodes P, such that $\varnothing \subsetneq P \subsetneq N$. We can then write $\ell(P,\overline{P})$ for $\sum_{a_{ij}:n_i \in P, n_j \in \overline{P}} \ell_{ij}$, and similarly for $x(P,\overline{P})$, and $\kappa(P,\overline{P})$. We can then state Hoffman's (1960)

Theorem III.8: For any circulatory network with lower bounds ℓ_{ij} and upper bounds κ_{ij}, a feasible solution exists if and only if

$$\ell(P,\overline{P}) \le \kappa(\overline{P},P) \ \forall \ \varnothing \subsetneq P \subsetneq N.$$

Proof: We first prove that if the condition is violated no circulation exists. In other words, assume that there exists at least one nonempty set $P \neq N$, such that

$$\ell(P,\overline{P}) > \kappa(\overline{P},P) \tag{33}$$

but a feasible circulation exists, requiring

$$x(P,\ \overline{P}) = x(\overline{P},P) \tag{34}$$

We have then

$$\begin{aligned} x(P,\overline{P}) &\ge \ell(P,\overline{P}) && \text{(due to feasibility)} \\ &> \kappa(\overline{P},P) && \text{(due to assumption (33))} \\ &\ge x(\overline{P},P) && \text{(due to feasibility)} \end{aligned}$$

so that $x(P,\ \overline{P}) > x(\overline{P},P)$ which contradicts the assumption (34).

We now prove that a circulatory network in which no feasible circulation exists violates the condition of the theorem. In order to do so, first set all flows equal to the lower bounds, i.e. set $x_{ij} = \ell_{ij} \ \forall \ a_{ij} \in A$. This flow will satisfy the lower and upper bound constraints, but it will not satisfy the conservation constraints, as otherwise a feasible circulation would exist. Define now the excess of a node n_j as $e_j = \sum_{i:a_{ij} \in A} x_{ij} - \sum_{k:a_{jk} \in A} x_{jk}$. A node with positive excess is called an *excess node*, a node with negative excess is called a *deficit node*, and a node with zero excess is said to be an *equilibrium node*. The sets E^+, E, and E^- are defined as the sets of excess, equilibrium, and deficit nodes, respectively. We now attempt to balance the flow by shipping as many flow units from nodes in E^+ to nodes in E^-, possibly via some intermediate nodes. By definition, it will not be possible to ship all excesses out of the nodes in E^+, since otherwise we would have a feasible circulation. Suppose that after as many excesses as possible are shipped, the remaining excess nodes are in some set E'^{+}. Furthermore, define U as the set of nodes to which at least one additional unit can be shipped from E'^{+}, including the nodes in E'^{+}. Defining $\overline{U}$ as the complement of U, i.e., $\overline{U} = N \setminus U$, then $(U, \overline{U})$ defines a cut in the network, such that

$x(U, \overline{U}) = \kappa(U, \overline{U})$ as otherwise additional units could be shipped to $\overline{U}$,

$x(\overline{U}, U) = \ell(\overline{U}, U)$ as otherwise the excess could be reduced by decreasing shipments to E'^{+}, and

$x(U, \overline{U}) < x(\overline{U}, U)$ as U includes at least one excess node, so that by definition the inflow into U must exceed the outflow. Hence,

$$\kappa(U, \overline{U}) = x(U, \overline{U}) < x(\overline{U}, U) = \ell(\overline{U}, U),$$

or simply $\kappa(U, \overline{U}) < \ell(\overline{U}, U)$ which violates the condition. □

CHAPTER 2 APPLICATIONS OF NETWORK FLOW MODELS

This section describes some of the many applications of network flow problems. For further examples and a detailed account of flow theory, the interested reader may consult Ahuja *et al.* (1993) or Murty (1992).

2.1 Building Evacuation

Consider the problem of evacuating people from public or office buildings, such as teachers and students from a school. An example of a single floor of a building is provided in Figure III.3 with five rooms A, ..., E and a common hall or foyer F.

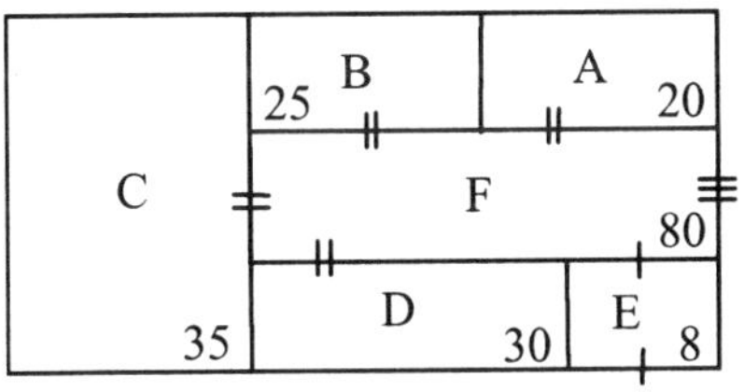

Figure III.3

The numbers in one of the lower corners of each room in the figure indicate the capacity of that room, and the lines connecting rooms indicate doors with different capacities: a door symbolized by a single line (such as that connecting rooms E and F) has a capacity of 15 people per minute, two lines (here, doors between all other rooms) symbolize doors with capacities of 50 people per minute. Finally, doors shown by three lines, such as the door connecting the foyer F to the outside of the building, have a capacity of 75 people per minute.

The question whether or not the building can be evacuated in one minute can be answered by constructing a graph as follows. Each room is represented as a node, and the outside is modeled as a sink. Undirected edges exist then between nodes n_i and n_j, if rooms i and j are connected by a door. Any such edge has a lower bound of zero and an upper bound given by the capacity of the door that connects them. In addition, each room is connected to a source and each arc a_{sj} connecting them has a lower and upper bound that equals the capacity of room j. In addition, all nodes representing rooms also have capacities that equal the capacities of the rooms. They are required to restrict the throughput to the room capacity. The graph $G = (N, A; \ell_{ij}, \kappa_{ij}; \kappa_i)$ of our example is shown in Figure III.4.

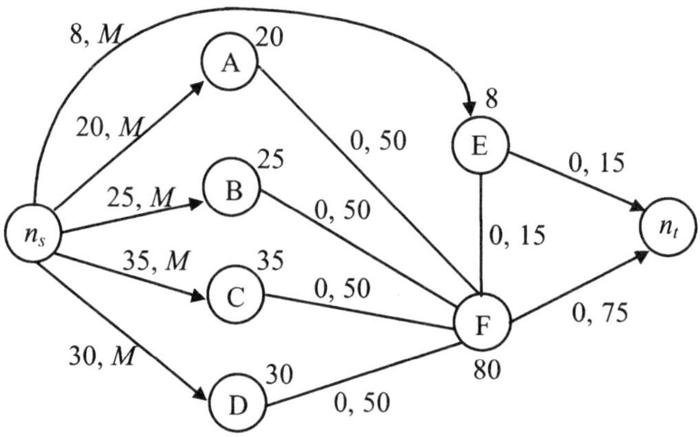

Figure III.4

It is apparent that there is no feasible flow, i.e., the building cannot be evacuated in a single minute with the foyer being the main bottleneck. However, the evacuation can be done in two minutes, which is easily demonstrated by resolving the problem with all arc and node capacities doubled, except those of the arcs a_{sj}.

Clearly, realistic problems will result in much more complicated networks that incorporate stairs, multiple levels or floors, and fire escapes, as well as "turn penalties", a well-known concept in traffic networks; see, e.g., Roberts (1978). It is also possible to perform interesting sensitivity analyses; e.g., room capacities may be determined based on a 1¼ or 1½ minute evacuation time. Furthermore, it is apparent that rooms are usually not filled to capacity. Hence, one could envisage a system that monitors the number of people in a room at any point in time and computes optimal escape routes in real time. These routes could then be indicated in the hallways by signs pointing the way. As room occupations change, so do optimal routes, so that the optimal escape route from any given room may change as well. If this is indeed efficient or just adds to the confusion as individuals tend to form habits, remains to be studied.

2.2 Flow Sharing Problems

Flow sharing problems were first described by Brown (1979). They have their roots in a real application, in which a strike at a central coal supplying plant rationed its shipments to power plants, so that the natural question arose how individual plants would share the available quantities. As such, the central task is to determine quantities that distribute the existing amount of coal and, similarly, the deficit supply to optimize some measure of "fairness". In general, this type of problem arises in cases where a scarce resource is to be shared among several consumers, as in the rationing of food and other essential supplies in a war or catastrophy situation.

Any solution to the problem will heavily depend on the underlying philosophy of fairness. For example, if there were two plants with demands of 100 and 60, respectively, and a total quantity of 120 were available, is the allocation of 75 and 45 (based on the idea that each plant receives the same proportion of the coal that it would normally receive; here $^{100}/_{160}$ = 62.5% and $^{60}/_{160}$ = 37.5%) fairer than an allocation of 80 and 40 (where each plant receives 20 units less than it normally does)? There is a rich literature on "equity" or fairness objectives, popularized by Rawls's (1971) "Theory of Justice" and Harsanyi's (1975) reply. In the context of operations research, a good discussion is provided by Savas (1978). "Equity" objectives often lead to meaningless and undesirable solutions if employed by themselves. Consider, for instance, a scenario in which n people earn \$100 each. This solution will be preferred by the standard "equity" objectives over a solution in which, say, $(n - 1)$ people earn \$1,000 each and the remaining one makes \$999, which is less equal but a solution preferred by each individual (given that all people concerned have the goal of maximizing their income). As a result, "equity" objectives are frequently coupled with efficiency objectives. Clearly, as in any bi- or multiple-objective problem, it is then required to specify a tradeoff function between efficiency and equity.

Consider now a problem in which p plants P_i, $i = 1, ..., p$ consume a certain fuel during the planning period that consists of weeks W_j, $j = 1, ..., t$. The demand of plant P_i in week W_j is d_{ij}, so that the total demand in week W_j can be calculated as $d_j = \sum_{i=1}^{p} d_{ij}$. In contrast, the supply available in week W_j is only s_j, $j = 1, ..., t$, with

$s_j < d_j$ for at least one j. The goal is now to determine values of variables x_{ij} that measure the quantity of the fuel shipped from the central supplier to plant P_i in week W_j. The objective in this model is to provide plants with fuel, so that the lowest proportion of average fuel received under shortage conditions to average fuel needed is as large as possible. For that purpose, define cumulative demands $D_{ij} = \sum_{\nu=1}^{j} d_{i\nu}$ and cumulative shipments $X_{ij} = \sum_{\nu=1}^{j} x_{i\nu}$ during the first j weeks. The objective of the problem can then be formulated as

$$\text{Max } z = \min_{\substack{i=1,\ldots,p \\ j=1,\ldots,t}} \left\{ \frac{X_{ij}}{D_{ij}} \right\}$$

subject to the availability of supplies.

The structural part of the problem can be expressed by the following network. Define a graph $G = (N, A, \kappa)$ with $N = \bigcup_{j=1}^{t} \{n_j\} \cup \bigcup_{i,j} \{n_{ij}\} \cup \{n\}$, so that each week W_j is modeled by a node n_j, each plant-week combination is represented by a node n_{ij}, and the node n is the general supplier. The set A consists of the following arcs:

- (n, n_j) $\forall j$ with capacities s_j and flows $\sum_i x_{ij}$,
- (n_j, n_{ij}) $\forall i, j$ with flows x_{ij},
- $(n_{ij}, n_{i,j+1})$ $\forall i$ and $j = 1, \ldots, t-1$ with flows X_{ij}, and
- (n_{it}, n) $\forall i$ with flows X_{it}.

If not otherwise specified, arc capacities are set to $M >> 0$.

Example: Consider a scenario in which two plants compete for fuel supplies during a planning period of three weeks. The individual demands for fuel are summarized in Table III.1.

Table III.1

	W_1	W_2	W_3
P_1	12.2	14.9	13.1
P_2	7.6	8.1	7.4

Furthermore, let the supplies available during the planning period be $\mathbf{s}$ = [17.1; 20.8; 20.2], clearly insufficient to satisfy the total weekly demands of 19.8, 23.0, and 20.5, the column sums in Table III.1. The structure of this problem is then expressed in the graph in Figure III.5 with capacities $M >> 0$ if not otherwise specified.

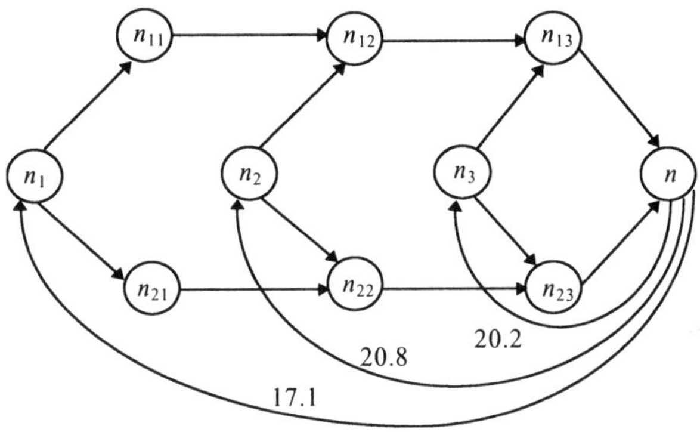

Figure III.5

Here, the flows on edges leading out of nodes n_i are individual fuel allocations x_{ij}, whereas the flows out of nodes n_{ij} are cumulative quantities X_{ij}.

Consider a specific solution in which the supplier ships 11.0, 12.0, and 13.2 units to plant P_1 and 6.1, 8.8, and 7.0 to plant P_2 during the three weeks of the planning period. This is shown in the network as arc flows of s_j in the arcs (n, n_j), the above individual allocations on arcs (n_j, n_{ij}) and cumulative quantities on the remaining arcs; e.g., 11.0 + 12.0 = 23.0 units on arc (n_{12}, n_{13}), 6.1 + 8.8 + 7.0 = 21.9 units on arc (n_{23}, n). The quantity of this specific solution is then z = $\min\left\{\frac{11.0}{12.2}; \frac{6.1}{7.6}; \frac{23.0}{27.1}; \frac{14.9}{15.7}; \frac{36.2}{40.2}; \frac{21.9}{23.1}\right\}$ = min{.9016; .8026; .8487; .9490; .9005; .9481} = .8026. In order to improve this solution, the central supplier should consider increasing the fuel shipment to P_2 in week 1.

A seemingly different, yet closely related, model arises in the hierarchical budget allocation process. Consider a state's budget for medical care. On the highest level in the hierarchy, the state planners will decide on money allocations (or flows in the context of network models) to hospitals and other health organizations. On the next, say hospital, level, the hospital administrators decide which units to allocate funds to and, within these units, the directors then decide which projects to fund. On each level and for each project, demands are formulated as well as the smallest amounts that are required to keep the unit functioning, albeit at a minimum level. Again, a variety of "equity" objectives may be applied to this situation.

2.3 A Worker Allocation Problem

The problem is to assign workers to shifts as well as departments, so as to satisfy some manpower requirements and minimize the overall cost of the assignment. More specifically, consider the following scenario. There are three 8-hour shifts given, *viz.*, shift 1 from 6 a.m. to 2 p.m., shift 2 from 2 p.m. to 10 p.m., and shift 3 from 10 p.m. to 6 a.m. Furthermore, there are four departments A, B, C, and D. The lower and upper bounds on the manpower requirements for all shift-department combinations are shown in Table III.2.

Table III.2

		Departments				All	Bonus
		A	B	C	D	depts.	factor
Shifts	1	(5, 7)	(11, 14)	(9, ∞)	(8, 11)	(38, ∞)	1.0
	2	(6, 9)	(14, 14)	(10, 16)	(10, 14)	(45, 51)	1.2
	3	(4, ∞)	(5, 14)	(8, ∞)	(7, ∞)	(24, ∞)	1.5
All shifts		(17, 18)	(34, 42)	(30, ∞)	(29, ∞)		
Bonus factor		1.0	1.1	1.3	1.0		

For example, Department D requires at least 8 and at most 11 workers in shift 1, Department A needs at least 4 (but no upper bound) workers in shift 3, and Department B needs exactly 14 workers in shift 2. In addition, there are requirements regarding the sum of workers being present in any given shift, and requirements regarding the number of workers present in departments. Here, a total of at least 45 and at most 51 workers should be available in shift 2, and no less than 34 and no more than 42 workers are needed in Department B; the requirements of the remaining departments are similar. Each worker receives a wage of \$175 per shift. Due to the hazardous, uncomfortable, or otherwise unpleasant or unhealthy conditions in some departments, a worker's wage is multiplied by a bonus factor. As an example, workers assigned to Department C receive a 30% bonus, so that their wage is (175)(1.3) = \$227.50 per shift. Similarly, there is a bonus for different shifts, so that the final wage a worker receives is the product of his base wage and the two bonus factors. Here, the wages per shift range from a low of (175)(1.0)(1.0) = \$175 for workers in shift 1 in Departments A and D to a high of (175)(1.3)(1.5) = \$341.25 for a worker in

Department C, working the graveyard shift. This problem can then be formulated as a minimum cost network flow problem as follows. From a source n_s, arcs lead to three nodes, each representing one of the shifts. The lower and upper bounds on these arcs are the requirements per shift, and the costs are zero. We now create one node for each department and connect each shift with all departments by an arc, so that an arc a_{ij} connects the node of shift j with the node of department j. The lower and upper bound on the flow in a_{ij} are the bounds provided in the table, and the costs are the amounts incurred if a worker is assigned to shift i and Department j. Finally, arcs connect each department with a sink n_t, so that the arc a_{jt} carries the lower and upper bounds that are specified for the j-th department, while the costs are zero. The network for the above example is shown in Figure III.6, where the values next to an arc a_{ij} are ℓ_{ij}, κ_{ij}; c_{ij}.

A cost-minimal feasible flow in the network of Figure III.6 provides an assignment of workers to shifts and departments. In particular, a flow of, say, 3, along the path $\pi = (n_s$, shift 2, B, $n_t)$ indicates that three workers are assigned to Department B during shift 2.

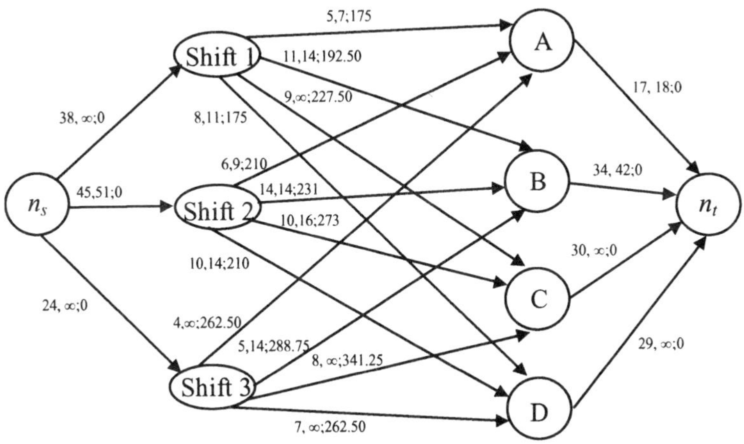

Figure III.6

The related problem of minimizing the number of workers required in the allocation problem regardless of the labor costs can be solved by using the same network. In that case, the costs are redefined as "1" for all arcs that connect shifts and departments. The cost-minimal feasible flow will then indicate the number of workers required (where workers working double/triple shifts will be counted as two or three workers, respectively.

2.4 Airline Crew Assignment

Assigning airline crews to flights is a multimillion dollar business. A poor schedule has numerous crews in the wrong places, so that these crews have to be flown as passengers to their flights, a process called *dead heading*, that loses money and time in the process. Many pertinent scheduling models exist in the industry. This section will develop two models that make a number of simplifying assumptions. For instance, we assume that there are no restrictions on work schedules, and that each crew can—at least in principle—be assigned to any of the flights. This is, of course, only possible if the crew is trained to fly the aircraft used on that particular flight. Here, we assume that each crew can be assigned to any flight under consideration, an assumption that would be satisfied if all aircraft were identical.

In this illustration, we consider two simple objectives: the first is to allocate crews to flights so as to minimize the number of crews used in the allocation, and the second objective allocates crews, so that a cost function is minimized. In both cases, the underlying structure of the network is very similar. Each flight will be represented by a node, and an arc a_{ij} from flight i to flight j indicates that the crew that operated flight i can be allocated to flight j as well. More formally, let t_i^d and t_i^a denote the scheduled departure and arrival time of flight i, respectively. Furthermore, let t denote the required transfer time between flights. We can then set up the (acyclic) network $G = (N, A; \ell, \kappa; c)$ with $N = \{n_i$: the i-th flight is considered$\} \cup \{n_s, n_t\}$ and $A = \{a_{ij}: n_i, n_j \in N;\ t_i^a + t \le t_j^d\} \cup \bigcup_j \{a_{sj}\} \cup \bigcup_i \{a_{it}\} \cup \{a_{ts}\}$. The lower and upper bounds on the arc flows are set to $\ell_{ij} = 0$ and $\kappa_{ij} = M >> 0\ \forall\ a_{ij} \in A$. However, each node n_i, $i \neq s, t$ has a lower bound of $\ell_i = 1$ associated with it that guarantees that each flight is operated by a crew. The arc costs will depend on the objective of the model.

Example: Consider a schedule of six flights between three cities shown in Table III.3 on the next page, where all departure and arrival times are given in eastern standard time.

Table III.3

Flight number	from	to	Departure time	Arrival time
1	Boston	Dallas	8:30 a.m.	11:00 a.m.
2	Boston	Los Angeles	9:00 a.m.	1:30 p.m.
3	Dallas	Boston	12:00 p.m.	2:30 p.m.
4	Dallas	Los Angeles	10:00 a.m.	2:00 p.m.
5	Los Angeles	Boston	12:00 p.m.	4:00 p.m.
6	Los Angeles	Dallas	2:30 p.m.	6:30 p.m.

If the required transfer time is $t = 1$ hour, then the graph in Figure III.7 results.

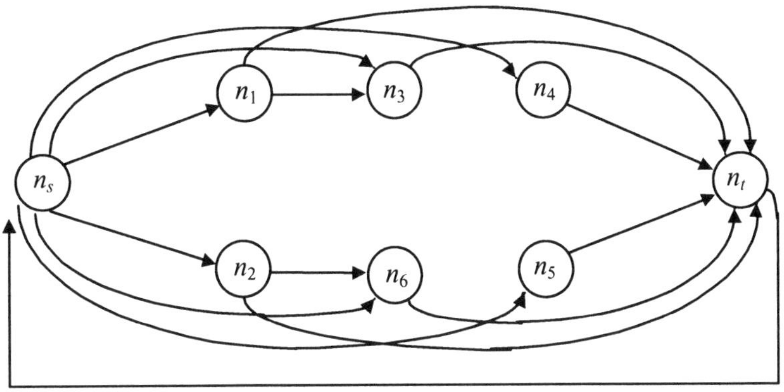

Figure III.7

Suppose now that our objective is to minimize the number of flight crews required to operate the six given flights. Assigning zero costs to all arcs except a_{ts} for which $c_{ts} = 1$, the unique optimal flow includes unit flows on paths $\pi_1 = (n_s, n_1, n_3, n_t)$, $\pi_2 = (n_s, n_4, n_t)$, $\pi_3 = (n_s, n_2, n_6, n_t)$, and $\pi_4 = (n_s, n_5, n_t)$ and a flow of $x_{ts} = 4$ to complete the circulation. The cost, i.e., the number of crews required is $c_{ts}x_{ts} = 4$, and one crew each will operate all flights on each of the four paths. Clearly, such a process may result in very uneven workloads, and additional constraints may have to be included to remedy the problem. Note that if the transfer time were cut to $t = \frac{1}{2}$ hour, an arc a_{46} would have to be added to the network. However, this addition would not result in a reduction of the number of crews required to operate the flights.

Instead of minimizing the number of flight crews used in the creation, we may instead wish to minimize the costs of maintaining operations with flight crews. While the general costs to support a crew could be expressed by an appropriate value c_{ts}, different accomodation and transfer costs at different airports could be expressed in the costs c_{sj} (or c_{it}). Regardless of the objective, while flow models can be solved more efficiently than general integer covering models, the latter are much more flexible and able to incorporate additional constraints. Even so, one of the uses of fairly simple flow models for this application is to quickly obtain lower bounds on the objective function.

2.5 Allocation of Representatives to Committees

An engineering faculty has q departments D_i, $i = 1, ..., p$ and r faculty members F_j, $j = 1, ..., r$ who belong to p ranks R_k, $k = 1, ..., p$. A faculty committee is to be created, consisting of q faculty members, one from each department. However, at least one but no more than u_k faculty members on the committee can belong to rank R_k, where the values of u_k, $k = 1, ..., p$ are predetermined numbers set to avoid the overrepresentation of any rank on the committee. Such a committee composition is said to be *balanced*. As an example, let the departments be Environmental, Civil, Electrical, Industrial and Mechanical engineering, so that $q = 5$. Furthermore, let $r = 10$ faculty members be considered for membership on the committee in question and assume that each individual belongs to exactly one rank, *viz*., Assistant, Associate and Full Professor, so that $p = 3$. The departmental and rank data for each faculty member are provided in Table III.4.

Table III.4

Department	Faculty member and rank
D_1: Environmental	F_1 (Associate)
D_2: Civil	F_2 (Assistant), F_3 (Assistant)
D_3: Electrical	F_4 (Assistant), F_5 (Associate), F_6 (Full)
D_4: Industrial	F_3 (Assistant), F_7 (Full), F_8 (Associate)
D_5: Mechanical	F_8 (Associate), F_9 (Associate), F_{10} (Full)

Here, faculty member F_3 is cross-appointed, belonging to the Departments of Civil and Industrial Engineering at the same time; similarly, F_8 belongs to both Industrial and Mechanical Engineering. Finally, no more than $(u_1, u_2, u_3) = (1, 3, 2)$ faculty members of the respective ranks should serve on the committee.

This situation can be modeled by introducing a source node n_s, a sink node n_t and arcs to form the network in Figure III.8. All arcs have upper bounds 1 except the ones leading into n_t, which have capacities $u_1, \ldots, u_p$. Finding a balanced committee is then equivalent to solving a maximum flow problem. If the value of the maximum flow is $f_{\bar{\mathbf{x}}} = q = 5$, then the corresponding flow indicates the composition of a balanced committee; if $f_{\bar{\mathbf{x}}} < 5$, no balanced committee exists. In this example, there are several possibilities for a balanced solution; e.g., $F_1, F_2, F_5, F_7, F_{10}$; F_1, F_3, F_6, F_8, F_9; $F_1, F_2, F_5, F_8, F_{10}$.

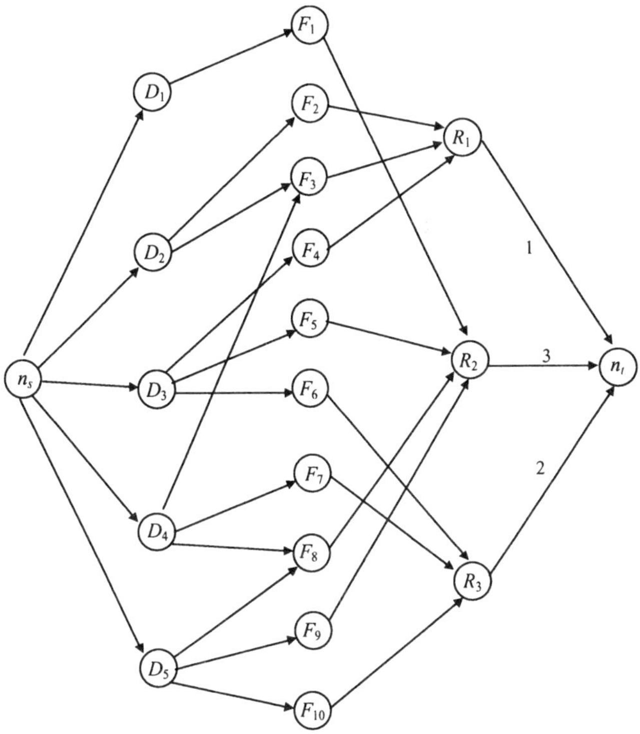

Figure III.8

Some general observations regarding the problem can be made. In particular, there are two extreme cases:

- If $\sum_{k=1}^{p} u_k < q$, no balanced solution exists, since for a balanced solution, the flow value $f_{\mathbf{x}}$ must equal q.

- If $\min\{u_k\} \geq q$, $1 \leq k \leq p$, balanced solutions trivially exist. One such solution includes one arbitrary member of each department.

Otherwise, a balanced solution may or may not exist. In the above example, these conditions imply that the problem with requirements $(u_1, u_2, u_3) = (1, 2, 1)$ has no balanced solution, while for the problem with $(u_1, u_2, u_3) = (6, 5, 9)$, any solution with a faculty member from each department has a balanced solution. Between these extremes, the problems with requirements $(u_1, u_2, u_3) = (1, 3, 2)$, $(1, 2, 2)$, $(3, 1, 1)$, $(1, 3, 1)$, $(1, 1, 3)$, $(2, 1, 2)$, and $(2, 2, 1)$ all have balanced solutions, whereas the problem with $(u_1, u_2, u_3) = (3, 0, 2)$ does not.

It is also worth mentioning that while the condition $\sum_{k=1}^{p} u_k < q$ prohibits balanced solutions from existing, the condition $\sum_{k=1}^{p} u_k \geq q$ does *not* imply that balanced solutions do exist. A counterexample is provided in Figure III.9.

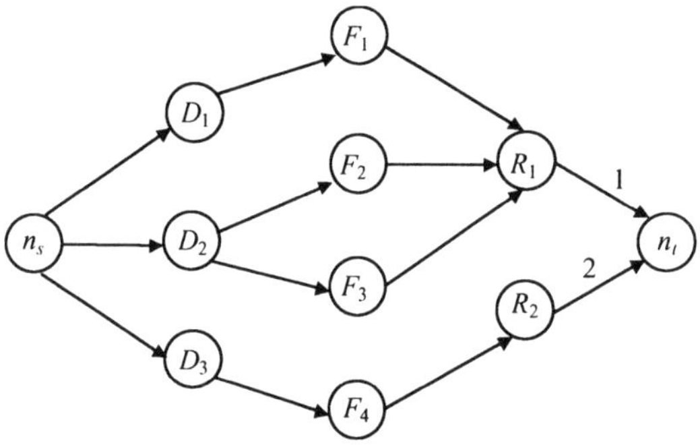

Figure III.9

A number of variants of the problem exist. For instance, instead of requiring that each department is represented by *exactly* one member on the committee, the requirement may demand that *at least* and/or *at most* t_i members from department D_i, $i = 1, ..., q$. Similarly, it may be required to have between ℓ_k and u_k faculty representation from rank R_k, $k = 1, ..., p$. Finally, the objective of the problem may be modified. Let x_k denote the flow through node R_k, $k = 1, ..., p$; then the original problem attempts to maximize $\sum_{k=1}^{p} x_k$. Denote now by $|R_k|$ the number of faculty members that have rank R_k regardless of their membership on the committee. Then we may employ the objective

$$\text{Min} \sum_{k=1}^{p} \left| x_k - q|R_k| \Big/ \sum_{\ell} |R_\ell| \right| \tag{1}$$

or

$$\text{Min} \max_{1 \le k \le p} \left| x_k - q|R_k| \Big/ \sum_{\ell} |R_\ell| \right|. \tag{2}$$

As x_k denotes the number of committee members of rank R_k and $|R_k| \Big/ \sum_{\ell} |R_\ell|$ denotes the proportion of faculty members of rank R_k, $q\ |R_k| \Big/ \sum_{\ell} |R_\ell|$ is the ideal number of representatives of rank R_k on the committee, as it reflects exactly the proportion of faculty members of that rank in the institution. The expression $\left| x_k - q|R_k| \Big/ \sum_{\ell} |R_\ell| \right|$ then measures the deviation of the actual committee assignment from that ideal. The objectives (1) and (2) then attempt to minimize the sum and the maximum of the imbalance, respectively. This is precisely the apportionment problem discussed in Section 3.1.7 of Part I.

2.6 Computer Program Testing

Each computer program whose steps consist of a sequence of instructions, should be tested on its validity. This requires that the consequence of each instruction is examined on at least one data set or other input. The question is now how to find the smallest number of inputs that guarantee that the consequence of each instruction is tested at least once. In order to explain the concept, consider the following fragment of a FORTRAN code.

```
READ (N)

READ (A)

IF N.LT.2 THEN

      IF N.LT.1 THEN

            S = - 1

      ELSE S = A(1) - 0.5

      ENDIF
```

```
      ELSE
            S = 0
            DO 10 I = 1, N
                  READ (A(I))
                  S = S + A(I)
10          CONTINUE
      ENDIF
      WRITE (S)
```

A flow chart and its more formal graph structure are displayed in Figures III.10a and b.

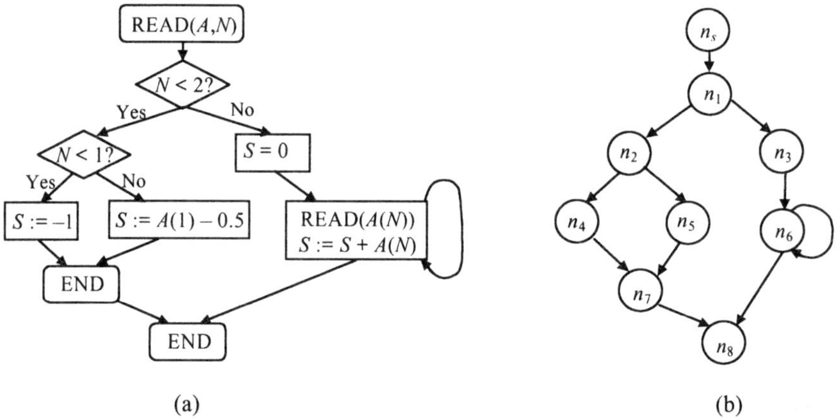

Figure III.10

Any input that is read at node n_s will follow a path to n_t. Hence, the task at hand can be reformulated to determine the smallest number of s-t paths that cover all arcs. This can be achieved by assigning a lower bound of 1 and capacities of $M >> 0$ to each arc, and determining an s-t flow of *minimal* rather than maximal value. This may be achieved by assigning arbitrary, but equal, costs to all arcs leading out of the source and zero costs to all other arcs. The minimal cost flow is then also the flow with the minimal flow value. The value of any minimal network flow from source to sink then indicates the minimal number of data sets required for testing purposes. For example, a minimal flow from n_s to n_t in the network shown in Figure III.10b is obtained by sending one unit of flow along each of the paths n_s, n_2, n_5, ..., n_5, n_t; n_s, n_1, n_3, n_6, n_t; and n_s, n_1, n_4, n_6, n_t, thus three data sets are needed to test the flow on each arc at least once. This result does, however, say nothing about what these input sets should include when they are used for testing purposes. Actually, such input sets may not even exist, as is the case in programs in which an arc will never be used. Consequently, the minimum flow value computed above may only serve as a lower bound on the number of test sets required in the process.

2.7 Distributed Computing

A "time critical" computer program that requires computing in a real-time environment, such as might arise in a military or manned satellite application, may take too long to be executed on a single processor. It may be possible to decompose the program into relatively independent parts, so that the individual parts can be run concurrently on different computers. This does, however, require that modules communicate with each other. Furthermore, since processors may differ, processing times of modules on processors may also be different as some processors may be more suited for running particular modules. Stone (1977) has examined and solved the case in which two processors are available.

Suppose a program has been decomposed into a number of modules M_1, ..., M_p and that the two available processors are S and T. Let the required volume of communication from module M_i to module M_j be d_{ij} $\forall$ i, j. These required volumes might have been estimated from previous runs or by prior simulations. Since communication between modules processed on the same processor can be handled efficiently with negligible cost, they can be neglected. However, communication between modules handled by different processors is expensive. Furthermore, let the cost of processing module M_i on processor S be a_i and b_i on T. The objective function is then to find a partition of modules that minimizes

$\sum_{i:M_i \in S} a_i + \sum_{j:M_j \in T} b_j + \sum_{i,j:M_i \in S, M_j \in T} d_{ij}$, where d_{ij} can be considered as the cost of communication between modules M_i and M_j, given that they are processed by different processors.

In order to develop the model further, we construct an undirected graph $G = (N, E, \kappa)$, with $N = \{n_i$: M_i is a module, $i = 1, ..., p\} \cup \{n_S, n_T\}$, $E = \{e_{ij}$: $n_i, n_j \in N\} \cup$

$\bigcup_{i=1}^{p}\{e_{si}\} \cup \bigcup_{i=1}^{p}\{e_{it}\}$, i.e., each module and the two processors are represented by nodes and, with the exception of e_{ST}, the graph is complete.

Example: Consider a problem with two computers S and T, and five modules M_1, ..., M_5. The cost of running the modules on the two computers is given in Table III.5.

Table III.5

		Modules				
		M_1	M_2	M_3	M_4	M_5
S	a_i	29	17	24	42	21
T	b_i	26	11	19	12	32

Furthermore, the communication costs d_{ij} between processors is shown in Table III.6, where $d_{ij} = 0$ indicates that modules M_i and M_j have no need to communicate.

Table III.6

i \ j	M_1	M_2	M_3	M_4	M_5
M_1	0	6	3	7	0
M_2	6	0	2	0	8
M_3	3	2	0	2	1
M_4	7	0	2	0	5
M_5	0	8	1	5	0

The graph constructed for this example is displayed in Figure III.11 on the next page.

To illustrate the situation, consider a specific assignment of modules to processors, e.g., assign M_1 and M_4 to computer S and modules M_2, M_3, and M_5 to computer T. The costs incurred in this solution are $a_1 + a_4 + b_2 + b_3 + b_5 + d_{12} + d_{13} + d_{15}\ (= 0) + d_{42}\ (= 0) + d_{43} + d_{45}$. Consider now the graph in Figure III.11. The cut (the broken line) separates the nodes that are processed by computer S, from those nodes representing modules processed by computer T. It includes the edges e_{12},

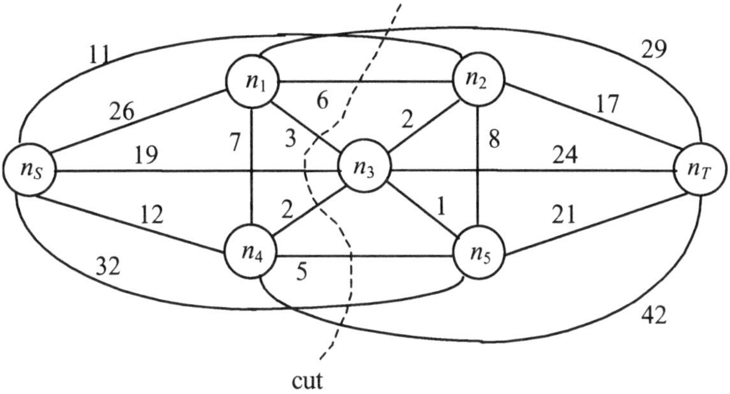

Figure III.11

e_{13}, e_{43}, e_{45}, whose processing times are counted towards the objective function. Moreover, the cut also intersects the edges e_{S2}, e_{S3}, and e_{S5} that connect modules with a processor. In the objective function, they are counted as b_2, b_3, and b_4. Similarly, the cut includes the edges e_{1T} and e_{4T} that connect a processor with modules which are counted in the objective function as a_1 and a_4. The above discussion allows us to define edge capacities κ_{ij}, so that $\kappa_{Si} = b_i \; \forall \; i$, $\kappa_{iT} = a_i \; \forall \; i$, and $\kappa_{ij} = d_{ij} \; \forall \; i, j \neq S, T$. The value of any cut will then express the total cost of communication if the modules are divided as indicated by the cut and processed in parallel. The objective is then to find a cut with minimal value, a task equivalent to determine the maximal flow from n_S to n_T. The resulting cut will specify (one of) the optimal division(s) of modules.

A few simple generalizations are easy to incorporate into the basic model. These include asymmetric communication costs with $d_{ij} \neq d_{ji}$ and cases in which a module may not be processed by one of the computers, in which case we set a_i (or b_i) $= M$ $>> 0$. Models that include more than two computers have been discussed in the literature, but are far more complicated.

2.8 Matrix Balancing Problems

Consider a problem in which a matrix $\mathbf{A} = (a_{ij})$ is given. This matrix must be updated periodically; however, the new matrix elements are not known, only its row and column sums. The goal is to provide as good an estimate of the updated matrix as possible. A typical application of matrix balancing occurs in migration matrices provided by the National Census Bureaus of many countries. In particular, given p regions R_1, R_2, ..., R_p, a matrix element a_{ij} indicates the migration from region R_i to region R_j in the year under consideration. Due to the high cost of data collection, a census is typically taken only every ten years. In any of the intervening years, only the number of people who moved within, into, or out of the regions are known. More specifically, let s'_i denote the number of people in R_i who have moved (possibly within the region), and let d'_j denote the number of people who have moved to region R_j (possibly from within the region). It is important to note that the parameters a_{ij} relate to the latest census figures, whereas the vectors $\mathbf{s}'$ and $\mathbf{d}'$ relate to some year between census takings.

The problem is now to estimate the relative migrations between countries by variables x_{ij}, so that the sums of in- and out-migrations coincide with the observed data s'_i and d'_j $\forall\ i, j$, and x_{ij} is "close" to a_{ij} / b, the observed relative migration from R_i to R_j in the past. One possibility is to solve the problem

$$\text{P: Min } z = \sum_{i=1}^{p}\sum_{j=1}^{p}(\tfrac{1}{b'}a_{ij} - x_{ij})^2$$

$$\text{s.t.} \quad \sum_{j=1}^{p} x_{ij} = \tfrac{1}{b'}s'_i \quad \forall\ i = 1, ..., p$$

$$\sum_{i=1}^{p} x_{ij} = \tfrac{1}{b'}d'_j \quad \forall\ j = 1, ..., p$$

$$x_{ij} \geq 0 \qquad \forall\ i, j = 1, ..., p.$$

Clearly, this problem formulation indicates that the structure is one of a standard transportation problem with a quadratic objective. The structure of the underlying graph $G = (N, A; \ell, \kappa; c)$ of the minimal cost flow problem is shown in Figure III.12.

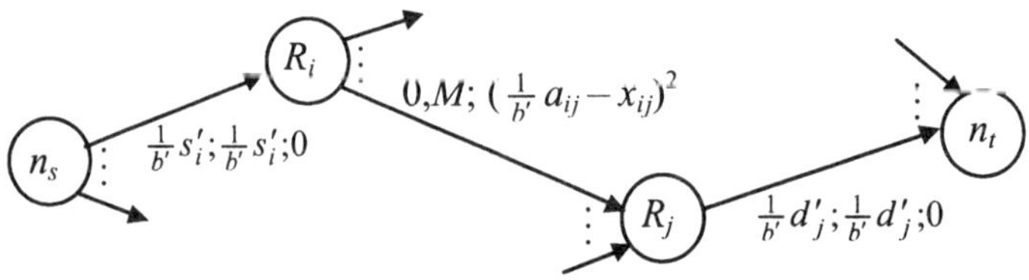

Figure III.12

Sometimes, it may by desired to assign weights w_{ij} to individual terms in the objective function, so that the weighted objective is then to minimize $z = \sum_{i=1}^{p}\sum_{j=1}^{p} w_{ij}(\frac{1}{b'}a_{ij} - x_{ij})^2$. As usual in the minimization of weighted objective functions, a large weight w_{ij} will force the variable x_{ij} as close as possible to the observed proportion $\frac{1}{b'}a_{ij}$.

Matrix balancing problems are also frequently used for updating population matrices in which a given population is categorized with respect to marital status (e.g., single, married, separated, divorced, widowed), age group, and/or other sociological data. As was the case with the migration matrices, cross-categorization data may be made available only infrequently, so that at intermediate points in time, estimations must be made on the basis of category totals.

2.9 Matrix Rounding Problems

Closely related to the problem of matrix balancing discussed in the previous section, is the problem of matrix rounding. Suppose that a $[m \times n]$-dimensional matrix $\mathbf{A} = (a_{ij})$ of real numbers is given. Its row sums are $s_i = \sum_{j=1}^{n} a_{ij}$ and its column sums are $d_j = \sum_{i=1}^{m} a_{ij}$. The problem is now to round all numbers a_{ij}, s_i, and d_j up or down to the nearest integer, so that the rounded s_i and d_j-values still represent the row and column sums of the rounded matrix, respectively. In other words, the problem is to find integers $\tilde{a}_{ij}, \tilde{s}_i,$ and $\tilde{d}_j$, where $\tilde{a}_{ij} = \lfloor a_{ij} \rfloor$ or $\lceil a_{ij} \rceil$, $\tilde{s}_i = \lfloor s_i \rfloor$ or $\lceil s_i \rceil$, and $\tilde{d}_j = \lfloor d_j \rfloor$ or $\lceil d_j \rceil$ $\forall\, i = 1, ..., m$ and $j = 1, ..., n$, such that $\sum_{j=1}^{n} \tilde{a}_{ij} = \tilde{s}_i$ $\forall\, i$ and $\sum_{i=1}^{m} \tilde{a}_{ij} = \tilde{d}_j$ $\forall\, j$. Such a rounding is called *consistent*.

Example: Consider the matrix

$$\mathbf{A} = \begin{bmatrix} 1.9 & 1.7 & 3.0 \\ 2.1 & 7.6 & 5.1 \end{bmatrix}$$

with $\mathbf{s} = [6.6, 14.8]^{\mathrm{T}}$ and $\mathbf{d} = [4.0, 9.3, 8.1]$. The rounded matrix

$$\tilde{\mathbf{A}} = \begin{bmatrix} 2 & 2 & 3 \\ 2 & 7 & 5 \end{bmatrix}$$

has $\tilde{\mathbf{s}} = [7, 14]^{\mathrm{T}}$ and $\tilde{\mathbf{d}} = [4, 9, 8]$, which are also rounded from the original row and column sums $\mathbf{s}$ and $\mathbf{d}$; hence, the rounding is consistent. On the other hand, consider the rounding

$$\tilde{\mathbf{A}} = \begin{bmatrix} 2 & 1 & 3 \\ 2 & 7 & 6 \end{bmatrix}$$

which is also a rounding of the original matrix. However, now $\tilde{\mathbf{s}} = [6, 15]^{\mathrm{T}}$ and $\tilde{\mathbf{d}} = [4, 8, 9]$, with $\tilde{d}_2 = 8$ which is not a rounding of $d_2 = 9.3$; hence, this rounding is not consistent.

The problem of consistent rounding can be formulated as a feasible flow problem in a capacitated network. For each row, we define a complete bipartite graph with one node R_i for each row of the matrix $\mathbf{A}$, $i = 1, ..., m$ and one node C_j for each column of $\mathbf{A}$, $j = 1, ..., n$. Add a source n_s and a sink n_t, and connect n_s to all nodes R_i, so that the arc a_{si} has bounds $(\ell_{si}, \kappa_{si}) = (\lfloor s_i \rfloor, \lceil s_i \rceil)$. Furthermore, connect all nodes C_j to the sink n_t, so that the arc a_{jt} has bounds $(\ell_{jt}, \kappa_{jt}) = (\lfloor d_j \rfloor, \lceil d_j \rceil)$. Furthermore, the arcs a_{ij} of the bipartite graph have bounds $(\ell_{ij}, \kappa_{ij}) = (\lfloor a_{ij} \rfloor, \lceil a_{ij} \rceil)$. The graph $G = (N, A; \ell, \kappa)$ associated with the above numerical example is shown in Figure III.13.

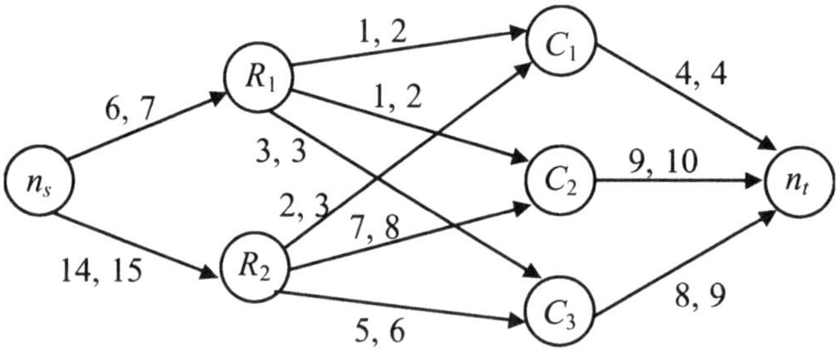

Figure III.13

Any feasible flow in the network will correspond to a consistent rounding. In case no feasible flow exists, then neither does a consistent rounding.

An application of the rounding problem is found in statistics, provided by the U.S. Census Bureau. The government agency must protect the privacy of individuals, which may require "fudging" the numbers, especially if some proportions are very small, which may otherwise point to specific individuals. In such a case, rounding may be required up and/or down to a multiple of some number α, not necessarily an integer.

CHAPTER 3 NETWORK FLOW ALGORITHMS

3.1 Maximal Flow Algorithms

This section first describes the general principle of augmenting flows as introduced by Ford and Fulkerson (1956) in their seminal contribution. Then a number of modifications and computational improvements are surveyed.

3.1.1 The Method of Ford and Fulkerson

The basic idea of all flow augmenting methods is to increase the flow by identifying paths from the source n_s to the sink n_t that have spare capacity on all arcs on the path. As an illustration, consider the following example.

Example: Let the graph $G = (N, A, \kappa)$ in Figure III.14 be given

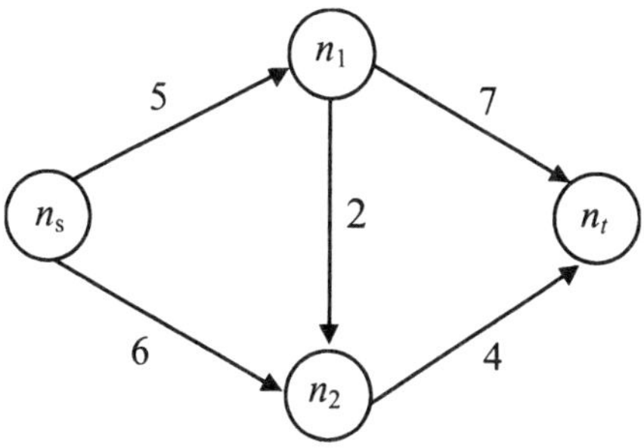

Figure III.14

At present the flow is zero in all areas. As the first path, we could choose $\pi_1 = (n_s, n_1, n_2, n_t)$ along which 2 units may be sent. Subsequently, an additional 3 units can be sent from source to sink on the path $\pi_2 = (n_s, n_1, n_t)$, and finally, 2 units can be sent on the path $\pi_3 = (n_s, n_2, n_t)$. This results in the flow pattern shown in the graph $G = (N, A, \kappa, x)$ in Figure III.15.

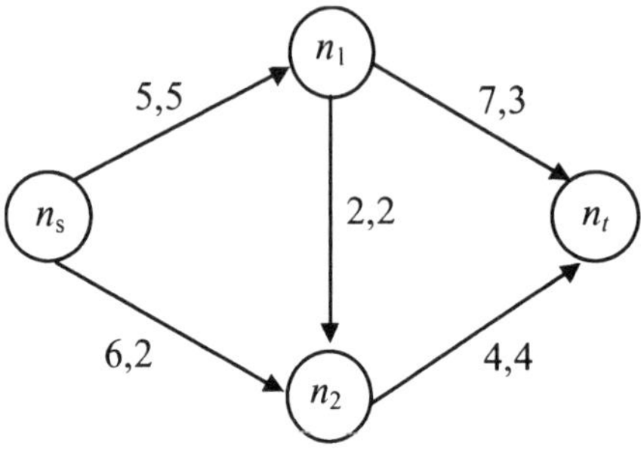

Figure III.15

At this point, it appears that no further flow augmenting path exists: leaving n_s towards n_1 is not possible as the flow on arc a_{s1} already equals its capacity. On the other hand, some flow may be shipped out of n_s to n_2, but there is no way for the flow to leave n_2 as the only outlet is the arc a_{2t}, and the current flow already equals its capacity. Hence, it seems that the present flow whose value is 7, is optimal. This is, however, not the case. The minimal cut is $S = \{n_s, n_2\}$ and $T = \{n_1, n_t\}$ with $A^{S,T} = \{a_{s1}, a_{2t}\}$ with a capacity of 9. In fact, if we were to reroute 2 flow units from the path $\pi_1 = (n_s, n_1, n_2, n_t)$ to (n_s, n_1, n_t), some capacity on arc a_{2t} would be freed up, allowing us to send 2 additional units on the path $\pi_3 = (n_s, n_2\ n_t)$, bringing to 9 the total flow, which is now maximal. The maximal flow pattern and the minimal cut of this problem are shown in Figure III.16.

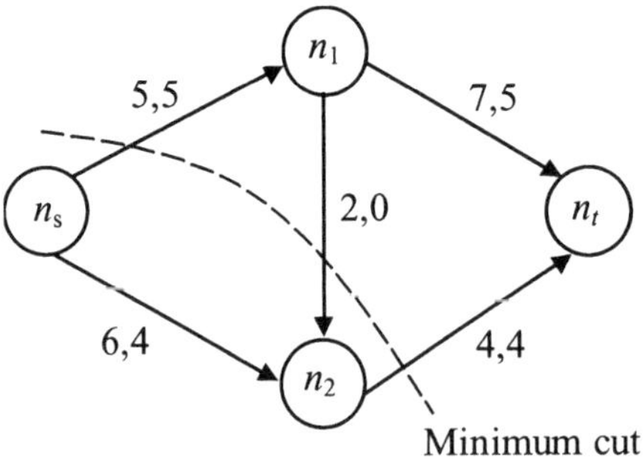

Figure III.16

The rerouting approach in the above example is, of course, haphazard and must be improved upon. One possibility to do so is to design a technique that reroutes flows by not only increasing *forward flows*, but also allowing to decrease *backward flows*. In particular, assume that it is known to be possible to ship some additional flow units from the source to some node n_i. We now attempt to ship those units onward to some node n_j that is connected to n_i by either an arc a_{ij} or a_{ji}. First, assume that $a_{ij} \in A$, a forward arc as seen from n_i. The free, slack, or residual, capacity of the arc a_{ij} is $\kappa'_{ij} = \kappa_{ij} - x_{ij}$ which is the maximal number of flow units that can be moved from n_i on to n_j. This is called forward routing. Assume now that $a_{ji} \in A$, a backward arc as seen from n_i. As opposed to forward routing which increases the flow in an arc, backward routing decreases the flow. Given that the current flow is x_{ji}, flows may be rerouted by decreasing the backward flow from n_j to n_i by at most $\kappa'_{ij} = x_{ji}$ units. In other words, we attempt to find a *flow augmenting chain* γ from the source to the sink, so that on each arc on that chain either the forward flow can be increased by a positive amount ε, or the backward flow can be decreased by some $\varepsilon > 0$. Such a path is shown by the bold lines in Figure III.17 with "+ε" and "–ε" indicating the change of the current flow.

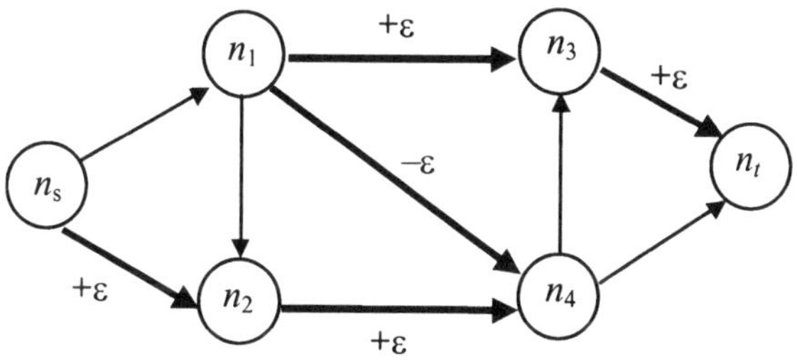

Figure III.17

Applying the principle of flow augmenting chains to the graph of Figure III.15, we find the chain $\gamma = (n_s, n_2, n_1, n_t)$ including the forward edge a_{s2}, the backward edge a_{12}, and the forward edge a_{1t}. The residual capacity of the forward arcs is $6 - 2 = 4$ and $7 - 3 = 4$, respectively, while the flow in the backward arc can be decreased by $x_{12} = 2$. The minimum change on all arcs in the chain determines the largest possible flow change on ε on that chain; here $\varepsilon = \min\{4, 2, 4\} = 2$. Modifying the flow on all arcs in the chain γ results in the flow shown in Figure III.16 which is known to be maximal.

We can now formalize the process. Define $R(\mathbf{x})$ as the *reachable set* given the current flow pattern $\mathbf{x}$, so that $R(\mathbf{x})$ includes all nodes n_i for which a flow-augmenting chain $\gamma = (n_s, ..., n_i)$ exists. We can then prove

Lemma III.9: Given a flow pattern $\mathbf{x}$ and a reachable set $R(\mathbf{x})$. Then either $n_t \in R(\mathbf{x})$, in which case a flow-augmenting chain γ from n_s to n_t exists (a *breakthrough*). In case of a *nonbreakthrough* with $n_t \notin R(\mathbf{x})$,

$$x_{ij} = \kappa_{ij} \quad \forall\ a_{ij}\colon n_i \in R(\mathbf{x}),\ n_j \notin R(\mathbf{x})$$

$$x_{ij} = 0 \quad \forall\ a_{ij}\colon n_i \notin R(\mathbf{x}),\ n_j \in R(\mathbf{x}), \text{ and}$$

$$f_{\bar{\mathbf{x}}} = \kappa\,(S, T) = \kappa\,(R(\mathbf{x}),\ \overline{R}\,(\mathbf{x}));$$

i.e., the flow pattern $\mathbf{x}$ is maximal and the value of the maximal flow equals the value of a minimal cut.

The constructive proof of the min cut = max flow theorem is straightforward and based on the principle of flow-augmenting chains. The flow-maximizing algorithm by Ford and Fulkerson (1956) is also based on that principle. It is initialized by any feasible flow pattern $\mathbf{x}$ (normally, but not necessarily, the zero flow), the set of labeled nodes $L = \{n_s\}$, so that in general, a node n_i is labeled as (ℓ_i, ε_i) where ε_i denotes the largest number of additional flow units that could be shipped from n_s to n_i on a flow-augmenting chain γ, and ℓ_i is n_i's predecessor node on that chain. Initially, we set $(\ell_s, \varepsilon_s) = (n_s, \infty)$. Finally, $J \subseteq L$ is the set of nodes from which we have already labeled. Initially, $J = \varnothing$.

The Ford and Fulkerson Algorithm

Step 1: Is $n_t \in L$?
If yes (breakthrough): Go to Step 3.
If no: Go to Step 2.

Step 2: Is $J = L$?
If yes (nonbreakthrough): The current flow is maximal and $(L, \overline{L}\,)$ is a minimal cut.
If no: Go to Step 4.

Step 3: Determine a flow-augmenting chain γ that includes forward arcs in $\gamma_F = \{a_{ij}\colon \ell_j = n_i\}$ and backward arcs $\gamma_B = \{a_{ij}\colon \ell_i = n_j\}$. The flow change is $\varepsilon = \varepsilon_t$ and the improved flow pattern $\mathbf{x} = (x_{ij})$ is determined as $x_{ij} := x_{ij} + \varepsilon\ \forall\ a_{ij} \in \gamma_F$ and $x_{ij} := x_{ij} - \varepsilon\ \forall\ a_{ij} \in \gamma_B$. Erase all labels except that of n_s, set $L := \{n_s\}$, $J := \varnothing$ and go to Step 1.

Step 4: Choose a node $n_i \in L \setminus J$. For each node $n_j \notin L$, label n_j with $(\ell_j, \varepsilon_j) = (n_i, \min\{\varepsilon_i;\ \kappa_{ij} - x_{ij}\})$ if $a_{ij} \in A$, and with $(\ell_j, \varepsilon_j) = (n_i; \min\{\varepsilon_i;\ x_{ji}\})$ if $a_{ji} \in A$. Set $J := J \cup \{n_i\}$, $L := L \cup (n_j\}$ and repeat as often as possible. Go to Step 1.

Example: Consider the network $G = (N, A, \kappa)$ in Figure III.18.

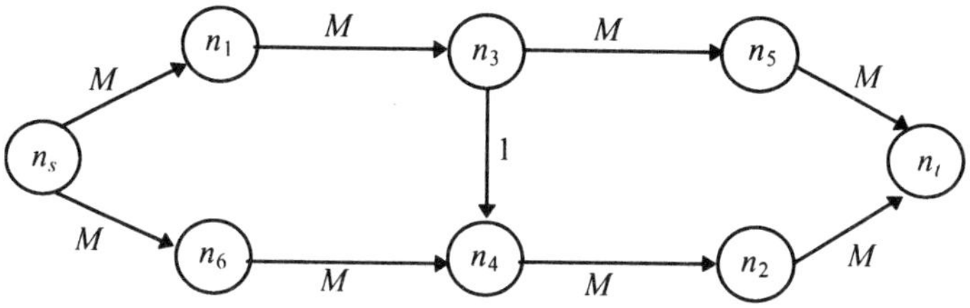

Figure III.18

The capacities in all arcs except a_{34} equal an arbitrarily large constant $M >> 0$. This example will not only demonstrate how the Ford-Fulkerson algorithm works, but will also show one of its shortcomings. Let the initial flow be $\mathbf{x} = 0$, and $L = \{n_s\}$ and $J = \emptyset$. Step 4 then uses n_s as the node to be labeled from (there is no choice), and labels the nodes n_1 and n_6, both with (n_s, M). At this point we update the sets J and L, so that $J = \{n_s\}$ and $L = \{n_s, n_1, n_6\}$. The next iteration commences with a node chosen from $L \setminus J = \{n_1, n_6\}$. As a real choice exists at this point, a rule must be devised for selecting a node. Here, we will choose the node with the smallest subscript. Labeling from n_1, we can label n_3 with (n_1, M), so that now $J = \{n_s, n_1\}$ and $L = \{n_s, n_1, n_3, n_6\}$. As $n_t \notin L$, the labeling procedure continues with n_3. From this node, we can label n_5 with (n_3, M) and n_4 with $(n_3, 1)$ so that now $J = \{n_s, n_1, n_3\}$ and $L = \{n_s, n_1, n_3, n_4, n_5, n_6\}$. Still, $n_t \notin L$ and the node with the smallest index in $L \setminus J$ is n_4. From n_4, we can label n_2 with $(n_4, 1)$ and update the sets J and L to $J = \{n_s, n_1, n_3, n_4\}$ and $L = \{n_s, n_1, n_2, n_3, n_4, n_5, n_6\}$. As $n_t \notin L$, the process continues with n_2. From n_2, we can label n_t with $(n_2, 1)$, and a breakthrough has occurred. We now determine the flow-augmenting chain γ by backtracking from n_t, i.e., $\ell_t = n_2$, $\ell_2 = n_4$, $\ell_4 = n_3$, $\ell_3 = n_1$, and $\ell_1 = n_s$. All arcs on this chain are forward arcs, so that $\gamma_B = \emptyset$ and with $\varepsilon = \varepsilon_t = 1$ the flows are modified to $x_{s1} = x_{13} = x_{34} = x_{42} = x_{2t} = 1$, and all other flows are unchanged (at the zero level). The resulting flow pattern is shown in the graph $G = (N, A, \kappa, x)$ in Figure III.19.

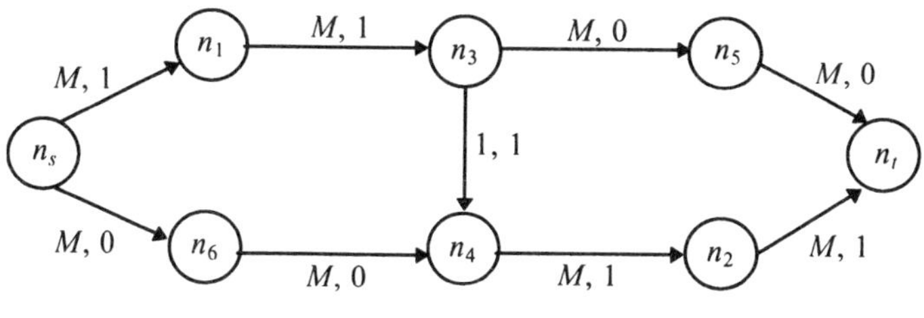

Figure III.19

The next iteration commences with $J = \varnothing$ and $L = \{n_s\}$. The first nodes to be labeled are n_1 and n_6 with the labels $(n_s, M-1)$ and (n_s, M), respectively. Suppose now that we change the choice rule in Step 4 to the node with the largest (rather than smallest) subscript. This means that we continue with n_6 and, from there, label n_4 with (n_6, M). Now $J = \{n_s, n_6\}$ and $L = \{n_s, n_1, n_4, n_6\}$ and we continue labeling from the node n_4. From n_4, we can label n_2 with $(n_4, M-1)$ and n_3 in backwards mode with $(n_4, 1)$, so that $J = \{n_s, n_4, n_6\}$ and $L = \{n_s, n_1, n_2, n_3, n_4, n_6\}$. Continuing the labeling procedure with n_3 leads to the labeling of n_5 with $(n_3, 1)$ and the updated sets $J = \{n_s, n_3, n_4, n_6\}$ and $L = \{n_s, n_1, n_2, n_3, n_4, n_5, n_6\}$. Finally, from n_5 it is possible to label n_t with $(n_5, 1)$, so that another breakthrough has occurred. Backtracking yields $\ell_t = n_5$, $\ell_5 = n_3$, $\ell_3 = n_4$, $\ell_4 = n_6$, $\ell_6 = n_s$, so that the flow-augmenting chain is $\gamma = (a_{s6}, a_{64}, a_{34}, a_{35}, a_{5t})$ with $\gamma_F = \{a_{s6}, a_{64}, a_{35}, a_{5t}\}$ and $\gamma_B = \{a_{34}\}$. Updating the flows yields a pattern shown in graph $G = (N, A, \kappa, x)$ in Figure III.20.

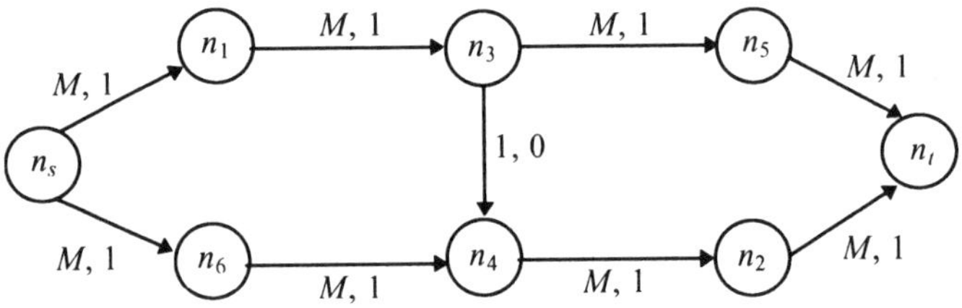

Figure III.20

Suppose now that in the succeeding iterations 3, 4, 5, ..., we use the "minimum subscript" rule in all odd-numbered iterations, and the "maximum subscript" rule in all even-numbered iterations. It soon becomes apparent that in odd-numbered iterations, we increase the flow by "1" along the path $\gamma = (a_{s1}, a_{13}, a_{34}, a_{42}, a_{2t})$, whereas we change the flow by "1"along the path $\gamma = (a_{s6}, a_{64}, a_{34}, a_{35}, a_{5t})$ in all even-numbered iterations. As the value of the flow through the network increases by one in each such iteration, the maximal flow of $f_{\bar{\mathbf{x}}} = 2M$ will be found only after M iterations, which can be made arbitrarily bad by choosing very large values of M. It is also worth noting that the "maximum subscript" rule, if applied from the beginning, finds the maximal flow in just two iterations. Also note that this problem exhibits multiple minimum cuts. As a matter of fact, any pair of arcs with one chosen from $\{a_{s1}, a_{13}, a_{35}, a_{5t}\}$ and the other from $\{a_{s6}, a_{64}, a_{42}, a_{2t}\}$ constitutes a minimal cut.

Before we discuss improvements of the Ford and Fulkerson algorithm that avoid slow convergence behavior as exhibited in the above example, we will introduce a modified version of the method. This version differs from the one shown above not by the way it proceeds, but by a simplified process. In particular, to each flow pattern $\mathbf{x}$ we can assign a *residual graph* $G(\mathbf{x}) = (N', A', \kappa')$, so that $N' = N$, $A' = \{ a'_{ij} \}$ with $a'_{ij} \in A'$ if $x_{ij} < \kappa_{ij}$, or if $x_{ji} > 0$. In other words, if the original arc a_{ij} presently carries a flow $x_{ij} < \kappa_{ij}$, then an arc a'_{ij} exists, and if the flow along a_{ij} is $x_{ij} > 0$, then an arc a'_{ji} exists. The residual capacity of an arc a'_{ij} is $\kappa'_{ij} = \kappa_{ij} - x_{ij}$ if $a_{ij} \in A$, and $\kappa'_{ij} = -x_{ji}$ if $a_{ji} \in A$. As an example, consider again the flow in the network of Figure III.15. The residual graph connected with this flow pattern is shown in Figure III.21.

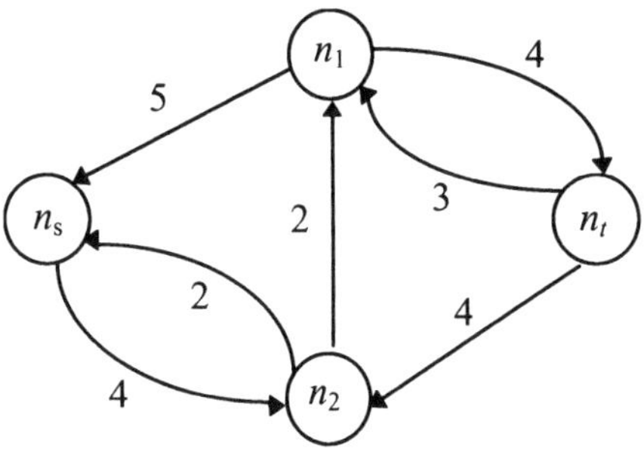

Figure III.21

Given the residual graph for any given graph and flow pattern, the Ford and Fulkerson algorithm can be modified so as to only consider forward labeling. If the residual graph includes a path from n_s to n_t (equivalent to a flow-augmenting chain in the original graph), then a breakthrough has occurred and the flow in the network can be increased. In the above example, the path $\pi = (n_s, n_2, n_1, n_t)$ is a flow-augmenting path with $\varepsilon = \min\{4, 2, 4\} = 2$, so that 2 additional flow units can be shipped on this path, i.e., 2 more units on the original arcs a_{s2} and a_{1t}, and 2 less units on the backward arc a_{12}. This modification of the flow pattern leads again to the maximal flow shown in Figure III.16.

One of the serious shortcomings of the original method by Ford and Fulkerson is that their algorithm, in case the capacities are irrational values, may not converge in a finite number of steps. Worse yet, the value it converges to may not be the maximal flow. A good description, based on an example by Ford and Fulkerson (1962), can be found in Chen (1990).

One way of trying to avoid many small flow augmentations was suggested by Edmonds and Karp (1972). The authors propose to always select a maximal capacity flow augmenting path. Gabow (1985) and later Ahuja and Orlin (1991) have demonstrated that it is not necessary to require that the capacity of the path be maximal, merely that it be "large".

Assume that all arc capacities are integers and that $U = \max_{a_{ij} \in A} \{\kappa_{ij}\}$. By $G(\mathbf{x}, \Delta)$ we denote the subnetwork of the residual network $G(\mathbf{x})$ that is determined by the current flow pattern $\mathbf{x}$ and where arcs all have a residual capacity of at least Δ. Initially, $\Delta = 2^k$ with $k = \min_{k \geq 0} \{2^k \geq U\}$. Ahuja and Orlin's (1991) *Capacity Scaling Algorithm* sets up the graph $G(\mathbf{x}, \Delta)$ and attempts to find an augmenting path in it. In case this is successful, a new flow pattern $\mathbf{x}$ with a higher flow value $f_\mathbf{x}$ has been found, and the process is repeated. In case no augmenting path was found, the value of Δ is redefined as $\Delta := ½\Delta$, and the process is repeated. The algorithm terminates with a maximal flow when $\Delta < 1$. The authors prove that the Capacity Scaling Algorithm solves the maximal flow problem in $O(m^2 \log U)$ time, which, with $m < n^2$, equals $O(n^4 \log U)$.

Edmonds and Karp (1972) suggested another polynomial implementation of Ford and Fulkerson's algorithm. The main feature of their work is a specific rule for the determination of the augmenting path. In particular, let any flow pattern $\mathbf{x}$ and its associated residual network $G(\mathbf{x})$ be given, and define the length of all arcs in $G(\mathbf{x})$ as one. Then the length of any path, and particularly a path from n_s to n_t, with respect to these arc lengths equals the number of arcs on that path. Note that the longest elementary path in $G(\mathbf{x})$ has length $(n - 1)$; any path whose length is n or more will visit at least one node more than once. The algorithm then goes through phases $k = 1, 2, \ldots, k - 1$, as follows. At the beginning of each phase, $G(\mathbf{x})$ is constructed and the flow is increased on all augmenting paths of length k. Once that is no longer possible, the new flow pattern is determined and its residual graph constructed, and a new phase begins. This version of Ford and Fulkerson's maximal flow algorithm has no more than $(n - 1) = O(n)$ phases, each of which consists of $O(m)$ flow augmentations, each of which, in turn, requires $O(m)$ operations. This results in an overall complexity of $O(m^2 n)$ that, as $m < n^2$, equals $O(n^5)$, which is polynomial, yet with a fairly high degree. A refinement of the algorithm that employs depth-first search was proposed by Dinic (1970); its complexity is $O(n^4)$.

3.1.2 Karzanov's Preflow Algorithm

As opposed to the techniques described in the previous subsection, Karzanov's (1974) algorithm does not merely specify certain rules of the selection of the augmenting path. In fact, there are two major differences between Karzonov's method and flow-augmenting algorithms. Firstly, while all flow-augmenting techniques first determine an augmenting path and end each iteration by augmenting the existing flow by sending additional flow units on that path, Karzanov's method ends each iteration with a marginal flow pattern (and not just a simple path) that is added to the existing flow. Secondly, each iteration of Karzanov's algorithm consists of two phases. The first phase, called "advance", is a forward sweep that pushes as many flow units as possible towards the sink. The result is called a preflow which may—and usually will—violate some of the conservation equations. The second phase is called "balance". It is a backward sweep that balances inflows and outflows wherever they were violated in the Advance phase. The result at the end of each phase is a marginal flow that, just as additional flow units on a flow-augmenting path, is added to the existing pattern.

In order to formalize the ideas, we first write

Definition III.10: A *preflow* $\mathbf{f} = (f_{ij})$ is a function that satisfies $f_{ij} \le \kappa_{ij}\ \forall\ a_{ij} \in A$ and $\sum_{j:a_{ji}\in A} f_{ji} - \sum_{j:a_{ij}\in A} f_{ij} \ge 0\ \forall\ n_i \in A$; i.e., the inflow into node n_i is equal to or exceeds the outflow out of n_i. The *excess* $e(n_i)$ of a node n_i is defined as the difference between in- and outflow of a node. A node is called *balanced* if $e(n_i) = 0$ and *unbalanced* if $e(n_i) > 0$.

Similar to the method by Edmonds and Karp, and that by Dinic, Karzanov's algorithm employs a distance function that indicates the length of the shortest path from the source to each node $n_i \in N$. Here, we formalize the idea as follows. As usual, we associate a residual graph $G(\mathbf{x}) = (N', A', \kappa')$ with any given flow pattern $\mathbf{x}$. We now define sets of nodes $N_0, N_1, \ldots, N_\ell$, so that $\{n_s\} = N_0$, and a node $n_i \in N_k$, if the length of the shortest path from n_s to n_i in $G(\mathbf{x})$ equals k. Each set of nodes N_k represents a layer of the graph. Clearly, if there exists no index ℓ, so that $n_t \in N_\ell$, then there exists no path from n_s to n_t in $G(\mathbf{x})$, and hence the existing flow pattern $\mathbf{x}$ is maximal. Otherwise we determine a *layered network* $L(\mathbf{x}) = (N'', A'', \kappa'')$ from $G(\mathbf{x})$ by

- deleting all arcs $a_{ij} \in A'$, for which $n_i \in N_k$, but $n_j \notin N_{k+1}$ (i.e., a_{ij} does not connect a node in one layer to a node in the next layer), and
- deleting all nodes and arcs that are not part of a path from n_s to n_t.

Given a layered network $L(\mathbf{x})$, the Advance part of an iteration commences by sending as many units out of the source as possible, i.e., by setting $f_{sj} = \kappa''_{sj} = \kappa'_{sj}\ \forall\ a_{sj} \in A''$. The general step is then to choose any unbalanced node n_i in the layer currently under consideration (the sequence in which the nodes in a layer are considered is one of the degrees of freedom in this algorithm) and sends as much flow out of n_i on arcs a''_{ij} as feasible until there is no longer any excess at n_i. The sequence in which the arcs a_{ij} are chosen represents another degree of freedom in this algorithm. Once no more flow can feasibly be sent out of the nodes of a layer, the Advance phase moves to the next layer of the layered network and repeats the procedure. The phase terminates once the method reaches layer N_ℓ, the layer that includes n_t.

The balance phase in an iteration commences with an unbalanced node in a layer with the highest subscript, i.e., the layer farthest from the source. The method achieves flow conservation by reducing the inflow into the node under consideration. This is accomplished in last in, first out (LIFO) fashion. Once the node is balanced, another node in the same layer is considered. Once all nodes in a layer arc balanced, the algorithm proceeds with the next layer of nodes closer to the source. The process terminates once all nodes in $L(\mathbf{x})$ are balanced. At that point in time, the iteration terminates with a marginal flow $\mathbf{y}$ in $L(\mathbf{x})$. The current flow $\mathbf{x}$ is then augmented to this marginal flow by setting $x_{ij} := x_{ij} + y_{ij}\ \forall\ a_{ij} \in A$ and $x_{ji} := x_{ji} - y_{ij}\ \forall\ a_{ji} \in A$, and the next iteration begins.

We are now able to formally describe the algorithm. It is initialized by setting $\mathbf{x} := \mathbf{0}$.

Karzanov's Preflow Algorithm

Step 1: Determine the residual graph $G(\mathbf{x})$ and the layered graph $L(\mathbf{x}) = (N'', A'', \kappa'')$ with layers of nodes $N_0, N_1, \ldots, N_\ell$.

Step 2: Is $n_\ell \in N_\ell$?
If yes: Go to Step 3.
If no: Stop; the current flow pattern $\mathbf{x}$ is maximal.

Step 3: Set $f_{sj} := \kappa''_{sj}\ \forall\ a_{sj} \in A''$. Set $k := 1$ and $S_k := \varnothing$, where S_k denotes the set of unbalanced nodes in layer k that have been scanned so far.

Step 4: Does there exist at least one node in $n_i \in N_k \setminus S_k$ with $e(n_i) > 0$?
If yes: Choose one such node n_i and go to Step 6.
If no: Go to Step 5.

Step 5: Is $k = \ell - 1$?
If yes: Go to Step 7.
If no: Set $k := k + 1$ and go to Step 4.

Step 6: Set $f_{ij} := \min\{\kappa''_{ij}; e(n_i)\}$ for some j with $a''_{ij} \in A''$. Repeat for other j until no further flow can be sent out of n_i. Set $S_k := S_k \cup \{n_i\}$ and go to Step 4.

Step 7: Is $e(n_i) = 0\ \forall\ n_i'' \in N''$?
If yes: Go to Step 10.
If no: Go to Step 8.

Step 8: Determine k as the highest subscript for which the current preflow **f** has an unbalanced node $n_i \in N_k$.

Step 9: Pick some unbalanced node $n_i \in N_k$, set $f_{vi} := \max\ \{0; f_{vi} - e(n_i)\}$, and go to Step 7.

Step 10: The present preflow **f** equals the new marginal flow $\mathbf{y} = (y_{ij})$. Modify the existing flow pattern $\mathbf{x}$ by setting $x_{ij} := x_{ij} + y_{ij}\ \forall\ a_{ij} \in A$ and $x_{ji} := x_{ji} - y_{ij}\ \forall\ a_{ji} \in A$. Go to Step 1.

It is apparent that the aforementioned Advance phase of an iteration comprises Steps 3-6, while the Balance module consists of Steps 7-9. The algorithm will now be illustrated by an

Example: Consider the graph $G = (N, A, \kappa)$ in Figure III.22.

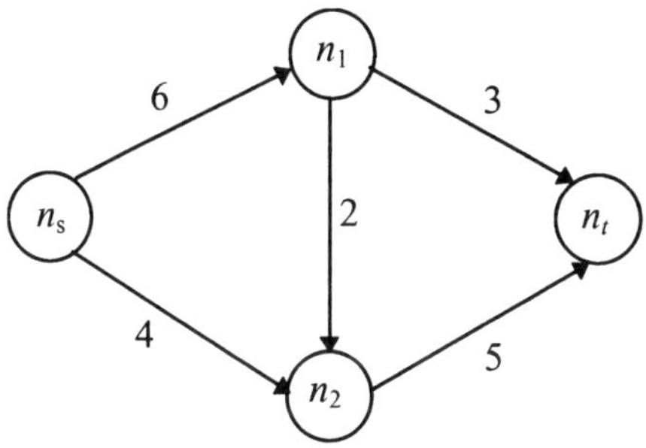

Figure III.22

Given an initial flow pattern $\mathbf{x} = \mathbf{0}$, the residual graph $G(\mathbf{x})$ equals the graph of Figure III.22. As the lengths of the shortest paths to the nodes n_s, n_1, n_2, and n_t are 0, 1, 1, and 2, respectively, we obtain layers with sets of nodes $N_0 = \{n_s\}$, $N_1 = \{n_1, n_2\}$, and $N_2 = \{n_t\}$. The layered graph $L(\mathbf{x}) = (N'', A'', \kappa'')$ is shown in Figure III.23 on the next page.

The Advance phase pushes 6 and 4 units out of the source on the arcs a''_{s1} and a''_{s2}, respectively. At this point, this preflow generates excesses of $e(n_1) = 6$ and $e(n_2) = 4$, respectively. First consider n_1 and push as many flow units as possible out of this node. As n_1's only outlet in $L(\mathbf{x})$ is the arc a''_{1t}, we set $f_{1t} = \min\ \{\kappa''_{1t}; e(n_1)\} = \min\{3; 6\} = 3$, thus leaving an excess of $e(n_1) = 3$. The second unbalanced node in

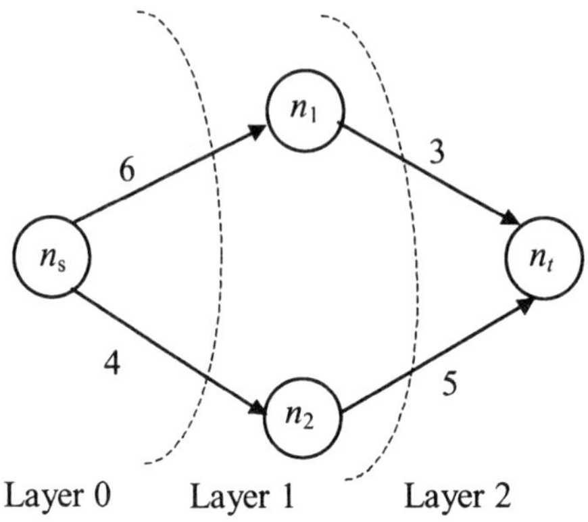

Figure III.23

this layer is n_2. Here, we manage to push the entire excess out of n_2 by setting f_{2t} = min{ κ''_{2t} ; $e(n_2)$} = min{5; 4} = 4. The Advance phase terminates with a preflow shown in Figure III.24a. The Balance phase then detects n_1 as the unbalanced node in the highest layer and balances it by reducing the flow into n_1 by 6 – 3 = 3 units. At the end of the phase we obtain the marginal flow **y** shown in Figure III.24b.

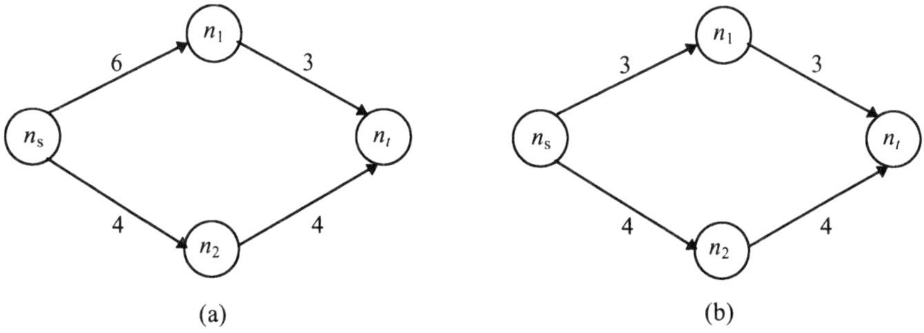

Figure III.24

Updating the current flow pattern $\mathbf{x} = \mathbf{0}$ by the marginal flow pattern **y** results in a new flow pattern **x** that equals that shown in Figure III.24b. The corresponding residual graph is shown in Figure III.25a and the associated layered network is given in Figure III.25b.

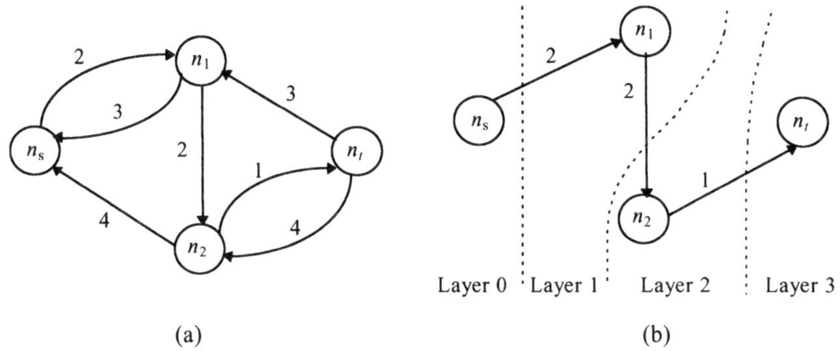

Figure III.25

The Advance phase pushes 2 flow units out of the source to n_1, pushes them on to n_2, and due to the capacity constraint $f_{2t} \leq \kappa''_{2t} = 1$, pushes one unit to the sink n_t. The preflow at the end of the Advance module is shown in Figure III.26a.

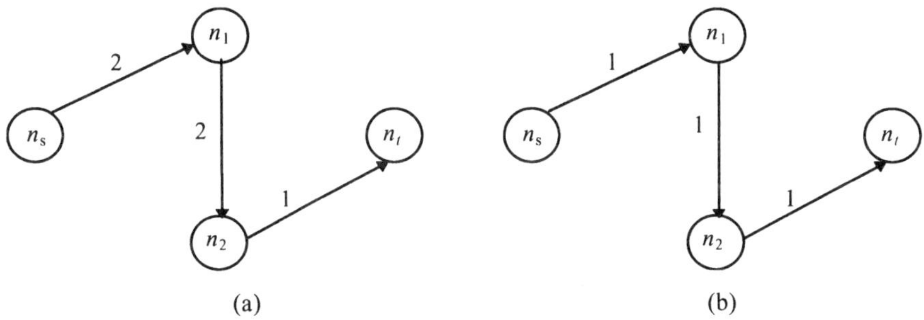

Figure III.26

The Balance phase identifies n_2 as an unbalanced node in the highest layer and attempts to balance it by pushing back flow. This is achieved by setting f_{12}:= 1. Now n_1 is an unbalanced node in the highest layer and, again, flow is pushed back by setting f_{s1}:= 1. The resulting marginal flow at the end of the Balance phase is shown in Figure III.26b. Augmenting the preset flow pattern **x** in Figure III.24b by the marginal flow in Figure III.26b, we obtain the new flow pattern **x** shown in

Figure III.27a together with its residual graph in Figure III.27b.

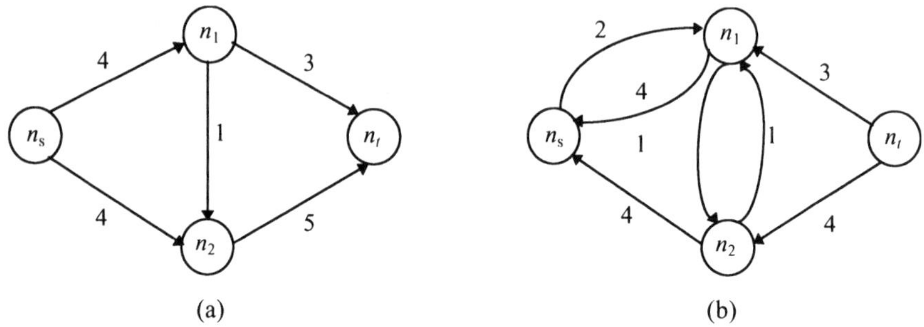

Figure III.27

The layered network $L(\mathbf{x})$ includes the nodes $n_s \in N_0$, $n_1 \in N_1$, and $n_2 \in N_2$. The sink n_t cannot be reached from the source; hence the current flow pattern in Figure III.27a is maximal.

It remains to be pointed out that a careful implementation of Karzanov's preflow algorithm has a computational complexity of $O(n^3)$. Another efficient preflow algorithm is due to Malhotra *et al.* (1978) which also runs in $O(n^3)$ time. Algorithms based on the concept of preflows are the most efficient maximum flow algorithms today.

Two algorithms worth mentioning were designed for finding maximal flows in graphs with special structures. One is an $O(n_1^2 m)$ algorithm for bipartite graphs with n_1 denoting the nodes in N_1. This algorithm is based on the concept of preflows. Secondly, there is an $O(n \log n)$ algorithm for *s-t* planar graphs, i.e., planar graphs with a planar representation that has the nodes n_s and n_t on the boundary of the external region. For details, readers are referred to Ahuja *et al.* (1993).

3.2 Feasible Flow Problems

The purpose of this section is to determine a technique that finds a feasible solution to the problem

$$\sum_j x_{ij} - \sum_j x_{ji} = 0 \qquad \forall\, i\colon n_i \in N \tag{1}$$

$$\left.\begin{array}{l} x_{ij} \geq \ell_{ij} \\ x_{ij} \leq \kappa_{ij} \end{array}\right\} \forall a_{ij} \in A \qquad (2),\ (3)$$

i.e., a flow in a network that satisfies the conservation equations (1), the lower bound constraints (2), and the upper bound or capacity constraints (3). For simplicity, consider a circulatory network, so that constraints (1) apply to all nodes n_i. Clearly, if $\ell_{ij} = 0\ \forall\ a_{ij} \in A$, then the zero flow $\mathbf{x} = \mathbf{0}$ is feasible, and solves the feasibility, or phase I, problem.

Following the standard practice in dealing with lower bound constraints, we define new variables x'_{ij}, so that $x'_{ij} = x_{ij} - \ell_{ij}$ or, equivalently,

$$x_{ij} = x'_{ij} + \ell_{ij} \tag{4}$$

Applying relation (4) to constraints (1), (2), and (3) leads to

$$\sum_j x'_{ij} - \sum_j x'_{ji} = \sum_j \ell_{ji} - \sum_j \ell_{ij} \tag{5}$$

$$x'_{ij} \geq 0 \tag{6}$$

and $$x'_{ij} \leq \kappa_{ij} - \ell_{ij}. \tag{7}$$

This process has eliminated the lower bounds which now all equal zero (6), but we have introduced artificial supplies and demands at the nodes as shown in (5). Define now $b_i = \sum_j \ell_{ji} - \sum_j \ell_{ij}\ \forall\ i$, so that $b_i > 0$ indicates that more lower bounds lead into a node than out of it, and similarly for $b_i < 0$. Recall that relation (4) implies that even for $x'_{ij} = 0$, there is a flow of ℓ_{ij} on the arc a_{ij}. Hence, if $b_i > 0$, more flow units enter than leave n_i. By artificially forcing a flow through a new arc into n_i, we allow this flow, in addition to the existing flow that equals the lower bounds, to satisfy the conservation equations. This leads to the following procedure. Given a network $G = (N, A, \ell, \kappa)$, define a new network $G^* = (N^*, A^*, \kappa^*)$, so that $N^* = N \cup \{n_s^*, n_t^*\}$ where n_s^* and n_t^* represent a "supersource" and a "supersink", respectively. Furthermore, let the set of arcs be $A^* = A \cup$

$\bigcup_{i:b_i>0} a_{s^*i} \cup \bigcup_{i:b_i<0} a_{it^*}$ and modified capacities $\kappa^*_{ij} = \begin{cases} \kappa_{ij} - \ell_{ij} \forall a_{ij} \in A \\ b_i \quad \forall a^*_{si} \in A^* \\ -b_i \quad \forall a^*_{it} \in A^*. \end{cases}$

We now determine the maximal flow from n_{s^*} to n_{t^*} in G^*. If the value of this flow is $f^*_{\bar{\mathbf{x}}} = \sum_i \kappa^*_{s^*i} = \sum_{i:b_i>0} b_i$, then a feasible flow exists. This flow can be reconstructed by virtue of relation (4), i.e., by adding the original lower bounds to the arc flows in the maximal flow pattern $\bar{\mathbf{x}}$. In case the maximal flow $f^*_{\bar{\mathbf{x}}} < \sum_{i:b_i>0} b_i$, no feasible flow exists.

Example: Consider the circulatory network $G = (N, A, \ell, \kappa)$ in Figure III.28.

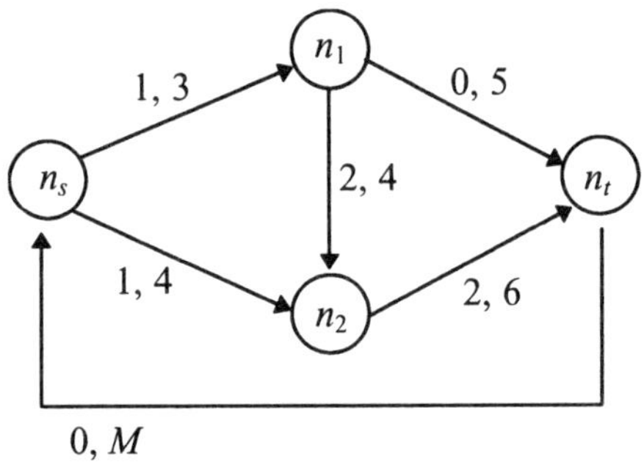

Figure III.28

Here, $b_s = -2$, $b_1 = -1$, $b_2 = 1$, and $b_t = 2$. Adding a supersource and supersink as well as the appropriate arcs (shown as broken lines), we obtain the graph $G^* = (N^*, A^*, \kappa^*)$ shown in Figure III.29 on the next page.

One maximal flow in G^* is obtained by sending one unit on the path $\pi_1 = (n_{s^*}, n_2, n_t, n_s, n_{t^*})$, one unit on $\pi_2 = (n_{s^*}, n_t, n_s, n_1, n_{t^*})$, and one unit on $\pi_3 = (n_{s^*}, n_t, n_s, n_{t^*})$. The resulting flow value is $f^*_{\bar{\mathbf{x}}'} = 3$, which equals the sum of positive surpluses b_i,

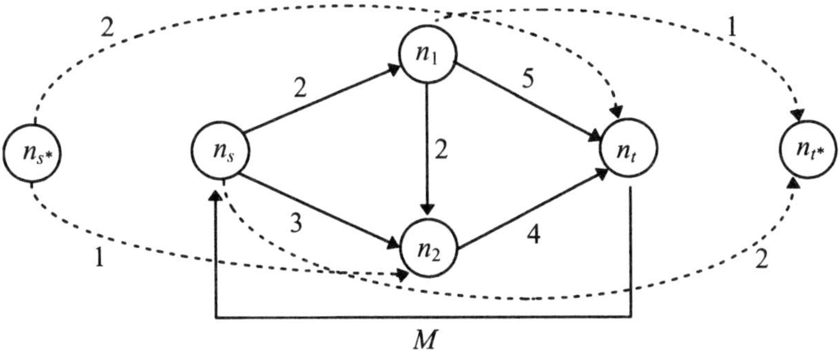

Figure III.29

so that the resulting flow pattern $\bar{\mathbf{x}}$ is feasible for the original problem. The arc flows in the maximal flow pattern $\bar{\mathbf{x}}'$ are $\bar{x}'_{s1} = \bar{x}'_{2t} = 1$ and $\bar{x}'_{ij} = 0$ otherwise. Applying relation (4), the feasible flow pattern in Figure III.30 results.

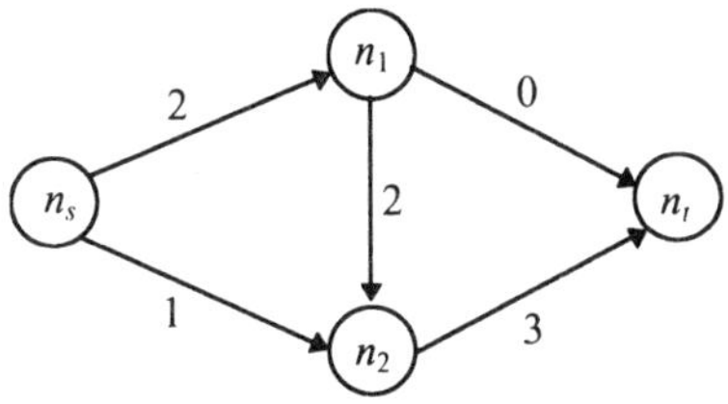

Figure III.30

Suppose now that the same problem as shown in Figure III.28 is given again, except that now $\ell_{1t} = 2$ instead of 0. The structure of the graph G^* is then the same as that in Figure III.35, except that now $b_s = -2$, $b_1, = -3$, $b_2 = 1$, and $b_t = 4$, so that $\kappa^*_{s^*t} = 4$, $\kappa^*_{1t} = 3$, and $\kappa^*_{1t^*} = 3$. The maximal flow value is then $f^*_{\bar{\mathbf{x}}'} = 4 < 5 = \sum_{i:b_i>0} b_i$, so that the problem has no feasible solution. This can easily be seen in this simple example, as no more than $\kappa_{s1} = 3$ flow units can enter the node n_1, but at least $\ell_{1t} + \ell_{12} = 2 + 2 = 4$ units must leave n_1.

This section has discussed a method that effectively reduces the task of finding a feasible flow to that of finding a maximal flow—a problem for which efficient algorithms are available, some of which are found in Section 3.1 of this chapter.

Given this result, the subsequent section can describe techniques that determine cost-minimal feasible flows.

3.3 Cost-Minimal Flow Problems

All algorithms in this section determine cost-minimal feasible flows. The first section describes a construction algorithm that determines an optimal flow pattern for a required flow value, the second section presents a primal improvement method that minimizes the cost starting with a given feasible flow pattern, the third section presents the primal-dual out-of-kilter algorithm, and the last section describes the network simplex method, which is probably the most effective solution method today.

3.3.1 An Augmenting Path Construction Algorithm

The method presented in this subsection determines a cost-minimal flow pattern in a network with capacities κ_{ij} and costs c_{ij} $\forall$ $a_{ij} \in A$, and a required flow from n_s to n_t of magnitude $\tilde{f}$. The idea of the algorithm was first put forward by Busacker and Gowen (1961) and is based on the following

Lemma III.11: Let $\mathbf{x}$ be a cost-minimal flow pattern with flow value $f_\mathbf{x}$. A cost-minimal flow pattern of value $f_\mathbf{x} + \varepsilon$ with $\varepsilon > 0$ but arbitrarily small is then found by adding to $\mathbf{x}$ a flow of ε units on the shortest path from n_s to n_t in the residual graph $G(\mathbf{x})$.

An algorithm based on this result can then be described as follows. It is initialized with a current flow pattern $\mathbf{x} = \mathbf{0}$ and the associated flow value $f_\mathbf{x} = 0$. The cost of this present flow pattern is $z_\mathbf{x}$.

Shortest Augmenting Path Algorithm

Step 1: Determine the residual graph $G(\mathbf{x})$ associated with the current flow pattern $\mathbf{x}$. Find P_{st}, the shortest path in $G(\mathbf{x})$ from n_s to n_t and denote its length by c_{st}.

Step 2: Determine $\varepsilon = \min\limits_{a_{ij} \in P_{st}} \{\kappa'_{ij} - x_{ij}; \tilde{f} - f_\mathbf{x}\}$. The new flow pattern is then $\mathbf{x}$

$= (x_{ij})$ with $x_{ij} := \begin{cases} x_{ij} + \varepsilon \text{ if } a_{ij} \in P_{st} \\ x_{ij} - \varepsilon \text{ if } a_{ji} \in P_{st} \\ x_{ij} \text{ otherwise} \end{cases}$ and an associated flow value $f_{\mathbf{x}}$.

The cost of this flow is $z_{\mathbf{x}} := z_{\mathbf{x}} + \varepsilon c_{st}$.

Step 3: Is $f_{\mathbf{x}} = \tilde{f}$?

If yes: Stop; the current flow pattern $\mathbf{x}$ is optimal.

If no: Go to Step 1.

Example: As a numerical illustration of the shortest augmenting path algorithm, consider the network $G = (N, A; \kappa; c)$ shown in Figure III.31. The desired flow in the network is $\tilde{f} = 6$.

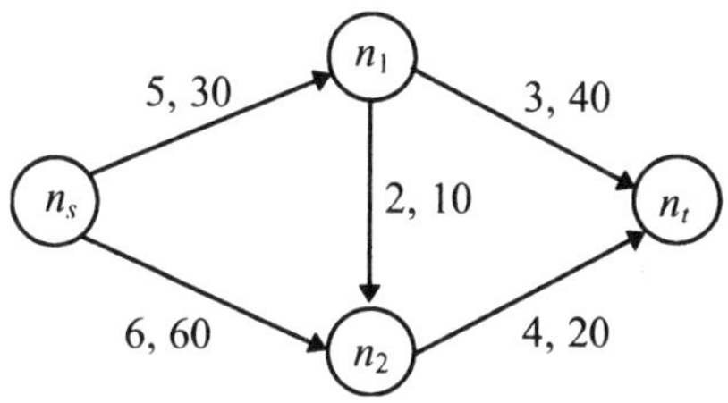

Figure III.31

The residual graph $G(\mathbf{x}) = (N, A'; \kappa'; c')$ associated with the initial flow pattern $\mathbf{x} = \mathbf{0}$ is the same as the original graph shown in Figure III.31. The shortest path is $P_{st} = (n_s, n_1, n_2, n_t)$ whose length is $c_{st} = 60$. The number of flow units to be shipped on this path is $\varepsilon = \min\{5, 2, 4; 6 - 0\} = 2$. The new flow pattern $\mathbf{x}$ with flow value $f_{\mathbf{x}} = 2$ and cost $z_{\mathbf{x}} = 120$ is shown in Figure III.32a and its associated residual graph is plotted in Figure III.32b.

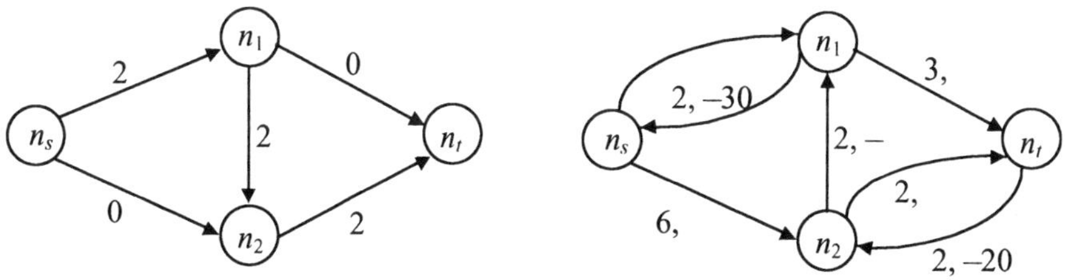

(a) (b)

Figure III.32

Since the required flow value has not yet been reached, another iteration is required. The shortest path in the current residual graph is $P_{st} = (n_s, n_1, n_t)$ with $c_{st} = 70$ and $\varepsilon = \min\{3, 3; 6-2\} = 3$. The new flow pattern $\mathbf{x}$ then has a flow value $f_\mathbf{x} = 5$ and cost $z_\mathbf{x} = 120 + 3(70) = 330$. It is shown in Figure III.33a together with its associated residual graph in III.33b.

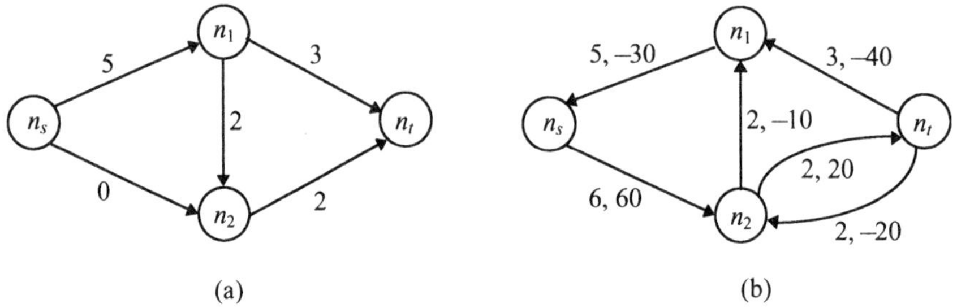

Figure III.33

As $f_\mathbf{x} = 5 < 6 = \tilde{f}$, another iteration is needed. The shortest (and, indeed, the only) path from n_s to n_t in the current residual graph is $P_{st} = (n_s, n_2, n_t)$ with length $c_{st} = 80$ and $\varepsilon = \min\{6, 2; 6-5\} = 1$. Increasing the current flow by $\varepsilon = 1$ will result in a flow $\mathbf{x}$ of the desired value $f_\mathbf{x} = \tilde{f} = 6$. The flow, whose cost is $z_\mathbf{x} = 330 + 1(80) = 410$ is shown in Figure III.34.

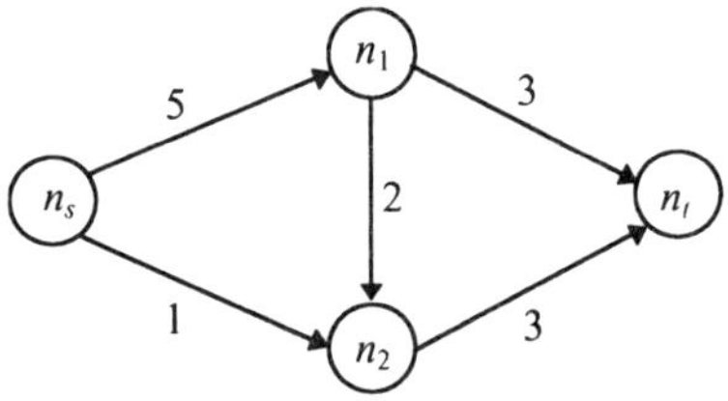

Figure III.34

It is obvious that the method requires no more than $\tilde{f}$ iterations, so its finiteness is easily proved. It should also be pointed out that the shortest augmenting path algorithm maintains dual feasibility throughout the process. As a result, when a primal feasible solution is obtained (i.e., one that satisfies the flow requirement) for the first time, it is optimal. The method's major drawback is that a change of parameters requires a "cold start", i.e., computation of the new optimal solution must start again with the zero flow. In such cases, the application of Klein's method, see Section 3.3.2, is recommended.

3.3.2 The Primal Improvement Algorithm of Klein

The method introduced in this section is a primal phase 2 method in that it commences with and maintains primal feasibility throughout the iterations while improving the value of the objective function. Suppose that the network under consideration is circulating and it is denoted by $G = (N, A; \ell, \kappa; c)$. Assume that an arbitrary feasible solution $\mathbf{x}$ is known and that $G'(\mathbf{x}) = (N, A'; \kappa'; c')$ is the corresponding residual network with $a'_{ij} \in A'$ if either $x_{ij} < \kappa_{ij}$, or $x_{ji} > \ell_{ji}$. For any $a'_{ij} \in A'$, if $a_{ij} \in A$, then $\kappa'_{ij} = \kappa_{ij} - x_{ij}$ and $c'_{ij} = c_{ij}$, if $a_{ji} \in A$, then $\kappa'_{ij} = x_{ji} - \ell_{ji}$ and $c'_{ij} = -c_{ji}$.

Consider now an arbitrary circuit in $G'(\mathbf{x})$ and denote it by $\rho = (a'_{i_1 i_2}, a'_{i_2 i_3}, \ldots, a'_{i_{r-1} i_r}, a'_{i_r i_1})$, so that $\rho_F = \{a'_{ij} : a'_{ij} \in \rho \text{ and } a_{ij} \in A\}$ and $\rho_B = \{a'_{ij} : a'_{ij} \in \rho \text{ and } a_{ji} \in A\}$; i.e., ρ_F is the set of forward arcs on the circuit, and ρ_B is the set of arcs that belong to backward arcs in the original graph G. Suppose now that the current flow $\mathbf{x}$ were modified by adding some arbitrarily small $\varepsilon > 0$ to all flows along arcs in ρ_F, and subtracting ε from the flows in all arcs in ρ_B. It is apparent that the new flow will still satisfy the conservation equations. Furthermore, as long as $\varepsilon \le \kappa'_{ij}$ $\forall$ $a'_{ij} \in \rho$, the new flow pattern will also satisfy the lower and upper bounding constraints. Consider now the effect of the flow change on the total costs. Any additional flow unit that is shipped on ρ will change the total costs by $\Delta_\rho = \sum_{a'_{ij} \in \rho_F} c_{ij} - \sum_{a'_{ij} \in \rho_B} c_{ij}$. This allows us to state

Lemma III.12: A flow pattern $\mathbf{x}$ in a network G is cost minimal, if it is feasible and $\Delta_\rho \ge 0$ for all circuits ρ in the associated residual graph $G'(\mathbf{x})$.

An immediate consequence of Lemma III.12 is the need for a method that detects circuits of negative length. The Floyd-Warshall algorithm of Section 2.3.3 of Part II could be employed for that purpose. We are now in a position to formally state the method of Klein (1967). It is initialized with a feasible flow pattern $\mathbf{x}$.

The Method of Klein

Step 1: Determine a circuit ρ in $G'(\mathbf{x})$, so that $\Delta_\rho = \min_i \{\Delta_{\rho_i}\}$.

Step 2: Is $\Delta_\rho < 0$?

If yes: Go to Step 3.

If no: Stop, the current flow pattern **x** is optimal.

Step 3: Determine $\varepsilon = \min_{a'_{ij} \in \rho} \{\kappa'_{ij}\}$ and define a new flow pattern

$$x'_{ij} = \begin{cases} x_{ij} + \varepsilon, \text{ if } a'_{ij} \in \rho_F \\ x_{ij} - \varepsilon, \text{ if } a'_{ij} \in \rho_B \\ x_{ij} \text{ otherwise} \end{cases} \text{. Go to Step 1.}$$

Example: Consider the minimum-cost feasible flow pattern shown in the graph $G = (N, A; \ell, \kappa; c)$ in Figure III.35.

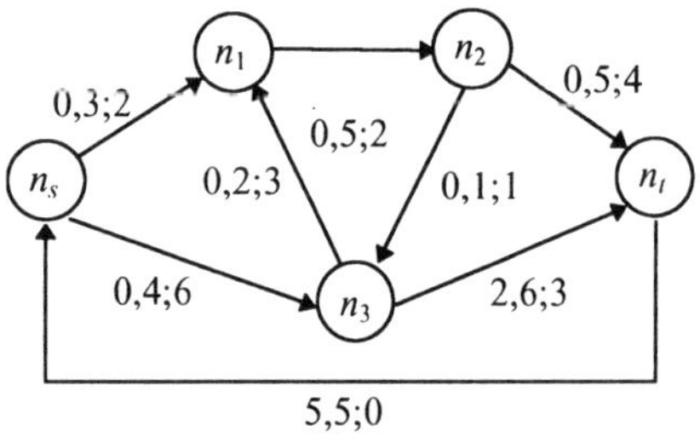

Figure III.35

A feasible flow pattern **x** and the associated residual graph $G'(\mathbf{x}) = (N, A'; \kappa', c')$ are shown in Figures III.36a and b, respectively.

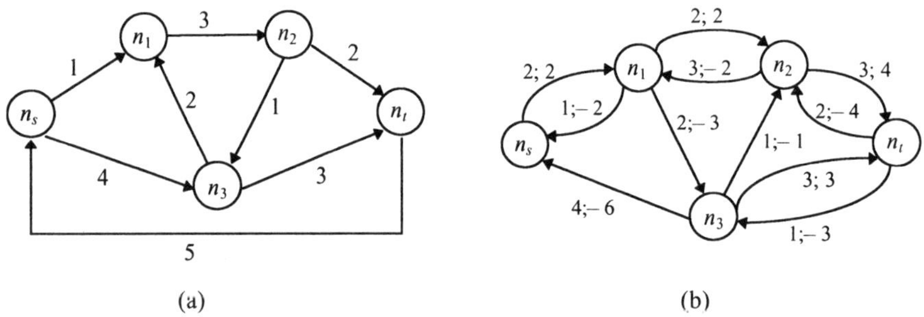

(a) (b)

Figure III.36

While it is sufficient to determine any circuit of negative length in $G'(\mathbf{x})$, all such circuits for this flow pattern **x** are shown in Table III.7.

Table III.7

Circuit ρ_i	Δ_{ρ_i}	ε_{ρ_i}	$\Delta_{\rho_i}\ \varepsilon_{\rho_i}$
$\rho_1 = (n_s, n_1, n_3, n_s)$	-7	2	-14
$\rho_2 = (n_1, n_3, n_2, n_1)$	-6	1	-6
$\rho_3 = (n_1, n_3, n_t, n_2, n_1)$	-6	2	-12
$\rho_4 = (n_s, n_1, n_2, n_t, n_3, n_1)$	-1	1	-1

In addition to the circuits, Table III.43 displays the cost change Δ_{ρ_i} if one flow unit were shipped on circuit ρ_i, the largest number of flow units ε_{ρ_i} that may be shipped on ρ_i while retaining feasibility, and the cost change if ε_{ρ_i} units are actually shipped on the circuit ρ_i. While a flow change on any circuit with negative length will improve the solution, the algorithm leaves the choice of the circuits to the user. One choice would be to use the "most negative" circuit, i.e., the circuit ρ_k with $\Delta_{\rho_k} = \min_i \{\Delta_{\rho_i}\}$. Another possibility would be to choose the circuit ρ_ℓ, so that the decrease of the total cost is maximized, i.e., $\Delta_{\rho_\ell}\varepsilon_{\rho_\ell} = \min_i \{\Delta_{\rho_i}\varepsilon_{\rho_i}\}$. Clearly, while the latter strategy results in the highest benefit in the most recent iteration, it also requires a higher computational effort. Whether or not this effort is justified is not known in general. We would like to point out the equivalence of the two selection rules for circuits and the "most negative" and "greatest change" rules for the selection of a pivot column in linear programming; see, e.g., Eiselt *et al.* (1987).

Choosing ρ_1 and shipping $\varepsilon_{\rho_1} = 2$ flow units on that circuit results in the flow pattern shown in Figure III.37a. The corresponding residual graph is shown in Figure III.37b.

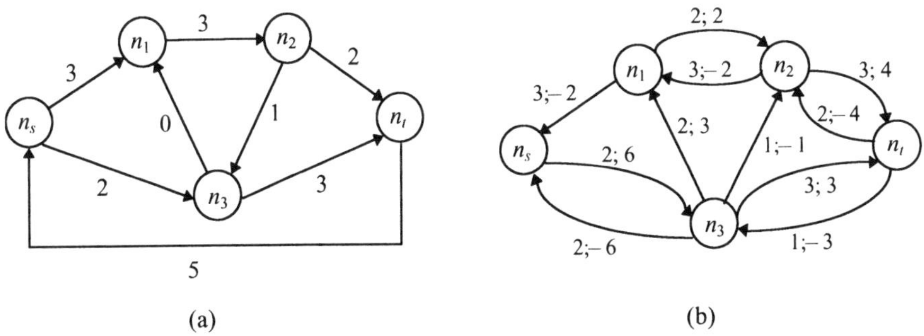

Figure III.37

As no circuit of negative length exists in the residual graph of Figure III.44b, the current flow pattern of Figure III.44a is optimal. Note, however, that a circuit of length 0 exists, *viz.*, $\rho = (n_3, n_2, n_t, n_3)$, indicating the existance of alternative optimal solutions. One of the alternative optimal flow patterns is shown in Figure III.38.

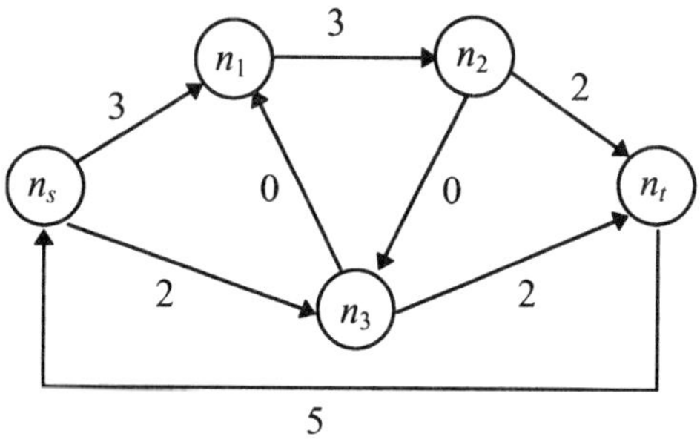

Figure III.38

3.3.3 The Primal-Dual Out-of-Kilter Algorithm

This section describes the first network algorithm to solve general minimum-cost network flow models. This primal-dual method is due to Fulkerson (1961). We will first describe the general idea of the method, then provide a formal statement of the algorithm, and then demonstrate the use of the algorithm by means of an example.

Recall the optimality conditions of Section 1.3

$\bar{c}_{ij} > 0$ and $x_{ij} = \ell_{ij}$,

$\bar{c}_{ij} < 0$ and $x_{ij} = \kappa_{ij}$, or

$\bar{c}_{ij} = 0$ and $x_{ij} \in [\ell_{ij}; \kappa_{ij}]$.

An arc that displays any of these three combinations of primal variables x_{ij} and dual variables $\bar{c}_{ij}$ is said to be "in kilter." More generally, each arc is associated with a *state* that depends on the current values of its primal and dual variables. There are two main classes of states: *in-kilter* states $\xi = 1^+, 2^+$, and 3^+ that correspond to the above three situations, and *out-of-kilter* states $\xi = 1^-, 2^-, \ldots, 6^-$. A summary of all possible states is shown in Table III.8 on the next page.

Table III.8

State ξ	Conditions	Kilter number K_{ij}
1^+	$\bar{c}_{ij} > 0$ and $x_{ij} = \ell_{ij}$	0
2^+	$\bar{c}_{ij} < 0$ and $x_{ij} = \kappa_{ij}$	0
3^+	$\bar{c}_{ij} = 0$ and $x_{ij} \in [\ell_{ij}; \kappa_{ij}]$	0
1^-	$\bar{c}_{ij} > 0$ and $x_{ij} < \ell_{ij}$	$\ell_{ij} - x_{ij}$
2^-	$\bar{c}_{ij} = 0$ and $x_{ij} < \ell_{ij}$	$\ell_{ij} - x_{ij}$
3^-	$\bar{c}_{ij} < 0$ and $x_{ij} < \ell_{ij}$	$\kappa_{ij} - x_{ij}$
4^-	$\bar{c}_{ij} > 0$ and $x_{ij} > \ell_{ij}$	$x_{ij} - \ell_{ij}$
5^-	$\bar{c}_{ij} = 0$ and $x_{ij} > \kappa_{ij}$	$x_{ij} - \kappa_{ij}$
6^-	$\bar{c}_{ij} < 0$ and $x_{ij} > \kappa_{ij}$	$x_{ij} - \kappa_{ij}$

Notice that the states 1^+, 1^-, and 4^-; 3^+, 2^-, and 5^-; and 2^+, 3^-, and 6^- span all the possibilities for $\bar{c}_{ij} > 0$, $\bar{c}_{ij} = 0$, and $\bar{c}_{ij} < 0$, respectively.

Considering the dual variables $\bar{c}_{ij}$ temporarily fixed, the flow in an arc in one of the states 1^-, 2^-, or 3^- should be increased, while the flow of an arc in any of the states 4^-, 5^-, and 6^- should be decreased. The reason is that the flow in an arc in states 1^- and 2^- is currently lower than the lower bound , while an arc in 3^- has $\bar{c}_{ij} < 0$ and $x_{ij} < \kappa_{ij}$; however, to be in kilter $x_{ij} = \kappa_{ij}$ is required (as per state 2^+), so that x_{ij} must again be increased. A similar argument shows the need to decrease the flow in arcs in states 4^-, 5^-, and 6^-.

Consider now the kilter numbers $K_{ij} \geq 0$ assigned to all arcs $a_{ij} \in A$. The kilter numbers for in-kilter arcs are zero, whereas the kilter numbers for all out-of-kilter arcs express the smallest flow change that transforms an arc to an in-kilter state, while keeping its $\bar{c}_{ij}$ value unchanged. As an example, consider state 1^-. Given its dual variable $\bar{c}_{ij} > 0$, the corresponding in-kilter state would be 1^+ with $x_{ij} = \ell_{ij}$. As in state 1^-, $x_{ij} < \ell_{ij}$, it takes a flow increase of $(\ell_{ij} - x_{ij})$ to bring the arc in kilter. The kilter numbers are frequently displayed in a *kilter diagram* such as that shown in Figure III.39 on the next page.

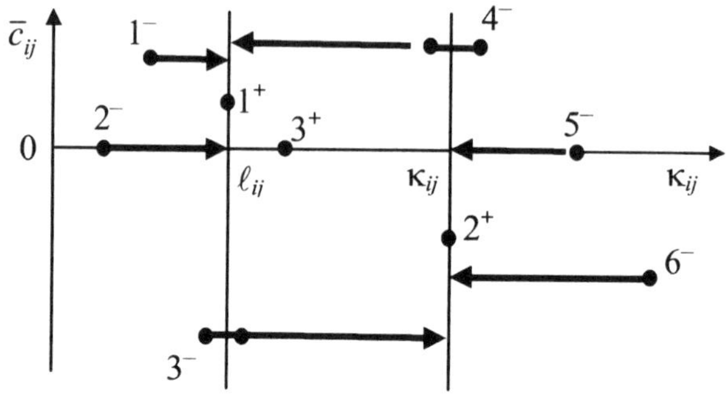

Figure III.39

Notice that the points representing the states 3^- and 4^- could be to the left on the right of the ℓ_{ij} and the κ_{ij} lines, respectively. The bold arrows show the magnitude of the kilter numbers. While the kilter numbers are not explicitly used in the algorithm presented below, the objective of the technique is to minimize $\sum_{a_{ij} \in A} K_{ij}$. If this sum equals zero at any stage of the algorithm, then the nonnegativity of the kilter numbers implies that $K_{ij} = 0 \ \forall \ a_{ij} \in A$, i.e., all arcs are in kilter, and an optimal solution has been determined. In case the sum of kilter numbers cannot be reduced to zero, the algorithm terminates with the message that no feasible solution exists. Furthermore, the kilter numbers are used to prove the finiteness of the algorithm by showing that their sum strictly decreases in each iteration.

In principle, the algorithm proceeds as follows. We first choose any out-of-kilter arc $a_{k\ell}$. This arc will be part of a cycle C, on which the flow will be modified. If the arc $a_{k\ell}$ is in one of the states 1^-, 2^-, or 3^-, its flow must be increased and $a_{k\ell}$ will be a forward arc on the cycle C. Similarly, if $a_{k\ell}$ is in one of the states 4^-, 5^-, or 6^-, its value must decrease and, in order to do so, it must be a backward arc on the cycle. To ensure that $a_{k\ell}$ is part of C in the proper direction, we start labeling at n_ℓ (if $a_{k\ell}$ is in 1^-, 2^-, or 3^-) or at n_k (if 4^-, 5^-, or 6^-). The node at which the labeling procedure commences is sometimes referred to as the *rerouting source*. The labeling process then commences at the rerouting source and attempts to label nodes so as to form a cycle (or a circuit in the associated residual graph). In case that task is completed successfully, we have a *breakthrough*, and the flow on the cycle is modified suitably. In case no cycle can be found, a *nonbreakthrough* has occured and the dual variables are modified.

We are now able to describe the out-of-kilter algorithm in detail. The method is initialized with arbitrary integer dual variables $u_i \; \forall \; n_i \in N$ and a flow pattern $\mathbf{x}$ that must satisfy the conservation equations, but may violate lower and upper bounding constraints. The algorithm can then be described as follows.

The Out-of-Kilter Algorithm

Step 1: Determine the dual costs $\bar{c}_{ij} = c_{ij} + u_i - u_j \; \forall \; a_{ij} \in A$ and the state ξ each arc is in.

Step 2: Are all arcs $a_{ij} \in A$ in one of the states 1^+, 2^+, or 3^+?
If yes: Stop; the current flow pattern $\mathbf{x}$ is optimal.
If no: Go to Step 3.

Step 3: Set up the residual graph $G'(\mathbf{x}) = (N', A', \kappa')$ for the present flow patterns $\mathbf{x}$ as follows:

Arc a_{ij} is in state ξ	Associated arc in residual graph G'	Capacity of the arc in G'
1^+	—	—
2^+	—	—
3^+ if $x_{ij} > \ell_{ij}$	a'_{ji}	$x_{ji} - \ell_{ji}$
if $x_{ij} < \kappa_{ij}$	a'_{ij}	$\kappa_{ij} - x_{ij}$
1^-	a'_{ij}	$\ell_{ij} - x_{ij}$
2^-	a'_{ij}	$\kappa_{ij} - x_{ij}$
3^-	a'_{ij}	$\kappa_{ij} - x_{ij}$
4^-	a'_{ji}	$x_{ij} - \ell_{ij}$
5^-	a'_{ji}	$x_{ij} - \ell_{ij}$
6^-	a'_{ji}	$x_{ij} - \kappa_{ij}$

Step 4: Let n_ℓ be a node for which there exists either an out-of-kilter arc $a_{k\ell} \in A \cap A'$ or an out-of-kilter arc $a_{\ell k} \in A$ with $a'_{k\ell} \in A'$. Let $G''_\ell = (N'', A'')$ be the subgraph of $G'(\mathbf{x})$ that includes all nodes and arcs in $G'(\mathbf{x})$ that can be reached on a path from u_ℓ.

Step 5: Is $n_k \in N''$?
If yes: Go to Step 6.
If no: Go to Step 7.

Step 6 (Breakthrough): Consider the cycle ρ that consists of a path from n_ℓ to n_k and the out-of-kilter arc for which $a'_{k\ell} \in A'$. Determine $\Delta_\rho = \min\limits_{a'_{ij} \in \rho}\{\kappa'_{ij}\}$

and define the new flow pattern $\mathbf{x} = \{x_{ij}\}$ with

$$x_{ij} := \begin{cases} x_{ij} + \Delta_\rho \ \forall\, a_{ij} \in A \text{ with } a'_{ij} \in \rho, \\ x_{ij} - \Delta_\rho \ \forall\, a_{ji} \in A \text{ with } a'_{ij} \in \rho, \text{and Go to Step 1.} \\ x_{ij} \text{ unchanged otherwise.} \end{cases}$$

Step 7 (Nonbreakthrough): Define the sets of arcs $A^+ = \{a_{ij} \in A: n_i \in N'', n_{ij} \notin N'', \bar{c}_{ij} > 0, x_{ij} \leq \kappa_{ij}\}$ and $A^- = \{a_{ij} \in A: n_i \notin N'', n_{ij} \in N'', \bar{c}_{ij} < 0, x_{ij} \geq \ell_{ij}\}$. Let $\delta^+ = \min\limits_{a_{ij} \in A^+} \{c'_{ij}\}$, $\delta^- = \min\limits_{a_{ij} \in A^-} \{-c'_{ij}\}$, and $\delta = \min\{\delta^+; \delta^-\}$, where $\delta^\nu := \infty$ if $A^\nu = \varnothing$ for $\nu = +, -$.

Step 8: Is $\delta = \infty$?
If yes: Stop; there exists no feasible solution to the problem.
If no: Set $u_i := \begin{cases} u_i + \delta \ \forall\ n_i \notin N'' \\ u_i \text{ otherwise} \end{cases}$ and go to Step 1.

Example: Consider a cost-minimal flow problem with the graph $G = (N, A; \ell, \kappa; c)$ shown in Figure III.40.

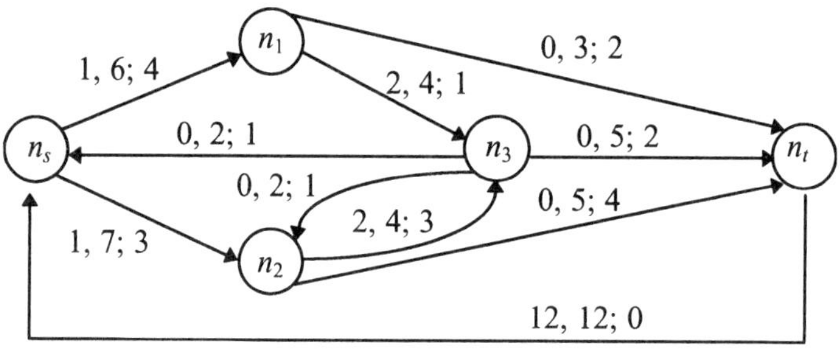

Figure III.40

Suppose that a flow pattern **x** and node variables **u** are known as shown in the

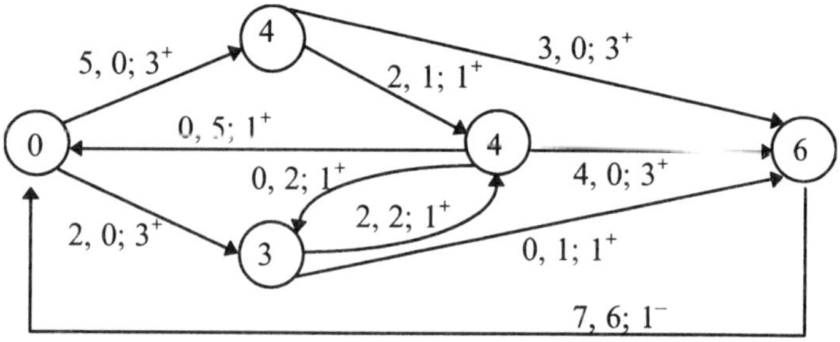

Figure III.41a

graph $G = (N, A; x, \bar{c}, \xi; u)$ in Figure III.41a, where the node variables are shown in the nodes they are associated with. The dual prices $\bar{c}_{ij}$ have been computed on that basis, and the states of all arcs have been determined as well.The associated residual graph $G'(\mathbf{x})$ is then shown in Figure III.41b.

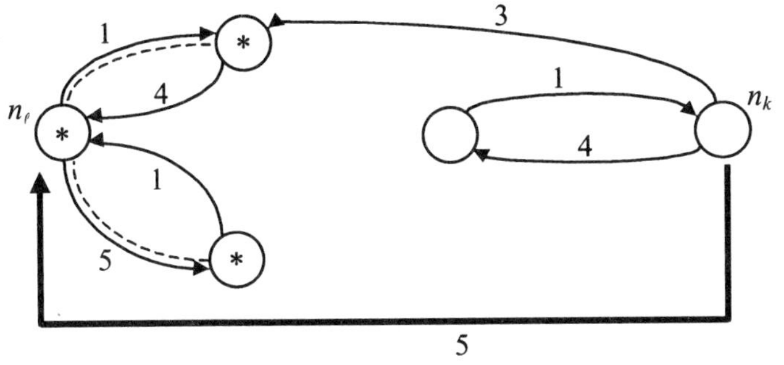

Figure III.41b

The bold arc a_{ts} is the chosen out-of-kilter arc with $n_s = n_\ell$. All nodes in $G'(x)$ that can be reached from n_ℓ are labeled with "*", and all reachable arcs are broken; together, they make up the subgraph G_2''. As $n_k \notin N''$, a nonbreakthrough has occurred and we will update the node variables. As $A^+ = \{a_{13}, a_{23}, a_{26}\}$ and $A^- = \emptyset$, we obtain $\delta^+ = \min\{1, 2, 1\} = 1$ and $\delta^- = \infty$, so that $\delta = 1$. The new solution with the same flow as that in the graph $G = (N, A; x, \bar{c}, \xi; u)$ Figure III.41a but new node variables and, consequently, now dual prices $\bar{c}_{ij}$, is shown in Figure III.42a.

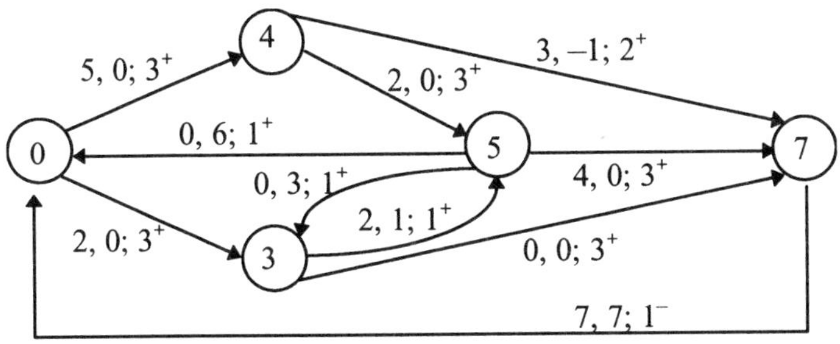

Figure III.42a

The associated residual graph $G'(x)$ is shown in Figure III.42b.

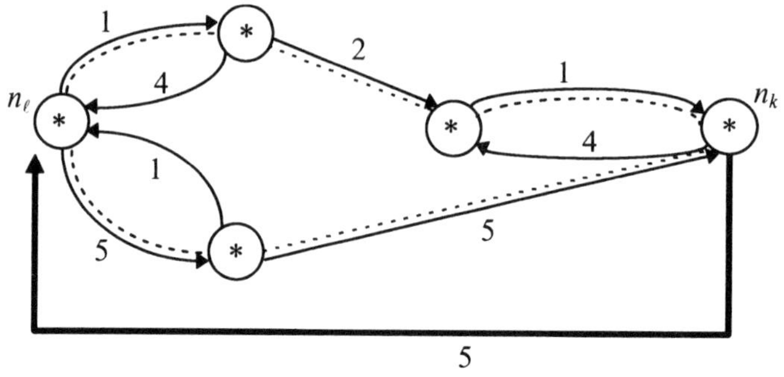

Figure III.42b

As before, the only out-of-kilter arc is a_{ts}. Choosing it, the node n_s is again labeled as n_ℓ and all nodes that can be reached from n_ℓ are labeled with "*". Here, n_k is labeled and a breakthrough has occurred.

One of the circuits is $\rho = (n_s, a_{s2}, n_2, a_{2t}, n_t, a_{ts})$, so that $\Delta_\rho = 5$ and $x_{s2} = 2 + 5 = 7$, $x_{2t} = 0 + 5 = 5$, and $x_{ts} = 7 + 5 = 12$. The new solution $G = (N, A; x, \bar{c}, \xi; u)$ is shown in Figure III.43.

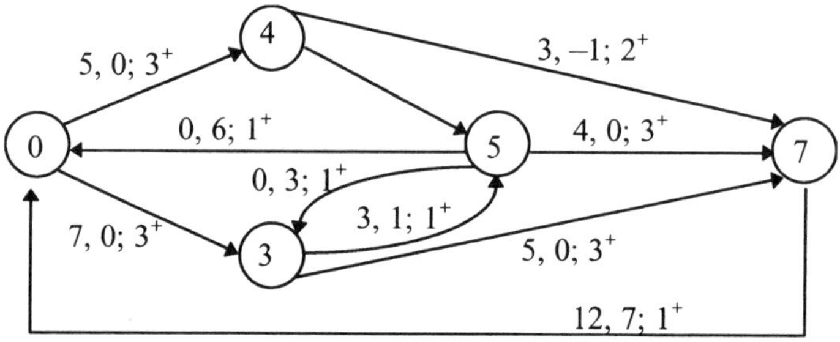

Figure III.43

As all arcs are in states 1^+, 2^+, and 3^+, the flow pattern in Figure III.43 is optimal.

It is apparent that the algorithm allows many degrees of freedom, e.g., in the choice of the out-of-kilter arc $a_{k\ell}$, and the choice of the path in case of a breakthrough. As Gondran and Minoux (1984) point out, the algorithm can be generalized to convex, piecewise linear functions.

3.3.4 The Network Simplex Method

While the out-of-kilter method of the previous section is very general in the type of problem it solves, its convergence tends to be fairly slow. In contrast, as shown earlier, any cost-minimal feasible network problem is nothing but a linear programming problem with a special structure, and the simplex method for linear programming, while not polynomial in the worst case, has efficiently solved linear programming problems for decades. The obvious idea is then to design a special implementation of the simplex method that takes the network structure into account. The basic ideas were first put forward by Johnson (1966). The method requires some new terminology as well as some transformations that align the network solutions with the simplex structure that necessitates basic solutions.

Definition III.13: Given a feasible flow pattern $\mathbf{x}$, an arc $a_{ij} \in A$ is said to be a *free arc*, if $\ell_{ij} < x_{ij} < \kappa_{ij}$. An arc that is not free is called *restricted*. Furthermore, a flow pattern is termed *cycle free*, if there exists no cycle that consists exclusively of free arcs.

Notice that on each cycle in G (or the equivalent circuit ρ in $G(\mathbf{x})$), the flows on all arcs can be increased or decreased by some positive, but arbitrarily small, amount ε. If a circuit ρ in $G(\mathbf{x})$ is of positive length Δ_ρ (defined as the sum of costs of all arcs it involves), then moving $\varepsilon > 0$ units on ρ will increase the total cost by $\Delta_\rho\varepsilon$; similarly, if $\Delta_\rho < 0$, then moving ε units on ρ will decrease the total cost by $\Delta_\rho\varepsilon$. This property can be used to prove

Lemma III.14: Given a minimum cost network flow problem with feasible, but no unbounded, solutions, there exists at least one optimal cycle free solution.

We can strengthen the requirement of cycle-free solutions further. Ignoring the directions of the arcs, we can define

Definition III.15: A feasible flow pattern $\mathbf{x}$ is called a *spanning tree solution* if there exists a spanning tree T as a subgraph of G, so that all arcs $a_{ij} \notin T$ are restricted.

The relation between cycle-free and spanning tree solutions can be demonstrated as follows. As per Definition 38, a cycle-free solution must define a forest in G. If this forest is connected, it is a tree as desired and the solution is called *nondegenerate*.

In case the forest is not connected, the disjoint subtrees can be connected one at a time until the desired tree T is obtained. In such a case, the solution is called *degenerate*. By constructing T in this fashion, we have not modified the flow pattern. In particular, any optimal cycle-free solution which, as per Lemma III.14, is guaranteed to exist can be transformed into a spanning tree solution. This proves

Theorem III.16: Given a maximum cost network flow problem with feasible but no unbounded solutions, there exists at least one optimal spanning tree solution.

For convenience, let now $L = \{a_{ij} \in G: a_{ij} \notin T, x_{ij} = \ell_{ij}\}$ and $K = \{a_{ij} \in G: a_{ij} \notin T, x_{ij} = \kappa_{ij}\}$ for any given flow pattern $\mathbf{x}$. It is apparent that the sets of arcs T, L, and K are mutually exclusive and collectively exhaustive. Given any spanning tree solution, it is trivial to determine the sets T, L, and K. Conversely, given sets T, L, and K, we can determine an associated solution by first setting $x_{ij} = \ell_{ij}\ \forall\ a_{ij} \in A$, and $x_{ij} = \kappa_{ij}\ \forall\ a_{ij} \in K$. As T forms a spanning tree, it includes $(n-1)$ arcs whose values have yet to be determined. This can be accomplished by use of the $(n-2)$ conservation equations for all nodes $n_i \neq n_s, n_t$ and the equation $f_{\mathbf{x}} = \sum_{j:a_{sj} \in A} x_{sj}$.

Given the definitions of the sets T, L, and K, and dual costs $\bar{c}_{ij} = c_{ij} + u_i - u_j$ with node potentials $u_i\ \forall\ n_i \in N$, the optimality conditions introduced in Section 1.3 can then be written as

$$\bar{c}_{ij} \begin{cases} = 0 & \forall\ a_{ij} \in T \\ \geq 0 & \forall\ a_{ij} \in L\ . \\ \leq 0 & \forall\ a_{ij} \in K \end{cases} \tag{8}$$

The idea of the network simplex method is then as follows. The procedure starts with a feasible spanning tree solution. Such a solution can always be obtained by applying a phase I algorithm such as that described in Section 3.2, followed by a procedure that generates a cycle-free solution (as described earlier in this section), and the successive addition of arcs to the forest of free arcs until a spanning tree is obtained. It is then possible to compute the node potentials $u_i\ \forall\ n_i \in N$ by using the conditions ($\bar{c}_{ij} =$) $c_{ij} + u_i - u_j = 0\ \forall\ a_{ij} \in T$. We are then able to calculate the dual costs $\bar{c}_{ij}\ \ \forall\ a_{ij} \in A$ and determine whether or not the optimality conditions (8) are satisfied. If this is the case, the algorithm terminates with an optimal flow pattern. Otherwise, a variable to enter the basis will have to be found. Such a variable will represent an arc $a_{ij} \in T$ which violates the optimality conditions; i.e., either an arc

$a_{ij} \in L$ with $\bar{c}_{ij} < 0$ or an arc $a_{ij} \in K$ with $\bar{c}_{ij} > 0$. Suppose first that $a_{ij} \in L$. Then, by assumption, a_{ij} violates the optimality condition (8) so that $\bar{c}_{ij} < 0$. Now $\{a_{ij}\} \cup T$ includes a unique cycle $\rho = (n_i, a_{ij}, n_j, ..., n_i)$ with $\sum_{a_{k\ell} \in \rho} \bar{c}_{k\ell} = \bar{c}_{ij} + \sum_{a_{k\ell} \in \rho \setminus \{a_{ij}\}} \bar{c}_{k\ell} = \bar{e}_{ij} < 0$ as all arcs $a_{k\ell} \in \rho, \{a_{ij}\}$ are in T and hence have $\bar{c}_{k\ell} = 0$. We can then prove

Lemma III.17: On any cycle ρ, $\sum_{a_{k\ell} \in \rho} \bar{c}_{k\ell} = \sum_{a_{k\ell} \in \rho} c_{k\ell}$.

The proof uses the fact that $\bar{c}_{k\ell} = c_{k\ell} + u_k - u_\ell$ and that the dual variables u_k and u_ℓ cancel out on a cycle. Clearly, in this case the flow along a_{ij} should be increased as a_{ij} is a part of a cycle ρ of negative length $\Delta_\rho = \bar{c}_{ij} < 0$; see Klein's method for a similar procedure. Increasing the flow on a_{ij} gives all arcs on ρ a direction, i.e., it classifies them as forward and backward arcs on ρ. In order to maintain the validity of the conservation equations, the flow on forward arcs will be increased, while the flow on backward arcs will be decreased. As usual, the value of the flow change ε is determined as the largest possible increase on forward arcs or decrease on backward arcs on ρ. If all arcs in $T \cap \rho$ are free, then $\varepsilon > 0$ and a nondegenerate pivot step results that is accompanied by a decrease of the total costs. As a general rule, the arc a_{ij} enters the basis, while the arc $a_{\nu\mu}$ that determines the value of ε will leave the basis. In case of ties, an arc $a_{\nu\mu}$ is arbitrarily chosen among those tied. The procedure for $a_{ij} \in K$ is similar. In this case, $\bar{c}_{ij} > 0$ as a_{ij} is assumed to violate the optimality conditions, and the flow in a_{ij} will be reduced. This is tantamount to defining the cycle as $\rho = (n_j, a_{ij}, n_i, ..., n_j)$ and again increasing the flows in all arcs in forward direction on ρ by some $\varepsilon > 0$, and decreasing the flows on all arcs in backward direction on ρ. The remaining arguments are the same as those in case $a_{ij} \in L$.

We are now able to formally state the network simplex method. The algorithm is initialized with a spanning tree solution **x** and can then be described as follows.

The Network Simplex Method

Step 1: Determine the sets T, L, and K defined by the current flow pattern **x** and compute the node potentials $u_i \; \forall \; n_i \in N$ as well as the reduced costs $\bar{c}_{ij} = c_{ij} + u_i - u_j$, by using $\bar{c}_{ij} = 0 \; \forall \; a_{ij} \in T$.

Step 2 (*Optimality test*): Is $\bar{c}_{ij} \geq 0 \; \forall \; a_{ij} \in L$ and $\bar{c}_{ij} \leq 0 \; \forall \; a_{ij} \in K$?

If yes: Stop; the current flow pattern is optimal.

If no: Go to Step 3.

Step 3 (*Select an incoming variable/arc*): Choose an arc a_{ij}, so that either $a_{ij} \in L$ and $\bar{c}_{ij} < 0$, or $a_{ij} \in K$ and $\bar{c}_{ij} > 0$. Determine the unique cycle ρ formed by a_{ij} with the arcs of T as well as the direction of the cycle ρ. Determine ε as the largest amount by which the flow on ρ can be modified.

Step 4 (*Select the departing variable/arc*): The arc $a_{\nu\mu}$ that determines ε in Step 3 will leave the basis. Update the flow pattern, the tree structure (T, L, K) and go to Step 1.

The above algorithm will be illustrated by means of the same example used for the out-of-kilter algorithm. To comply with the required format, the problem is restated here as a flow, rather than a circulation, problem.

Example: Consider the network $G = (N, A; \ell, \kappa; c)$ shown in Figure III.44.

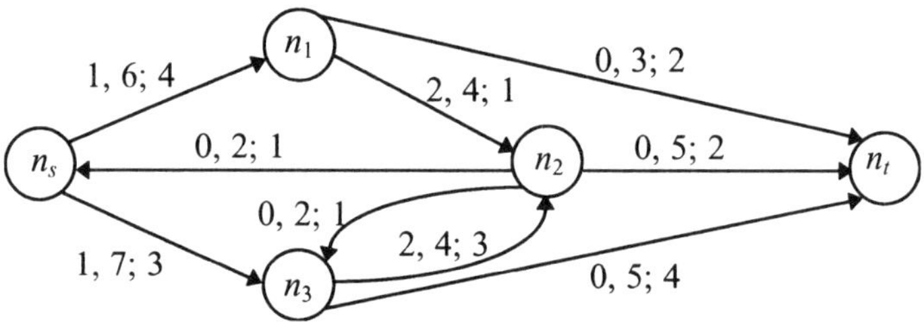

Figure III.44

The objective is to send $\tilde{f} = 12$ units of flow from n_s to n_t at minimal cost. After applying a phase I procedure, a feasible solution is determined, which is subsequently modified so as to represent a spanning tree (or basic) solution. This initial solution together with its dual costs is shown in the graph $G = (N, A; u, x; \bar{c})$ in Figure III.45 on the next page in which the arcs in T are shown as bold lines. Note that the solution displayed in Figure III.45 has associated costs of $z = 92$.

As a_{32} has a flow $x_{32} = \kappa_{32} = 4$, but its dual costs are $\bar{c}_{32} = 4 > 0$, this arc violates the optimality conditions and is chosen for introduction into the basis. This is

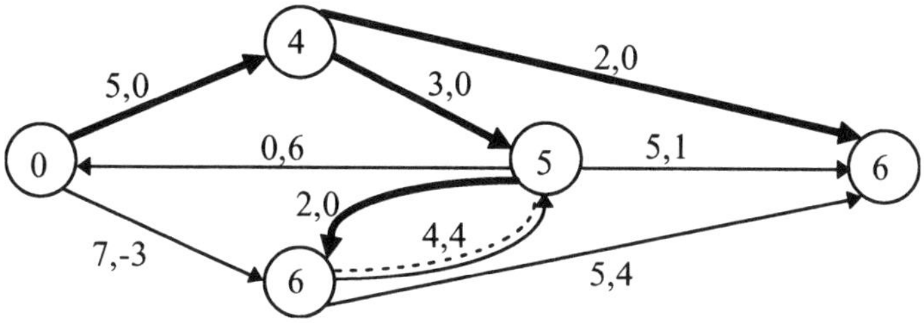

Figure III.45

indicated by the broken line in Figure III.45. The unique cycle created by this arc in conjunction with the arcs in T is $\rho = (a_{32}, a_{23})$, so that the arc a_{32} is in backward direction on that cycle. As the arc a_{23} leaves the basis but is not free, a degenerate pivot step results. The new solution is shown in Figure III.46 with the dual variables u_i in the nodes and arc values x_{ij} and $\bar{c}_{ij}$ along the arcs.

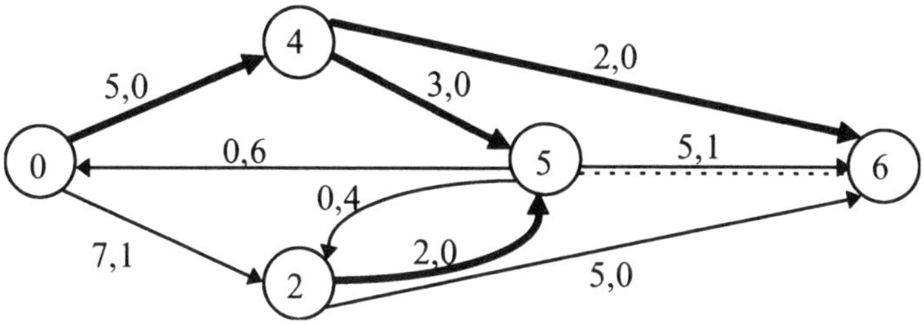

Figure III.46

The solution in Figure III.46 includes two arcs that violate the optimality conditions, *viz.*, a_{s3} and a_{2t}; both have $\bar{c}_{ij} > 0$ while $x_{ij} > \ell_{ij}$. Arbitrarily choose a_{2t} for introduction into the basis. The resulting cycle ρ is shown in Figure III.46 by the broken line and some bold lines, i.e., $\rho = (a_{2t}, a_{12}, a_{2t})$ so that a_{2t} is on this cycle in a backward direction. The flow change is then $\varepsilon = \min\{x_{2t} - \ell_{2t}; x_{12} - \ell_{12}; \kappa_{1t} - x_{1t}\} = \min\{5 - 0; 3 - 2; 3 - 2\} = 1$. Here, either a_{12} or a_{1t} leaves the basis; we arbitrarily choose a_{1t}. Changing the flow and the basis accordingly results in the solution in graph $G = (N, A; u; x, \bar{c})$ in Figure III.47.

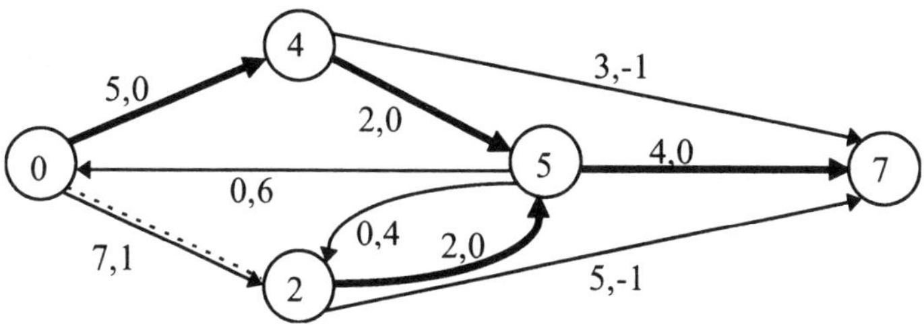

Figure III.47

At this point, all arcs except a_{s3} satisfy the optimality conditions. Introducing a_{s3} into the basis results in the cycle $\rho = (a_{s3}, a_{s1}, a_{12}, a_{32})$ on which a_{s3} is in a backward orientation. The resulting flow change is $\varepsilon = \min\{x_{s3} - \ell_{s3};\ \kappa_{s1} - x_{s1};\ \kappa_{12} - x_{12};\ x_{32} - \ell_{32}\} = \min\{7-1;\ 6-5;\ 4-2;\ 2-2\} = 0$, so that the arc a_{32} leaves the basis and another degenerate step results. This solution is shown in Figure III.48.

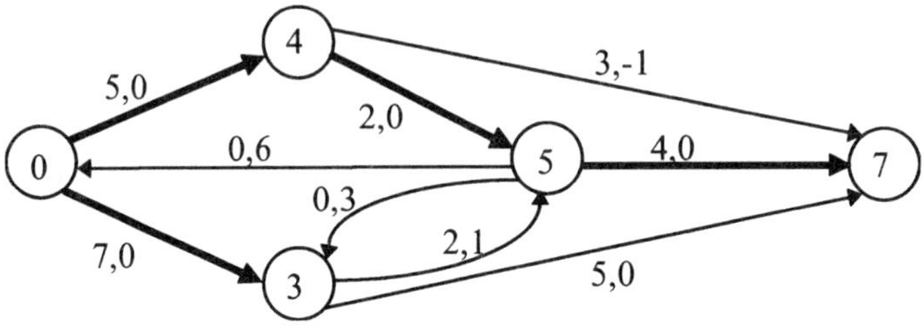

Figure III.48

Now all arcs satisfy the optimality conditions (8) and the flow pattern **x** in the graph of Figure III.48 is optimal. Its cost is 83 and alternative optimal solutions exist, as the nonbasic variable a_{3t} has reduced costs of $\bar{c}_{3t} = 0$. The adjacent optimal basic solution would introduce a_{3t} into the basis, use the cycle $\rho = (a_{st}, a_{2t}, a_{12}, a_{s1}, a_{13})$ with a_{s3} and a_{3t} in backward direction and the remaining arcs of ρ in forward direction, so that $\varepsilon = 1$ and either a_{s1} or a_{2t} leaves the basis.

The degeneracy inherent in the above example allows some degree of freedom during the execution of the algorithm. For example, for the flow pattern of the initial solution in Figure III.44, we could have used the spanning tree $T = \{a_{s1}, a_{12}, a_{1t}, a_{23}\}$, or $T = \{a_{s1}, a_{12}, a_{1t}, a_{3t}\}$ instead of $T = \{a_{s1}, a_{12}, a_{1t}, a_{23}\}$. Another degree of freedom involves the choice of the variable that enters the basis. Just as there is a multitude of pivot column selection rules (see, e.g., Eiselt *et al.* (1987)), any arc that violates the optimality conditions (8) can be chosen as for introduction into the basis. One popular rule is Dantzig's "steepest unit ascent" rule which introduces the arc a_{ij} into the basis that has $a_{ij} = \underset{a_{k\ell} \in L \cup K}{\arg\max} \{|\bar{c}_{k\ell}|\}$. Yet another choice is possible in cases of primal degeneracy when the leaving variable is not uniquely determined. Again, a variety of specific rules are available. For a detailed discussion of efficient implementations and other simplex-based approaches to general network problems, readers are referred to Ahuja *et al.* (1993).

Another development should also briefly be mentioned. Similar to the capacity scaling algorithm of Section 3.1, Röck (1980), Goldberg and Tarjan (1990) and others have developed *Cost Scaling Algorithms*. Their major attraction is that they have polynomial (some even strongly polynomial) running time.

CHAPTER 4 MULTICOMMODITY NETWORK FLOWS

One of the assumptions in the previous chapters was that the flow units sent through the network under consideration were all of the same commodity. In this chapter, we drop this assumption. Clearly, doing so requires assurance that the inflow equals the outflow at all nodes for each commodity separately. Incorporating these requirements in a network flow model is the subject of this chapter.

4.1 The Model, its Formulation and Properties

In order to understand the basic issues involved in multicommodity flow problems, consider two passengers of an airline. The first passenger wants to fly from Los Angeles to New York, and the second passenger's plan involves a trip from Boston to San Francisco. Using the standard airline codes for the airports Los Angeles International (LAX), Newark, NJ (EWR) serving New York, Boston Logan (BOS) and San Francisco (SFO), the approximate distances between the airports (in miles) are shown in Table III.9.

Table III.9

from \ to	LAX	EWR	BOS	SFO
LAX	0	2,450	2,610	340
EWR	2,450	0	160	2,560
BOS	2,610	160	0	2,700
SFO	340	2,560	2,700	0

Augmenting the corresponding undirected graph by a source n_s and sink n_t which are connected to the passenger origins and destinations by arcs with zero costs, respectively, the graph $G = (N, A; c)$ in Figure III.49 on the next page results.

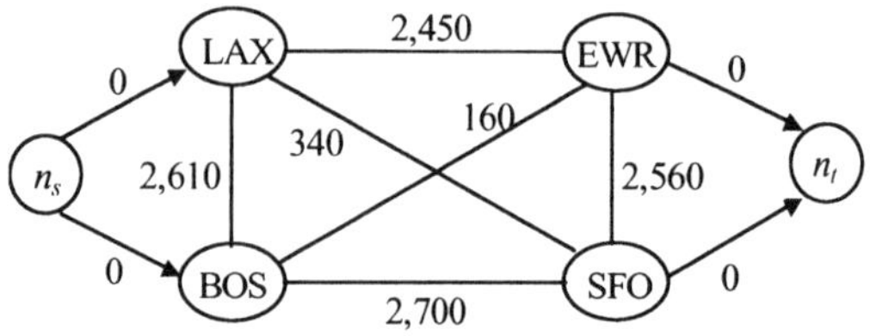

Figure III.49

The upper bounds on the flows of all arcs are set sufficiently high, and all lower bounds are zero, except for arcs incident to n_s or n_t, which have a lower bound of one. This design is supposed to ensure one passenger trip through LAX and to EWR, and similarly one passenger trip from BOS to SFO. This approach, however, has a cost-minimal feasible flow that routes Passenger 1 from LAX to SFO, and Passenger 2 from BOS to EWR—cheap, but not where the passengers want to go. The problem with this single-commodity approach applied to a multicommodity flow problem is that flow units, once they have left the source, can no longer be identified. In other words, to the graph in Figure III.58, all passengers are alike and once one of them arrives in, say, BOS, the model has no way of knowing whether this customer starts his trip in BOS or transfers here from somewhere else.

One way to overcome this modeling difficulty is to assign to each passenger (or, in general, commodity) a specific origin-destination pair. In this example, we could identify Passenger 1 with the source-sink pair (n_{s^1}, n_{t^1}), while the second passenger is associated with the pair (n_{s^2}, n_{t^2}). The graph that models this situation is shown in Figure III.50.

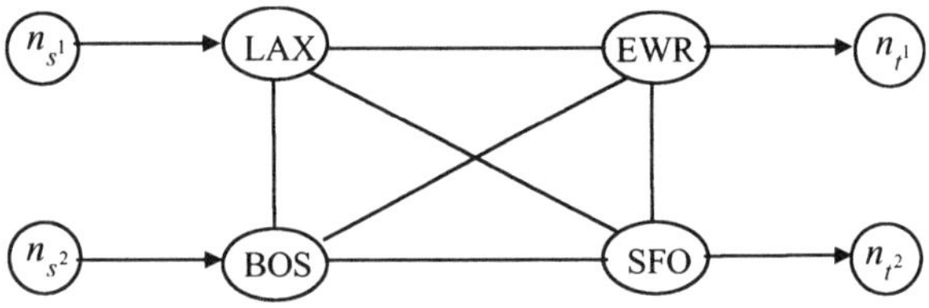

Figure III.50

The important difference between the graphs in Figures III.49 and III.50 is not so much their structure, but the requirement and possibility of identifying different passengers by their origin-destination pairs at any point in the network.

The problem with this approach is that each commodity needs its own source-sink pair. In the case of airline passengers, there are typically thousands within the course of a few hours. However, some aggregation is possible as is shown below. Consider again the case of two passengers as above, and assume that a third passenger wants to fly from BOS to LAX. Modeling each passenger separately, we obtain the graph in Figure III.51.

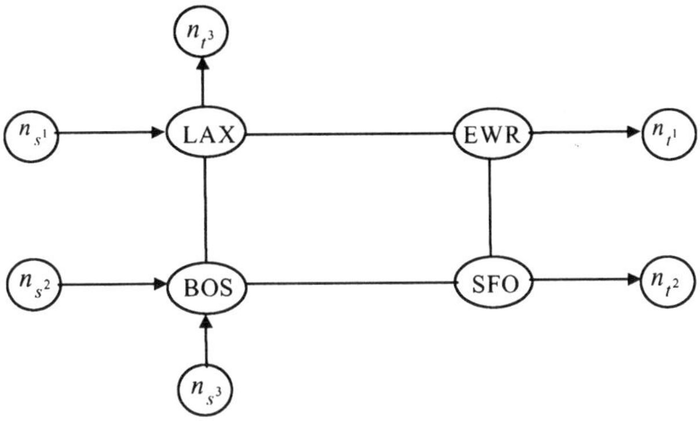

Figure III.51

The capacities of all directed arcs are arbitrarily high (or possibly restricted by the capacity of the airplane scheduled on that route), and the lower bounds of all edges is zero, whereas the arcs incident to sources or sinks have lower bounds of "1" each. (In case our "Passenger 1" is actually not a single passenger, but a collection of, say, b passengers all of whom want to fly from LAX to EWR, we may use lower bounds of b for the arcs (n_{s^1}, LAX) and (EWR, n_{t^1}).) It is now possible to aggregate all passengers whose trips commence at the same origin or those passengers whose trips end at the same destination. These two types of aggregation are commonly referred to as *origin aggregation* and *destination aggregation*, respectively. Since the destinations of the three passengers in our example are all different, whereas two passengers have BOS as their origin, only origin aggregation is possible. The sources n_{s^2} and n_{s^3} are constracted to n_{s^2} that is connected to BOS by an arc with lower bound 2. Similarly, the sinks n_{t^2} and n_{t^3} are contracted to a single sink n_{t^2} which can be reached from SFO and LAX on arcs with a lower bound of 1 each. The structure of the network is shown in Figure III.52 on the next page.

Similarly, close attention must be paid if it is desired to represent a multicommodity maximal flow problem as an equivalent minimal cost problem. As

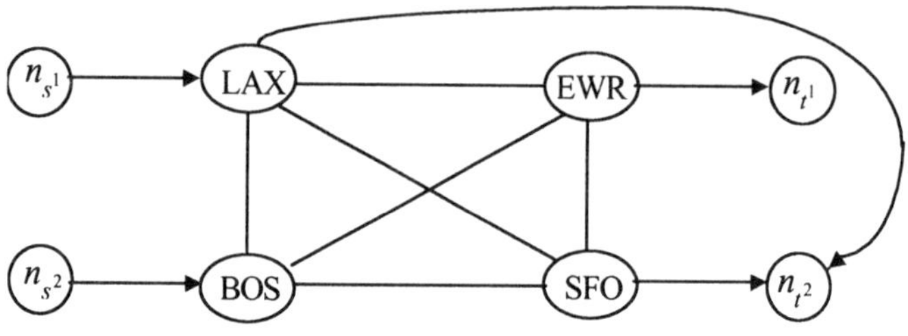

Figure III.52

an illustration, consider the graph $G = (N, A; \kappa)$ in Figure III.53, in which two commodities are to be shipped; Commodity 1 originates at n_1 and terminates at n_2, while Commodity 2 originates at n_3 and is destined for n_1.

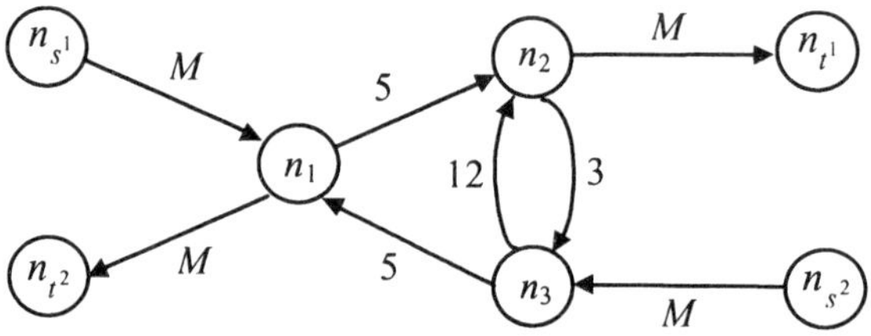

Figure III.53

The maximal 2-commodity flow is of value 10, achieved by shipping five flow units of the first commodity on the path $\pi_1 = (n_{s^1}, n_1, n_2, n_{t^1})$ and five units of the second commodity on the path $\pi_2 = (n_{s^2}, n_3, n_1, n_{t^2})$. Casting this maximal flow problem into the mold of a minimal cost circulation problem could be achieved by introducing additional arcs (n_{t^1}, n_{s^1}) and (n_{t^2}, n_{s^2}) with capacity $M >> 0$ and cost $= -1$, while the costs of all other arcs would be zero. However, this transformation above is not sufficient; in addition to the (maximal) flow on π_1 and π_2, it also allows a flow on the path $\pi_3 = (n_{s^1}, n_1, n_{t^2}, n_{s^2}, n_3, n_2, n_{t^1})$ with a capacity of 12. A flow of this (illegal) path could only be curbed by introducing *individual capacities* in addition to the common capacities shown in Figure III.53. In particular, define c_{ij}^k as the largest number of flow units of the k-th commodity allowed on arc a_{ij}, the above maximal flow problem could be properly solved as a minimal cost flow problem, if we were to associate individual upper bounds $c_{t^1 s^1}^2 = c_{t^2 s^2}^1 = 0$ with the two respective circulatory arcs.

We will now provide a mathematical formulation for multicommodity flow problems. For that purpose, define x_{ij}^k as the flow of commodity $k = 1, ..., p$ on the arc a_{ij}, and similarly let c_{ij}^k denote the unit cost of a flow of the k-th commodity on a_{ij}. Furthermore, define κ_{ij}^k (ℓ_{ij}^k) as the upper (lower) bound on the flow of the k-th commodity on arc a_{ij}, and let κ_{ij} denote the common upper bound on the flow along a_{ij}. The common lower bound ℓ_{ij} is usually assumed to be zero. We can then formulate the p-commodity problem as follows.

$$\text{P: Max } f_x = \sum_k \sum_{j: a_{s^k j} \in A} x_{s^k j} \quad \text{or} \quad \text{Min } z = \sum_k \sum_{a_{ij} \in A} c_{ij}^k x_{ij}^k$$

$$\text{s.t.} \quad \sum_j x_{ij}^k - \sum_j x_{ji}^k = 0 \quad \forall\, k = 1, ..., p;\ n_i \neq n_{s^k}, n_{t^k}$$

$$\ell_{ij}^k \leq x_{ij}^k \leq \kappa_{ij}^k \quad \forall\, a_{ij} \in A;\ k = 1, ..., p$$

$$\sum_{k=1}^p x_{ij}^k \leq \kappa_{ij} \quad \forall\, a_{ij} \in A$$

$$x_{ij}^k \geq 0 \quad \forall\, a_{ij} \in A;\ k = 1, ..., p$$

The first set of constraints are the usual conservation equations which must be satisfied for all nodes that are not sources or sinks, and for all commodities separately. The next set of constraints provides lower and upper bounds on the flows of the individual commodities, and the third set of constraints provides common upper bounds. The nonnegativity constraints were added even though they are redundant since $x_{ij}^k \geq \ell_{ij}^k \geq 0$ was already required. The first objective maximizes the sum of flows of all commodities (which could be extended to maximizing the weighted sum), and the second objective minimizes the costs of the flows.

First and foremost, it is important to realize that this formulation is a linear program and as such can be solved with any appropriate method or code. Furthermore, two features of the formulation are readily apparent: firstly, and not unexpectedly, the size of a multicommodity flow problem with p commodities is about the size of p single-commodity problems, and secondly, the only component that connects the p individual single-commodity problems are the capacity constraints $\sum_{k=1}^p x_{ij}^k \leq \kappa_{ij}$. Without them, the problem could be solved by optimizing p single-commodity problems. This loose connection is used algorithmically in Section 4.2 of this chapter.

So far in this section, we have shown that multicommodity network flow problems are not simple extensions of their single-commodity counterparts. The difference is, however, not overcome by small tricks in the formulation or the graphical representation. As the following example by Bellmore *et al.* (1970) shows, there are deep structural differences. Consider the three-commodity problem shown in Figure III.54, where the common upper bounds are $\kappa_{ij} = 1 \ \forall \ a_{ij} \in A$.

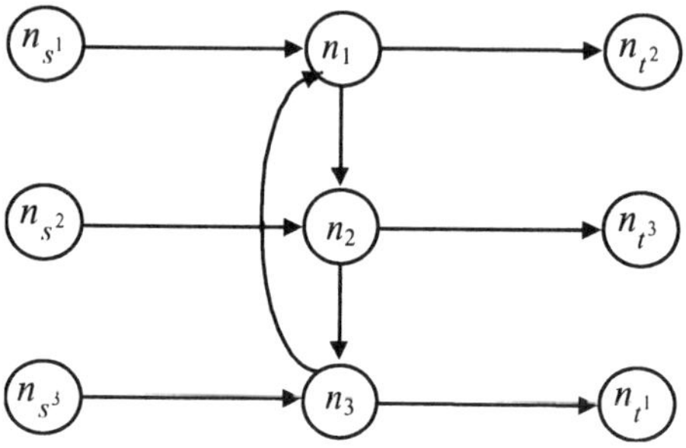

Figure III.54

Any of the sets $\{a_{12}, a_{23}\}$, $\{a_{23}, a_{31}\}$, and $\{a_{31}, a_{12}\}$ is a *disconnecting set*, i.e., a set of arcs whose removal disconnects *all* sources from their respective sinks. (Note that a disconnecting set is the straightforward generalization of cuts to multicommodity flow problems). Furthermore, each of these three disconnecting sets is minimal with capacity 2. The unique maximal flow has ½ flow unit from each of the three sources on the unique simple path to its associated sink. Hence, we can make two main observations:

- The maximal multicommodity flow is not necessarily integer, even if all capacities are integer, and
- The value of the maximal flow is not necessarily equal to the value of the smallest disconnecting set.

4.2 Solution Methods

This section will survey some of the methods commonly used to solve multicommodity flow problems. As became clear in the previous section, multicommodity flow problems can be written as linear programs, albeit very large ones. However, these programs have a special structure. This can be seen by defining vectors of variables $\mathbf{x}^k$, $k = 1, \ldots, p$ and writing the mathematical problem in matrix notation as

$$\text{Max } f_{\mathbf{x}} = \sum_{k=1}^{p} \mathbf{c}^k \mathbf{x}^k \qquad \text{or} \qquad \text{Min } z = \sum_{k=1}^{p} \tilde{\mathbf{c}}^k x^k \quad (1)$$

$$\text{s.t.} \qquad \mathbf{A}^k \mathbf{x}^k = \mathbf{0} \ \ \forall \ k = 1, \ldots, p \qquad (2)$$

$$\sum_{k=1}^{p} \mathbf{x}^k \leq \boldsymbol{\kappa} \qquad (3)$$

$$\mathbf{x}^k \geq \mathbf{0} \;\; \forall\, k = 1, ..., p \qquad (4)$$

for the minimal cost and maximal flow problems, respectively, where $\mathbf{A}^k$ denotes the node-arc incidence matrix without the rows for the nodes n_{s^k} and n_{t^k}, $\mathbf{c}^k$ is a vector of zeroes and ones with ones in all positions that correspond to arcs leading out of n_{s^k}, and a vector $\tilde{\mathbf{c}}^k$ that expresses in its ℓ-th position the cost incurred of a single unit of the k-th commodity shipped on arc a_ℓ. It is apparent that without the common upper bounding constraints (3), the problem is block-angular and can be decomposed into p single-commodity problems, even if we were to add individual capacity constraints $\mathbf{x}^k \leq \boldsymbol{\kappa}^k$ for all commodities. Hence, the constraints (3) are frequently referred to as *complicating constraints*, or *bundling constraints*, as they bundle the commodities together. Given this structure, it is not surprising that most solution methods use decomposition techniques. In this section, we will describe price-directive (i.e., Dantzig-Wolfe) and resource-directive (i.e., subgradient) decompositions.

4.2.1 Price-Directive Decomposition

The main idea of price-directive decomposition is to solve p separate single-commodity problems, whose objective functions are modified in an iterative fashion, so that ultimately the bundling constraints are satisfied. For that purpose, we will adopt the Dantzig-Wolfe decomposition method; for a description of the general technique, see, e.g., Eiselt *et al.* (1987). Define the feasible set for the k-th subproblem as $S^k = \{\mathbf{x}^k\colon \mathbf{A}^k\mathbf{x}^k = \mathbf{0};\ \mathbf{0} \leq \mathbf{x}^k \leq \boldsymbol{\kappa}\}$. As $\mathbf{0} \in S^k$, all sets S^k are nonempty and as $\kappa_{ij} < \infty \;\forall\, a_{ij} \in A$, the sets are bounded as well. For simplicity, consider the maximal flow problem. Let now $\mathbf{x}^{k\nu}$, $\nu = 1, ..., p^k$ denote the extreme points of the set S^k, $k = 1, ..., p$. Then each single-commodity solution $\mathbf{x}^k$ can be written as a linear convex combination of the extreme points $\mathbf{x}^{k\nu}$ with $\mathbf{x}^k = \sum_{\nu=1}^{p^k} \lambda_\nu^k \mathbf{x}^{k\nu}$ with weights λ_ν^k that satisfy $\sum_{\nu=1}^{p^k} \lambda_\nu^k = 1$ and $\lambda_\nu^k \geq 0 \;\forall\, \nu = 1, ..., p^k$. Suppose now that after r iterations of the Dantzig-Wolfe method, we have iteratively generated the extreme points $\mathbf{x}^{k\nu}$, $\nu \in P^k(r) \subseteqq P^k = \{1, ..., p^k\}$, $k = 1, ..., p$. While the method can be initialized with any set of known extreme points, $\mathbf{x}^{k1} = \mathbf{0}$ is always known and may serve as an initial solution. More generally, we set the iteration counter $r := 1$ and generate $\mathbf{x}^{k\nu}$, $\nu \in P^k(1)$ for $k = 1, ..., p$ and set $\tilde{\mathbf{c}}_\nu^k := \mathbf{c}^k\mathbf{x}^{k\nu}$. The method for the multicommodity maximal flow problem can then be described as follows.

Dantzig-Wolfe Decomposition Method

Step 1: Set up and solve the restricted master problem

$$\mathrm{P_{RM}}: \max_{\lambda_v^k} z^{\mathrm{RM}} = \sum_{k=1}^{p} \sum_{v \in P^k(r)} \tilde{\mathbf{c}}_v^k \lambda_v^k$$

$$\text{s.t.} \quad \sum_{k=1}^{p} \sum_{v \in P^k(r)} \lambda_v^k \mathbf{x}^{kv} \le \kappa$$

$$\sum_{v \in P^k(r)} \lambda_v^k = 1 \;\; \forall\, k = 1, \ldots, p$$

$$\lambda_v^k \ge 0 \;\; \forall\, v \in P^k(r);\, k = 1, \ldots, p$$

Let the solution be $\tilde{\lambda}_v^k$, $\tilde{\mathbf{u}}$, and $\tilde{\alpha}_k$, where $\mathbf{u}$ and α_k denote the dual variables associated with the constraints relating to the capacity and the linear convex combination constraints, respectively.

Step 2: Set up and solve the subproblems

$$\mathrm{P}_k^{\mathrm{SUB}}: \max_{x^k} z_k^{\mathrm{SUB}} = (\mathbf{c}^k - \tilde{\mathbf{u}})\mathbf{x}^k$$

$$\text{s.t.} \quad \mathbf{A}^k \mathbf{x}^k = \mathbf{0}$$

$$\mathbf{x}^k \ge \mathbf{0}$$

for $k = 1, \ldots, p$. Denote the optimal solution by $\tilde{\mathbf{x}}^k$ with an objective function value of $\tilde{z}_k^{\mathrm{SUB}}$.

Step 3 (*Optimality test*): Is $\tilde{z}_k^{\mathrm{SUB}} \le \tilde{\alpha}_k \;\; \forall\, k = 1, \ldots, p$?

If yes: Stop; $\tilde{x}^k = \sum_{v \in P^k(r)} \tilde{\lambda}_v^k \mathbf{x}^{kv}$ is an optimal solution.

If no: For any $\tilde{z}_k^{\mathrm{SUB}} > \tilde{\alpha}_k$, $x^{kv'} := \tilde{x}^k$ is a new extreme point of P^k, so that $P^k(r + 1) := P^k(r) \cup \{v'\}$. Set $\tilde{\mathbf{c}}_{v'}^k := \mathbf{c}^k \mathbf{x}^{kv'}$, $v := v'$, $r := r + 1$, and go to Step 1.

Notice that in this procedure, the network structure is retained in the subproblems, so that efficient network algorithms can be used for their solution. However, if the original multicommodity flow problem has the objective to maximize the flow (i.e., the vectors $\mathbf{c}^k$ consists entirely of zeroes and ones), the subproblems have objective function coefficients $(\mathbf{c}^k - \tilde{\mathbf{u}})$, making them standard minimal cost flow problems. On the other hand, if the original multicommodity problem had a minimal cost objective, then so do the subproblems.

Example: Consider the network $G = (N, A, \kappa)$ in Figure III.55.

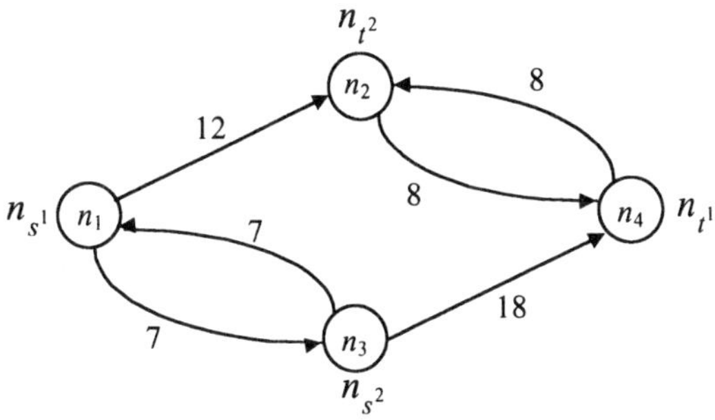

Figure III.55

As indicated $(n_{s^1}, n_{t^1}) = (n_1, n_4)$, and $(n_{s^2}, n_{t^2}) = (n_3, n_2)$. The problem can then be formulated as follows.

$$\text{P: Max } x_{12}^1 + x_{13}^1 + x_{31}^2 + x_{34}^2$$

$$\begin{aligned}
\text{s.t} \quad x_{12}^1 + x_{42}^1 + x_{24}^1 &= 0 \\
x_{13}^1 - x_{31}^1 - x_{34}^1 &= 0 \\
x_{31}^2 - x_{13}^2 - x_{12}^2 &= 0 \\
x_{34}^2 + x_{24}^2 - x_{42}^2 &= 0 \\
x_{12}^1 + x_{12}^2 &\le 12 \\
x_{13}^1 + x_{13}^2 &\le 7 \\
x_{24}^1 + x_{24}^2 &\le 8 \\
x_{31}^1 + x_{31}^2 &\le 7 \\
x_{34}^1 + x_{34}^2 &\le 18 \\
x_{42}^1 + x_{42}^2 &\le 8 \\
x_{ij}^k &\ge 0 \quad \forall\, a_{ij} \in A;\, k = 1, 2
\end{aligned}$$

Consider now the fact that $x_{13}^1 = x_{13}^2 = 0$ as flows on arcs leading into the source node of a commodity can be set to zero; similarly $x_{42}^1 = x_{24}^2 = 0$. The constraints then reduce to

$$\begin{aligned} x_{12}^1 &= x_{24}^1 & x_{12}^1 + x_{12}^2 &\le 12 \\ x_{13}^1 &= x_{34}^1 & x_{13}^1 &\le 7 \\ x_{31}^2 &= x_{12}^2 & x_{24}^1 &\le 8 \\ x_{34}^2 &= x_{42}^2 & x_{31}^2 &\le 7 \\ & & x_{34}^1 + x_{34}^2 &\le 18 \\ & & x_{42}^2 &\le 8 \\ & & x_{ij}^k &\ge 0 \end{aligned}$$

It can be seen that Kirchhoff's node equations have reduced to simple identities that can be used to eliminate the variables x_{24}^1, x_{34}^1, x_{12}^2, and x_{42}^2. The remaining problem is then

$$\begin{aligned} \text{P: Max } z = \ & x_{12}^1 + x_{13}^1 + x_{31}^2 + x_{34}^2 \\ \text{s.t.} \quad & x_{12}^1 + x_{31}^2 \le 12 \\ & x_{13}^1 \le 7 \\ & x_{12}^1 \le 8 \\ & x_{31}^2 \le 7 \\ & x_{13}^1 + x_{34}^2 \le 18 \\ & x_{34}^2 \le 8 \\ & x_{12}^1,\ x_{13}^1,\ x_{31}^2,\ x_{34}^2 \ge 0 \end{aligned}$$

With the node equations being implicitly included and satisfied, the single-commodity problems are characterized by feasible sets $S^1 = \{(x_{12}^1,\ x_{13}^1): 0 \le x_{12}^1 \le 8;\ 0 \le x_{13}^1 \le 7\}$ and $S^2 = \{(x_{31}^2,\ x_{34}^2): 0 \le x_{31}^2 \le 7;\ 0 \le x_{34}^2 \le 8\}$.

We initialize the price-directive decomposition with $\mathrm{P}^1(0) = [x_{12}^1, x_{13}^1] = \mathrm{P}^2(0) = [x_{31}^2,\ x_{34}^2] = [0,\ 0]$, so that $\tilde{c}_1^1 = \mathbf{c}^1\begin{bmatrix}0\\0\end{bmatrix} = [1,\ 1]\begin{bmatrix}0\\0\end{bmatrix} = 0$ and $\tilde{c}_1^2 = \mathbf{c}^2\begin{bmatrix}0\\0\end{bmatrix} = [1,\ 1]\begin{bmatrix}0\\0\end{bmatrix} = 0$. The restricted master problem is then

$$\begin{aligned} \mathrm{P^{RM}}\text{: Max } z^{\mathrm{RM}} = \ & 0\lambda_1^1 + 0\lambda_1^2 \\ \text{s.t.} \quad & 0\lambda_1^1 + 0\lambda_1^2 \le 12 \\ & 0\lambda_1^1 + 0\lambda_1^2 \le 18 \end{aligned}$$

$$\begin{aligned} \lambda_1^1 \quad & = 1 \\ \lambda_1^2 & = 1 \\ \lambda_1^1, \ \lambda_1^2 & \geq 0 \end{aligned}$$

With the dual variables u_1, u_2, α_1, and α_2 assigned to the respective primal structural constraints, the optimal solution of the problem is $\tilde{\lambda}_1^1 = \tilde{\lambda}_1^2 = 1$ and $\tilde{u}_1 = \tilde{u}_2 = \tilde{\alpha}_1 = \tilde{\alpha}_2 = 0$. The subproblems are then

$$\mathrm{P}_1^{\mathrm{SUB}}: \text{ Max } z_1^{\mathrm{SUB}} = ([1, 1] - [0, 0]) \begin{bmatrix} x_{12}^1 \\ x_{13}^1 \end{bmatrix} = x_{12}^1 + x_{13}^1$$

$$\begin{aligned} \text{s.t.} \quad & 0 \leq x_{12}^1 \leq 8 \\ & 0 \leq x_{13}^1 \leq 7 \end{aligned}$$

with the optimal solution $\tilde{x}_{12}^1 = 8$, $\tilde{x}_{13}^1 = 7$, and $\tilde{z}_1^{\mathrm{SUB}} = 15$, and

$$\mathrm{P}_2^{\mathrm{SUB}}: \text{ Max } z_2^{\mathrm{SUB}} = ([1, 1] - [0, 0]) \begin{bmatrix} x_{31}^2 \\ x_{34}^2 \end{bmatrix} = x_{31}^2 + x_{34}^2$$

$$\begin{aligned} \text{s.t.} \quad & 0 \leq x_{31}^2 \leq 7 \\ & 0 \leq x_{34}^2 \leq 8 \end{aligned}$$

with the optimal solution $\tilde{x}_{31}^2 = 7$, $\tilde{x}_{34}^2 = 8$, and $\tilde{z}_2^{\mathrm{SUB}} = 15$.

Since $\tilde{z}_1^{\mathrm{SUB}} = 15 > 0 = \tilde{\alpha}_1$ and $\tilde{z}_2^{\mathrm{SUB}} = 15 > 0 = \tilde{\alpha}_2$, we add the extreme points $\mathbf{x}^{12'} = [8, 7]^{\mathrm{T}}$ and $\mathbf{x}^{22'} = [7, 8]^{\mathrm{T}}$ to P^1 and P^2, respectively. We also obtain $\tilde{c}_2^1 = [1, 1]\begin{bmatrix} 8 \\ 7 \end{bmatrix} = 15$ and $\tilde{c}_2^2 = [1, 1]\begin{bmatrix} 7 \\ 8 \end{bmatrix} = 15$, so that the new restricted master problem is

$$\begin{aligned} \mathrm{P}^{\mathrm{RM}}: \text{ Max } z^{\mathrm{RM}} = \ & 15\lambda_2^1 + 15\lambda_2^2 \\ \text{s.t.} \quad & 8\lambda_2^1 + 7\lambda_2^2 \leq 12 \\ & 7\lambda_2^1 + 8\lambda_2^2 \leq 18 \\ & \lambda_1^1 + \lambda_2^1 = 1 \\ & \lambda_1^2 \quad + \lambda_2^2 = 1 \\ & \lambda_2^1, \ \lambda_2^2 \geq 0 \end{aligned}$$

which has an optimal solution $\tilde{\lambda}_1^1 = 3/8$, $\tilde{\lambda}_2^1 = 5/8$, $\tilde{\lambda}_1^2 = 0$, and $\tilde{\lambda}_2^2 = 1$, so that $[\tilde{x}_{12}^1, \tilde{x}_{13}^1] = \tilde{\lambda}_1^1 [0, 0] + \tilde{\lambda}_2^1 [8, 7] = [5, 4\,3/8]$, and $[\tilde{x}_{31}^2, \tilde{x}_{34}^2] = \tilde{\lambda}_1^2 [0, 0] + \tilde{\lambda}_2^2 [7, 8] = [7, 8]$. Furthermore, $\tilde{u}_1 = 1\,7/8$, $\tilde{u}_2 = 0$, $\tilde{\alpha}_1 = 0$, and $\tilde{\alpha}_2 = 1\,7/8$. The subproblems are then

$$P_1^{SUB}: \text{Max } z_1^{SUB} = ([1, 1] - [{}^{15}\!/\!_8, 0]) \begin{bmatrix} x_{12}^1 \\ x_{13}^1 \end{bmatrix} = -{}^7\!/\!_8\, x_{12}^1 + x_{13}^1$$

$$\text{s.t.} \quad 0 \le x_{12}^1 \le 8$$

$$0 \le x_{13}^1 \le 7$$

with an optimal solution $\tilde{x}_{12}^1 = 0$, $\tilde{x}_{13}^1 = 7$ and $\tilde{z}_1^{SUB} = 7$, and

$$P_2^{SUB}: \text{Max } z_2^{SUB} = ([1, 1] - [{}^{15}\!/\!_8, 0]) \begin{bmatrix} x_{31}^2 \\ x_{34}^2 \end{bmatrix} = -{}^7\!/\!_8\, x_{31}^2 + x_{34}^2$$

$$\text{s.t.} \quad 0 \le x_{31}^2 \le 7$$

$$0 \le x_{34}^2 \le 8$$

with the optimal solution $\tilde{x}_{31}^2 = 0$ and $\tilde{x}_{34}^2 = 8$, with objective value $\tilde{z}_2^{SUB} = 8$.

As $\tilde{z}_1^{SUB} = 7 > 0 = \tilde{\alpha}_1$ and $\tilde{z}_2^{SUB} = 8 > 1\,{}^7\!/\!_8 = \tilde{\alpha}_2$, we add the extreme points $\mathbf{x}^{13'} = [0, 7]^T$ and $\mathbf{x}^{23'} = [0, 8]^T$ to P^1 and P^2, respectively, and obtain $\tilde{c}_3^1 = [1, 1]\begin{bmatrix} 0 \\ 7 \end{bmatrix} = 7$, and $\tilde{c}_3^2 = [1, 1]\begin{bmatrix} 0 \\ 8 \end{bmatrix} = 8$, so that the restricted master problem is now

$$\begin{aligned} P^{RM}: \text{Max } z^{RM} = \; & 15\lambda_2^1 + 15\lambda_2^2 + 7\lambda_3^1 + 8\lambda_3^2 \\ \text{s.t. } & 8\lambda_2^1 + 7\lambda_2^2 \le 12 \\ & 7\lambda_2^1 + 8\lambda_2^2 + 7\lambda_3^1 + 8\lambda_3^2 \le 18 \\ & \lambda_2^1 + \lambda_3^1 \le 1 \\ & \lambda_2^2 + \lambda_3^2 \le 1 \\ & \lambda_2^1,\ \lambda_2^2,\ \lambda_3^1,\ \lambda_3^2 \ge 0. \end{aligned}$$

The optimal solution of the restricted master problem is $\tilde{\lambda}_1^1 = \tilde{\lambda}_1^2 = \tilde{\lambda}_3^1 = 0$, $\tilde{\lambda}_2^1 = 1$, $\tilde{\lambda}_2^2 = {}^4\!/\!_7$, and $\lambda_3^2 = {}^3\!/\!_7$, with $\tilde{z}^{RM} = 27$, so that $[\tilde{x}_{12}^1, \tilde{x}_{13}^1] = \tilde{\lambda}_1^1[0, 0] + \tilde{\lambda}_2^1[8, 7] + \lambda_3^1[0, 7] = 0[0, 0] + 1[8, 7] + 0[0, 7] = [8, 7]$, and $[\tilde{x}_{31}^2, \tilde{x}_{34}^2] = \tilde{\lambda}_1^2[0, 0] + \tilde{\lambda}_2^2[7, 8] + \tilde{\lambda}_2^3[0, 8] = 0[0, 0] + {}^4\!/\!_7[7, 8] + {}^3\!/\!_7[0, 8] = [4, 8]$. The optimal values of the dual variables of P^{RM} are $\tilde{u}_1 = 1$, $\tilde{u}_2 = 0$, $\tilde{\alpha}_1 = 7$, and $\tilde{\alpha}_2 = 8$. The two subproblems are then

$$P_1^{SUB}: \text{Max } z_1^{SUB} = ([1, 1] - [1, 0]) \begin{bmatrix} x_{12}^1 \\ x_{13}^1 \end{bmatrix} = x_{13}^1$$

$$\text{s.t.} \quad 0 \le x_{12}^1 \le 8$$

$$0 \le x_{13}^1 \le 7$$

with an optimal solution $\tilde{x}_{12}^1 = 0$, $\tilde{x}_{13}^1 = 7$ and $\tilde{z}_1^{\text{SUB}} = 7$ (as before) and

$$\text{P}_2^{\text{SUB}} : \text{Max } z_2^{\text{SUB}} = ([1, 1] - [1, 0]) \begin{bmatrix} x_{31}^2 \\ x_{34}^2 \end{bmatrix} = x_{34}^2$$

$$\text{s.t.} \quad 0 \leq x_{31}^2 \leq 7$$

$$0 \leq x_{34}^2 \leq 8$$

with the optimal solution $\tilde{x}_{31}^2 = 0$, $\tilde{x}_{34}^2 = 8$, and $\tilde{z}_2^{\text{SUB}} = 8$ (as before).

As $\tilde{z}_1^{\text{SUB}} = 7 = \tilde{\alpha}_1$ and $\tilde{z}_2^{\text{SUB}} = 8 = \tilde{\alpha}_2$, the stop criterion is satisfied, and the algorithm terminates with the optimal solution $[\bar{x}_{12}^1, \bar{x}_{13}^1] = [8, 7]$, and $[\bar{x}_{31}^2, \bar{x}_{34}^2] = [4, 8]$.

In summary, $\bar{x}_{12}^1 = 8$, $\bar{x}_{13}^1 = 7$, $\bar{x}_{24}^1 = \bar{x}_{31}^1 = \bar{x}_{42}^1 = 0$, and $\bar{x}_{34}^1 = 7$ for commodity 1, and $\bar{x}_{12}^2 = 4$, $\bar{x}_{13}^2 = \bar{x}_{24}^2 = 0$, $\bar{x}_{31}^2 = 4$, $\bar{x}_{34}^2 = \bar{x}_{42}^2 = 8$ for commodity 2. The flow values of the two commodities are 15 and 12, respectively, for a total flow of 27.

It should be pointed out that while this optimal solution is already integer, this will not necessarily be the case in general. Even in this example, the flow $\bar{x}_{12}^1 = 6\,^2\!/_5$, $\bar{x}_{12}^2 = 5\,^3\!/_5$, $\bar{x}_{24}^1 = 6\,^2\!/_5$, $\bar{x}_{31}^2 = 5\,^3\!/_5$, and all other variables the same as in the above solution, also represents a noninteger flow with a flow value of 27; hence, it is also maximal.

4.2.2 Resource-Directive Decomposition

As opposed to the price-directive decomposition discussed in the previous section, resource-directive (or subgradient) decomposition considers a series of problems with parametrically changing right-hand side values. Consider again a maximal flow or minimal cost multicommodity flow problem whose complicating constraints are $\sum_{k=1}^{p} x_{ij}^k \leq \kappa_{ij}\ \forall\ a_{ij} \in A$. Let now $\bar{x}_{ij}^k$, $a_{ij} \in A$, $k = 1, \ldots, p$ represent an optimal solution of the original problem P, whose objective function value is $\bar{z}^P$. Consider now a problem P′, which is identical to the original problem P, except that the complicating capacity constraints in P are replaced by individual upper bounding constraints $x_{ij}^k < \bar{x}_{ij}^k\ \forall\ a_{ij} \in A$ and $k = 1, \ldots, p$, assuming that the values $\bar{x}_{ij}^k$ were known. Let $\tilde{x}_{ij}^k$; $a_{ij} \in A$, $k = 1, \ldots, p$ denote an optimal solution of P′ with a value of the objective function of $\tilde{z}^P$. It is apparent that $\tilde{x}_{ij}^k$ is also a feasible solution to the original problem P as it satisfies all conservation equations (which are also included in P′) and as $\tilde{x}_{ij}^k \le \bar{x}_{ij}^k$, $\sum_{k=1}^{p} \tilde{x}_{ij}^k \le \sum_{k=1}^{p} \bar{x}_{ij}^k \le \kappa_{ij}$. Hence, $z^{P'} \le \bar{z}^{P}$ in case of maximal flow problems, and $\tilde{z}^{P'} \ge \bar{z}^{P}$ in case of minimal cost problems. Furthermore, since $(\bar{x}_{ij}^k)$ is a feasible solution to the problem P′ as well, $\tilde{z}^{P'} = \bar{z}^{P}$. The advantage of solving the problem P′ instead of P is that whereas P includes the complicating capacity constraints, P′ can be decomposed into p independent single-commodity problems, thus allowing it to be solved very efficiently. For the ease of the exposition, we consider maximal flow problems and without loss of generality assume that P has finite optimal solutions.

The problem with the transformation from P to P′ is that the values $\bar{x}_{ij}^k$ are not known beforehand. Instead, we define resource availabilities r_{ij}^k $\forall\, a_{ij} \in A$, $k = 1, \ldots, p$ which, in essence, are guesses of the unknown values $\bar{x}_{ij}^k$. Clearly, the guesses must satisfy $\sum_{k=1}^{p} r_{ij}^k \le \kappa_{ij}$. Define now a problem $\mathrm{P}''(\mathbf{r})$ which is identical to P′, except that it replaces $\bar{x}_{ij}^k$ by r_{ij}^k. Any solution $(\tilde{x}_{ij}^k)$ that is feasible for $\mathrm{P}''(\mathbf{r})$ will, by virtue of an argument similar to that made above for $(\tilde{x}_{ij}^k)$ and P′, also be feasible for P. Define now $\hat{z}^{P''}(\mathbf{r})$ as the objective function value of the problem $\mathrm{P}''(\mathbf{r})$ at optimum, it follows that $\hat{z}^{P''}(\mathbf{r}) \le \bar{z}^{P}$ $\forall\, \mathbf{r}$ that satisfy $\sum_{k=1}^{p} r_{ij}^k \le \kappa_{ij}$ $\forall\, a_{ij} \in A$. Since $\hat{z}^{P''}(\bar{\mathbf{x}}) = \bar{z}^{P}$, we can attempt to maximize $z^{P''}(\mathbf{r})$ as a function of the right-hand side (resource) vector $\mathbf{r}$; this procedure gives the approach its name "resource-directive decomposition". Again, the problem $\mathrm{P}''(\mathbf{r})$ decomposes into p single-commodity problems $\mathrm{P}^k(\mathbf{r}^k)$ which are optimized parametrically with respect to the right-hand sides $\mathbf{r}^k = (r_{ij}^k)$. This resource-directive problem can be written as

$$\mathrm{P}^{\mathrm{RD}}\text{: } \max_{\mathbf{r}}\ z^{\mathrm{RD}}(\mathbf{r}) = \sum_{k-1}^{p} \hat{z}^k(\mathbf{r}^k)$$

$$\text{s.t. } \sum_{k=1}^{p} r_{ij}^k \le \kappa_{ij}\ \forall\, a_{ij} \in A$$

where $\hat{z}^k(\mathbf{r}^k)$ is the optimal objective function value of the k-th single-commodity problem with resource vector $\mathbf{r}^k$.

Each individual single-commodity flow problem $P^k(\mathbf{r}^k)$ is a linear programming problem with node-balancing constraints and parametrized upper bounding constraints $x_{ij}^k \leq r_{ij}^k \; \forall \; a_{ij} \in A$; $k = 1, ..., p$. It can be shown that $\hat{z}^k(\mathbf{r}^k)$ is a piecewise linear concave function of $\mathbf{r}^k$ for maximal flow problems; for minimal cost problems, it is piecewise linear and convex. Furthermore, defining dual variables $\mathbf{s}^k = (s_{ij}^k)$ to the primal capacity constraints $x_{ij}^k \leq r_{ij}^k$ in $P^k(\mathbf{r}^k)$, these dual variables turn out to be subgradients of $\hat{z}^k(\mathbf{r}^k)$.

Hence, we can solve the problem P^{RD} by using subgradient optimization. Given some allocation $\mathbf{r}^k$ as right-hand side values, the problem $P^k(\mathbf{r}^k)$ is solved, resulting in optimal values of the dual variables $\hat{\mathbf{s}}^k$ and a value of the objective function $\hat{z}^k(\mathbf{r}^k)$. The allocation to the right-hand side values is then modified by taking a step of length θ into the direction of $\hat{\mathbf{s}}^k$. However, since it is possible that $\sum_{k=1}^{p} r_{ij}^k + \theta \hat{s}_{ij}^k > \kappa_{ij}$ for some $a_{ij} \in A$, we may need to modify the projection $\mathbf{r}^k + \theta \hat{\mathbf{s}}^k$, so that it is feasible. The projection used here simply reduces the allocation $r_{ij}^k + \theta \hat{s}_{ij}^k$ by the same amount for all commodities. Formally, if $\sum_{k=1}^{p} r_{ij}^k + \theta \hat{s}_{ij}^k > \kappa_{ij}$ for some arc $a_{ij} \in A$, then the infeasible allocations $r_{ij}^k + \theta \hat{s}_{ij}^k$ are reduced by the term $\frac{1}{p}\left[\left(\sum_{k=1}^{p} r_{ij}^k + \theta \hat{s}_{ij}^k\right) - \kappa_{ij}\right]$.

Consider now the step length θ and define the length of a step taken in iteration ν as θ_ν. A commonly used sufficient condition for convergence is that θ_ν tends to 0 as ν tends to ∞, while $\sum_{\nu=1}^{\infty} \theta_\nu \to \infty$. Here we use $\theta_\nu = {}^4/_\nu$. Another feature of the algorithm that needs to be decided upon is the stop criterion. The usual possibilities exist, e.g., a threshold value on the change of the solution (i.e., the resource allocation $\mathbf{r}$) or the change of the value of the objective function, a limit on the number of iterations, computer run time, or any combination thereof. Since the convergence of the method is not necessarily monotonic, it is required to store the best known solution $\mathbf{x}^*$ and its objective value z^{RD*}.

For simplicity, we describe the resource-directive decomposition algorithm for maximal flows. It is initialized with the initial capacity/resource allocation $r_{ij}^{k1} = \frac{1}{p}\kappa_{ij}$ $\forall$ $a_{ij} \in A$, $k = 1, ..., p$. Set the iteration counter $\nu := 1$ and let the best known objective value be $z^{\text{RD*}} := -\infty$. The algorithm can then be described as follows.

Resource-Directive Decomposition Algorithm

Step 1: Given the current resource allocations $r_{ij}^{k\nu}$, solve the single-commodity subproblems $\text{P}^{\nu}(\mathbf{r}^{\nu})$. The optimal primal solution is ($\hat{x}_{ij}^{k\nu}$), the optimal values of the dual variables are $\hat{\mathbf{s}}^{k\nu}$, and the objective value is $\hat{z}^k(\mathbf{r}^{k\nu})$. Determine the corresponding $\hat{z}^{\text{RD}}(\mathbf{r}^{\nu})$ and set $z^{\text{RD*}} = \max\{\hat{z}^{\text{RD}}(\mathbf{r}^{\nu}); z^{\text{RD*}}\}$.

Step 2: Is the chosen stop criterion satisfied?

If yes: Stop; the optimal value of the objective function is $z^{\text{RD*}}$ and the associated solution is optimal.

If no: Go to Step 3.

Step 3: Update the resource allocations to

$$r_{ij}^{k,\nu+1} = r_{ij}^{k\nu} + \tfrac{4}{\nu}\hat{s}_{ij}^{k\nu} - \tfrac{1}{p}\max\{0; \sum_{k=1}^{p} r_{ij}^{k\nu} + \tfrac{4}{\nu}\hat{s}_{ij}^{k\nu} - \kappa_{ij}\} \;\; \forall\, a_{ij} \in A$$

$k = 1, ..., p$. Set $\nu := \nu + 1$ and go to Step 1.

Example: Consider again the two-commodity maximal flow problem that relates to the graph in Figure III.55. Unlike in the previous subsection, though, we retain all twelve flow variables x_{ij}^1, x_{ij}^2, $a_{ij} \in A$. As a stop criterion, we employ a limit of 3 on the number of iterations. The flow variables are sorted, so that $\mathbf{x}^{1\nu} = [x_{12}^{1\nu}, x_{13}^{1\nu}, x_{24}^{1\nu}, x_{31}^{1\nu}, x_{34}^{1\nu}, x_{42}^{1\nu}]$ and $\mathbf{x}^{2\nu} = [x_{12}^{2\nu}, x_{13}^{2\nu}, x_{24}^{2\nu}, x_{31}^{2\nu}, x_{34}^{2\nu}, x_{42}^{2\nu}]$, and similar for the arc capacities $\kappa = [12, 7, 8, 7, 18, 8]$. With $\nu = 1$, the initial resource allocation is $(r_{ij}^{11}) = (r_{ij}^{21}) = [6, 3½, 4, 3½, 9, 4]$.

Solving the two single-commodity maximal flow problems with these capacities results in the unique primal solution $\hat{\mathbf{x}}^{11} = [4, 3½, 4, 0, 3½, 0]$ and $\hat{\mathbf{x}}^{21} = [3½, 0, 0, 3½, 4, 4]$, an associated dual solution $\mathbf{s}^{11} = [0, 1, 1, 0, 0, 0]$, and $\mathbf{s}^{21} = [0, 0, 0, 1, 0, 1]$, and objective value $\hat{z}^1(\mathbf{r}^{11}) = \hat{z}^2(\mathbf{r}^{21}) = 7½$. Therefore, $z^{\text{RD}}(\mathbf{r}^1) = 7½ + 7½ = 15$, and $z^{\text{RD*}} = \max\{15; -\infty\} = 15$.

The capacities are now updated to $(\mathbf{r}^{11} + \frac{4}{1}\hat{\mathbf{s}}^{11}) = (r_{ij}^{11} + 4\hat{s}_{ij}^{11}) = [5 + 4(0), 3½ + 4(1), 4 + 4(1), 3½ + 4(0), 9 + 4(0), 4+4(0)] = [6, 7½, 8, 3½, 9, 4]$, and $(\mathbf{r}^{21} + \frac{4}{1}\hat{\mathbf{s}}^{21}) = (r_{ij}^{21} + 4\hat{s}_{ij}^{21}) = [6 + 4(0), 3½ + 4(0), 4 + 4(0), 3½ + 4(1), 9 + 4(0), 4 + 4(1)] = [6, 3½, 4, 7½, 9, 8]$. As the sum of these two resource vectors equals [12, 11, 12, 11, 18, 12], and as such exceeds the available capacities κ in the positions 2, 3, 4 and 6, we have to make the appropriate adjustments to $\mathbf{r}^{12} = [6, 7½ - \frac{11-7}{2}, 8 - \frac{12-8}{2}, 3½ - \frac{11-7}{2}, 9, 4 - \frac{12-8}{2}] = [6, 5½, 6, 1½, 9, 2]$, and $\mathbf{r}^{22} = [6, 3½ - \frac{11-7}{2}, 4 -, 7½ - \frac{11-7}{2}, 9, 8 - \frac{12-8}{2}] = [6, 1½, 2, 5½, 9, 6]$, and set $\nu := 2$.

In the next interation, the subproblems have the optimal primal solutions $\hat{\mathbf{x}}^{12} = [6, 5½, 6, 0, 5½, 0]$, and $\hat{\mathbf{x}}^{22} = [5½, 0, 0, 5½, 6, 6]$, and the associated dual solutions $\hat{\mathbf{s}}^{12} = [1, 1, 0, 0, 0, 0]$ and $\hat{\mathbf{s}}^{22} = [0, 0, 0, 1, 0, 1]$ with a value of the objective function of $\hat{z}^1(\mathbf{r}^{12}) = \hat{z}^2(\mathbf{r}^{22}) = 11½$, so that $z^{RD}(\mathbf{r}^2) = 11½ + 11½ = 23$ and $\hat{z}^{RD*} = \max\{23; 15\} = 23$. The capacity bounds are then $(\mathbf{r}^{12} + \frac{4}{2}\hat{\mathbf{s}}^{12}) = [6 + 2(1), 5½ + 2(1), 6 + 2(0), 1½ + 2(0), 9 + 2(0), 2 + 2(0)] = [8, 7½, 6, 1½, 9, 2]$, and $(\mathbf{r}^{22} + \frac{4}{2}\hat{\mathbf{s}}^{22}) = [6 + 2(0), 1½ + 2(0), 2 + 2(0), 5½ + 2(1), 9 + 2(0), 6 + 2(1)] = [6, 1½, 2, 7½, 9, 8]$, whose sum equals [14, 9, 8, 9, 18, 10] and as such exceeds the given capacity vector κ in positions 1, 2, 4, and 6. The adjusted resource allocations are then $\mathbf{r}^{13} = [8 - \frac{14-12}{2}, 7½ - \frac{9-7}{2}, 6, 1½ - \frac{9-7}{2}, 9, 2 - \frac{10-8}{2}] = [7, 6½, 6, ½, 9, 1]$, and $\mathbf{r}^{23} = [6 - \frac{14-12}{2}, 1½ - \frac{9-7}{2}, 2, 7½ - \frac{9-7}{2}, 9, 8 - \frac{10-8}{2}] = [5, ½, 2, 6½, 9, 7]$.

In the third iteration, we solve the subproblems and obtain the optimal primal solutions $\hat{\mathbf{x}}^{13} = [6, 6½, 6, 0, 6½, 0]$, and $\hat{\mathbf{x}}^{23} = [5, 0, 0, 5, 7, 7]$. The associated dual solutions are $\hat{\mathbf{s}}^{13} = [0, 1, 1, 0, 0, 0]$ and $\hat{\mathbf{s}}^{23} = [1, 0, 0, 0, 0, 1]$, and the value of the objective function is $\hat{z}^{RD}(\mathbf{r}^3) = 12½ + 12 = 24½$, so that $z^{RD*} = \max\{24½, 23\} = 24½$. At this point the process terminates, as three iterations are completed. The best known solution is $(\hat{\mathbf{x}}^{13}, \hat{\mathbf{x}}^{23})$ with a toal flow value of 24½. This value is not very close to the true optimal value of 27; however, one more iteration with the resource-directive decomposition method finds a solution with a flow value of $26^2/_3$, reasonably close to the optimum.

It is worth pointing out that there exist solution techniques other than the two decomposition methods described in this chapter. Most prominent among them are Lagrangean relaxation, column generation, and basis partitioning. For a full description of these methods and multi-commodity flows in general, the reader is referred to Ahuja *et al.* (1993) and Kennington and Helgason (1980).

4.3 Network Design Problems

So far in this chapter, we have assumed that a network exists and the required flow has to be routed through it. The problems in this section are of the design type, meaning that for a given network $G = (N, A)$ of *potential* connections, we are to choose or design a subnetwork $G' = (N, A')$ with $A' \subset A$, so that a required multicommodity flow through the network is still feasible, while some function of the costs of establishing and/or generating the network and its flow are minimized. In order to determine such a subnetwork, define zero-one variables y_{ij} which assume a value of 1, if the arc a_{ij} is included in A', and 0 otherwise. In addition to the arc selection variables y_{ij}, we have the usual arc flow variables x_{ij}^k that denote the flow of commodity k along arc a_{ij}. The parameters of this problem include p_{ij}, the cost of constructing the arc a_{ij} (i.e., including it in A'), the (operating) cost of shipping one flow unit along a_{ij} as c_{ij}, the given budget b, an individual upper bound on x_{ij}^k denoted by κ_{ij}^k, and a required flow value of thc k-th comodity n_{s^k} to n_{t^k}, denoted by $\tilde{f}_k$. One version of the optimal network design problem can then be written as

$$\text{P:} \quad \text{Min } z = \sum_{a_{ij} \in A} c_{ij} \sum_{k=1}^{p} x_{ij}^k$$

$$\text{s.t.} \quad \sum_{j: a_{ij} \in A} x_{ij}^k - \sum_{j: a_{ji} \in A} x_{ji}^k = 0 \quad \forall\, k;\ i \neq s^k, t^k \tag{5}$$

$$\sum_{a_{s^k j} \in A} x_{s^k j} = \tilde{f}_k \quad \forall\, k = 1, \ldots, p \tag{6}$$

$$x_{ij}^k \leq \kappa_{ij}^k y_{ij} \quad \forall\, a_{ij} \in A;\ k = 1, \ldots, p \tag{7}$$

$$\sum_{a_{ij} \in A} p_{ij} y_{ij} \leq b \tag{8}$$

$$x_{ij}^k \geq 0 \quad \forall\, a_{ij} \in A;\ k = 1, \ldots, p \tag{9}$$

$$y_{ij} = 0 \vee 1 \quad \forall\, a_{ij} \in A \tag{10}$$

where the constraints (5) are the usual conservation equations, constraints (6) regulate the required flow, constraints (7) allow flows only in existing arcs, and (8) is the budget constraint. Notice that relations (6) are the only constraints that connect the variables x_{ij}^k and y_{ij}; without them, the problem p would decompose into two subproblems, one with variables x_{ij}^k and the other with variables y_{ij}.

Clearly, if $\sum_{a_{ij} \in A} p_{ij} \le b$, all arcs in A can be established and the problem reduces to a (much easier) multicommodity feasible flow problem. On the other hand, if the budget is so tight that it does not allow to connect n_{s^k} with n_{t^k} $\forall\, k = 1, \ldots, p$, then no feasible solution exists (note that this is not the same as a spanning tree of G). The problems with budget b between these two extremes constitute the interesting problems. In many applications the objective includes both, construction and operating costs. Furthermore, if individual capacities $\kappa_{ij}^k = \tilde{f}_k$ $\forall\, a_{ij}$ that are part of some path from n_{s^k} to n_{t^k}, we can dualize the complicating constraints (6) with multipliers λ_{ij}^k $\forall\, a_{ij} \in A$; $k = 1, \ldots, p$ and obtain the Lagrangean function

$$L(\lambda) = \sum_{a_{ij} \in A} \sum_{k=1}^{p} \left[(c_{ij} + \lambda_{ij}^k) x_{ij}^k - \lambda_{ij}^k \kappa_{ij}^k y_{ij}\right]$$

which is to be minimized subject to the constraints (1), (2), (4), (5), and (6). The problem now decomposes into two subproblems, one that involves the flow variables x_{ij}^k and the other that comprises the arc selection variables y_{ij}. We get:

$$\text{P}_{ij}\text{: Min } z_{ij} = \sum_{a_{ij} \in A} \sum_{k=1}^{p} (c_{ij} + \lambda_{ij}^k) x_{ij}^k$$

$$\text{s.t.} \quad \sum_{j: a_{ij} \in A} x_{ij}^k - \sum_{j: a_{ji} \in A} x_{ji}^k = 0 \qquad \forall\, k = 1, \ldots, p;\ i \neq s^k, t^k$$

$$\sum_{j: a_{s^k j} \in A} x_{s^k j} = \tilde{f}_k \qquad \forall\, k = 1, \ldots, p$$

$$x_{ij}^k \ge 0 \qquad \forall\, a_{ij} \in A;\ k = 1, \ldots, p$$

and

$$\text{P}_k\text{: Max } z_k = \sum_{a_{ij} \in A} \sum_{k=1}^{p} \lambda_{ij}^k \kappa_{ij}^k y_{ij}$$

$$\text{s.t.} \quad \sum_{a_{ij} \in A} p_{ij} y_{ij} \le b$$

$$y_{ij} = 0 \vee 1 \quad \forall\, a_{ij} \in A$$

Note that in the problem P_{ij}, the variables of the different commodities are not linked to each other, so that the problem can be solved for each commodity separately. In particular, the optimal solution will include flows on the shortest path from n_{s^k} to n_{t^k} with respect to the arc values ($c_{ij} + \lambda_{ij}^k$), for each $k = 1, ..., p$. The problem P_k is nothing but a zero-one knapsack problem with objective function coefficients $\sum_{k=1}^{p} \lambda_{ij}^k \kappa_{ij}^k$. A solution procedure using the Lagrangean $L(\lambda)$ to determine lower bounds on z will commence with some suitably chosen values λ_{ij}^k, compute the bound, and modify the λ-values based on the result. Such a procedure may be visualized in the following

Example: Consider the undirected graph $G = (N, E)$ in Figure III.65, where each adjacent pair of nodes (n_i, n_j) with $e_{ij} \in E$ represents a commodity with a required flow of $\tilde{f}_k$. The values next to the edges are the required flow values $\tilde{f}_k$ and the construction costs p_{ij}; the operating costs c_{ij} are assumed to be 1 for all $e_{ij} \in E$. There is a budget of $b = 680$, where construction costs for an edge are counted twice, one in each direction, effectively leaving a budget of $b = 340$.

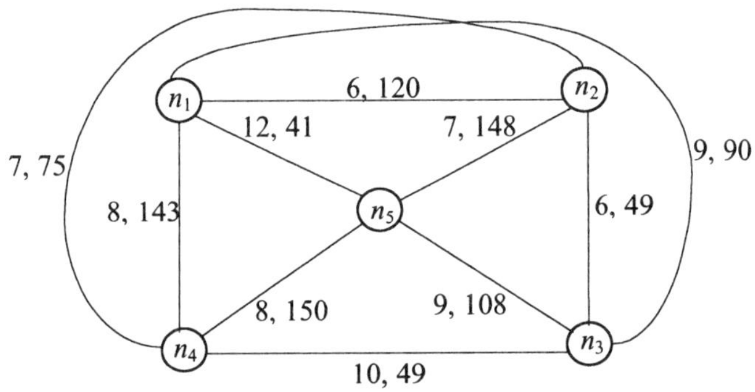

Figure III.65

First compute a lower bound with the above Lagrangean function by arbitrarily choosing $\lambda_{ij}^k = 0 \ \forall \ e_{ij} \in E, k = 1, ..., p$. As a result, $x_{ij} = \tilde{f}_k \ \forall \ e_{ij} \in E$ so that z_{ij} = 164; and $z_k = 0$, resulting in $L(0) = 164$. A better bound can be obtained by considering the fact that flows on the direct connections e_{ij} are still optimal for all commodities as long as $\lambda_{ij}^k \leq 1 \ \ \forall \ e_{ij} \in E$; $k = 1, ..., p$ as the length of direct connection will be ≤ 2, whereas each detour uses at least two edges and with $c_{ij} = 1$

$\forall\ e_{ij} \in E$ will have a length of at least 2. Setting $\lambda_{ij}^k = 1\ \ \forall\ e_{ij} \in E$; $k = 1, ..., p$ we obtain $z_{ij} = 328$. The now nontrivial knapsack problem P_k can either be solved exactly or an upper bound can be found for the computation of the lower bound on z by using the linear programming relaxation with $0 \le y_{ij} \le 1\ \ \forall\ e_{ij} \in E$ and $\kappa_{ij}^k = \tilde{f}_k\ \ \forall\ k = 1, ..., p$ that sets variables equal to 1 based on the ratio $\lambda_{ij}\tilde{f}_k/p_{ij}$ as long as possible. In this example, we obtain $y_{15} = y_{34} = y_{23} = y_{13} = y_{24} = 1$, $y_{35} = {}^1/_3$, and $y_{ij} = 0$ otherwise for $z_k = 94$, so that the improved lower bound on z is $z_{ij} - z_k = 328 - 94 = 234$.

For an upper bound , take the above solution of the knapsack problem and set $y_{35} := 0$, so that the network to be established has the edge set $E' = \{e_{15}, e_{34}, e_{23}, e_{13}, e_{24}\}$ for a used budget of $\sum_{e_{ij} \in E'} p_{ij} = 304$. The flows along the edges of E' are 36, 26, 19, 47, and 7, so that an upper bound on z is 270, i.e., $234 \le \bar{z} \le 270$. Further appropriate modification of the Lagrangean multipliers results in the optimal network with edge set $E' = \{ e_{13}, e_{15}, e_{23}, e_{34}, e_{35}\}$ with edge flows 23, 12, 26, 33, and 24 for an objective value of $\bar{z} = 236$.

CHAPTER 5 NETWORKS WITH CONGESTION

This chapter examines a number of problems that frequently occur in traffic networks. The main difference between the models in this chapter and those discussed so far is that in traffic models, flows are related to time. In particular, the average amount of time taken by each of the x_{ij} flow units along arc a_{ij} is defined as $t_{ij}(x_{ij})$. Sometimes, $t_{ij}(x_{ij})$ is referred to as *delay function*. Typically, $t_{ij}(x_{ij})$ is assumed to be increasing at an increasing rate, i.e., $\frac{dt_{ij}(x_{ij})}{dx_{ij}} > 0$ and $\frac{d^2t_{ij}(x_{ij})}{dx_{ij}^2} > 0$. Obvious applications of network flow problems with delay functions are those involving road traffic, and telecommunication networks. Delay functions in road networks are frequently of the type $t_{ij}(x_{ij}) = t_{ij}^0[1+\alpha(x_{ij}/\kappa_{ij})^\beta]$ with a capacity κ_{ij}, a transmission time under light traffic t_{ij}^0, and constants α and β; a value of $\beta = 4$ is typical in practical applications. In telecommunication networks, we frequently encounter delay functions of the type $t_{ij}(x_{ij}) = t_{ij}^0\,\kappa_{ij}/(\kappa_{ij} - x_{ij})$ which can be derived from $M/M/1$ queues, given random arrivals and message lengths that follow a negative exponential distribution.

There are three main problems related to traffic networks. They are

(1) Given delay functions, find an overall optimum to the traffic flow from an overall and an individual point of view.
(2) Reconsider the problem under (1); how should flow units be routed if it is no longer allowed to break up supplies?
(3) Given delay functions and traffic flows, and a finite budget, how should the network and the capacities of its arcs be designed, so as to minimize the overall time required to ship all units through the network?

These three questions are dealt with in the next three sections of this chapter.

5.1 System-Optimal and User-Optimal Network Flows

This section describes methods that optimally distribute traffic flows in a network, given delay functions. We distinguish between two objectives, an individual and a collective objective. We can write

Definition III.18: A *system-optimal* network flow minimizes the sum of delays on the entire network, whereas a *user-optimal* flow has the property that none of its flow units can be rerouted unilaterally, such that its delay decreases.

Loosely speaking, the two concepts differ in that the former represents an overall optimum, whereas the latter flow is in a state of equilibrium. In order to formulate the problem, define the parameters and variables of the multicommodity flow problem as follows. Let P_ℓ^k denote the ℓ-th path of commodity k (from n_{s^k} to n_{t^k}) and let $y^{k\ell}$ denote the flow on P_ℓ^k. The parameter $\delta_{ij}^{k\ell} = 1$, if $a_{ij} \in P_\ell^k$, and 0 otherwise. This parameter provides the connection between the arc a_{ij} and its flow x_{ij} and the path P_ℓ^k. Furthermore, define $\tilde{f}_k$ as the desired flow value of the k-th commodity (from n_{s^k} to n_{t^k}), and, as before, let the delay function for each unit of flow be given by $t_{ij}(x_{ij})$ so that the total delay on arc a_{ij} is then $c_{ij}(x_{ij}) = x_{ij}t_{ij}(x_{ij})$. The flow on arc a_{ij} is then $x_{ij} = \sum_{k=1}^{p}\sum_{\ell}\delta_{ij}^{k\ell}y^{k\ell}$, so that the system-optimizing *Flow Assignment Problem* can be written as

$$\text{P}_{\text{FAP}}\text{: Min } z(y^{k\ell}) = \sum_{a_{ij}\in A} c_{ij}\left(\sum_{k=1}^{p}\sum_{\ell}\delta_{ij}^{k\ell}y^{k\ell}\right)$$

$$\text{s.t.} \quad \sum_{\ell} y^{k\ell} = \tilde{f}_k \ \forall\, k = 1, \dots, p$$

$$y^{k\ell} \geq 0 \ \forall\, k, \ell.$$

The Lagrangean of this problem can then be written as

$$L(y^{k\ell}, \lambda_k) = z(y^{k\ell}) + \sum_{k=1}^{p}\lambda_k\left(\tilde{f}_k - \sum_{\ell\in L_\lambda} y^{k\ell}\right),$$

where L_k is the set of subscripts of all paths from n_{s^k} to n_{t^k}. The Kuhn-Tucker conditions (see, e.g., Eiselt *et al.* (1987)) then require that at optimality $\frac{\partial L}{\partial y^{k\ell}} \geq 0$, $y^{k\ell}\frac{\partial L}{\partial y^{k\ell}} = 0$, $y^{k\ell} \geq 0 \ \forall\, k, \ell$; and $\frac{\partial L}{\partial \lambda_k} = 0 \ \forall\, k$. This requires that $\frac{\partial L}{\partial y^{k\ell}} = 0 \ \forall\, y^{k\ell}$

> 0 and $\frac{\partial L}{\partial y^{k\ell}} \geq 0 \ \forall \ y^{k\ell} = 0$. As $\frac{\partial L}{\partial y^{k\ell}} = \frac{\partial z_{\text{sys}}}{\partial y^{k\ell}} + \frac{\partial}{\partial y^{k\ell}}\left[\sum_{k=1}^{p} \lambda_k \left(\tilde{f}_k - \sum_{\ell} y^{k\ell}\right)\right] =$

$\sum_{a_{ij} \in A} \frac{\partial c_{ij}}{\partial x_{ij}} \frac{\partial x_{ij}}{\partial y^{k\ell}} - \lambda_k = \left(\sum_{a_{ij} \in A} \delta_{ij}^{k\ell} \frac{\partial c_{ij}}{\partial x_{ij}}\right) - \lambda_k$, so that

$$\sum_{a_{ij} \in A} \delta_{ij}^{k\ell} \frac{\partial c_{ij}}{\partial x_{ij}} = \lambda_k \quad \forall \ y^{k\ell} > 0, \text{ and} \tag{1}$$

$$\sum_{a_{ij} \in A} \delta_{ij}^{k\ell} \frac{\partial c_{ij}}{\partial x_{ij}} \geq \lambda_k \quad \forall \ y^{k\ell} = 0. \tag{2}$$

Due to the convexity of $z(y^{k\ell})$, the conditions (1) and (2) are also sufficient for a minimum. Recall that $t_{ij}(x_{ij})$ and $c_{ij}(x_{ij})$ are defined as the average individual time requirement and the total time requirement on arc a_{ij}, respectively. Then $\frac{\partial t_{ij}(x_{ij})}{\partial x_{ij}}$ denotes the time an additional flow unit would require to move along a_{ij}, whereas $\frac{\partial c_{ij}(x_{ij})}{\partial x_{ij}}$ denotes the increase in the total time required by all flow units to move on a_{ij} caused by an additional flow unit. Given this interpretation, the optimality conditions (1) and (2) can be summarized as

Theorem III.19 (Wardrop's first principle): At *system optimum*, for each source-sink pair (n_{s^k}, n_{t^k}),

- all used routes $(n_{s^k}, \ldots, n_{t^k})$ have the same marginal total duration λ_k, and
- all unused routes $(n_{s^k}, \ldots, n_{t^k})$ have marginal total durations of at least λ_k, $\forall \ k$.

Consider now an individual user's perspective. It is usually assumed that users, who are each supposed to decentrally optimize their own utility, do so in the presence of complete information. In traffic networks, this assumption may be justified by a driver's access to up-to-date broadcast traffic reports. Given that, a user will then choose a route of minimal length $\sum_{a_{ij} \in A} \delta_{ij}^{k\ell} t_{ij}(x_{ij})$, resulting in

Theorem III.20 (Wardrop's second principle): At *user optimum*, for each source-sink pair (n_{s^k}, n_{t^k}),

- all used routes $(n_{s^k}, \ldots, n_{t^k})$ have the same individual duration, and
- all unused routes $(n_{s^k}, \ldots, n_{t^k})$ have an individual duration that is at least as long as any of that of the used routes.

Example: Consider a network with two nodes n_s and n_t, and two parallel arcs a_1 and a_2. The respective durations of the two arcs are $t_1(x_1) = 1/(16 - x_1)$ and $t_2(x_2) = 1/(64 - x_2)$. It is required to move a total of 50 units from n_s to n_t, i.e., $x_1 + x_2 = 50$.

First consider system optimality. With $c_1(x_1) = x_1t_1(x_1) = x_1/(16 - x_1)$ and $c_2(x_2) = x_2t_2(x_2) = x_2/(64 - x_2)$, the optimality conditions (1) and (2) are $\frac{\partial c_1(x_1)}{\partial x_1} = \frac{16}{(16-x_1)^2} = \lambda$ and $\frac{\partial c_2(x_2)}{\partial x_2} = \frac{64}{(64-x_2)^2} = \lambda$. The system of simultaneous equations is then $16/(16 - x_1)^2 = \lambda$, $64/(64 - x_2)^2 = \lambda$, and $x_1 + x_2 = 50$ and its solution is $x_1 = 6$ and $x_2 = 44$. Shipping 6 and 44 flow units on a_1 and a_2, respectively, results in average durations of $t_1(6) = \frac{1}{10}$ and $t_2(44) = \frac{1}{20}$ per flow unit and a total duration of $z = c_1(x_1) + c_2(x_2) = \frac{6}{10} + \frac{44}{20} = 2.8$.

Consider now user optimality. The optimality conditions are $t_1(x_1) = (16 - x_1)^{-1} = t_2(x_2) = (64 - x_2)^{-1}$ which, with $x_1 + x_2 = 50$, results in $x_1 = 1$ and $x_2 = 49$. In this solution, the average durations are $t_1(x_1) = t_2(x_2) = \frac{1}{15}$ and the total duration is $\frac{1}{15} + \frac{49}{15} = 3.\dot{3}\dot{3}$. Comparing this user optimum with the system optimal solution, it is apparent that the latter has a fairly long duration on arc a_1, which would prompt users, if they were able to choose themselves, to switch to arc a_2 instead. However, the system optimum provides a solution with a total duration (i.e., lost time) of only 2.8, whereas the total duration in the decentralized optimum is almost 20% higher at $3.\dot{3}\dot{3}$.

It is worth mentioning the similarity between centralized optima and system optimality on the one hand, and competitive equilibrium and user optimality on the other. The fact that decentralized or user optimality does not generally optimize the overall system is well known. In this context, Braess (1968) have provided a pertinent example known as the *Braess paradox*. It demonstrates that user optimization, even when coupled with the addition of an arc, may not improve the solution from a systems point of view. In fact, as the following example demonstrates, the solution may deteriorate.

Example: Consider the graph in Figure III.57. The duration functions $t_{ij}(x_{ij})$ are shown next to the edges, and the required flow from n_s to n_t equals 6 units.

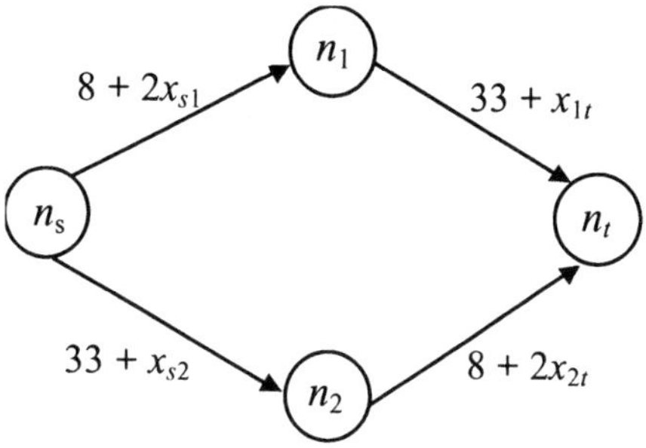

Figure III.57

Due to the inherent symmetry, the user optimum and system optimum coincide with 3 users taking the upper path (n_s, n_1, n_t), and the other 3 users choosing the lower path (n_s, n_2, n_t), so that the flows $x_{s1} = x_{1t} = x_{s2} = x_{2t} = 3$ result. Each user on the upper path faces a duration of $(8 + 2(3)) + (33 + 1(3)) = 50$, while each user on the lower path deals with a duration of $(33 + 1(3)) + (8 + 2(3)) = 50$. The sum of durations is then $\bar{z}_{\text{sys}} = 6(50) = 300$.

Suppose now that an additional arc a_{12} is introduced as shown in Figure III.58. Its duration is $13 + 3x_{12}$, and all other durations remain unchanged.

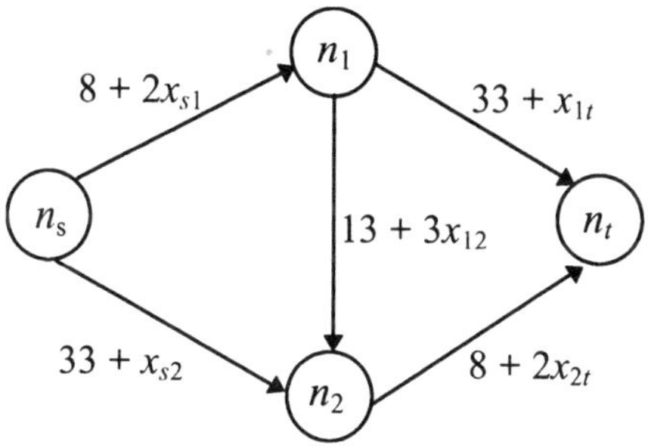

Figure III.58

Some users on the upper path will choose to take the intermediate path (n_s, n_1, n_2, n_t) instead, as its total duration is lower, i.e., as $(8 +2(3)) + (13 + 3(1)) + (8 + 2(3)) = 46 < 50$. This will, of course, add to the congestion in the arc a_{2t}. A user optimum is achieved for $(8 + 2x_{s1}) + (33 + x_{1t}) = (8 + 2x_{s1})\ (13 + 3x_{12}) + (8 + 2x_{2t}) = (33 + x_{s2}) + (8 + 2x_{2t})$, i.e, all paths have the same durations, plus the Kirchhoff node equations for n_1 and n_2, and the required flow value $x_{s1} + x_{s2} = 6$. The optimal solution has 2 users each take the upper, intermediate, and lower path, respectively. The result is a duration of 51 for each user for a total of 306 time units. Clearly, user optimization has not led to an overall optimum. The system optimum is achieved via conditions (1) and (2) that are $(8 + 4x_{s1}) + (33 + 2x_{1t}) = \lambda$,

$(8 + 4x_{s1}) + (13 + 6x_{12}) + (8 + 4x_{2t}) = \lambda$, $(33 + 2x_{s2}) + (8 + 4x_{2t}) = \lambda$, the Kirchhoff node equations for n_1 and n_2, and the flow requirement $x_{s1} + x_{s2} = 6$. In the resulting flow pattern, $^8/_3$ units each flow on the upper and lower path, respectively, while $^2/_3$ units take the intermediate path. The durations on these paths are $50^1/_3$ for the upper and the lower path, and $44^1/_3$ for the intermediate path for a total duration of $\bar{z} = 298$.

5.2 Solving Flow Assignment Problems

The flow assignment problem introduced in the previous section had the objective Min $z = \sum_{a_{ij} \in A} c_{ij}(x_{ij})$. In case of system optimization, we have defined $c_{ij}(x_{ij}) = x_{ij}t_{ij}(x_{ij})$. In case a user optimum is desired, we can define $c_{ij}(x_{ij}) = \int_0^{x_{ij}} t_{ij}(\xi)d\xi$. In network terms, for system optimization, we can define arc lengths $d_{ij} = \frac{\partial c_{ij}(x_{ij})}{\partial x_{ij}}$, so that the relations (1) and (2) indicate that all flow units will move on the shortest paths from the source to the sink. Similarly, in case of user optimization, arc lengths can be defined as $d_{ij} = t_{ij}(x_{ij})$ and again, units will flow on the shortest source-sink paths. This property is exploited by the algorithms described in this section. Given that system and user optimization differ only by the definition of the parameters d_{ij}, we assume generic arc lengths that cover both versions of the problem.

Popular solution methods for the flow assignment problem are shortest augmenting path algorithms, primal improvement methods, and the network simplex method. Without modification, Busacker and Gowen's (1961) successive shortest path method would assign the entire required flow to the route that has the least duration of the zero flow level. Such a procedure would generally lead to solutions of very poor quality. A modified approach is based on the idea behind user optima. It assigns a small increment of the required flow to the source-sink path of minimal duration, then recomputes the durations on the basis of the new flows, finds the shortest path again, and repeats this process until the required flow levels have been reached. Given that the increments ε by which the flow is increased, are sufficiently small, one may believe that an optimal solution can be found by this procedure. However, as the following example demonstrates, this is not the case.

Example: Consider the graph $G = (N, A, \kappa, t^{\circ})$ in Figure III.59 whose arc durations are $t_{ij}(x_{ij}) = t_{ij}^{\circ}\,[1 + 0.15(x_{ij}/\kappa_{ij})^4]\ \forall\ a_{ij} \in A$. For simplicity, we consider the values κ_{ij} as thresholds in the delay functions, but not restrictive as capacities on the flows.

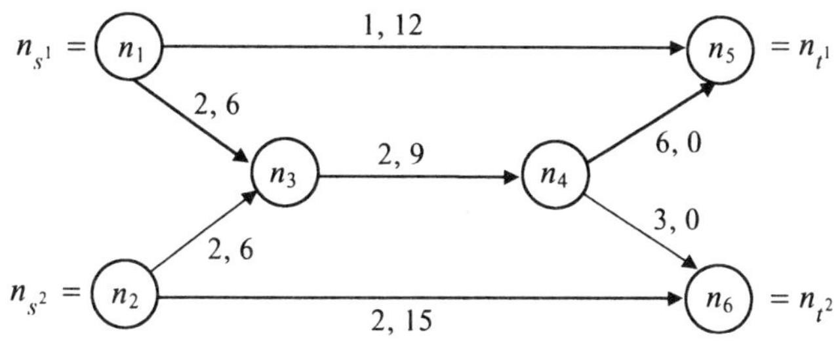

Figure III.59

Let the required flow values of the two commodities be $\tilde{f}_1 = 4$ and $\tilde{f}_2 = 2$, respectively. The four source-sink paths are then $P_1 = (n_1, n_5)$, $P_2 = (n_1, n_3, n_4, n_5)$, $P_3 = (n_2, n_3, n_4, n_6)$, and $P_4 = (n_2, n_6)$, where P_1 and P_2 are used by the first, and P_3 and P_4 are used by the second commodity. The duration on path P_k is denoted by t_k for $k = 1, ..., 4$, so that $t_1 = 12[1 + 0.15(x_{15})^4]$, $t_2 = 6[1 + 0.15(\frac{1}{2}x_{13})^4] + 9[1 + 0.15(\frac{1}{2}x_{34})^4] + 0$, $t_3 = 6[1 + 0.15(\frac{1}{2}x_{23})^4] + 9[1 + 0.15(\frac{1}{2}x_{34})^4] + 0$, and $t_4 = 15[1 + 0.15(\frac{1}{2}x_{26})^4]$. Initially, all flows are zero, so that the durations on the four paths are 12, 15, 15, and 15, respectively. Initially, flow pertaining to the first commodity is shipped on P_1 until the $t_1 = t_2$, while equal numbers of units of the second commodity are shipped on P_3 and P_4. At some point the flows on paths P_1, P_2, P_3 and P_4 are $x_1 = 1.136$, $x_2 = 0$, $x_3 = 0.284$ and $x_4 = 0.284$ with $t_1 = t_2 = t_3 = t_4 \approx 15$, so that 28.4% of the required flows are shipped from the respective sources to their sinks. Continuing with this procedure, the flow in all paths will either remain unchanged or increase. In particular, notice that $x_3 > 0$ when the procedure terminates. In contrast, the system optimum has $x_1 = 1.481$, $x_2 = 2.519$, $x_3 = 0$ and $x_4 = 2$ with $t_1 = t_2 = 20.66$, $t_3 = 18.40$ and $t_4 = 17.25$. The method of augmenting shortest paths will not find this solution.

Another solution method was used and popularized by Weintraub (1974). His approach is based on the method of Klein, described in Section 3.3.2. In essence, it starts with a feasible multicommodity flow pattern $\mathbf{x}$ and then determines the circuit of minimal length in $G(\mathbf{x})$. If the length of this circuit is negative, the objective is improved if some number of flow units $\varepsilon > 0$ are moved on that circuit, thus increasing the flow in all arcs $a'_{ij} \in A'$ on the circuit if that belongs to arcs $a_{ij} \in A$ (forward arcs) while decreasing the flow in arcs $a'_{ij} \in A'$ with $a_{ji} \in A$ (backward arcs). The main problem is how the value of ε is to be determined. One possibility would be to apply any one-dimensional search technique such as Golden Section search, see, e.g., Eiselt *et al.* (1987). This approach can be illustrated by an example.

Example: Consider again the graph in Figure III.59 and suppose that a feasible flow was determined with $x_1 = 1.532$, $x_2 = 2.468$, $x_3 = 0.284$ and $x_4 = 1.716$, so that $t_1(x_1) = 21.915$, $t_2(x_2) = 21.927$, $t_3(x_3) = 19.840$ and $t_4(x_4) = 16.219$. It is apparent that the circuit that consists of the arcs of P_4 and P_3 has length $t_4 - t_3 = 16.219 - 19.840 = -3.621 < 0$, so that ε units should be moved on that circuit, thus increasing the flow in a_{26} and decreasing the flow in a_{23}, a_{34}, and a_{46} by ε units each. The flow change ε is bounded by the flow change for which $t_4(x_4 - \varepsilon) = t_3(x_3 - \varepsilon)$, and $x_3 - \varepsilon \geq 0$, whichever comes first. Here, $\varepsilon = 0.284$ and the new flow pattern has $x_1 = 1.532$, $x_2 = 2.468$, $x_3 = 0$ and $x_4 = 2$, so that $t_1(x_1) = 21.915$ (unchanged), $t_2(x_2) = 20.217$, $t_3(x_3) = 18.130$ and $t_4(x_4) = 17.250$. At this point, P_2 and P_1 form another cycle with t_2 $t_1 = 20.217 - 21.915 = -1.698 < 0$, requiring another iteration. It turns out that moving $\varepsilon = 0.051$ units on this cycle leads to the system optimal solution with $x_1 = 1.481$, $x_2 = 2.519$, $x_3 = 0$ and $x_4 = 2$. In general, the exact optimal solution will only be approached asymptotically by this method.

One problem with this and similar approaches is that they must keep track of the arc flows of the individual commodities. In large-scale real-life networks, this may be an arduous task that requires large amounts of memory. In the following, we describe an approach that requires only the knowledge of the aggregated flows $x_{ij} = \sum_{k=1}^{p} x_{ij}^k$. The method is based on Frank and Wolfe's (1956) optimization method that is designed to find the minimum of a convex function subject to linear constraints. The method starts with a feasible solution $\hat{\mathbf{x}}^\nu$ with the iteration counter ν. It then performs a gradient minimization step that results in a solution $\tilde{\mathbf{x}}^\nu$. Then an interpolation step is performed, so that the next iteration commences with a solution $\hat{\mathbf{x}}^{\nu+1}$ that is a linear convex combination of $\hat{\mathbf{x}}^\nu$ and $\tilde{\mathbf{x}}^\nu$.

The application of the ν-th iteration of the Frank and Wolfe method can be described as follows. Given a feasible flow pattern $\mathbf{x}^\nu$, where the superscript does not specify the commodity, but is an iteration counter, compute the partial derivatives $\eta_{ij}^\nu = \dfrac{\partial c_{ij}(x_{ij})}{\partial x_{ij}}$ evaluated at x_{ij}^ν. For each commodity k, we then solve the shortest path problem from n_{s^k} to n_{t^k} with arc lengths η_{ij}^ν, resulting in the flow pattern $\tilde{\mathbf{x}}^\nu$. We then perform an interpolation search $\underset{\alpha\in[0;1]}{\text{Min}}\, z((1-\alpha)\hat{\mathbf{x}}^\nu + \alpha\tilde{\mathbf{x}}^\nu)$. This is a function of the single variable α. The result is an optimal value $\bar{\alpha}$, which is used to compute the new solution $\hat{\mathbf{x}}^{\nu+1} = (1-\bar{\alpha})\hat{\mathbf{x}}^\nu + \bar{\alpha}\tilde{\mathbf{x}}^\nu$. Due to the convexity of the feasible set, the new solution $\hat{\mathbf{x}}^{\nu+1}$ is feasible.

This procedure is repeated until some stop criterion is satisfied. Since the objective functions of flow assignment problems tend to be shallow in the vicinity of the optimum, we may replace the usual stop criterion "Stop if $[z(\hat{\mathbf{x}}^{\nu}) - z(\hat{\mathbf{x}}^{\nu+1})]/z(\hat{\mathbf{x}}^{\nu}) < \varepsilon$ for some predetermined $\varepsilon > 0$" by "Stop if $\left\|\hat{\mathbf{x}}^{\nu+1} - \hat{\mathbf{x}}^{\nu}\right\| / \left\|\hat{\mathbf{x}}^{\nu}\right\| < \varepsilon$". This is particularly relevant if the arc flows are more of a concern than the objective value $z(\mathbf{x})$.

We are now able to formally describe the algorithm. It is initialized with a feasible flow pattern $\hat{\mathbf{x}}^{0}$ (that is possibly obtained with the incremental method described above). Set the iteration counter $\nu := 0$ and define some $\varepsilon > 0$.

The Minimum Delay Flow Deviation Algorithm

Step 1: Given the feasible flow pattern $\hat{\mathbf{x}}^{\nu}$, determine the partial derivatives
$\eta_{ij}^{\nu} = \dfrac{\partial c_{ij}(x_{ij})}{\partial x_{ij}}$ evaluated at $\hat{x}_{ij}^{\nu}$.

Step 2: For each $k = 1, \ldots, p$, determine the shortest path from n_{s^k} to n_{t^k} based on the arc values η_{ij}^{ν}. Send $\tilde{f}_k$ units on the shortest path from n_{s^k} to n_{t^k}, $k = 1, \ldots, p$. The resulting flow pattern is denoted by $\tilde{\mathbf{x}}^{\nu}$.

Step 3: Perform an interpolation search to solve the single-variable problem
Min $z((1-\alpha)\hat{\mathbf{x}}^{\nu} + \alpha\tilde{\mathbf{x}}^{\nu})$, s.t. $\alpha \in [0; 1]$.
Denote the result by $\overline{\alpha}$ and compute $\hat{\mathbf{x}}^{\nu+1} = (1-\overline{\alpha})\hat{\mathbf{x}}^{\nu} + \overline{\alpha}\,\tilde{\mathbf{x}}^{\nu}$.

Step 4: Is the chosen stop criterion satisfied?
If yes: Stop; a near-optimal solution has been found.
If no: Set $\nu := \nu + 1$ and go to Step 1.

It is apparent that the flows of the individual commodities do not have to be stored. Recall that the initial flow pattern $\hat{\mathbf{x}}^{0}$ is assumed to satisfy the conservation equations for all commodities at all nodes except at the sources and sinks. Similarily, the flow pattern $\tilde{\mathbf{x}}^{\nu}$ sends flows on a simple path, thus automatically satisfying Kirchhoff's node equations. As a result, any linear convex combination such as the one that determines the new solution $\hat{\mathbf{x}}^{\nu+1}$ will also satisfy the conservation equations.

Example: Consider the graph in Figure III.60 where the sources of the two commodities are $n_{s^1} = n_1$ and $n_{s^2} = n_2$ and the two sinks coincide at $n_{t^1} = n_{t^2} = n_4$. The functions $t_{ij}(x_{ij})$ are shown next to the arcs.

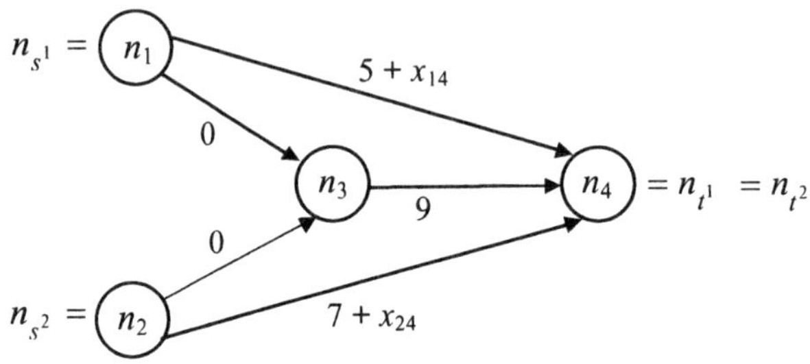

Figure III.60

Furthermore, suppose that the required flow values are $\tilde{f}_1 = \tilde{f}_2 = 5$, and let the initial solution be $\hat{\mathbf{x}}^0 = [\hat{x}_{14}^0, \hat{x}_{13}^0, \hat{x}_{34}^0, \hat{x}_{23}^0, \hat{x}_{24}^0] = [5, 0, 0, 0, 5]$. This implies individual durations of $t(\hat{\mathbf{x}}^0) = [10, 0, 9, 0, 12]$ and total durations $c(\hat{\mathbf{x}}^0) = \hat{\mathbf{x}}^0 t(\hat{\mathbf{x}}^0) = [50, 0, 0, 0, 60]$, so that $z(\hat{\mathbf{x}}^0) = \sum_{a_{ij} \in A} c_{ij}(\hat{x}_{ij}^0) = 50 + 0 + 0 + 0 + 60 = 110$. Then $\eta^0 = [5 + 2\hat{x}_{14}^0, 0, 9, 0, 7 + 2\hat{x}_{24}^0] = [15, 0, 9, 0, 17]$. Given these arc lengths, the shortest paths for the first and second commodity are (n_1, n_3, n_4) and (n_2, n_3, n_4), respectively. Shipping $\tilde{f}_1 = 5$ and $\tilde{f}_2 = 5$ flow units on these paths results in the flow pattern $\tilde{\mathbf{x}}^0 = [0, 5, 10, 5, 0]$. We now have to minimize $z[(1 - \alpha)\hat{\mathbf{x}}^0 + \alpha\tilde{\mathbf{x}}^0] = z[(5 - 5\alpha, 0, 0, 0, 5 - 5\alpha) + (0, 5\alpha, 10\alpha, 5\alpha, 0)] = z(5 - 5\alpha, 5\alpha, 10\alpha, 5\alpha, 5 - 5\alpha) = c(\mathbf{x}) = \mathbf{x}t(\mathbf{x}) = 5(5 - 5\alpha) + (5 - 5\alpha)^2 + 90\alpha + 7(5 - 5\alpha) = (5 - 5\alpha)^2 = 50\alpha^2 - 70\alpha + 85$. This function is optimized by setting $\frac{dz}{d\alpha} = 100\alpha - 70 = 0$, so that $\bar{\alpha} = 0.7$, and $\hat{\mathbf{x}}^1 = (1 - \bar{\alpha})\hat{\mathbf{x}}^0 + \bar{\alpha}\tilde{\mathbf{x}}^0 = (0.3)[5, 0, 0, 0, 5] = (0.7)[0, 5, 10, 5, 0] = [1.5, 3.5, 7, 3.5, 1.5]$ with $z(\hat{\mathbf{x}}^1) = 9.75 + 0 + 63 + 0 + 12.75 = 85.5$. The progression of the algorithm in the first three iterations is summarized in Table III.10.on the next page.

As the table demonstrates, the quality of the solutions improves quickly; the deviation of the solutions from the unique exact optimum $\bar{\mathbf{x}} = [2, 3, 7, 4, 1]$ with $z(\bar{\mathbf{x}}) = 85$ is only 0.08% after two and 0.0282% after three iterations. However, the flow patterns, while close to optimum as well, deviate by higher percentages.

Table III.10

Iteration ν	$\hat{\mathbf{x}}^{\nu}$	$z(\hat{\mathbf{x}}^{\nu})$	$\tilde{\mathbf{x}}^{\nu}$	$z(\tilde{\mathbf{x}}^{\nu})$	$\overline{\alpha}$
0	5, 0, 0, 0, 5	110	0, 5, 10, 5, 0	90	0.7
1	1.5, 3.5, 7, 3.5, 1.5	85.5	5, 0, 5, 5, 0	95	0.1724
2	2.1034, 2.8966, 6.6552, 3.7586, 1.2414	85.0690	0, 5, 10, 5, 0	90	0.0867
3	1.921, 3.079, 6.945, 3.866, 1.134	85.024	–	–	–

In general, the convergence of this method tends to be slow as the solutions approach the optimum. The convergence may be speeded up by alternating between the Frank and Wolfe method and the PARTAN method; for references regarding the latter, see, e.g., Luenberger (1984) or Boffey (1992). Wolfe (1970) suggested another technique regarding improvements in convergence.

5.3 Discrete Route Assignment

The flow assignment of Section 5.1 minimized the total transportation of all $\sum_{k=1}^{p} \tilde{f}_k$ flow units required to be shipped from their sources to the associated sinks. That problem allowed supplies to be split up on their way through the network. Sometimes, however, it is not feasible or desirable to do so. For example, customers may not wish to separate their $\tilde{f}_k$ containers that are to be moved from n_{s^k} to n_{t^k}, but may require that they be shipped on the same train, so as to ensure that all containers are available at the same time. Similarly, it may be required to submit packets of information through a computer network, and sending all of them on the same route will generally require resequencing. Formally, the *discrete route assignment* problem is a standard flow assignment problem with the added requirement that all units of one commodity are shipped on the same route.

In order to formulate the problem, define zero-one variables $y^{k\ell} = 1$, if all units of the k-th commodity are shipped on the ℓ-th path, and 0 otherwise. Recall that L_k denotes the set of paths from n_{s^k} to n_{t^k} and $\tilde{f}_k$ is the number of flow units available at n_{s^k} and in demand at n_{t^k}. The problem can then be written as

$$\text{P}_{\text{DRA}}: \text{ Min } z = \sum_{a_{ij} \in A} c_{ij}(x_{ij})$$

$$\text{s.t. } x_{ij} = \sum_{k=1}^{p} \sum_{\ell \in L_k} \delta_{ij}^{k\ell} y^{k\ell} \tilde{f}_k$$

$$\sum_{\ell \in L_k} y^{k\ell} = 1 \;\; \forall\, k = 1, \ldots, p$$

$$y^{k\ell} = 0 \vee 1 \;\; \forall\, k = 1, \ldots, p;\; \ell \in L_k$$

If desired, capacity constraints $x_{ij} \le \kappa_{ij}\ \forall\ a_{ij} \in A$ can also be added. It is easy to show that the flow assignment problem P_{FAP} is a special case of the discrete route assignment problem P_{DRA}. In order to do so, define in P_{FAP} new variables $\tilde{y}^{k\ell} = y^{k\ell}/\tilde{f}_k \;\; \forall\ \ell \in L_k;\ k = 1, \ldots, p$. Then the structural constraints in P_{FAP} are $\sum_{\ell \in L_k} \tilde{y}^{k\ell} = \frac{1}{\tilde{f}_k} \sum_{\ell \in L_k} y^{k\ell} = 1\ \forall\ k = 1, \ldots, p$, which shows that the flow assignment problem P_{FAP} corresponds to the discrete route assignment problem P_{DRA} but without integrality requirements on the variables $y^{k\ell}$.

Given that the already difficult flow assignment problem is a special case of the discrete route assignment problem, the latter is obviously difficult. In order to solve the problem, some lower bounding procedure may be helpful. Therefore, we dualize the first set of constraints of P_{DRA} and obtain the Lagrangean function

$$L(x, \lambda) = \sum_{a_{ij} \in A} c_{ij}(x_{ij}) - \sum_{a_{ij} \in A} \lambda_{ij} \left[x_{ij} - \sum_{k=1}^{p} \sum_{\ell \in L_k} \delta_{ij}^{k\ell} y^{k\ell} \tilde{f}_k \right]$$

$$\text{s.t. } \sum_{\ell \in L_k} y^{k\ell} = 1 \;\; \forall\, k = 1, \ldots, p$$

$$x_{ij} \in [0; \kappa_{ij}] \;\; \forall\, a_{ij} \in A$$

$$y^{k\ell} = 0 \vee 1 \;\; \forall\, k = 1, \ldots, p;\; \ell \in L_k.$$

The Lagrangean can be written as

$$L(x, \lambda) = z_1 + z_2 = \sum_{a_{ij} \in A} [c_{ij}(x_{ij}) - \lambda_{ij} x_{ij}] + \sum_{a_{ij} \in A} \left[\lambda_{ij} \sum_{k=1}^{p} \sum_{\ell \in L_k} \delta_{ij}^{k\ell} y^{k\ell} \tilde{f}_k \right].$$

It is now apparent that the problem decomposes into two sets of independent problems; one that involves the variables x_{ij}, and another that depends on the variables $y^{k\ell}$. Furthermore, the problems that include the variables x_{ij} are independent of each other, so that we can independently solve $m = |A|$ problems

$$P_1^{ij}: \text{ Min } z_1^{ij} = c_{ij}(x_{ij}) - \lambda_{ij} x_{ij}$$

$$\text{s.t. } x_{ij} \in [0; \kappa_{ij}],$$

so that $z_1 = \sum_{a_{ij} \in A} z_1^{ij}$. A similar argument applies to the set of problems that comprise variables $y^{k\ell}$. That set consists of p independent problems of the type

$$P_2^k: \text{ Min } z_2^k = \sum_{a_{ij}\in A} \lambda_{ij} \sum_{\ell\in L_k} \delta_{ij}^{k\ell} y^{k\ell} \tilde{f}_k$$

$$\sum_{\ell\in L_k} y^{k\ell} = 1$$

$$y^{k\ell} = 0 \vee 1 \quad \forall\, \ell \in L_k,$$

so that $z_2 = \sum_{k=1}^{p} z_2^k$. Clearly, the dual variables λ_{ij} connect the two types of problems. A possible approach to this problem is examined in the following

Example: Consider the graph in Figure III.61 in which travel along an edge a_{ij} takes $t_{ij}(x_{ij}) = t_{ij}^0 + h_{ij}x_{ij}$.

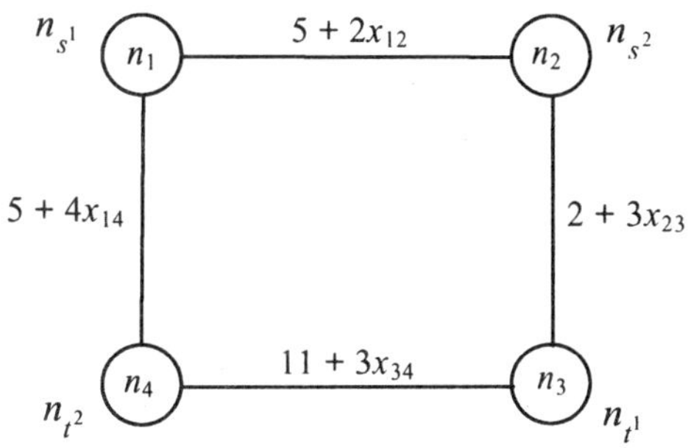

Figure III.61

As indicated, $n_{s^1} = n_1$, $n_{s^2} = n_2$, $n_{t^1} = n_3$, and $n_{t^2} = n_4$. Furthermore, $\tilde{f}_1 = \tilde{f}_2 = 2$ and, for computational convenience, we assume that if $\tilde{f}_k$ units are shipped from n_{s^k} to n_{t^k} on a path p_k, then $\tilde{f}_k$ units are returned from n_{t^k} to n_{s^k} on the same path. Given the durations $t_{ij}(x_{ij}) = t_{ij}^0 + h_{ij}x_{ij}$ and sufficiently high capacities κ_{ij} $\forall$ $a_{ij} \in A$, the problem P_1^{ij} can be written as Min $z_1^{ij} = t_{ij}^0 x_{ij} + h_{ij} x_{ij}^2 - \lambda_{ij}x_{ij}$, s.t. $x_{ij} \geq 0$. Differentiating with respect to x_{ij} results in $\frac{dz_1^{ij}}{dx_{ij}} = t_{ij}^0 + 2h_{ij}\bar{x}_{ij} - \lambda_{ij} = 0$ or simply $\bar{x}_{ij} = \frac{\lambda_{ij} - t_{ij}^0}{2h_{ij}}$, if $\lambda_{ij} \geq t_{ij}^0$, and 0 otherwise so that $c_{ij}(\bar{x}_{ij}) = \frac{1}{4h_{ij}}(\lambda_{ij}^2 - t_{ij}^{0^2})$ if $\lambda_{ij} \geq t_{ij}^0$, and 0 otherwise. This results in $\bar{z}_1^{ij} = \frac{-1}{4h_{ij}}(\lambda_{ij} - t_{ij}^0)^2$

if $\lambda_{ij} \geq t_{ij}^0$, and 0 otherwise, and $\bar{z}_1 = 2 \sum\limits_{a_{ij} \in A} \bar{z}_1^{ij}$.

Consider now the problem P_2^k. Define y^{11}, y^{12}, y^{21}, and y^{22} as the proportion of the two commodities on the paths (n_1, n_2, n_3), (n_1, n_4, n_3), (n_2, n_3, n_4), and (n_2, n_1, n_4), respectively. Suppose now that $\lambda_{ij} = \lambda \ \forall \ a_{ij}$, then the problem P_2^1 is Min $z_2^1(\lambda) = 2\lambda \tilde{f}_1 [(\delta_{12}^{11} y^{11} + \delta_{23}^{11} y^{11}) + (\delta_{14}^{12} y^{12} + \delta_{43}^{12} y^{12})]$, s.t. $y^{11} + y^{12} = 1$ and $y^{11}, y^{12} = 0 \vee 1$, where the constant "2" in the objective function is due to the return flow. Similarly, P_2^2 is Min $z_2^2(\lambda) = 2\lambda \tilde{f}_2 [(\delta_{23}^{21} y^{21} + \delta_{34}^{21} y^{21}) + (\delta_{21}^{22} y^{22} + \delta_{14}^{22} y^{22})]$, s.t. $y^{21} + y^{22} = 1$ and $y^{21}, y^{22} = 0 \vee 1$. As a result, we obtain $z_2(\lambda) = z_2^1(\lambda) + z_2^2(\lambda) = 2[2\lambda(y^{11} + y^{11}) + (y^{12} + y^{12}) + (y^{21} + y^{21}) + (y^{22} + y^{22})] = 8\lambda[y^{11} + y^{12} + y^{21} + y^{22}]$ which, as $y^{11} + y^{12} = y^{21} + y^{22} = 1$, reduces to $z_2(\lambda) = 16\lambda$.

Assuming that $\lambda_{ij} \geq t_{ij}^0 \ \forall \ a_{ij} \in A$, we can now compute a lower bound on the objective function value of P_{DRA} as $\underline{z} = z_1 + z_2 = -2 \sum\limits_{a_{ij} \in A} \frac{1}{4h_{ij}} (\lambda_{ij} - t_{ij}^0)^2 + 16\lambda = -2[\frac{1}{8}(\lambda - 5)^2 + \frac{1}{12}(\lambda - 2)^2 + \frac{1}{12}(\lambda - 11)^2 + \frac{1}{16}(\lambda - 5)^2 + 16\lambda = -\frac{17}{24}\lambda^2 + \frac{578}{24}\lambda - \frac{725}{24}$. Differentiating with respect to λ and setting the resulting expression equal to zero results in $\bar{\lambda} = 17$. This value satisfies $\bar{\lambda}_{ij} \geq t_{ij}^0 \ \forall \ a_{ij} \in A$ and results in $\bar{z}_1(\lambda) = -97.5$ and $\bar{z}_2(\lambda) = 272$ for a lower bound of $\bar{z}(\lambda) = 174.5$.

Furthermore, given $\bar{\lambda} = 17$ we obtain $x_{12} = 3$, $x_{23} = 2.5$, $x_{34} = 1$, and $x_{41} = 1.5$ as well as $(y^{11}, y^{12}, y^{21}, y^{22}) = (1, 0, 1, 0), (1, 0, 0, 1), (0, 1, 1, 0)$, or $(0, 1, 0, 1)$.

At this point we may proceed by improving the value of λ by means of subgradient techniques for minimizing the Lagrangean function. Doing so in our example reveals that the value of $\lambda = 17$ is already optimal. For details regarding subgradient optimization of the Lagrangean function, see, e.g., Eiselt *et al.* (1987) or Boffey (1992).

5.4 Network Design Problems

Whereas the problems in the previous sections of this chapter have dealt with the assignment of traffic flows to the links of a given network, design models take the problem one step further. In particular, they add the design of the network to the routing issues of the standard problems. In the most general case, network design problems include decisions concerning

- the selection of links to be included in the network,
- the capacities of the links that are included in the network, and
- the traditional issue of how to route the traffic in the newly designed network.

Clearly, as network design problems address the design issues in addition to all the traditional routing issues, they are much harder than, say, flow assignment problems.

As far as the capacities of the links are concerned, we normally have the choice among a discrete set of possibilities. Typical examples are single or double tracks in railroad networks, single, double, triple,... lanes in the design of highway networks, or ¼", ½", ¾", or 1" pipe diameters. However, LeBlanc and Reddoch (1990) argue that in some telecommunications networks arc capacities can be considered as continuous variables. More specifically, such instances include cases in which standard leased communication lines can be supplemented by telephone lines with a dial-up modem when the demand requires it. Problems of this type are considered in the next section.

5.4.1 Continuous Network Design

In order to concentrate on the aspect under consideration within the limited space available in this book, consider a problem in which it is already determined which links will be introduced and where the flow is routed through the network. In this *link size determination problem*, the only variables are the capacities κ_{ij} associated with the existing arcs in A. This problem can be formulated as

$$\text{P}_{\text{LDP}}: \text{ Min } z = \sum_{a_{ij}\in A} c_{ij}(\bar{x}_{ij}, \kappa_{ij})$$

$$\text{s.t.} \quad \sum_{a_{ij}\in A} g_{ij}(\kappa_{ij}) \le b$$

$$\kappa_{ij} \ge 0 \quad \forall\, a_{ij} \in A,$$

where $\bar{x}_{ij}$ is the parametric flow in arc a_{ij}, κ_{ij} is the variable capacity of arc a_{ij}, $c_{ij}(\bar{x}_{ij}, \kappa_{ij})$ is the total duration or delay on the arc a_{ij} as a function of the known flow $\bar{x}_{ij}$ and the capacity κ_{ij}, $g_{ij}(\kappa_{ij})$ are the costs of constructing a link a_{ij} of capacity κ_{ij}, and b is the available budget. As stated, the problem P_{LDP} is a continuous nonlinear knapsack problem. As the duration function $c_{ij}(\bar{x}_{ij}, \kappa_{ij})$ may be assumed to be decreasing for increasing capacities, whereas the costs $g_{ij}(\kappa_{ij})$ are increasing with increasing capacity, an optimal solution will use the entire budget b, i.e., the constraint will be satisfied as equality at optimum. Given that, we can write the Lagrangean as $L(\kappa, \lambda) = \sum_{a_{ij}\in A} c_{ij}(\bar{x}_{ij}, \kappa_{ij}) - \lambda\,[b - \sum_{a_{ij}\in A} g_{ij}(\kappa_{ij})]$, so that the optimality conditions are $\dfrac{\partial L(\kappa,\lambda)}{\partial \kappa_{ij}} = 0 \; \forall \; a_{ij} \in A$ and $\dfrac{\partial L(\kappa,\lambda)}{\partial \lambda} = 0$, resulting in

$$\frac{\partial c_{ij}(\bar{x}_{ij},\kappa_{ij})}{\partial \kappa_{ij}} + \lambda \frac{\partial g_{ij}(\kappa_{ij})}{\partial \kappa_{ij}} = 0, \text{ and} \tag{3}$$

$$\sum_{a_{ij}\in A} g_{ij}(\kappa_{ij}) = b. \tag{4}$$

There is little that can be said about the problem without further knowledge of the functions $c_{ij}(\bar{x}_{ij},\kappa_{ij})$ and $g_{ij}(\kappa_{ij})$. In order to further explore the problem, we consider two examples with different delay functions $c_{ij}(\bar{x}_{ij},\kappa_{ij})$.

Example 1: Let $c_{ij}(\bar{x}_{ij},\kappa_{ij}) = \bar{x}_{ij}/(\kappa_{ij} - \bar{x}_{ij})$ and $g_{ij}(\kappa_{ij}) = p_{ij}\,\kappa_{ij} \; \forall \; a_{ij} \in A$ with given parameters p_{ij}. The optimality condition (3) is then $-\bar{x}_{ij}\big/(\kappa_{ij} - \bar{x}_{ij})^2 + \lambda p_{ij} = 0 \; \forall \; a_{ij} \in A$ or, after some routine calculations, $\kappa_{ij} = \bar{x}_{ij}\left[1 + 1\big/\sqrt{\lambda p_{ij}\bar{x}_{ij}}\right]$. Substituting this expression in the constraints (4) results in $\lambda = [(\sum_{a_{ij}\in A}\sqrt{p_{ij}\bar{x}_{ij}}\,) / (b - \sum_{a_{ij}\in A} p_{ij}\bar{x}_{ij}\,)]^2$. Once the optimal value for λ is calculated, the optimal capacities can be determined as well.

Example 2: Consider a link size determination problem with $c_{ij}(\bar{x}_{ij},\kappa_{ij}) = t_{ij}^0\bar{x}_{ij} + \alpha_{ij}(\bar{x}_{ij}/\kappa_{ij})^\beta$ and $g_{ij}(\kappa_{ij}) = p_{ij}\kappa_{ij} \; \forall \; a_{ij} \in A$. The optimality conditions applied to these functions are $-\alpha_{ij}\beta\,\bar{x}_{ij}^\beta\,\kappa_{ij}^{-\beta-1} + \lambda p_{ij} = 0 \; \forall \; a_{ij} \in A$, or $\kappa_{ij}(\lambda) = \sqrt[\beta+1]{\dfrac{\alpha_{ij}\,\beta\,\bar{x}_{ij}^\beta}{\lambda p_{ij}}} \; \forall \; a_{ij} \in A$, and $\sum_{a_{ij}\in A} p_{ij}\,\kappa_{ij}(\lambda) = b$. Some standard algebraic transformations reveal that

$$\lambda = \frac{1}{b^{\beta+1}}\left[\sum_{a_{ij}\in A}\sqrt[\beta+1]{\alpha_{ij}\,\beta\,p_{ij}^\beta\,\bar{x}_{ij}^\beta}\right]^{\beta+1} \tag{5}$$

which can be used to calculate the values $\kappa_{ij}(\lambda)$. To illustrate the process, consider the network of an airline that operates flights between $n_1, n_2, \ldots, n_5$ as shown in the graph $G = (N, A, x, p)$ in Figure III.62 on the next page.

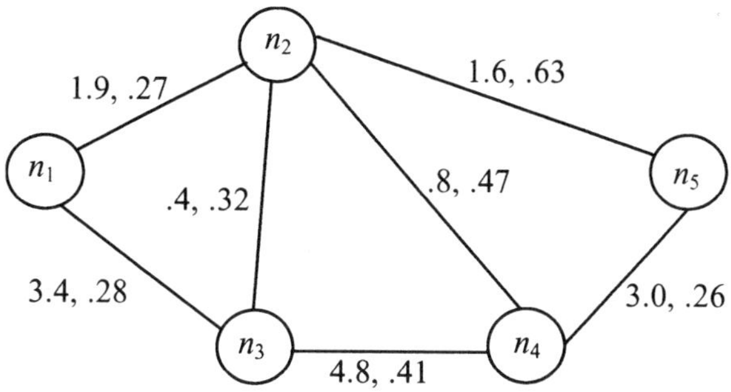

Figure III.62

Suppose that the budget is $b = 10$ and let $\alpha_{ij} = ½ \ \forall \ a_{ij} \in A$ and $\beta = 2$. We then obtain $\lambda \frac{1}{1{,}000} (5.806599)^3 = .195779$ which yields capacities as shown in the graph $G = (N, A, \kappa)$ shown in Figure III.63.

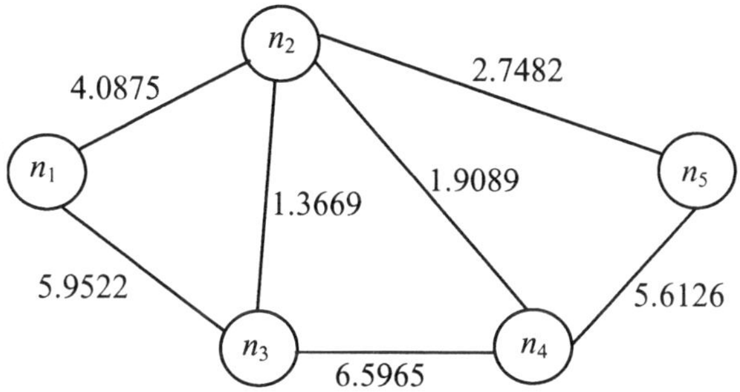

Figure III.63

It can be seen that the total costs of the capacities equal the budget.

In the link size determination problem, we have assumed that the flows through the network are known, even before the arc capacities have been determined. In order to incorporate capacity *and* flow optimization, one could assume some initial flow pattern, optimize the capacities as shown above, use these capacities as temporarily fixed and optimize the flows in a flow assignment problem, and continue to alternate between capacity and flow optimization.

Suppose that the optimality conditions (3) and (4) can be solved analytically for any given flow pattern $\mathbf{x}$. Then the capacities κ_{ij} will be expressed as functions of x_{ij}, say, $\kappa_{ij}(x_{ij})$. It then remains to solve the problem

$$\text{P: Min } z = \sum_{a_{ij} \in A} c_{ij} \, (x_{ij}, \kappa_{ij}(x_{ij}))$$

subject to the usual conservation and nonnegativity constraints. This problem may be solved with the Frank and Wolfe algorithm. Provided that c_{ij} is a convex function of x_{ij}, it is assumed that a local minimum is also a global minimum. The delay function $c_{ij} = x_{ij}/(\kappa_{ij} - x_{ij})$ does satisfy this requirement. However, this delay function is not convex in both, x_{ij} and κ_{ij} simultaneously, so that the flow deviation algorithm may not lead to a global optimum.

5.4.2 Discrete Network Design

The problem in this section is the same as that in the previous section, except that the capacities of the links are no longer continuous but discrete, i.e., we can choose the capacity κ_{ij} of an arc $a_{ij} \in A$ from a finite set K_{ij}, i.e., $\kappa_{ij} \in K_{ij} \; \forall \; a_{ij} \in A$. For simplicity, we assume here that the sets K_{ij} are the same for all arcs; we will denote them by $K = \{\kappa_0, \kappa_1, ..., \kappa_r\}$. We can then formulate the *discrete link size determination problem* as

$$\text{P}_{\text{DLDP}}\text{: Min } z = \sum_{a_{ij} \in A} c_{ij}(\bar{x}_{ij}; \kappa_{ij})$$

$$\text{s.t.} \quad \sum_{a_{ij} \in A} g_{ij}(\kappa_{ij}) \leq b$$

$$\kappa_{ij} \in K \quad \forall \; a_{ij} \in A.$$

A lower bound to the problem is provided by the Lagrangean function

$$L(\kappa, \lambda) = \sum_{a_{ij} \in A} c_{ij}(\bar{x}_{ij}; \kappa_{ij}) - \lambda[b - \sum_{a_{ij} \in A} g_{ij}(\kappa_{ij})]$$

$$\text{s.t. } \kappa_{ij} \in K \quad \forall \; a_{ij} \in A.$$

It is readily apparent that the Lagrangean function is decomposable into $m = |A|$ independent subproblems, one for each arc. For each $a_{ij} \in A$, the subproblem is

$$\text{P}_{ij}\text{: Min } z_{ij} = c_{ij}(\bar{x}_{ij}, \kappa_{ij}) + \lambda g_{ij}(\kappa_{ij})$$

$$\text{s.t.} \quad \kappa_{ij} \in K$$

Given that $|K| \ll \infty$, each problem P_{ij} is easy to solve. A lower bound on the objective value z of P is then $\underline{z} = \sum_{a_{ij} \in A} \bar{z}_{ij} - \lambda b$. Details concerning the computation of this lower bound are provided below.

Example: Consider again the problem of Example 2 in the previous section. Assuming that $K = \{2, 4, 6, 8\}$, each of the subproblems is

$$P_{ij}\text{: Min } z_{ij} = ½(\bar{x}_{ij} / \kappa_{ij})^2 + \lambda p_{ij}\kappa_{ij}$$
$$\text{s.t.} \quad \kappa_{ij} = 2 \vee 4 \vee 6 \vee 8.$$

Assuming that $\lambda = 1.0$, the problems P_{ij} and their objective values for $\kappa_{ij} = 2, 4, 6$, and 8 are shown in Table III.11, where the optimal value $\bar{z}_{ij}$ is indicated by a "*".

Table III.11

e_{ij}	Problem P_{ij}	z_{ij} value for κ_{ij} =				$\hat{\lambda}$
		2	4	6	8	
e_{12}	min z_{12} = $1.805/\kappa_{12}^2 + .27\kappa_{12}$	.9913*	1.1928	1.6701	2.1882	.6267
e_{13}	min z_{13} = $5.78/\kappa_{13}^2 + .28\kappa_{13}$	2.0050	1.4813*	1.8406	2.3303	.3584
e_{23}	min z_{23} = $.08/\kappa_{23}^2 + .32\kappa_{23}$	.6600*	1.2850	1.9222	2.5613	.0234
e_{24}	min z_{24} = $.32/\kappa_{24}^2 + .47\kappa_{24}$	1.0200*	1.9000	2.8289	3.7650	.0638
e_{25}	min z_{25} = $1.28/\kappa_{25}^2 + .63\kappa_{25}$	1.5800*	2.6000	3.8156	5.0600	.1905
e_{34}	min z_{34} = $11.52/\kappa_{34}^2 + .41\kappa_{34}$	3.7000	2.3600*	2.7800	3.4600	.4878
e_{45}	min z_{45} = $4.5/\kappa_{45}^2 + .26\kappa_{45}$	1.6450	1.3213*	1.6850	2.1503	.3005

Setting the capacities κ_{ij} to their optimal values results in $\kappa_{12} = \kappa_{23} = \kappa_{24} = \kappa_{25} = 2$, and $\kappa_{13} = \kappa_{34} = \kappa_{45} = 4$, so that the expenditures for the capacities are $\sum_{a_{ij} \in A} p_{ij}\kappa_{ij} = 7.18$, well short of the budget $b = 10$. In order to decide which capacities are to increase, we decrease the value of λ. Note that the solution will remain optimal for all problems P_{ij} until a break point is reached. Suppose that for some arc a_{ij} the present capacity is $\kappa_{ij} = \kappa_\ell$. The break point is then reached for $\lambda = \hat{\lambda}$, where the optimal z_{ij} values for two capacities equal each other. In this example, a breakpoint occurs for

$$½\frac{\bar{x}_{ij}^2}{\kappa_\ell^2} + \hat{\lambda} p_{ij}\kappa_\ell = ½\frac{\bar{x}_{ij}^2}{\kappa_{\ell+1}} + \hat{\lambda} p_{ij}\kappa_{\ell+1}, \text{ or}$$

$$\hat{\lambda} = \frac{\bar{x}_{ij}^2}{2p_{ij}} \frac{\kappa_\ell + \kappa_{\ell+1}}{\kappa_\ell^2 \kappa_{\ell+1}^2}$$

The break points $\hat{\lambda}$ for all arcs in the graph are shown in Table III.11.

The procedure now decreases λ from its present value to the nearest break point, increases the capacity of the corresponding arc (if the budget allows it, otherwise the process terminates with a value of λ which can be used to compute a lower bound on the original problem), and computes the next break point for the arc under consideration. This procedure is repeated until the budget is used. In this example, the current capacities cost 7.18 and the present value of $\lambda = 1.0$. The highest break point is $\hat{\lambda}_{12} = .6267$, so that the capacity of the arc a_{12} is increased from $\kappa_{12} = 2$ to 4. This costs an additional $2p_{ij} = 0.54$, so that the costs for the capacities is 7.18 + .54 = 7.72, still well within budget. The new break point for a_{12} is then $\hat{\lambda}_{12} = .1161$. The next highest break point is $\hat{\lambda}_{34} = .4878$. Increasing the capacity of a_{34} from $\kappa_{34} = 4$ to 6 results in additional costs of .82 for a budget consumption of 7.72 + .82 = 8.54. The new break point for a_{34} is $\hat{\lambda}_{34} = .1707$. The next lowest break point is $\hat{\lambda}_{13} = .3584$, so that κ_{13} is raised from 4 to 6. The additional costs are .54 for a new total of 8.54 + .56 = 9.1. The new break point for a_{13} is $\hat{\lambda}_{13} = .1254$. The next highest break point in the list is $\hat{\lambda}_{45} = .3005$, hence κ_{45} is raised from 4 to 6. The additional costs are .52 for a new total of 9.1 +.52 = 9.62; the new break point is $\hat{\lambda}_{45} = .1052$. The next highest break point is $\hat{\lambda}_{25} = .1905$. However, raising κ_{25} from 2 to 4 would cost an additional 1.26 and exceed the budget. Hence the process terminates with $\lambda = .1905$ and $\kappa_{23} = \kappa_{24} = \kappa_{25} = 2$, $\kappa_{12} = 4$, and $\kappa_{13} = \kappa_{45} = 6$. With these values, the lower bound on the objective value of the original problem can be computed.

5.4.3 Combined Routing and Discrete Link-Size Determination

In this section we drop the assumption that either the edge capacities or the flow patterns are fixed. Clearly, this problem is much more difficult than the (already *NP-hard*) problems discussed in the previous sections. Assuming that the edge capacities have to be chosen from a discrete and finite set, the combined routing and discrete link size determination problem can be formulated as

$$\text{P}_{\text{DLDRP}}: \quad \text{Min } z = \sum_{a_{ij} \in A} c_{ij}(x_{ij}, \kappa_{ij})$$

$$\text{s.t.} \quad \sum_{a_{ij} \in A} g_{ij}(\kappa_{ij}) \le b$$

$$x_{ij} = \sum_{k=1}^{p} \sum_{\ell = L_k} \delta_{ij}^{k\ell} y^{k\ell} \tilde{f}_k \quad \forall\, a_{ij} \in A$$

$$\sum_{\ell \in L_k} y^{k\ell} = 1 \;\forall\, k = 1, \ldots, p$$

$$x_{ij} \in [0; \kappa_{ij}] \quad \forall \; a_{ij} \in A$$
$$y^{k\ell} = 0 \vee 1 \quad \forall \; \ell \in L_k; \; k = 1, \ldots, p$$
$$\kappa_{ij} \in K = \{\kappa_0, \kappa_1, \ldots, \kappa_r\}.$$

With $\kappa_0 = 0$, we allow a link not to be built. Dualizing the budget constraint, the *Discrete Design Problem* results; a problem that is still difficult. Dualizing the constraints that define the arc flows x_{ij}, two subproblems result; one that includes the variables x_{ij} and the other that involves the path flow variable $y^{k\ell}$. This situation is not unlike that encountered in the Discrete Route Assignment problem discussed earlier. Methods that alternate between determining flows and capacities in one phase and improved dual variables in the next suggest themselves for the solution of this difficult class of problems.

REFERENCES

Agarwal, Y., Mathur, K., Salkin, H.M. (1989): A Set-Partitioning-Based Algorithm for the Vehicle Routing Problem. Networks 19, 731-750

Ahuja, R.K., Magnanti, T.L., Orlin, J.B. (1991): Some Recent Advances in Network Flows. SIAM Review 33, 175-219

Ahuja, R.K., Magnanti, T.L., Orlin, J.B. (1993): Network Flows: Theory, Algorithms and Applications. Prentice-Hall, Englewood Cliffs, NJ

Ahuja, R.K., Orlin, J.B. (1991): Distance-Directed Augmenting Path Algorithms for Maximum Flow and Parametric Maximum Flow Problems. Naval Research Logistics Quarterly 38, 413-430

Anbil, R., Gelman, E., Patty, B., Tanga, R (1991): Recent Advances in Crew-Pairing Optimization at American Airlines. Interfaces 21, 62-74

Aneja, Y.P. (1980): An Integer Linear Programming Approach to the Steiner Problem in Graphs. Networks 10, 167-178

Balas, E. (1965): Discrete Programming by the Filter Method, Operations Research 13, 915-957

Balas, E., Christofides, N. (1981): A Restricted Lagrangean Approach to the Traveling Salesman Problem. Mathematical Programming 21, 19-46

Balas, E., Toth, P. (1985): Branch and Bound Methods. In: Lawler, E.L., Lenstra, J.K., Rinnooy Kan, A.H.G., Shmoys, D.B. (Eds.): The Traveling Salesman Problem: A Guided Tour of Combinatorial Optimization. Wiley, Chichester, 361-401

Balas, E., Zemel, E. (1984): Lifting and Complementing Yields all the Facets of Positive Zero-One Programming Polytopes. In: Cottle, R.W. *et al.* (Eds.):

Mathematical Programming, Proceedings of the International Conference in Mathematical Programming, 13-24

Balinski, M.L., Quandt, R. (1964): On an Integer Program for a Delivery Problem. Operations Research 12, 300-304

Balinski, M.L., Young, P (1982): Fair Representation. Yale University Press, New Haven, CT

Beale, E.M., Tomlin J. (1970): Special Facilities in a General Mathematical Programming System for Nonconvex Problems Using Ordered Sets of Variables. In: Lawrence, J. (Ed.): Proceedings of the 5th International Conference on Operational Research.Tavistock, London, 447-454

Beasley, J.E. (1984): An Algorithm for the Steiner Problem in Graphs. Networks 14, 147-159

Beasley, J.E., Christofides, N. (1989): An Algorithm for the Resource Constrained Shortest Path Problem. Networks 19, 379-394

Belenguer, J.M., Benavent, E. (1991): Polyhedral Results on the Capacitated Arc Routing Problem. Working Paper, Departamento de Estadística e Investigación Operativa, Universidad de Valencia

Bellman, R. (1957): Dynamic Programming. Princeton Unversity Press Princeton, NJ

Bellman, R., Dreyfus, S.E. (1962): Applied Dynamic Programming. Princeton University Press, Princeton, N.J

Bellmore, M., Greenberg, H.J., Jarris, J.J. (1970): Multi-Commodity Disconnecting Sets. Management Science 16B, 427-433

Bellmore, M., Malone, J.C. (1971): Pathology of Traveling Salesman Subtour Elimination Algorithms. Operations Research 19, 278-307

Ben-Israel, A., Charnes, A. (1962): On Some Problems of Diophantine Programming. Cahiers du C.E.R.O. 4, 215-280

Berge, C. (1958): La Théorie des Graphes et ses Applications. Dunod, Paris. Translated as: The Theory of Graphs and its Applications, Wiley, New York, 1962

Berge, C. (1962): The Theory of Graphs and its Application. J. Wiley and Sons, New York

Berge, C. (1970): Graphes et Hypergraphes. Dunod, Paris. Translated as: Graphs and Hypergraphs. Amsterdam

Berge, C. (1973): Graphs and Hypergraphs. Amsterdam, North Holland

Bertsekas, D.P. (1991): Linear Network Optimization: Algorithms and Codes. MIT Press, Cambridge, MA

Biggs, N.L., Lloyd, E.K., Wilson, R.L. (1976): Graph Theory 1736-1936. Clarendon Press, Oxford

Bland, R.G., Shallcross, D.F. (1989): Large Traveling Salesman Problems Arising from Experiments in X-ray Crystallography: A Preliminary Report on Computation. Operations Research Letters, 8, 125-128

Bodin, L.D., Golden, B.L., Assad, A.A., Ball, M.O. (1983): Routing and Scheduling of Vehicles and Crews, the State of the Art. Computers & Operations Research, 10, 69-211

Boffey, T.B. (1991): Computer Network Design Problems. Working paper, Dept. of Statistics and Computational Mathematics, University of Liverpool, GB

Boffey, T.B. (1992): Distributed Computing: Associated Combinatorial Problems. Blackwell Scientific Publications, Oxford

Borgwardt, K, -H. (1982): The Average Number of Pivot Steps Required by the Simplex Method is Polynomial. Zeitschrift für Operations Research 26, 157-177

Bowman, V.J., Nemhauser, G.L. (1970): A Finiteness Proof for Modified Dantzig Cuts in Integer Programming. Naval Research Logistics Quarterly 17, 309-313

Boyce, W.M. (1977): An Improved Program for the Full Steiner Tree problem. ACM Transactions on Mathematical Software 3, 359-385

Bradley, G.H. (1971): Transformations of Integer Programs to Knapsack Problems. Discrete Mathematics 1, 29-45

Brearley, A.L., Mitra, G., and Williams, H.P. (1975): An Analysis of Mathematical Programming Problems Prior to Applying the Simplex Method. Mathematical Programming 8, 54-83

Brown, J.R. (1979): The Sharing Problem. Operations Research 27, 324-340

Brucker, P. (1981): Minimizing Maximum Lateness in a Two-Machine Unit-Time

Job Shop. Computing 27, 367-370

Bullnheiner, B., Harte, R.F., Strauss, C. (1999): A New Rank Based Version of the Ant System: A Computational Study. Central European Journal of Operations Research 7, 25-38

Busacker, R.G., Gowen, P.J. (1961): A Procedure for Determining Minimal Cost Network Flow Patterns. Technical Report 15, Operational Research Office, Johns Hopkins University, U.S.A.

Busacker, R.G., Saaty, T.L. (1965): Finite Graphs and Networks: An Introduction with Applications. McGraw-Hill, New York

Camerini, P.M., Galbiati, G., Maffioli, F. (1988): Algorithms for Finding Optimum Trees: Description, Use and Evaluation. Annals of Operations Research 13, 265-397

Carpaneto, G., Martello, S., Toth, P. (1988): Algorithms and Codes for the Assignment Problem. In: Simeone, B., Toth, P., Gallo, G., Maffioli, F., Pallottino, S. (Eds.): FORTRAN Codes for Network Optimization. Annals of Operations Research 13, 193-223

Carpaneto, G., Toth, P. (1980): Some New Branching and Bounding Criteria for the Asymmetric Travelling Salesman Problem. Management Science 26, 736-743

Cayley, A. (1875): On the Analytical Forms Called Trees, with Application to the Theory of Chemical Combinations. Report of the British Association for the Advancement of Science, 257-305; also Mathematical Papers IX, 427-460

Cerny, V. (1985): A Thermodynamical Approach to the Travelling Salesman Problem. An Efficient Simulated Annealing Algorithm. Journal of Optimization Theory and Applications 45, 41-51

Chandy, K.M., Lo, T. (1973): The Capacitated Minimum Spanning Tree. Networks 3, 173-181

Chen, W. K. (1990): Theory of Nets: Flows in Networks. Wiley, New York

Christofides, N. (1973): The Optimum Traversal of a Graph. Omega 1, 719-732

Christofides, N. (1976): Worst-Case Analysis of a New Heuristic for the Travelling Salesman Problem. Report No 388, Graduate School of Industrial Administration, Carnegie Mellon University, U.S.A.

Christofides, N. (1985a): Vehicle routing. In: Lawler, E.L., Lenstra, J.K., Rinnooy

Kan, A.H.G., Shmoys, D.B. (Eds.): The Traveling Salesman Problem: A Guided Tour of Combinatorial Optimization. Wiley, Chichester, 431-448

Christofides, N., Benavent, E., Campos, V., Corberan, A., Mota, E. (1984): An Optimal Method for the Mixed Postman Problem. In: Thoft-Christensen, P. (Ed.): System Modelling and Optimization, Lecture Notes in Control and Information Sciences 59. Springer, Berlin

Christofides, N., Campos, V., Corberán, A., Mota, E. (1986): An Algorithm for the Rural Postman Problem on a Directed Graph, Mathematical Programming Study 26, 155-166

Clarke, G., Wright, J.W. (1964): Scheduling of Vehicles from a Central Depot to a Number of Delivery Points. Operations Research 12, 568-581

Colorni, A., Dorigo, M., Maniezzo, V. (1991): Distributed Optimization by Ant Colonies. In: Varela, F., Bourgines, P. (Eds.): Proceedings of the European Conference on Artificial Life (ECAL '91, Paris, France). Elsevier, Amsterdam, 134-142

Colorni, A., Dorigo, M., Maniezzo, V. (1992): An Investigation of Some Properties of an Ant Algorithm. In: Maenner, R., Manderick, B. (Eds.): Proceedings of the Second Conference on Parallel Problem Solving from Nature (PPSN II, Brussels, Belgium). North Holland, Amsterdam, 509-520

Cook, S.A. (1971): The Complexity of Theorem-Proving Procedures. Proceedings of the 3rd Annual ACM Symposium on the Theory of Computing, Association for Computing Machinery, New York, 151-158

Crowder, H.P., E.L. Johnson, M.W. Padberg (1983): Solving Large-Scale Zero-One Linear Programming Problems. Operations Research 31, 803-834

Dakin, R.J. (1965): A Tree-Search Algorithm for Mixed Integer Programming Problems. The Computer Journal 8, 250-255

Dantzig, G. B. (1959): Note on Solving Linear Programs in Integers. Naval Research Logistics Quarterly 6, 75-76

Dantzig, G.B. (1957): Discrete Variable Extremum Problems. Operations Research 5, 266-277

Dantzig, G.B., Fulkerson, D.R., Johnson, S.M. (1954): Solution of a Large-Scale Traveling Salesman Problem. Operations Research 2, 393-410

Dantzig, G.B., Fulkerson, D.R., Johnson, S.M. (1959): On a Linear Programming Combinatorial Approach to the Traveling Salesman Problem. Operations Research 7, 58-66

Dantzig, G.B., van Slyke, R.M. (1967): Generalized Upper Bounded Techniques for Linear Programming. Journal of Computer and Systems Sciences 1, 213-226

Darwin, C. (1859): On the Origin of Species, 1st Edition (facsimile - 1964). Harvard University Press, Cambridge, MA

Dawkins, R. (1986): The Blind Watchmaker, Penguin, London

De Werra, D. and Hertz, A. (1989): Tabu search techniques. A Tutorial and an Application to Neural Networks. OR Spektrum 11 131-141

Denardo, E.V. (1982): Dynamic Programming: Models and Applications. Prentice Hall, Englewood Cliffs, NJ

Dennis, J.B. (1959): Mathematical Programming and Electrical Networks. J. Wiley and Sons, New York

Deo, N., Pang, C. (1984): Shortest Path Algorithms: Taxonomy and Annotation. Networks 14, 275-323

Desrochers, M., Desrosiers, J., Solomon, M.M. (1992): A New Optimization Algorithm for the Vehicle Routing Problem with Time Windows. Operations Research 40, 342-354

Desrochers, M., Laporte, G. (1991): Improvements and Extensions to the Miller-Tucker-Zemlin Subtour Elimination Constraints. Operations Research Letters 10, 27-36

Desrochers, M., Lenstra, J.K., Savelsbergh, M.W.P. (1990): A Classification Scheme for Vehicle Routing and Scheduling Problems. European Journal of Operational Research 46, 322-332

Desrosiers, J., Soumis, F., Desrochers, M. (1984): Routing with Time Windows by Column Generation. Networks 14, 545-565

Dial, R. (1965): Algorithm 360: Shortest Path Forest with Topological Ordering. Communications of the ACM 12, 632-633

Dijkstra, E.W. (1959): A Note on Two Problems in Connection with Graphs. Numerische Mathematik 1, 269-271

Dinic, E.A. (1970): Algorithm for Solution of a Problem of Maximum Flow in Networks with Power Estimation. Soviet Mathematics Doklady II, 1277-1280

Dreyfus, S.E. (1969): An Appraisal of Some Shortest Path Algorithms. Operations Research 17, 395-412

Dreyfus, S.E., Law, A.M. (1977): The Art and Theory of Dynamic Programming. Acedemic Press, New York

Dror, M., Gavish, B., Choquette, J. (1990): Directed Steiner Tree Problem on A Graph: Models, Relaxations, and Algorithms. INFOR 28, 266-281

Eastman, W.L. (1958): Linear Programming with Pattern Constraints. Ph.D. Thesis, Harvard University, Cambridge, MA

Edmonds, J. (1965): Paths, Trees and Flowers. Canadian Journal of Mathematics 17, 449-467

Edmonds, J., Johnson, E.L. (1973): Matching, Euler Tours and the Chinese Postman Problem. Mathematical Programming 5, 88-124

Edmonds, J., Karp, R.M. (1972): Theoretical Improvements in Algorithmic Efficiency for Network Flow Problems. Journal of the ACM 19, 248-264

Eilon, S., Watson-Gandy, C.D.T., Christofides, N. (1971): Distribution Management: Mathematical Modelling and Practical Analysis. Griffin, London

Eiselt, H.A., Gendreau, M., Laporte, G. (1995a): Arc Routing Problems, Part I: The Chinese Postman Problem. Operations Research 43, 231-242

Eiselt, H.A., Gendreau, M., Laporte, G. (1995b): Arc Routing Problems, Part II: The Rural Postman Problem. Operations Research 43, 399-414

Eiselt, H.A., Laporte, G. (1991): A Combinatorial Optimization Problem Arising in Dartboard Design. Journal of the Operational Research Society 42, 113-118

Eiselt, H.A., Sandblom, C.-L., Pederzoli, G. (1987): Continuous Optimization Models. W. de Gruyter, Berlin

Esau, L.R., Williams, K.C. (1966): On Teleprocessing System Design, Part II. IBM Systems Journal 5, 142-147

Euler, L. (Ed. Newman, J.R.) (1953): Leonhard Euler and the Königsberg Bridges. Scientific American 189, 66-70

Evans, J.R., Minieka, E. (1992): Optimization Algorithms for Networks and Graphs. Marcel Dekker, New York

Feuermen, M., Weiss, H. (1973): A Mathematical Programming Model for Test Construction and Scoring. Management Science 19, 961-966

Fischetti, M., Toth, P. (1992): An Additive Bounding Procedure for the Asymmetric Travelling Salesman Problem. Mathematical Programming 53, 173-197

Fisher, M.L., Jaikumar, R. (1978): A Decomposition Algorithm for Large-Scale Vehicle Routing. Working paper 78-11-05, Department of Decision Sciences, University of Pennsylvania, U.S.A.

Fisher, M.L., Jaikumar, R. (1981): A Generalized Assignment Heuristic for Vehicle Routing. Networks 11, 109-124

Fleischner, H. (1990): Eulerian Graphs and Related Topics (Part 1, Volume 1). Annals of Discrete Mathematics 45, North Holland, Amsterdam

Fleischner, H. (1991): Eulerian Graphs and Related Topics (Part 1, Volume 2), Annals of Discrete Mathematics 50, North-Holland, Amsterdam

Fleury, M. (1885): Deux Problèmes de Géométrie de Situation. Journal de Mathématiques Elémentaire, 157

Florian, M., Nguyen, S., Pallottino, S. (1981): A Dual Simplex Algorithm for Finding all Shortest Paths. Networks 11, 367-378

Floyd, R.W. (1962): Algorithm 97: Shortest Path. Communications of the ACM 5, 345

Ford, L.R., Fulkerson, D.R. (1956): Maximal Flow Through a Network. Canadian Journal of Mathematics 8, 399-404

Ford, L.R., Fulkerson, D.R. (1962): Flows in Networks. Princeton University Press, Princeton

Foster, B., Ryan, D. (1976): An Integer Programming Approach to the Vehicle Scheduling Problem. Operational Research Quarterly 27, 367-384

Frank, M., Wolfe, P. (1956): An Algorithm for Quadratic Programming. Naval Research Logistics Quarterly 3, 95-110

Frederickson, G.N. (1979): Approximation Algorithms for Some Postman Problems. Journal of the Association for Computing Machinery 26, 538-554

Frederickson, G.N., Hecht, M.S., Kim, C.E. (1978): Approximation Algorithms for Some Postman Problems. SIAM Journal on Computing 7, 178-193

Gabow, H.N. (1985): Scaling Algorithm for Network Problems. Journal of Computer and System Sciences 31, 148-168

Gallo, G., Pallottino, S. (1986): Shortest Path Methods: A Unifying Approach. Mathematical Programming Study 26, 38-64

Gallo, G., Pallottino, S. (1988): Shortest Path algorithms. Annals of Operations Research 13, 3-79

Garey, M.R., Johnson, D.S. (1977): Two-Processor Scheduling with Start-Times and Deadlines. SIAM Journal on Computing 6, 416-426

Garey, M.R., Johnson, D.S. (1979): Computers and Intractability: A Guide to the Theory of NP-Completeness. Freeman, New York

Garfinkel, R.S. (1973): On Partitioning the Feasible Set in a Branch and Bound Algorithm for the Asymmetric Traveling Salesman Problem. Operations Research 21, 340-343

Garfinkel, R.S. (1977): Minimizing Wallpaper Waste: Part I: A Class of Traveling Salesman Problems. Operations Research 25, 741-751

Garfinkel, R.S. Nemhauser, G.L. (1972): Integer Programming. Wiley, New York

Gaskell, T. (1967): Bases for Vehicle Fleet Scheduling. Operations Research Quarterly 18, 281-295

Gendreau, M., Laporte, G., Zhao, Y. (1990): The Windy-Postman Problem on General Graphs. Publication #698, Centre de recherche sur les transports, Montreal

Gershkoff, I. (1989): Optimizing Flight Crew Schedules. Interfaces, 19, 29-43

Ghiani, G., Laporte, G. (2000): A Branch-and-Cut Algorithm for the Undirected Rural Postman Problem. Mathematical Programming, (forthcoming)

Gilmore, P.C., Gomory, R.E. (1961): A Linear Programming Approach to the Cutting Stock Problem:Part I. Operations Research 9, 849-859

Gilmore, P.C., Gomory, R.E. (1963): A Linear Programming Approach to the Cutting Stock Problem:Part II. Operations Research 11, 863-888

Gilmore, P.C., Gomory, R.E. (1964): Sequencing a One State-Variable Machine: A Solvable Case of the Traveling Salesman Problem. Operations Research 12, 655-679

Glover, F. (1968): A New Foundation for a Simplified Primal Integer Programming Algorithm. Operations Research 16, 727-740

Glover, F. (1975): New Results on Equivalent Integer Programming Formulations. Mathematical Programming 8, 84-90

Glover, F. (1986): New Approaches for Heuristic Search. A Bilateral Linkage With Artificial Intelligence. European Journal of Operational Research 39, 119-130

Glover, F. (1990a): Tabu Search: A Tutorial. Interfaces 20, 74-94

Glover, F., Woolsey, R.E. (1970): Aggregating Diophantine Equations. Report #70-4, University of Colorado, Boulder

Goldberg, A.V., Tarjan, R.E. (1990): Solving Maximum Cost Flow Problems by Successive Approximation. Mathematics of Operations Research 15, 430-466

Goldberg, D. (1989): Genetic Algorithms in Search: Optimization and Machine Learning. Addison Wesley, Reading, MA

Golden, B.L., Assad, A.A. (1988): Vehicle Routing: Methods and Studies. North Holland, Amsterdam

Golden, B.L., DeArmon, J.S., Baker, E.K. (1983): Computational Experiments with Algorithms for a Class of Routing Problems. Computers & Operations Research 10, 47-59

Golden, B.L., Magnanti, T.L. (1977): Deterministic Network Optimization—A Bibliography. Networks 7, 149-183

Golden, B.L., Magnanti, T.L., Nguyen, H.Q. (1977): Implementing Vehicle Routing Algorithms. Networks 7, 113-148

Golden, B.L., Wong, R.T. (1981): Capacitated Arc Routing Problems. Networks 11, 305-315

Gomory, R.E. (1958): Outline of an Algorithm for Integer Solutions to Linear Programs. Bulletin of the American Mathematical Society 64, 275-278

Gomory, R.E. (1963): An Algorithm for Integer Solutions to Linear Programs. In: Graves, R.L. and Wolfe, P. (Eds.): Recent Advances in Mathematical Proramming. McGraw-Hill, New York, 269-302

Gomory, R.E. (1969): Some Polyhedra Related to Combinatorial Problems. Linear Algebra and its Applications 2, 451-558

Gomory, R.E., Hu, T.C. (1961): Multi-Terminal Network Flows. SIAM Journal on Applied Mathematics 9, 551-556

Gondran, M., Minoux, M. (1984): Graphs and Algorithms. Wiley, New York

Grötschel, M., Win, Z. (1992): A Cutting Plane Algorithm for the Windy Postman Problem. Mathematical Programming 55, 339-358

Guan, M. (1984a): A Survey of the Chinese Postman Problem. Journal of Mathematical Research and Exposition 4, 113-119 (in Chinese)

Guan, M. (1984b): On the Windy Postman Problem. Discrete Applied Mathematics 9, 41-46

Guignard, M. (1995): Lagrangean Relaxation: A Short Course. Belgian Journal of Operations Research, Statistics and Computer Science 35, 5-21

Guignard, M., Spielberg, K. (1981): Logical Reduction Methods in Zero-One Programming (Minimal Preferred Variables). Operations Research 29, 49-74

Hadley, G. (1964): Nonlinear and Dynamic Programming. Addison-Wesley, Reading, MA

Hakimi, S.L. (1971): Steiner's Problem in Graphs and its Implications. Networks 1, 113-133

Hamilton, W.R. (1856): Memorandum Respecting a New System of Roots of Unity (the Icosian Calculus), Philosophy Magazine 12, 446

Hansen, P. (1986): The Steepest Ascent Mildest Descent Heuristic for Combinatorial Programming. Congress on Numerical Methods in Combinatorial Optimization, Capri, Italy

Harsanyi, S.C. (1975): Can the Maximin Principle Serve as a Basis for Morality?

A Critique of John Rawl's Theory. American Political Science Review 14, 594-606

Heller, I., Thompkins, C.B. (1956): An Extension of a Theorem of Dantzig. In: Kuhn, H.W., Tucker, A.W. (Eds.): Linear Inequalities and Related Systems. Annals of Mathematical Studies, Princeton University Press, NJ

Hertz, A., Laporte, G., Nanchen-Hugo, P. (1999): Improvement Procedures for the Undirected Rural Postman Problem. INFORMS Journal on Computing 11, 53-62

Hess, S., Weaver, J., Siegfeldt, H., Whelan, J., Zitlau, P. (1965): Nonpartisan Political Redistricting by Computer. Operations Research 13, 993-1006

Hierholzer, C. (1873): Über die Möglichkeit, einen Linienzug ohne Wiederholung und ohne Unterbrechung zu umfahren. Mathematische Annalen 6, 30-32

Hillier, F.S., Lieberman, G.J. (1995): Introduction to Operations Research, 6th ed. Holden-Day, San Francisco, CA

Hoffman, A.J. (1960): Some Recent Applications of the Theory of Linear Inequalities to Extreme Combinatorial Analysis. Proceedings of the Symposium on Applied Mathematics 10, 113-128

Hoffman, A.J., Kruskal, J.G. (1956): Internal Boundary Points of Convex Polyhedra. In: Kuhn, H.W., Tucker, A.W. (Eds.): Linear Inequalities and Related Systems. Annals of Mathematics Study 38, 223-246, Princeton University Press, Princeton

Hoffman, K., Padberg, M. (1985): LP-Based Combinatorial Problem Solving. Annals of Operations Research 4, 145-194

Hojati, M. (1996): Optimal Political Districting. Computers & Operations Research 23, 1147-1161

Holland, J.H. (1975): Adaptation in Natural and Artificial Systems. The University of Michigan Press, Ann Arbor, MI

Hopcroft, J.E., Tarjan, R.E. (1974): Efficient Planarity Testing. Journal of the ACM 21, 549-568

Hopfield, J.J., Tank, D (1986): Computing With Neural Circuits: A Model. Science 233, 625-633

Hopfield, J.J., Tank, D. (1985): Neural Computations of Decisions in Optimization Problems. Biological Cybernetics 52, 141-152

Howard, R.A. (1960): Dynamic Programming and Markov Processes. MIT Press, Cambridge, MA

Jacobsen, H.D., Mayne, D.Q. (1970): Differential Dynamic Programming. Elsevier-North Holland, New York

Jeroslow, R.G. (1973): There Cannot Be any Algorithm for Integer Programming with Quadratic Constraints. Operations Research 21, 221-224

Jeroslow, R.G. (1974): Trivial Integer Programs Unsolvable by Branch and Bound. Mathematical Programming 6, 105-109

Johnson, D., Aragon, C., McGeoch, L., Schevon, C. (1989): Optimization by Simulated Annealing: An Experimental Evaluation; Part I, Graph Partitioning. Operations Research 37, 865-892

Johnson, D., Aragon, C., McGeoch, L., Schevon, C. (1991): Optimization by Simulated Annealing: An Experimental Evaluation; Part II, Graph Coloring and Number Partitioning. Operations Research 39, 378-406

Johnson, E.L. (1966): Networks and Basic Solutions. Operations Research 14, 619-624

Karmarkar, N. (1984): A New Polynomial-Time Algorithm for Linear Programming. Combinatorica 4, 373-395

Karp, R.M. (1972): Reducibility Among Combinatorial Problems. In: Miller, R.E., Thatcher, J.W. (Eds.): Complexity of Computer Computation. Plenum Press, New York, 85-101

Karp, R.M. (1979): A Patching Algorithm for the Nonsymmetric Traveling Salesman Problem. SIAM Journal on Computing 8, 561-573

Karzanov (1974): Determining the Maximal Flow in a Network by the Method of Preflows. Soviet Mathematics Doklady 15, 434-437

Kaufmann, A. (1967): Graphs, Dynamic Programming and Finite Games. Academic Press, New York

Kendall, K.E., Zionts, S. (1977): Solving Integer Programming Problems by Aggregating Constraints. Operations Research 25, 346-351

Kennington, J.L., Helgason, R.V. (1980): Algorithms for Network Programming.

Wiley, New York

Kirchhoff, G.R. (1847): Über die Auflösung der Gleichungen, auf welche man bei der Untersuchung der linearen Verteilung galvanischer Ströme geführt wird. [On the Solution of the Equations Obtained from the Investigation of the Linear Distribution of Galvanic Currents]. Annalen der Physik und Chemie 72, 497-508

Kirkpatrick, S., Gerlatt, C.D. Jr., Vecchi, M.P. (1982): Optimization by Simulated Annealing. IBM Research Report RC 9355

Klee, V., Minty, G.J. (1972): How Good is the Simplex Algoithm? In: Shisha, O. (Ed.): Inequalities III. Academic Press, New York, 159-175

Klein, M. (1967): A Primal Method for Minimal Cost Flows, with Applications to the Assignment and Transportation Problems. Management Science 14, 205-220

Knuth, D. (1973): The Art of Computer Programming, Volume 1: Fundamental Algorithms. Addison-Wesley, Reading, MA

König, D. (1936): Theorie der Endlichen und Unendlichen Graphen. Reprinted by Chelsea Publishing Co., New York, 1950

Korte, B. (1989): Applications of Combinatorial Optimization. In: Iri, M., Tanabe, K. (Eds.): Mathematical Programming, Recent Developments and Applications. Kluwer, Tokyo, 1-55

Kruskal, J.B. (1956): On the Shortest Spanning Subtree of a Graph and the Travelling Salesman Problem. Proceedings of the American Mathematical Society 7, 48-50

Kuratowski, K. (1930): Sur le Probleme des Courbes Gauches en Topologie. Fundamenta Mathematicae 15, 271-283

Ladner, R.E. (1977): The Computational Complexity of Provability in Systems of Modal Propositional Logics. SIAM Journal of Computing 6, 467-480

Land, A.H. (1979): The Solution of Some 100-City Travelling Salesman Problems. Working paper, London School of Economics

Land, A.H., Doig, A.G. (1960): An Automatic Method of Solving Discrete Linear Programming Problems. Econometrica 28, 496-520

Langevin, A., Soumis, F., Desrosiers, J. (1990): Classification of Traveling Salesman Problem Formulations. Operations Research Letters 9, 127-132

Laporte, G. (1990): Développements Algorithmiques Récents et Perspectives de Recherche en Distributique. Les Cahiers Scientifiques du Transport 21, 61-84

Laporte, G., Mercure, H., Nobert, Y. (1986): An Exact Algorithm for the Asymmetrical Capacitated Vehicle Routing Problem. Networks 16, 33-46

Laporte, G., Nobert, Y. (1987): Exact Algorithms for the Vehicle Routing Problem. In Martello, S., Laporte, G., Minoux, M., Ribeiro, C. (Eds.): Surveys of Combinatorial Optimization. Volume 31 in Annals of Discrete Mathematics, Volume 32 in Mathematics Studies. North-Holland, Amsterdam

Laporte, G., Nobert, Y., Desrochers, M. (1985): Optimal Routing Under Capacity and Distance Restrictions. Operations Research 33, 1050-1073

Lawler, E.L. (1976): Combinatorial Optimizaton: Networks and Matroids. Holt, Rinehart and Winston, New York

Lawler, E.L., Lenstra, J.K., Rinnooy Kan, A.H.G., Shmoys, D.B. (1985): The Traveling Salesman Problem: A Guided Tour of Combinatorial Optimization. Wiley, Chichester

LeBlanc, L.J., Reddoch, R. (1990): Backbone Telecommunication Network Design with Decreasing Marginal Capacity Costs. Working Paper 90-43, Owen Graduate School of Management, Vanderbilt University, Nashville, Tennessee

Lenstra, J.K., Rinnooy Kan, A.H.G. (1975): Some Simple Applications of the Traveling Salesman Problem. Operational Research Quarterly 26, 717-733.

Lenstra, J.K., Rinnooy Kan, A.H.G. (1976): On General Routing Problems. Networks 6, 273-280

Lenstra, J.K., Rinnooy Kan, A.H.G. (1979): Complexity of Vehicle Routing and Scheduling Problem. Report BW 111, Mathematisch Centrum, Amsterdam

Lin, S. (1965): Computer Solution of the Traveling Salesman Problem. Bell System Technical Journal 44, 2245-2269

Little, J.D.C., Murty, K.G., Sweeney, D.W., Karel, C. (1963): An Algorithm for the Traveling Salesman Problem. Operations Research 11, 972-989

Loberman, H., Weinberger, A. (1957): Formal Procedures for Connecting Terminals with a Minimum Total Wire Length. Journal of the ACM 4, 428-438

Lockwood, A., Moore, T. (1992): Harvest Scheduling With Spatial Constraints: A

Simulated Annealing Approach. Canadian Journal of Forest Research 23, 468-478

Luenberger, D.L. (1984): Linear and Nonlinear Programming, 2nd ed. Addison-Wesley, Reading, MA

Maffioli, F. (1981): Complexity of Optimum Undirected Tree Problems: A Survey of Recent Results. In: Analysis and Design of Algorithms in Combinatorial Optimization. International Center for Mechanical Sciences, Courses and Lectures 266. Springer, New York

Magnanti, T.L., Wolsey, L.A. (1995): Optimal Trees. In: Ball, M.O., Magnanti, T.L., Monma, C.L., Nemhauser, G.L. (Eds.): Handbooks in Operations Research and Management Science. North-Holland, Amsterdam

Malhotra, V.M., Pramodh-Kumar, M., Mahashwari, S.N. (1978): An $O(V^3)$ Algorithm for Finding Maximum Flows in Networks. Information Processing Letters 7, 277-278

Martello, S., Toth, P. (1990): Generalized Assignment Problem. In: Martello, S.; Toth, P. (Eds.): Knapsack Problems: Algorithms and Computer Implementations. Wiley, Chichester, 189-220

Martin, G.T. (1966): Solving the Traveling Salesman Problem by Integer Programming. Working paper, CEIR, New York

Martin, J. (1969): Design of Real-Time Computer Systems. Prentice Hall, Englewood Cliffs, NJ

Mathews, G.B. (1897): On the Partition of Numbers. Proceedings of the London Mathematical Society 28, 486-490

McMullen, P. (1970): The Maximum Number of Faces of a Convex Polytope. Mathematika 17, 179-184

Melzak, Z.A. (1961): On the Problem of Steiner. Canadian Mathematics Bulletin 4, 143-148

Michalewicz, Z., Fogel, D.B. (2000): How to Solve It: Modern Heuristics. Springer, Berlin Heidelberg New York

Miliotis, P. (1976): Integer Programming Approaches to the Travelling Salesman Problem. Mathematical Programming 10, 367-378

Miliotis, P. (1978): Using Cutting Planes to Solve the Symmetric Travelling Salesman Problem. Mathematical Programming 15, 177-188

Miller, C.E., Tucker, A.W., Zemlin, R.A. (1960): Integer Programming Formulations and Traveling Salesman Problems. Journal of the ACM 7, 326-329

Miller, D.L., Pekny, J.F. (1991): Exact Solution of Large Scale Asymmetric Traveling Salesman Problems. Science 251, 754-761

Minieka, E. (1979): The Chinese Postman Problem for Mixed Networks. Management Science 25, 643-648

Monma, C.L., Shallcress, D.F. (1989): Methods for Designing Communications Networks with Certain Two-Connected Survivability Constraints. Operations Research 37, 531-541

Müller-Merbach, H. (1981): Heuristics and their Design: A Survey. European Journal of Operational Research 8, 1-23

Murty, K.G. (1968): An Algorithm for Ranking all the Assignments in Order of Increasing Cost. Operations Research 16, 682-687

Murty, K.G. (1992): Network Programming. Prentice Hall, Englewood Cliffs, NJ

Nemhauser, G.L. (1966): Introduction to Dynamic Programming. Wiley, New York

Nemhauser, G.L., Wolsey, L.A. (1988): Integer and Combinatorial Optimization. Wiley, New York

Norback, J., Love, R. (1977): Geometric Approaches to Solving the Traveling Salesman Problem. Management Science 23, 1208-1223

Norback, J.; Love, R. (1979): Heuristic for the Hamiltonian Path Problem in Euclidean Two Space. Journal of the Operational Research Society 30, 363-368

Or, I. (1976): Traveling Salesman-Type Combinatorial Problems and their Relation to the Logistics of Regional Blood Banking. Ph.D. Dissertation, Northwestern University, Evanston, IL

Orloff, C. (1976): Route Constrained Fleet Scheduling. Transportation Science 10, 149-168

Osman, I.H., Laporte, G. (1996): Metaheuristics: A Bibliography. Annals of Operations Research 63, 513-623

Padberg, M.W., Rinaldi, G. (1990): Facet Identification for the Symmetric Traveling Salesman Problem. Mathematical Programming 47, 219-257

Padberg, M.W., Rinaldi, G. (1991): A Branch-and-Cut Algorithm for the Resolution of Large-Scale Symmetric Traveling Salesman Problems. SIAM Review 33, 60-100

Paessens, H. (1988): The Savings Algorithm for the Vehicle Routing Problem. European Journal of Operational Research 34, 336-344

Parker, R.G., Rardin, R.L. (1988): Discrete Optimization. Academic Press

Pearn, W.-L. (1989): Approximate Solutions for the Capacitated Arc Routing Problem. Computers & Operations Research 16, 589-600

Picard, J.-C., Queyranne, M. (1980): On the Structure of all Minimum Cuts in a Network and Applications. Mathematical Programming Study 13, 8-16

Pierce, A.R. (1975): Bibliography on Algorithms for Shortest Path, Shortest Spanning Tree, and Related Circuit Routing Problems. Networks 5, 73-82

Pirlot, M. (1993): General Local Search Heuristics in Combinatorial Optimization: A Tutorial. Belgian Journal of Operations Research, Statistics and Computer Science 32, 7-67

Plesnik, J. (1981): A Bound for the Steiner Tree Problem in Graphs. Math. Slovaca 31, 155-163

Prim, R.C. (1957): Shortest Connection Networks and Some Generalizations. Bell System Technical Journal 36, 1389-1401

Ram, B., Karwan, M.H., Babu, A.J.G. (1988): Aggregation of Constraints in Integer Programming. European Journal of Operational Research 35, 216-227

Rao, M.R., Zionts, S. (1968): Allocation of Transportation Units to Alternative Trips — A Column Generation Scheme with Out-of-Kilter Subproblems. Operations Research 16, 52-63

Rardin, Ronald R. (1998): Optimization in Operations Research. Prentice Hall, New Jersey

Ravindran, A., Phillips, D.T., Solberg, J.J. (1987): Operations Research: Principles and Practice, 2nd ed. Wiley, New York

Rawls, J. (1971): A Theory of Justice. Harvard University Press, Cambridge, MA

Rayward-Smith, V.J. (1983): The Computation of Nearly Minimal Steiner Trees in Graphs. International Journal of Mathematical Education in Science and Technology 14, 15-23

Rayward-Smith, V.J., Osman, I.H., Reeves, C.R., Smith, G.D. (Eds.) (1996): Modern Heuristic Search Methods. Wiley, Chichester

Reeves, C.R. (1993): Modern Heuristic Techniques for Combinatorial Problems. Blackwell, Oxford

Reinelt, G. (1992): Fast Heuristics for Large Geometric Traveling Salesman Problems. ORSA Journal on Computing 4, 206-217

Roberts, F.S. (1978): Graph Theory and Its Applications to Problems of Society. Volume 29 in CBMS-NSF: Regional Conference Series in Applied Mathematics, Society for Industrial and Applied Mathematics, Philadelphia,

Röck, H. (1980): Scaling Techniques for Minimal Cost Network Flows. In: Page, V. (Ed.): Discrete Structures and Algorithms. Carl Hanser, Munich, 181-191

Rosenkrantz, D.J., Stearns, R.E., Lewis, P.M. II (1977): An Analysis of Several Heuristics for the Traveling Salesman Problem. SIAM Journal on Computing 6, 563-581

Saaty, T.L. (1970): Optimization in Integers and Related External Problems. McGraw Hill, New York

Salkin, H.M. (1975): Integer Programming. Addison-Wesley, Reading, MA

Savas, E.S. (1978): On Inequality in Providing Public Services. Management Science 24, 800-808

Schrijver, A. (1986): Theory of Linear and Integer Programming. Wiley, Chichester

Shapiro, D.M. (1966): Algorithms for the Solution of the Optimal Cost and Bottleneck Traveling Salesman Problems. Sc.D. Thesis, Washington University, St. Louis, MO

Sharda, R. (1994): Neural Networks for the MS/OR Analyst: An Application Bibliography. Interfaces 24, 116-130

Shier, D.R. (1976): Iterative Methods for Determining the k Shortest Paths in a

Network. Networks 6, 205-229

Singh, S. (1997): Fermat's Enigma: The Epic Quest to Solve the World's Greatest Mathematical Problem. Walker, New York

Sinha, P., Zoltners, A.A. (1979): The Multiple Choice Knapsack Problem. Operations Research 27, 503-515

Smale, S. (1982): The Problem of the Average Speed of the Simplex Method. In: Bachem, A., Grötschel, M., Korte, B. (Eds): Mathematical Programming: The State of the Art. Springer, Bonn, 530-539

Smith, T.H.C., Srinivasan, V., Thompson, G.L. (1977): Computational Performance of Three Subtour Elimination Algorithms for Solving Asymmetric Traveling Salesman Problems. Annals of Discrete Mathematics 1, 495-506

Snyder, S, ReVelle, C. (1997): Multiobjective Grid Packing Model: An Application in Forest Management. Location Science 5, 165-180

Steenbrink, P.A. (1974): Optimization of Transport Networks. Wiley, London

Stewart, W.R. Jr. (1977): A Computationally Efficient Heuristic for the Traveling Salesman Problem. Proceedings of the 13th Annual Meeting of S.E. TIMS, 75-85

Stone, L.D. (1977): Search for Targets with Generalized Conditionally Deterministic Motion. SIAM Journal of Applied Mathematics 33, 456-468

Strassen, V. (1969): Gaussian Elimination is Not Optimal. Numerische Mathematik 13, 354-356

Taha, H.A. (1975): Integer Programming: Theory, Applications, and Computations. Academic Press, New York.

Taha, H.A. (1997): Operations Research-An Introduction, 6th ed. Prentice Hall, New Jersey

Tarjan, R. E. (1983): Data Structures and Network Algorithms. Volume 44 in CBMS-NSF: Regional Conference Series in Applied Mathematics, Society for Industrial and Applied Mathematics, Philadelphia, PA

Toregas, C., ReVelle, C.S. (1972): Optimal Location Under Time or Distance Constraints. Papers of the Regional Science Association 28, 133-143

van Aardenne-Ehrenfest, T., Bruijn, N.G. (1951): Circuits and Trees in Oriented Linear Graphs. Simon Stevin 28, 203-217

Veinott, A.F. Jr., Dantzig, G.B. (1968): Integral Extreme Point. SIAM Review 10, 371-372

Wagner, H.M. (1975): Principles of Operations Research, 2nd ed. Prentice Hall, Englewood Cliffs, NJ

Warshall, S. (1962): A Theorem on Boolean Matrices. Journal of the ACM 9, 11-12

Williams, H. P. (1974): Experiments in the Formulation of Integer Programming Problems. Mathematical Programming Study 2, 180-197

Williams, H.P. (1978): Model Building in Mathematical Programming. Wiley, Chichester

Williams, H.P. (1993): Model Solving in Mathematical Programming, 3rd Edition. Wiley, Chichester

Win, Z. (1989): On the Windy Postman Problem on Eulerian Graphs. Mathematical Programming 44, 97-112

Winter, P. (1985): Generalized Steiner Problem in Outerplanar Networks. BIT 25, 485-496

Wolfe, P. (1970): Convergence Theory in Nonlinear Programming. In: Abadie, J. (Ed.): Integer and Nonlinear Programming. North Holland, Amsterdam, 1-36

Wolsey, L.A. (1989): Strong Formulations for Mixed Integer Programming: A Survey. Mathematical Programming 45, 173-191

Wong, R.T. (1980): Integer Programming Formulations of the Traveling Salesman Problem. Proceedings of the IEEE International Conference on Circuits and Computers, 149-152

Wong, R.T. (1984): A Dual Ascent Approach for Steiner Tree Problems on a Directed Graph. Mathematical Programming 28, 271-287

Wythoff, B.J. (1993): Backpropagation Neural Networks. A Tutorial. Chemometrics and Intelligent Laboratory Systems 18, 115-155

Yao, A.C.C. (1975): An O(|E|log log |V|) Algorithm for Finding Minimum Spanning Trees. Information Processing Letters 4, 21-25

Yellow, P. (1970): A Computational Modification to the Savings Method of Vehicle Scheduling. Operational Research Quarterly 21, 281-283

Young, R.D. (1965). A Primal (All-Integer) Integer Programming Algorithm. Journal of Research of the Natural Bureau of Standards, Section B69, 213-250

Young, R.D. (1968): A Simplified Primal (All-Integer) Integer Programming Algorithm. Operations Research 16, 750-782

Zahn, C.T. (1971): Graph-Theoretical Methods for Detecting and Describing Gestalt Clusters. IEEE Transactions on Computers C-20, 68-86

SUBJECT INDEX

K

L

M

N

O

P

R

S

T

U

V

W